AF247589

Hematology: The Red Cell and Its Diseases

Hematology: The Red Cell and Its Diseases

Editor: Corbin Mitchell

MURPHY & MOORE
www.murphy-moorepublishing.com

www.murphy-moorepublishing.com

ⓂMURPHY & MOORE

Cataloging-in-Publication Data

Hematology : the red cell and its diseases / edited by Corbin Mitchell.
    p. cm.
Includes bibliographical references and index.
ISBN 978-1-63987-766-9
1. Hematology. 2. Blood cells. 3. Blood--Diseases. I. Mitchell, Corbin.
RC636 .H46 2023
616.15--dc23

Murphy & Moore Publishing
1 Rockefeller Plaza,
New York City,
NY 10020, USA

ISBN 978-1-63987-766-9

Contents

Preface

Hematology is a branch of medicine that deals with the study of blood, and the causes, prognosis, treatments and prevention of blood-related diseases. Red blood cells are a specific type of blood cells, which are the primary means of transporting oxygen to the tissues in the body through blood flow. They transport oxygen to the lungs through the capillaries and then release it in the tissues. The cytoplasm of these cells consists of hemoglobin, which contains the iron biomolecule having the capability to bind oxygen. Hemoglobin also provides the red color to the cells and blood. The cell membrane of the red blood cells comprises lipids and proteins, which provides essential properties for various physiological functions of the cells. These functions include stability and deformability of the blood cells, while traversing the capillary network and circulatory system. The red blood cells are created by the process of erythropoiesis, in which stem cells are transformed into red blood cells. Red blood cells have clinical significance as diseases such as anemia, hemolysis, and polycythemia can occur due to low or high red blood cell count. This book contains some path-breaking studies on red blood cells and the diseases associated with them. It will serve as a resource guide for experts as well as students.

The information contained in this book is the result of intensive hard work done by researchers in this field. All due efforts have been made to make this book serve as a complete guiding source for students and researchers. The topics in this book have been comprehensively explained to help readers understand the growing trends in the field.

I would like to thank the entire group of writers who made sincere efforts in this book and my family who supported me in my efforts of working on this book. I take this opportunity to thank all those who have been a guiding force throughout my life.

Editor

ZOOMICS: Comparative Metabolomics of Red Blood Cells from Old World Monkeys and Humans

Lorenzo Bertolone[1], Hye K. Shin[2], Davide Stefanoni[1], Jin Hyen Baek[2], Yamei Gao[2], Evan J. Morrison[1], Travis Nemkov[1], Tiffany Thomas[3], Richard O. Francis[3], Eldad A. Hod[3], James C. Zimring[4], Tatsuro Yoshida[5], Matthew Karafin[6,7], Joseph Schwartz[3], Krystalyn E. Hudson[3], Steven L. Spitalnik[3], Paul W. Buehler[8,9]*[†] and Angelo D'Alessandro[1,10]*[†]

*Correspondence:
Paul W. Buehler
pbuehler@som.umaryland.edu
Angelo D'Alessandro
angelo.dalessandro@ucdenver.edu

[†] These authors have contributed equally to this work and share last/senior authorship

[1] Department of Biochemistry and Molecular Genetics, University of Colorado Denver – Anschutz Medical Campus, Aurora, CO, United States, [2] Center for Biologics Evaluation and Research, Food and Drug Administration, Silver Spring, MD, United States, [3] Department of Pathology and Cell Biology, Columbia University, New York, NY, United States, [4] Department of Pathology, University of Virginia, Charloteseville, VA, United States, [5] Hemanext, Inc., Lexington, MA, United States, [6] Blood Center of Wisconsin, Milwaukee, WI, United States, [7] Department of Pathology and Laboratory Medicine, Milwaukee, WI, United States, [8] Department of Pathology, University of Maryland School of Medicine, Baltimore, MD, United States, [9] Department of Pediatrics, Center for Blood Oxygen Transport and Hemostasis, University of Maryland School of Medicine, Baltimore, MD, United States, [10] Division of Hematology, Department of Medicine, University of Colorado Denver – Anschutz Medical Campus, Aurora, CO, United States

As part of the ZOOMICS project, we set out to investigate common and diverging metabolic traits in the blood metabolome across various species by taking advantage of recent developments in high-throughput metabolomics. Here we provide the first comparative metabolomics analysis of fresh and stored human (n = 21, 10 males, 11 females), olive baboon (n = 20), and rhesus macaque (n = 20) red blood cells at baseline and upon 42 days of storage under blood bank conditions. The results indicated similarities and differences across species, which ultimately resulted in a differential propensity to undergo morphological alterations and lyse as a function of the duration of refrigerated storage. Focusing on purine oxidation, carboxylic acid, fatty acid, and arginine metabolism further highlighted species-specific metabolic wiring. For example, through a combination of steady state measurements and $^{13}C_6{}^{15}N_4$-arginine tracing experiments, we report an increase in arginine catabolism into ornithine in humans, suggestive of species-specific arginase 1 activity and nitric oxide synthesis — an observation that may impact the translatability of cardiovascular disease studies carried out in non-human primates (NHPs). Finally, we correlated metabolic measurements to storage-induced morphological alterations via scanning electron microscopy and hemolysis, which were significantly lower in human red cells compared to both NHPs.

Keywords: comparative biology, red blood cell, metabolomics, blood storage, transfusion

INTRODUCTION

Mere scientific curiosity has historically led science to the discovery of novel natural phenomena that turned out to be mechanistically relevant to our understanding of human biology. From Linnaeus' taxonomical efforts to order nature into separate categories, to modern day DNA-based phylogenetic classifications, scientists have leveraged comparative biology to unravel Nature's mysteries and learn more about humankind as a species. Embracing this philosophy, we embarked on the ZOOMICS project: by taking advantage of recent developments in high-throughput metabolomics (Nemkov et al., 2017; Reisz et al., 2019), we set out to investigate common and diverging metabolic traits in the blood metabolome across various species. The choice to focus on this matrix and, specifically, on red blood cells (RBCs), stems from an appreciation that blood offers a window into systems metabolism. From classical clinical biochemistry to early clinical metabolomics approaches (D'Alessandro et al., 2012), studies on RBC metabolism have furthered our understanding of systemic responses to aging, inflammation, and physiological (e.g., high altitude) or pathological (e.g., hemorrhage, ischemia) hypoxia. RBC metabolism modulates hemoglobin oxygen binding and off-loading and, in so doing, modulates metabolic activity of the rest of bodily cells that, unlike RBCs, contain mitochondria (Nemkov et al., 2018a). As such, it is unlikely that understanding RBC metabolic variance across multiple species will turn out to be nothing more than a sophisticated exercise in technical metabolomics proficiency. Even if ignoring the relevance of such studies to the standpoint of veterinary medicine, we propose that a metabolomics effort in cataloging RBC metabolomes across species will be relevant to the design and interpretation of research studies that rely on these animals as models for human health and disease.

Animal models have contributed significantly to the fields of hematology and transfusion medicine. Rodent [e.g., mice (Howie et al., 2019), rats (Williams et al., 2019), guinea pigs (Baek et al., 2017)], canine (Klein, 2017), and swine (Clendenen et al., 2017) models have been extensively adopted in blood research for years. However, being genotypically closest to humans, non-human primate (NHP) biology is most phenotypically comparable to human biology. Among all NHPs, baboons (e.g., *Papio anubis*, olive baboon) and macaques (e.g., *Macaca mulatta*, Rhesus macaque, hereon referred to as macaques) are the most frequently studied for biomedical research (VandeBerg et al., 2009; Cox et al., 2013). While both species are equally distant phylogenetically from *Homo sapiens* (Siepel, 2009) (~93% of DNA sequence homology) (Rhesus Macaque Genome Sequencing and Analysis Consortium, Gibbs et al., 2007; Cox et al., 2013), baboon size and anatomy makes them more phenotypically similar to humans and, thus, preferred for some comparative research studies, including blood research (Valeri and Ragno, 2006). For example, similar to humans, the hematocrit is 39–45% in macaques and 33–46% in baboons (Valeri et al., 1981b), with corresponding Hb levels of 13.1 ± 0.9 and 12.5 ± 0.2 g/dl in male and female

macaques, and 12.6 ± 1.2 and 12.5 ± 1.0 g/dl in male and female baboons (Harewood et al., 1999), respectively (Chen et al., 2009). In addition, RBC distribution widths are 13.0 ± 0.7 and 12.9 ± 1.0% in macaques and baboons (Mahaney et al., 2005), respectively, similar to human RBCs (Chen et al., 2009). The mean circulatory life span of macaque RBCs is 98 ± 21 days (Fonseca et al., 2016, 2018), comparable to humans (100–120 days), and significantly longer than mice (55–60 days) (Kaestner and Minetti, 2017). Fresh baboon RBCs also have a similar lifespan to humans (~100 days) (Valeri et al., 2002), and hydrogen peroxide-damaged baboon RBCs have increased levels of spectrin–hemoglobin complexes facilitating more rapid removal from circulation (McKenney et al., 1990). Further, RBCs in both humans and baboons respond to altitude (D'Alessandro et al., 2016) or hemorrhagic hypoxia (Reisz et al., 2017) by promoting synthesis of 2,3-diphosphoglycerate, thereby enhancing hemoglobin transition state responsiveness to pH and carbon dioxide; this process promotes hemoglobin's T state conformation (right shift of the hemoglobin oxygen-dissociation curve), oxygen off-loading, and restoration of tissue oxygen homeostasis (Herman et al., 1971).

Studying baboon RBCs in the context of transfusion medicine is not novel. For example, when stored for 3 weeks in citrate phosphate dextrose (CPD) followed by washing, baboon and human RBCs had comparable post-transfusion recoveries (i.e., ~77%) (Valeri et al., 1981a). This suggested that baboon RBCs were a candidate animal model for blood storage, *in vivo* recovery, and, potentially, for transfusion outcomes. Subsequent studies in 2005 suggested that storing baboon RBCs for 42 days in CPD combined with Additive Solution-1 (i.e., CPD-ADSOL) yielded inferior post-transfusion recoveries, compared to human RBCs. The interspecies differences were even more significant after 49 days of storage, but not at earlier time points (i.e., 74% for both species at storage day 35) (Valeri and Ragno, 2005), suggesting some limitations regarding the translational relevance of baboons as a model for human RBC storage. Further, both young and older baboons demonstrated comparable RBC storage and post-transfusion recoveries based on ^{59}Fe-labeling studies (Valeri et al., 1985); interestingly, this observation is not consistent with human data (Tuo et al., 2014) suggesting that RBCs from older donors are more susceptible to the "storage lesion" when compared to those from younger donors. Although not directly related to RBC storage, recent studies have explored the differentiation of baboon-induced pluripotent stem cells (iPSCs) into enucleated mature RBCs (Olivier et al., 2019), as a novel therapeutic approach that may become relevant for transfusion medicine.

Despite several studies, little is known about the metabolism of baboon RBCs, especially how they compare to fresh and stored macaque and human RBCs. Recently, we extensively described the metabolic phenotypes of fresh RBCs from humans and macaques (*Macaca mulatta*), as well as along a weekly continuum in 42-day storage studies (Stefanoni et al., 2020). These experiments identified significant metabolic differences between macaque and human RBCs regarding purine deamination, glutathione metabolism (especially the gamma-glutamyl cycle), arginine metabolism, and membrane phosphatidylserines as a function of storage duration (Stefanoni et al., 2020). Herein, we provide a comparative metabolomics analysis of fresh and stored RBCs from humans and two old-world monkeys, olive baboons (*Papio anubis*) and macaques.

Based on defining RBC metabolic processes that may affect blood storage, disease progression, and translational research, the current work identifies parallel and divergent metabolomics in three primate species (n = 20 per group, equally distributed by sex), with several unique and storage-dependent similarities and differences in non- and human–primate RBC metabolism. Notably, a cross-species dimorphism in arginine metabolism was identified in a non-targeted analysis and further validated using stable isotope-labeled arginine tracing. We highlight this pathway, given the importance of nitric oxide (NO) as an RBC-transported signaling molecule that modulates vascular responsiveness to hypoxia (Doctor and Stamler, 2011); in addition, NO depletion may be a relevant component of the storage lesion that affects transfusion efficacy (Bennett-Guerrero et al., 2007; Kanias et al., 2013; Reynolds et al., 2018). Further, in the context of cardiovascular disease (CVD), RBC imbalances in arginase and NO synthase suggest that increased RBC arginase-1 activity decreases NO bioavailability, superoxide production, endothelial dysfunction, and enhances post-ischemic cardiac failure (Yang et al., 2018; Mahdi et al., 2019; Pernow et al., 2019). Nonetheless, to our knowledge no studies have examined RBC metabolomes of multiple, closely related, primate species, particularly regarding arginine metabolism. This "ZOOmics" study of RBC metabolism may critically affect the design and interpretation of CVD studies focusing on NO signaling and metabolism in humans and NHP models (Havel et al., 2017). Such models may be important for studying communication between RBCs and the vasculature in the progression of CVD, and of RBC transfusion quality-based outcomes that are affected by dysregulated metabolic RBC communication.

MATERIALS AND METHODS

Since all the methods used in this study have been described in prior work, extensive analytical details and related references to methodological papers and their application to recent RBC storage studies are provided in the **Supplementary File—Materials** and Methods extended.

Ethical Statement

All experimental protocols were approved by named institutional committees. Specifically, animal studies were performed according to FDA White Oak Animal Care and Use protocol 2018-31. Human blood was collected under informed consent according to NIH study IRB #99-CC-0168 "Collection and Distribution of Blood Components from Healthy Donors for In Vitro Research Use" under an NIH-FDA material transfer agreement and in compliance with the Declaration of Helsinki.

Blood Collection, Processing, and Storage

Blood was collected into a syringe using a 20-G needle from the femoral vein of 5-year-old rhesus macaques (*Macaca mulatta*— n = 20; 10 males/10 females) and olive baboons (*Papio anubis*— n = 20; 10 males/10 females) under ketamine/dexmedotomidine (7 mg/kg/0.2 mg/kg) anesthesia according to FDA White Oak Animal Care and Use protocol 2018-31. All blood donor macaques originated from the same colony located at Morgan Island, South Carolina, while blood donor olive baboons originated from Southwest National Primate Research Center, San Antonio, Texas, prior to arrival at FDA's White Oak Campus, Silver Spring, Maryland. Donor blood collections for both species were obtained from basal animals prior to their allocation in other, unrelated studies. Human donor blood was collected into a syringe using a 16-G needle from the median cubital vein of 30- to 75-year-old human volunteers (n = 21; 11 males/10 females) under informed consent according to NIH study IRB #99-CC-0168 "Collection and Distribution of Blood Components from Healthy Donors for In Vitro Research Use" under an NIH-FDA material transfer agreement. Blood was collected into acid citrate dextrose, leukofiltered, and stored in AS-3 in pediatric-sized bags designed to hold 20-ml volumes and mimicking the composition of standard full-sized units [i.e., incorporating polyvinylchloride (PVC) and phthalate plasticizers]. RBCs were stored at 4–6°C for 42 days. RBCs and supernatants were separated via centrifugation upon sterile sampling of each unit on days 0, 7, 14, 21, 28, 35, and 42.

Tracing Experiments With $^{13}C_6{}^{15}N_4$-Arginine

All available RBC lysates from the three species were incubated for 5 min and 24 h at 37°C in AS-3 supplemented with 5 mM stable isotope-labeled $^{13}C_6{}^{15}N_4$-arginine (product no: CNLM-539-H-0.05, Cambridge Isotopes).

Ultrahigh-Pressure Liquid Chromatography-Mass Spectrometry Metabolomics, Lipidomics, and Arginine Tracing Experiments

A volume of 50 µl of frozen RBC aliquots was extracted 1:10 in ice cold extraction solution (methanol:acetonitrile:water 5:3:2 *v/v/v*) (Reisz et al., 2019). Samples were vortexed and insoluble material pelleted, as described (Nemkov et al., 2016). Analyses were performed using a Vanquish UHPLC coupled online to a Q Exactive mass spectrometer (Thermo Fisher, Bremen, Germany). Samples were analyzed using a 3-min isocratic condition (Nemkov et al., 2017) or a 5-, 9-, and 17-min gradient, as described (Fu et al., 2016; D'Alessandro et al., 2017b). While this method is not directly comparable to classic methods for quantitation of high-energy phosphate compounds (e.g., ATP and DPG), which require acidic extraction, it still affords ac comprehensive overview of (stored) RBC metabolism (Nemkov et al., 2016). For targeted quantitative experiments, extraction solutions were supplemented with stable isotope-labeled standards, and endogenous metabolite concentrations were quantified against the areas calculated for heavy isotopologs for each internal standard (Fu et al., 2016; D'Alessandro et al., 2017b). For arginine tracing experiments, ^{13}C and ^{15}N tracing into arginine, ornithine, and citrulline was performed as previously described (Seim et al., 2019), through the auxilium of the software El-MAVEN (Agrawal et al., 2019). Graphs and statistical analyses (either t-test or repeated measures ANOVA) were prepared with GraphPad Prism 8.0 (GraphPad Software, Inc, La Jolla, CA, United States), GENE E (Broad Institute, Cambridge, MA, United States), and MetaboAnalyst 4.0 (Chong et al., 2018). Extensive details for this section are provided in the **Supplementary File—Materials** and Methods extended.

Hemolysis Measurements

Percent hemolysis was measured based on % hematocrit, supernatant hemoglobin (Hb) (g/dl), and total (Hb) (supernatant + RBC, g/dl) in 50-μl samples obtained weekly from storage bags. Supernatant and RBCs were separated using a hematocrocrit centrifuge (Thermo Fisher, Frederick, MD, United States). Hematocrit was recorded, and supernatant was separated from RBCs. Supernatant and lysed RBC Hb levels were measured using a Carey 60 UV-visible spectrophotometer (Agilent Technologies, Santa Clara, CA, United States). Oxy ferrous Hb ($HbFe^{2+}O_2$) and ferric Hb ($HbFe^{3+}$) concentrations were determined based on the extinction coefficients for each species. Molar extinction coefficients used to calculate Hb concentrations in heme equivalents were: 15.2 mM^{-1} cm^{-1} at 576 nm for $Hb(O_2)$ and 4.4 mM^{-1} cm^{-1} at 631 nm for ferric Hb using 50 mM potassium phosphate buffer, pH 7.0 at ambient temperature, in both cases. Total heme was calculated by adding these values and converting (heme) (microM) to total (Hb) (g/dl).

Red Blood Cell Morphological Evaluation

Red blood cells were fixed (1% glutaraldehyde in 0.1M phosphate buffer) and post-fixed with 1% osmium tetroxide for 1 h at room temperature, prior to further preparation and evaluation by scanning electron microscopy, as described (Baek et al., 2018).

RESULTS

The Metabolic Phenotypes of Fresh RBCs From NHPs Are More Similar to Each Other Than to Humans

Metabolomics analyses were performed on freshly drawn human (n = 21), baboon (n = 20), and macaque (n = 20) RBCs (**Figure 1A**). All the raw data are extensively reported in tabulated form as **Supplementary Table 1** and as a heat map in **Supplementary Figure 1**. Partial least square-discriminant analysis (PLS-DA; **Figure 1B—Supplementary Table 1** also includes a comparison between PCA and PLS-DA, details of PLS-DA performances, and validation) and hierarchical clustering analyses (**Figure 1C**) were performed on these data, showing that RBC metabolomes of the NHPs were separated

from those of humans across principal component (PC) 1, which explained 35.5% of the total variance (**Figure 1B**). This observation is clearer in the heat map in **Figure 1C**, which shows the top 50 significant metabolites by ANOVA from cross-species comparisons. Some metabolites (alanine, threonine, ornithine, $NADP^+$, and urate) were consistently higher in human RBCs, whereas AMP and carnosine were significantly lower in humans compared to NHPs (**Figure 1D**). We also observed metabolic divergences between baboons and macaques. For example, pyridoxal was significantly lower in baboons, whereas macaques had comparable levels to humans (**Figure 1D**). In other cases, metabolic differences that we had previously reported in the comparison of macaques and humans (Stefanoni et al., 2020) (e.g., arginine), were further increased in baboon RBCs and significantly greater than macaques (**Figure 1D**). Other metabolites (e.g., UDP-glucose in the hexosamine pathway) were higher in macaque RBCs than in humans or baboons, which had similar levels of this metabolite (**Figure 1D**).

Species-Specific Metabolic Differences Were Enhanced by Storage Under Blood Bank Conditions

Red blood cell storage in blood banks is critically important for therapeutic purposes; however, RBC storage induces many metabolic changes affecting energy and redox metabolism, membrane function, tissue oxygenation, and cellular communication (Yoshida et al., 2019). Because these pathways differed in humans, compared to NHPs, in fresh RBCs, we hypothesized that they would be further aggravated by storage for 42 days, the shelf-life of packed RBCs in the United States and most European countries. Metabolomics analyses were performed on samples collected weekly by sterile docking of units from Storage Day 0 through 42 (**Figure 2A**). Analyses were performed on 854 samples [n = 61 (20 baboons, 20 macaques, and 21 humans) each for seven storage time points for RBCs and supernatants], which were clustered on the basis of their metabolic phenotypes in the PCA (**Figure 2B**). PC1 showed a significant impact of storage duration (from left to right—storage days 0–42; **Figure 2B**), explaining 24.5% of the total variance. However, PC2 and PC3 still explained > 14% of the residual variance, with human RBCs clustering closer to macaques than baboons throughout storage.

In the heat map in **Figure 2C**, the most significant metabolites and time-dependent %hemolysis comparisons were evaluated by repeated measures two-way ANOVA (**Supplementary Table 1**) and were grouped by general metabolic pathways. Further comparisons between baboon and human %hemolysis were evaluated using a Student's t-test or a by one-way ANOVA with Holm–Sidak's multiple-comparison test to compare %hemolysis across the three species. A vectorial version of this figure is provided as **Supplementary Figures 2, 3** for RBCs and supernatants, along with metabolite names. Major patterns in these time-series data as a function of the species were identified via ANOVA simultaneous component analysis (ASCA—**Supplementary Table 1**). Some of the metabolites that are differentially impacted across

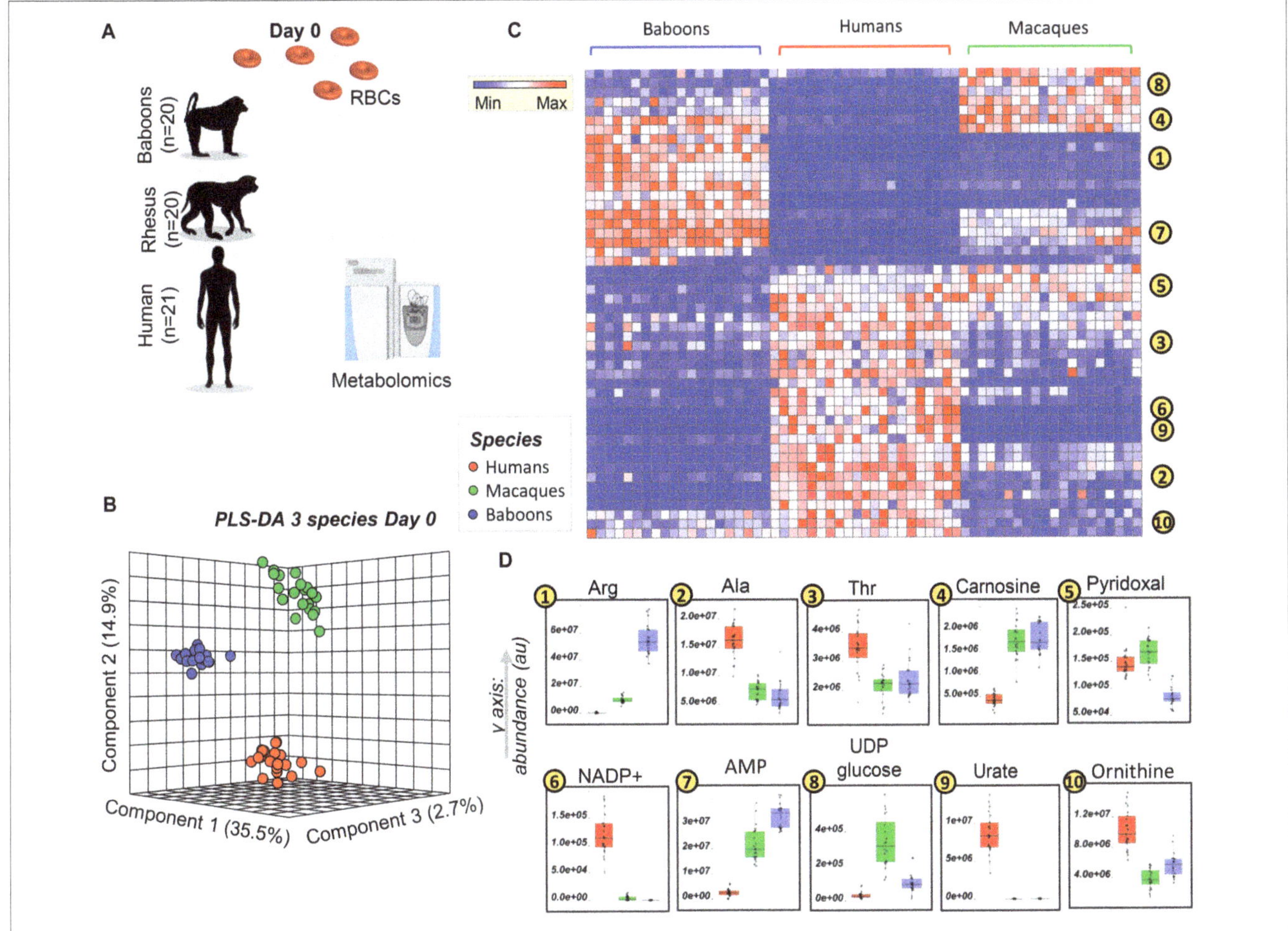

FIGURE 1 | Metabolic phenotypes of freshly drawn human (red), macaque (green), and baboon (blue) red blood cells (RBCs) were tested via ultrahigh-pressure liquid chromatography-mass spectrometry (UHPLC-MS)-based metabolomics approaches **(A)**. Partially supervised partial least square-discriminant analysis (PLS-DA; **B)** and hierarchical clustering analysis **(C)** of the metabolomics data indicate a significant difference between human and NHP RBCs (PC1: 35.5% of the total variance), and between macaque and baboon RBCs (PC2: 14.9% of the total variance; **B**). **(D)** Several representative metabolites, among the top significant changes at baseline, across species, as determined by ANOVA.

species as a function of storage are representatively grouped into classes in **Figure 2C**. While intracellular fatty acids and phosphatidylethanolamines were comparable across species during storage, supernatant levels of lipids were the lowest in humans and highest (and increased during storage) in macaques; nonetheless, baboons unexpectedly had the highest levels of carnitine-conjugated fatty acids in the supernatants (**Figure 2C**).

Species-Specific Metabolic Changes Throughout Storage

To expand on the unsupervised analyses described above, line plots (median $\pm$ quartile ranges) were plotted for human (red), baboon (blue), and macaque (green) RBCs for different pathways, including glycolysis and glutamine/glutathione metabolism (**Figure 3**), purine, arginine, and carboxylic acid/transaminase metabolism (**Figures 4A–C**, respectively), and fatty acids and acyl-carnitines (**Figure 5**). Detailed statistics including

fold changes, ANOVA, and ASCA analyses are reported in **Supplementary Table 1**.

Baboon RBCs had the lowest levels of glucose, glycolytic, and pentose phosphate pathway intermediates and byproducts (lactate and ribose phosphate isomers) throughout storage, whereas humans had the highest (though not significantly higher than macaques; **Figure 3**). Human RBCs had the highest levels of reduced glutathione (GSH) and the lowest of oxidized glutathione (GSSG), which was significantly higher in macaques compared to baboons (**Figure 3**). In contrast, baboons had the highest levels of 5-oxoproline, an activation marker of the gamma-glutamyl cycle (**Figure 3**). Compared to humans, baboons and macaques both had significantly higher markers of glutaminolysis (low extracellular glutamine, high intracellular or extracellular glutamate for baboons and macaques, respectively). Further alterations in sulfur metabolism were noted in NHPs. For example, levels of the antioxidant taurine were highest in baboons throughout storage; however, methionine and S-adenosylmethionine levels were significantly lower in baboons,

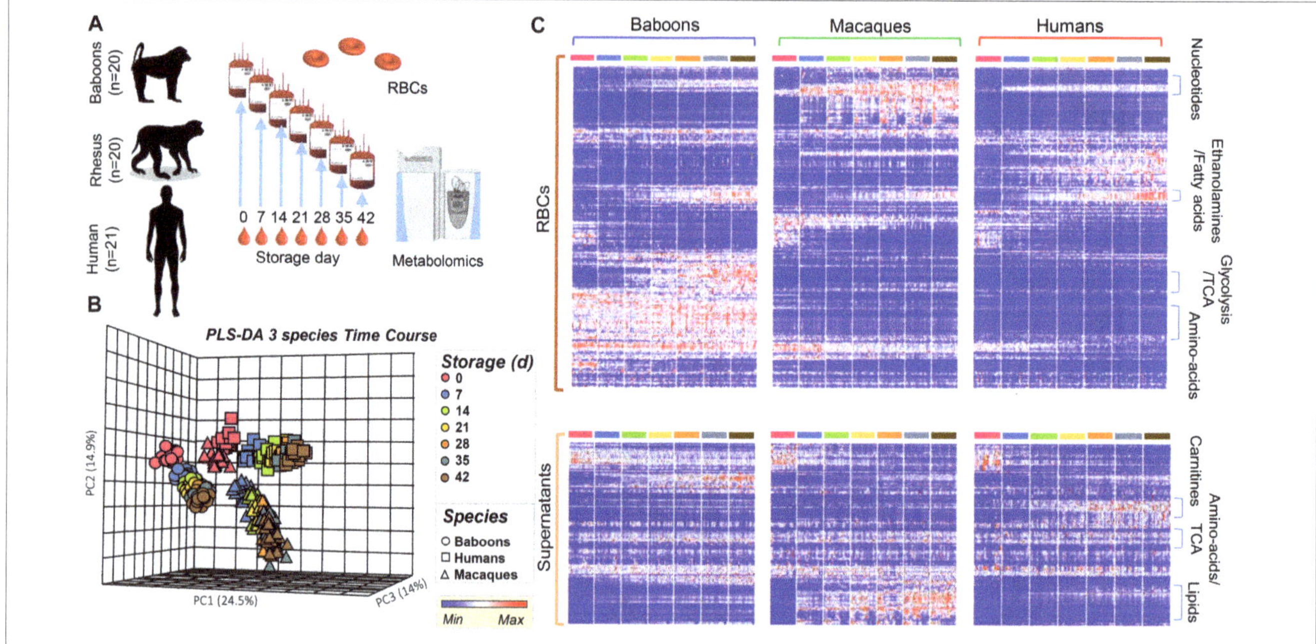

FIGURE 2 | Metabolic phenotypes of human (red), macaque (green), and baboon (blue) RBCs and supernatants during refrigerated storage under blood bank conditions **(A)**. Units were tested weekly, from storage day 0 until the end of the shelf-life (storage day 42). Partially supervised PLS-DA **(B)** and hierarchical clustering analysis **(C)** of the metabolomics data indicate a significant difference between human and non-human primate (NHP) RBCs (PC1: 24.5% of the total variance), and between macaque and baboon RBCs (PC2: 14.9%) and storage duration (14%; **B**). In **(C)**, heat maps indicate significant metabolic changes during storage across species, as determined by repeated measures ANOVA, for RBCs (top) and supernatants (bottom).

suggesting a decreased capacity to repair isoaspartyl damage compared to human and macaque RBCs (**Figure 3**). Analysis of purine metabolism revealed significantly higher levels of methylthioadenosine in baboons, compared to the other species, suggesting an increased activation of salvage pathway S-adenosyl metabolites. Conversely, humans had the highest levels of deaminated purine products: IMP, urate [a potent antioxidant (Tzounakas et al., 2015), which was barely detectable in NHPs], 5-hydroxyisourate, and allantoate, with the notable exception of hypoxanthine, which was highest in NHPs (**Figure 4A**). Although we previously reported an effect of sex on this pathway in macaques and humans (Stefanoni et al., 2020) (e.g., the hypoxanthine guanosine phosphoribosyl transferase gene X-linked), only minor, but significant, sex-associated differences were noted in hypoxanthine and allantoate levels in baboons (female > male) during storage (**Supplementary Figure 4A**).

The most notable metabolic change was that baboons had the highest levels of arginine, citrulline, ornithine, and creatine (**Figure 4B**), with ornithine/arginine (a marker of arginase activity) (D'Alessandro et al., 2019a) and citrulline/arginine ratios (a marker of nitric oxide synthase activity) (Kleinbongard et al., 2006) being higher in females and males, respectively (**Supplementary Figure 4B**). Despite having the lowest levels of arginine, human RBCs consistently had the second highest levels of citrulline throughout storage and the highest citrulline/arginine ratios (a marker of nitric oxide synthase activity). To validate this observation further, fresh RBC lysates from these species were incubated with 5 mM $^{13}C_6{}^{15}N_4$-arginine for 5 min and 24 h at 37°C (**Supplementary Figure 5A**). As

expected, no residual urea cycle activity was observed in any species (i.e., no isotopolog $^{13}C_5{}^{15}N_2$-citrulline was detected), consistent with the absence of mitochondria in mature RBCs. However, significant arginase activity was seen in humans ($^{13}C_5{}^{15}N_2$-ornithine detected as > 80% of the total after 24 h in humans—**Supplementary Figure 5C**) and nitric oxide synthase activity in all species—in particular, macaques (**Supplementary Figure 5C**), were observed, resulting in the generation of $^{13}C_6{}^{15}N_3$-citrulline. Overall, when normalized to the total levels of labeled arginine detected as a percentage of the total (**Supplementary Figure 5C**), our results indicate that human RBCs have significantly higher arginase activity, and lower nitric oxide synthase activity, than macaques and baboons.

Human RBCs also had the highest transaminase activity (as inferred by the levels of the metabolic products of the activity of these enzymes—e.g., alanine, aspartate—and consumption of glutamate), but the lowest levels of carboxylic acid products of transamination reactions (e.g., alpha-ketoglutarate and related citrate and succinate), which were comparable between NHPs (**Figure 4C**). On the other hand, the highest levels of fumarate and malate in human RBCs are consistent with increased purine deamination and, perhaps, salvage in comparison to NHPs (**Figure 4C**). Finally, human RBCs had the highest levels of free carnitine and long-chain acyl-carnitines (14 carbon atoms or longer) throughout storage, followed by baboons (**Figure 5A**), especially males (**Supplementary Figure 6**). Although macaque RBCs had significantly lower levels of acyl-carnitines throughout storage, they had the highest levels of free fatty acids, from 7 to 18 carbon atoms (**Figure 5A**). In

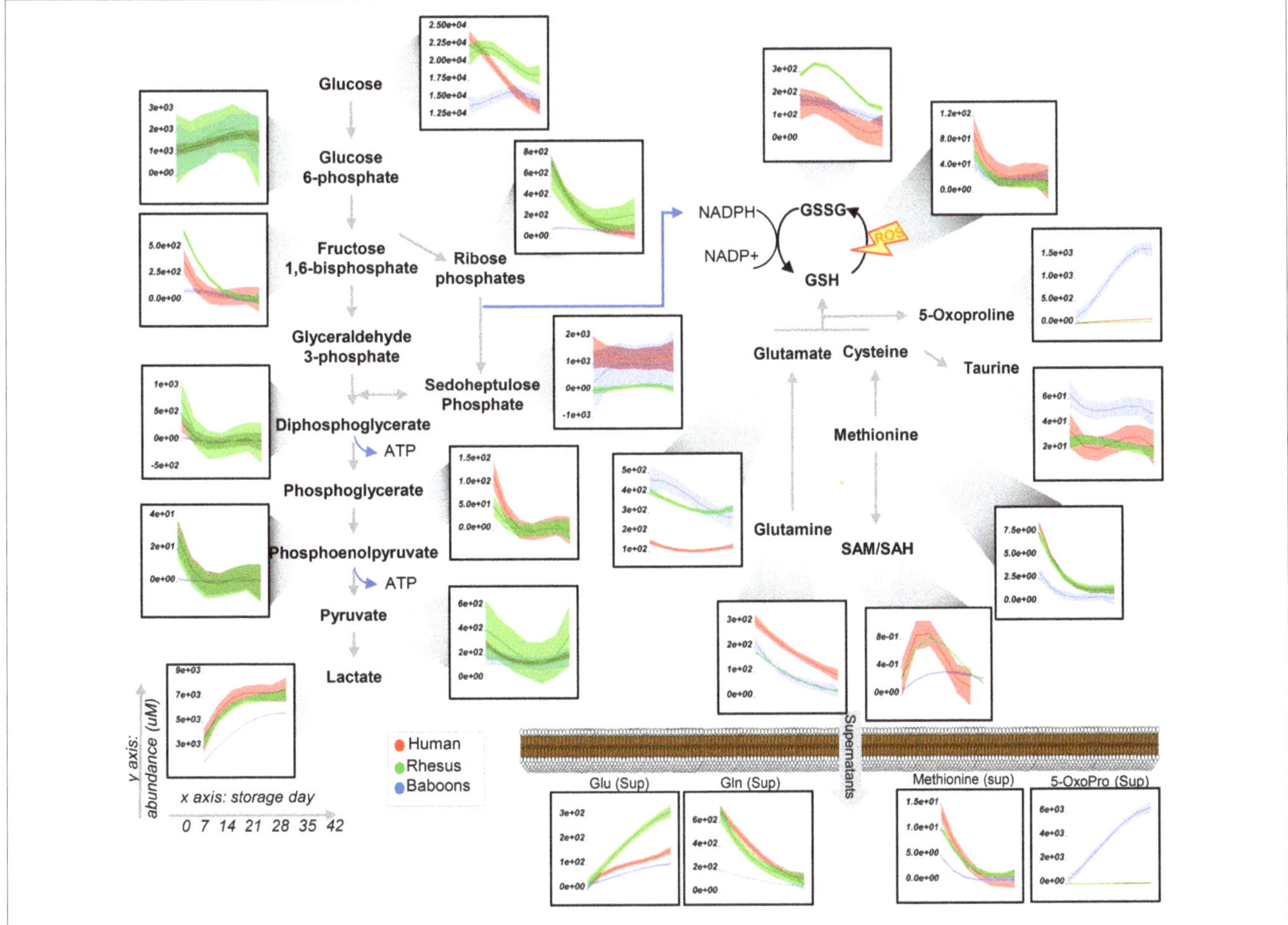

FIGURE 3 | Glycolysis and pentose phosphate pathway, glutathione, and methionine metabolism in human (red), macaques (green), or baboon (blue) RBCs from storage days 0 to 42. Line plots indicate median ($n = 20$) $\pm$ quartile ranges for all the groups. Y axis represents peak areas (arbitrary units) normalized to the median of week 0 measurement for each independent species. Extensive statistical analyses, including fold changes and repeated measures ANOVA, are provided in **Supplementary Table 1**.

contrast, human RBCs had increased levels of polyunsaturated fatty acids and lower levels of oxylipins, which were highest in baboons [hydroxyeicosatetraenoic acids (HETEs)] and macaques [hydroxyoctadecenoic acids (HODEs)] (**Figure 5B**). Macaque and human RBCs had the highest levels of sphingosine 1-phosphate (**Figure 5C**).

Human RBCs Are Characterized by Lower Storage Hemolysis and Morphological Alterations

Considering the differences in lipid metabolism in stored RBCs and supernatants across these species, we anticipated a lower rate of vesiculation and a lower severity of the storage lesion (based on RBC morphology and hemolysis *in vitro*) with human RBCscompared to NHPs. Storage hemolysis increased progressively in baboons, with more than half demonstrating hemolysis levels $\geq$ 1% at storage day 42 (i.e., above the Food and Drug Administration quality threshold; **Figure 6A**). This was similar in baboons and macaques, whereas all human RBC

samples had 42-day hemolysis values of < 1% (**Figure 6B**). By scanning electron microscopy, by storage day 42, > 70% of baboon RBCs had lost their discocytic phenotype (**Figure 6C**). Macaque RBCs were similar, with 80% loss of the discocytic phenotype by storage day 42 (**Supplementary Figure 7**). Unlike the NHPs, human RBCs had a better-preserved morphology, with > 70% discocytes on storage day 42 (**Supplementary Figure 7**), consistent with the patterns of storage hemolysis across these three species.

Common and Divergent Metabolic Correlates to Species-Specific Hemolysis

We performed correlation analyses (**Figure 7A**) of end of storage metabolites with storage hemolysis (**Figure 7B**). This metabolic linkage analysis (D'Alessandro et al., 2017a) provides further, unsupervised, data-driven insights into the metabolic wiring of RBCs evaluated in this study and provides additional correlative evidence of their potential contributions to RBC

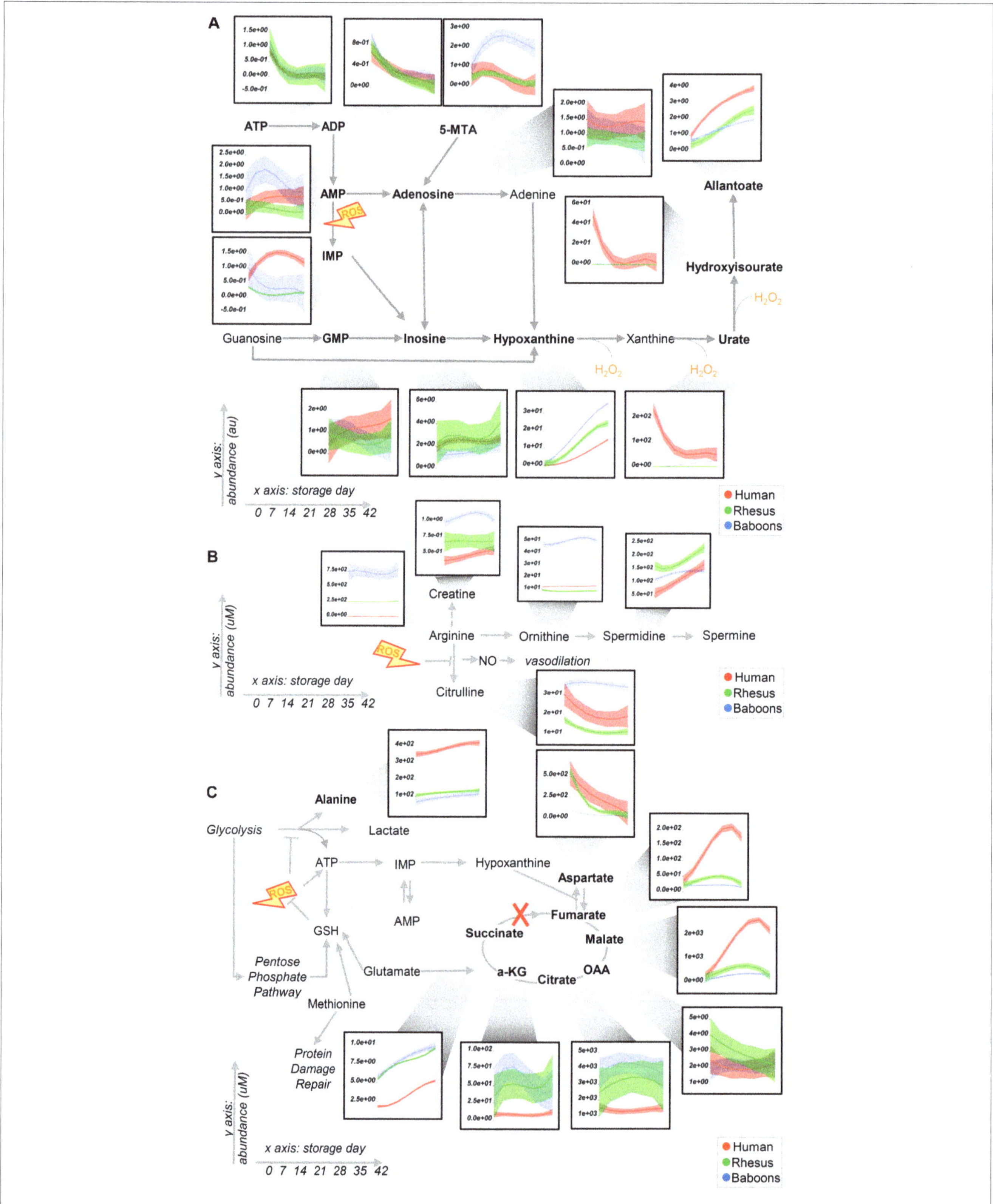

FIGURE 4 | Purine **(A)** and arginine **(B)** metabolism, transaminases, and carboxylic acids **(C)** in human (red), macaque (green), or baboon (blue) RBCs from storage days 0 to 42. Line plots indicate median ($n = 20$) $\pm$ quartile ranges for all the groups. Y axis represents peak areas (arbitrary units) normalized to the median of week 0 measurement for each independent species. Extensive statistical analyses, including fold changes and repeated measures ANOVA are provided in **Supplementary Table 1**.

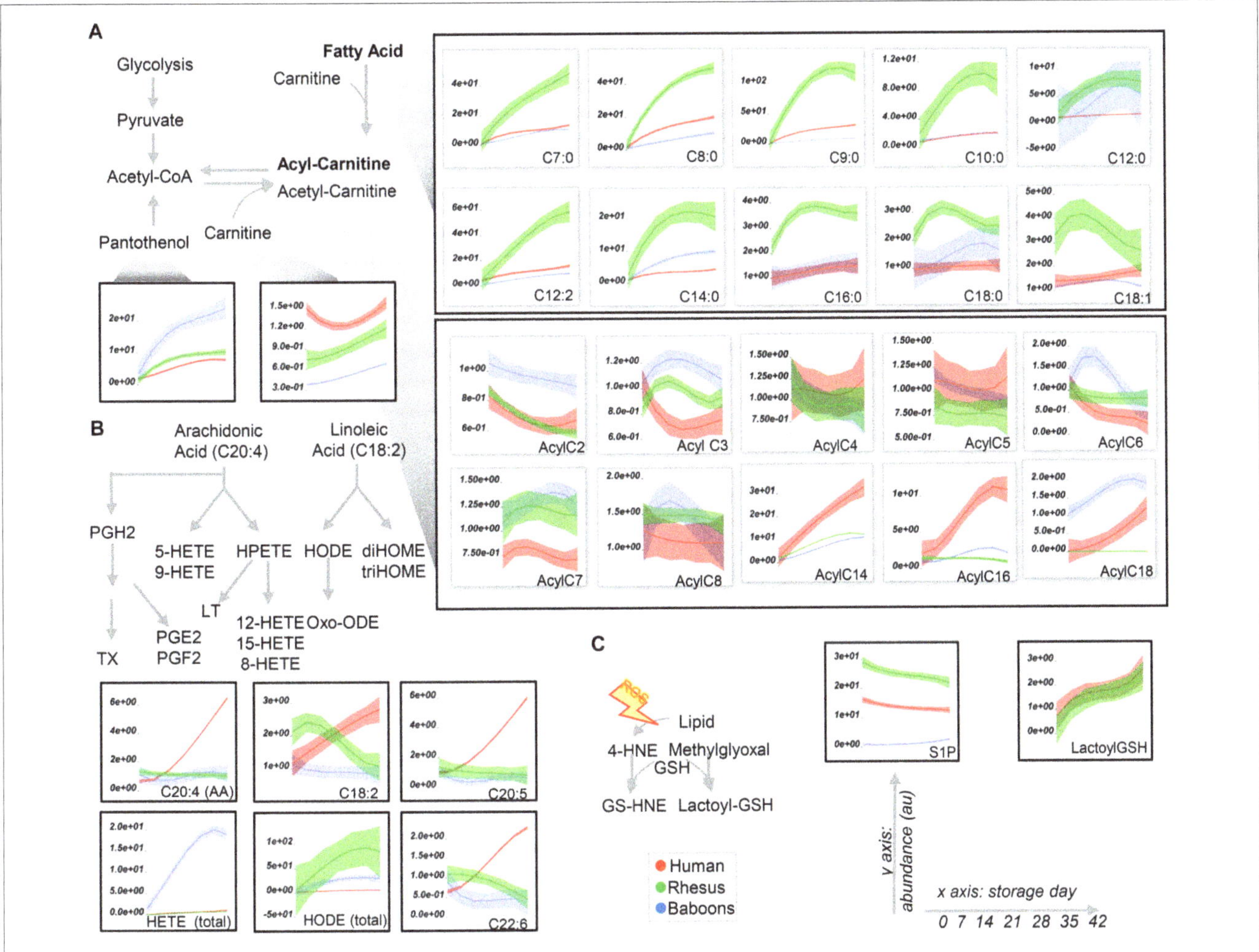

FIGURE 5 | Fatty acid (top), acyl-carnitine (bottom; **A**), eicosanoids **(B)**, and glyoxylate and sphingosine 1-phosphate (S1P) metabolism in human (red), macaque (green), or baboon (blue) RBs from storage days 0 to 42. In **(A,B)**, abbreviations indicate the length in carbon atoms **(C)** of the fatty acid chain and the number of desaturations. The prefix "Acyl-" is used to denote acyl-carnitines in **(A)**. Line plots indicate median (n = 20) ± quartile ranges for all the groups. Y axis represent peak areas (arbitrary units) normalized to the median of week 0 measurement for each independent species. Extensive statistical analyses, including fold changes and repeated measures ANOVA are provided in **Supplementary Table 1**.

hemolysis mechanisms. Thus, in human and macaque RBCs, strong correlations were preserved between the metabolite levels involved in S-adenosylmethionine-dependent protein damage repair mechanisms, purine metabolism (except for urate), and glutathione and pentose phosphate pathway homeostasis (**Figure 7C**). In contrast, human RBCs were more comparable to baboons regarding short-chain fatty acid metabolism, whereas long-chain and polyunsaturated fatty acid metabolism was similar between macaques and baboons (**Figure 7C**, bottom half). These observations are consistent with our prior findings identifying several polyunsaturated fatty acids among the top negative correlates of storage hemolysis (**Figure 7D**). Therefore, this was further leveraged in **Figure 7C** to rank species-specific metabolic correlates to hemolysis (in tabular form in **Supplementary Table 1**). Notably, as in our prior studies, deaminated purines, including IMP, positively correlated with hemolysis in human RBCs, while urate negatively correlated (**Figure 7E**). However, this

observation was only partially recapitulated in baboons and macaques, with a positive correlation of hypoxanthine with hemolysis in both species, but a negative correlation with the hypoxanthine precursor, IMP, or the downstream oxidation product, urate, in baboon and macaque RBCs, respectively (**Figure 7E**). This suggests species-specific biochemistry (at the structural, expression, or functional level) in xanthine dehydrogenase oxidase, AMP deaminase 3 (AMDP3), or hypoxanthine guanosine-phosphoribosyltransferase (HGPRT). Although no species-specific polymorphisms were noted around the active sites in AMPD3 (99% homology across species) and HGPRT (100% sequence homology), some potentially relevant polymorphisms around the active site (residues 803E and 881R) were seen in XDH, despite ∼97% sequence homology across the three species (**Supplementary Figure 8**). Similar changes in correlation trends to storage hemolysis across these species were seen with carboxylic acids, such as malate and fumarate (positive correlation with hemolysis in humans, but negative correlations

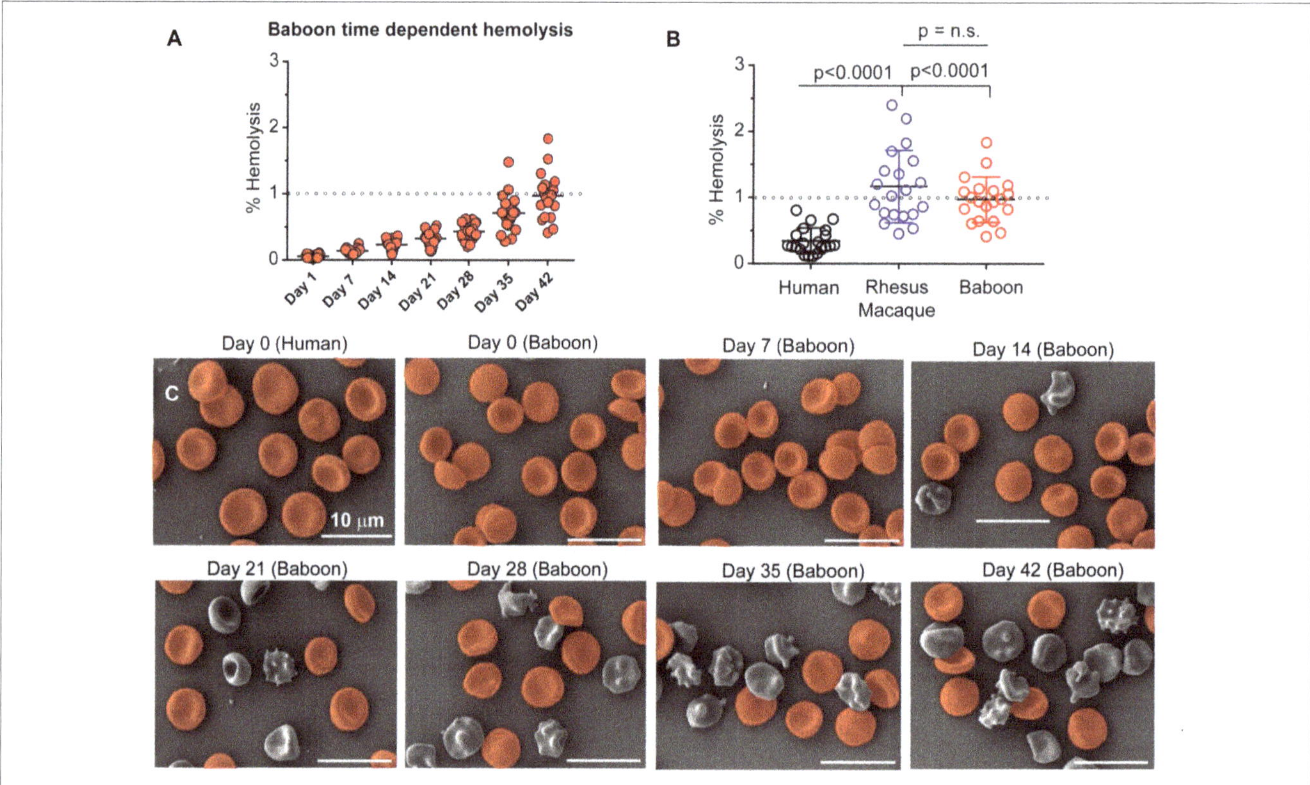

FIGURE 6 | Spontaneous hemolysis as a function of storage duration in baboons **(A)** and an interspecies comparison of end of storage hemolysis **(B)**. NHPs have significantly higher end-of-storage hemolysis than humans, with individual macaques showing the most variability and the greatest maximum individual hemolysis levels. Collectively, mean hemolysis levels between baboons and macaques were similar (p = 0.231, n.s.). Scanning electron micrographs **(C)** are shown for human (day 0) and baboon (weekly) RBCs as a function of storage duration (discocytes in red; echinocytes, spheroechinocytes, acanthocytes, stomatocytes, and spherocytes in gray). Days 0 and 42 equivalent micrographs for human and macaque RBCs are provided as **Supplementary Figure 6**.

in NHPs). Additionally, polyunsaturated fatty acids negatively correlated with hemolysis in human and baboon RBCs, but were positively correlated in macaques (**Figure 7F**).

DISCUSSION

The present study describes significant metabolic, morphological, and storage hemolysis changes between fresh and stored RBCs from human donors and old-world monkeys, specifically olive baboons and rhesus macaques. Baseline metabolic changes were further enhanced by refrigerated storage under standard blood bank conditions. In prior studies, we showed that RBCs undergo alterations to energy and redox metabolism during refrigerated storage (Cancelas et al., 2015; Rolfsson et al., 2017; D'Alessandro et al., 2018). The so-called metabolic "storage lesion" varies across blood donors (D'Alessandro et al., 2019b), with RBCs from some donors being more susceptible to hemolysis *in vitro* or upon oxidant, mechanical, or osmotic challenge applied at the end of their shelf-life (Kanias et al., 2017). The metabolic storage lesion ultimately impacts the tissue metabolome of healthy autologous transfusion recipients (D'Alessandro et al., 2019a) or in sickle-cell disease patients undergoing RBC exchange transfusion (Gehrke et al., 2019). These observations were phenocopied in rodent models of

blood storage and transfusion (Howie et al., 2019), providing evidence *in vivo* of the dysregulation of RBC viability, recovery, systemic iron exposure, and redox metabolism in the initial 24 h after transfusion (Gehrke et al., 2019). Further, storage-induced alteration of purine deamination, and subsequent metabolism by AMP deaminase 3 and xanthine dehydrogenase/oxidase, predicted the ability of transfused RBCs to circulate in the recipient (Nemkov et al., 2018b). Although prestorage levels of urate, a major circulating antioxidant, were associated with improved RBC storage quality (Tzounakas et al., 2015), the levels of its precursor (i.e., hypoxanthine) were negatively associated with the preservation of energy metabolism in stored RBCs and predict accelerated RBC removal post-transfusion (Nemkov et al., 2018b). In the present study, metabolites of the purine deamination pathway were the most divergent when comparing human and NHP RBCs; human RBCs had an order of magnitude higher levels of urate both at baseline and after refrigerator storage. Of note, although such differences had been reported in plasma studies (Misra et al., 2018) involving baboons, the impact of this RBC phenotype change had not been described in relation to RBC storage quality. This observation can be explained by species-specific peculiarities in urate metabolism, which would end up affecting the circulating levels of this metabolite and, as a consequence, RBC urate concentration

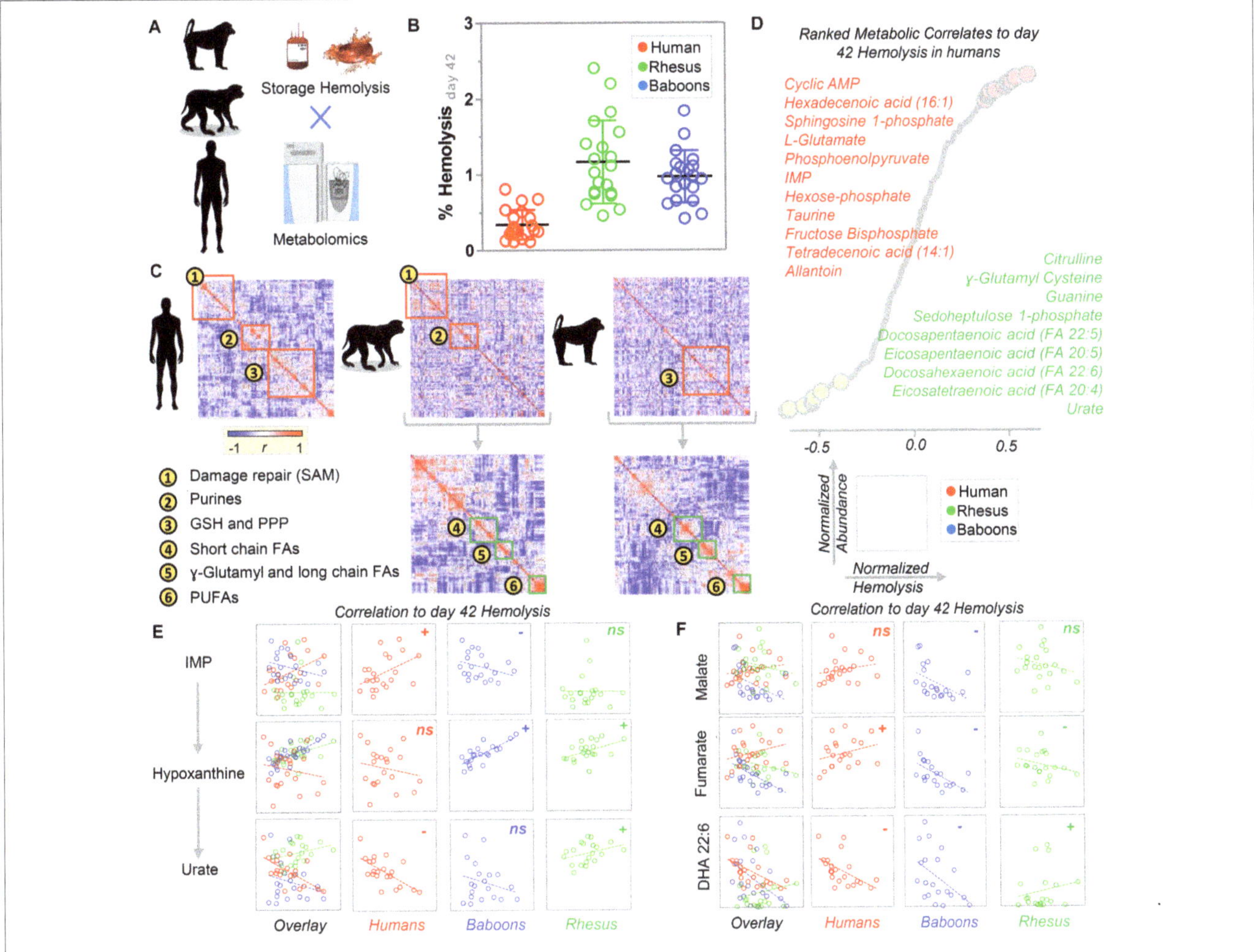

FIGURE 7 | Determination of species-specific metabolic correlates to end of storage hemolysis in baboons, macaques, and humans **(A)**. **(B)** A side-by-side comparison of end of storage hemolysis across the three species. Significantly lower hemolysis was observed in human RBCs. **(C)** Metabolic linkage analysis (i.e., correlation of metabolites to metabolites) in humans, macaques, and baboons [the latter two maintaining the clustered order from human metabolic linkage analyses (i.e., top) or performing species-specific correlations (i.e., bottom)]. These analyses highlighted groups of metabolites and related metabolic pathways that are preserved between humans and macaques, humans and baboons, or baboons and humans. **(D)** A ranked list of metabolic correlates to hemolysis in humans. Based on the analyses described above, a few representative metabolites were plotted that significantly correlated to hemolysis in one species, but not (or inversely correlated) in both of the other two species tested, including purine oxidation metabolites **(E)**, carboxylic acids, and polyunsaturated fatty acids **(F)**.

at the time of blood drawn. In addition, since one of the enzymes involved in hypoxanthine recycling into guanosine (i.e., HGPRT) is encoded by an X-linked gene, it is interesting to note a marginal, but significant, effect of sex on hypoxanthine accumulation across the primate species evaluated here and in previous comparisons between human and macaque RBCs (Stefanoni et al., 2020).

A possible explanation of the divergencies in this pathway across these species is provided by appreciating the polymorphisms in the genes coding for xanthine oxidase in old-world monkeys, compared to humans, as indicated by polymorphisms that mapped within 20 amino acids the active sites of this enzyme (E803 and R881). Increases in carboxylic acid metabolites in macaque and baboon RBCs, compared to humans, along with increases in 5-methylthioadenosine, are consistent with greater purine salvage reactions in

NHPs, which may explain the observed decreases in purine deamination products.

Prior studies of macaque RBC storage observed alterations in free fatty acids, acyl-carnitines, and phosphatidylserines, compared in stored human RBCs. Herein, we expanded on these findings by confirming increases in storage-induced hemolysis and morphologic changes in this species, compared to humans. In addition to confirming similar metabolic, hemolytic, and morphological dysregulation in baboon RBCs, we identified unique alterations of lipid oxidation and remodeling (e.g., via carnitine conjugation and the Lands cycle) (Wu et al., 2016) between baboons and macaques. Alterations in this pathway (especially with respect to fatty acyl-compositions) may at least be in part explained by species-specific diets (Stefanoni et al., 2020).

Perhaps one of the most translationally relevant observations in this study is the divergent metabolism of arginine in human RBCs, compared to baboons and macaques. Arginine conversion to citrulline by an RBC-specific nitric oxide synthase (Kleinbongard et al., 2006) was proposed to contribute to system-wide NO metabolism and, thus, vasodilation. Conversely, arginine metabolism by arginase 1, which is also present and functional in mature RBCs (D'Alessandro et al., 2019a), generates ornithine without the net production of vasoactive NO. Increased arginase-1 activity in the RBCs of patients with Type 2 diabetes is associated with superoxide production and accelerated endothelial dysfunction (Yang et al., 2018; Mahdi et al., 2019; Pernow et al., 2019), suggesting a link between increased RBC arginase-1 activity and CVD. We previously showed that macaque RBCs had significantly higher levels of arginine and its catabolites, compared to human RBCs. To understand this observation further, herein we combined steady state measurements and metabolic tracing with $^{13}C^{15}N$-arginine to show that baboon RBCs have an order of magnitude higher levels of arginine, creatine, and ornithine than macaque RBCs. However, human RBCs had significantly higher steady state levels of citrulline than macaques and significantly higher labeled ornithine (suggestive of higher arginase activity), compared to both NHPs. This observation is critical in light of the central role of this pathway in cardiovascular studies, suggesting caution when interpreting (Havel et al., 2017) CVD studies focusing on NO signaling in baboons and macaques. In addition, these observations cannot be explained by polymorphisms in arginase 1 (>98% sequence homology across the three species) or NO synthase. On the other hand, expression levels and/or activity of arginase may vary across these species as a result of regulatory alterations that has evolved under positive pressure, as classic comparative biology studies report a 957 $\pm$ 206 μmol urea/g of hemoglobin/h arginase activity in human RBCs compared to < 1 in *Papio* baboon species (Spector et al., 1985). Moreover, sex dimorphisms in this pathway that were seen previously in human tissues (Reckelhoff et al., 1998) and RBCs (Contreras-Zentella et al., 2019), and macaque RBCs (Stefanoni et al., 2020), are also observed in baboons.

Finally, levels of amino acids involved in transamination reactions were increased in human RBCs, compared to NHPs, with increased glutaminolysis and glutamate consumption in stored human RBCs as storage progressed. Similarly, human and macaque RBCs showed comparable levels of methionine (a marker of oxidant stress and protein isoaspartyl damage-repair in mature RBCs) (Reisz et al., 2018). In contrast, baboon RBCs coped with oxidant stress through activating the gamma-glutamyl cycle (5-oxoproline levels were two orders of magnitude higher in this species compared to macaques and humans) and taurine, a dietary antioxidant that only lower mammals (e.g., rodents, cats) are thought to be capable of synthesizing (Ripps and Shen, 2012).

REFERENCES

Agrawal, S., Kumar, S., Sehgal, R., George, S., Gupta, R., Poddar, S., et al. (2019). El-MAVEN: a fast, robust, and user-friendly mass spectrometry data processing engine for metabolomics. *Methods Mol. Biol. Clifton N. J.* 1978, 301–321. doi: 10.1007/978-1-4939-9236-2_19

CONCLUSION

Despite the observational nature of the study, we provide the first comparative metabolomics analysis of fresh and stored human, baboon, and macaque RBCs. The results indicated similarities and differences across species, which ultimately resulted in a differential propensity to undergo morphological alterations and lyse as a function of the duration of refrigerated storage. Focusing on purine oxidation and carboxylic acids, fatty acids, and arginine metabolism, we further highlighted species-specific metabolic wiring (e.g., increased arginine catabolism into citrulline and ornithine in humans) and correlations with storage-induced hemolysis. Notably, while RBCs from these NHPs had been reported to have similar *in vivo* lifespan to human RBCs (Valeri et al., 2002; Kaestner and Minetti, 2017), the increased susceptibility to hemolysis during storage may result from the appreciation that current storage additives have been optimized for human RBCs, a caveat that may be relevant for future transfusion medicine studies using NHPs as a model.

AUTHOR CONTRIBUTIONS

HS, JB, and PB collected and stored the samples. TY, SS, PB, and AD'A provided the essential materials and methods to perform the study. PB and YG performed the hemolysis and SEM evaluation of RBCs. LB, EM, DS, TN, and AD'A performed the metabolomics analyses (untargeted and targeted quantitative) and tracing experiments. LB and AD'A performed the data analysis and prepared the figures and tables. AD'A, PB, and SS wrote and modified the first draft of the manuscript, which was revised by all the other authors, including MK, JS, TT, JZ, EH, RF, and KH. All the authors contributed to the finalizing of the manuscript.

FUNDING

Research reported in this publication was supported by funds from the Boettcher Webb-Waring Investigator Award (AD'A, JZ, and SS), RM1GM131968 by the National Institute of General Medical Sciences (AD'A), R01HL146442, R01HL149714, and R01HL148151 by the National Heart, Lung and Blood Institute (AA, JZ, and SS). Zoomics was part of the project MIRAGES: Metabolic Investigation of Red blood cells as a function of Aging, Genetics and Environment sponsored by the National Heart, Lung and Blood Institute (R21HL150032 to AD'A).

Baek, J. H., Yalamanoglu, A., Gao, Y., Guenster, R., Spahn, D. R., Schaer, D. J., et al. (2017). Iron accelerates hemoglobin oxidation increasing mortality in vascular diseased guinea pigs following transfusion of stored blood. *JCI Insight* 2:e93577. doi: 10.1172/jci.insight.93577

Baek, J. H., Yalamanoglu, A., Moon, S.-E., Gao, Y., and Buehler, P. W. (2018). Evaluation of renal oxygen homeostasis in a preclinical animal model to elucidate difference in blood quality after transfusion. *Transfusion* 58, 1474–1485. doi: 10.1111/trf.14560

Bennett-Guerrero, E., Veldman, T. H., Doctor, A., Telen, M. J., Ortel, T. L., Reid, T. S., et al. (2007). Evolution of adverse changes in stored RBCs. *Proc. Natl. Acad. Sci. U.S.A.* 104, 17063–17068. doi: 10.1073/pnas.0708160104

Cancelas, J. A., Dumont, L. J., Maes, L. A., Rugg, N., Herschel, L., Whitley, P. H., et al. (2015). Additive solution-7 reduces the red blood cell cold storage lesion. *Transfusion* 55, 491–498. doi: 10.1111/trf.12867

Chen, Y., Qin, S., Ding, Y., Wei, L., Zhang, J., Li, H., et al. (2009). Reference values of clinical chemistry and hematology parameters in rhesus monkeys (*Macaca mulatta*). *Xenotransplantation* 16, 496–501. doi: 10.1111/j.1399-3089.2009.00554.x

Chong, J., Soufan, O., Li, C., Caraus, I., Li, S., Bourque, G., et al. (2018). MetaboAnalyst 4.0: towards more transparent and integrative metabolomics analysis. *Nucleic Acids Res.* 46, W486–W494. doi: 10.1093/nar/gky310

Clendenen, N., Nunns, G. R., Moore, E. E., Reisz, J. A., Gonzalez, E., Peltz, E., et al. (2017). Hemorrhagic shock and tissue injury drive distinct plasma metabolome derangements in swine. *J. Trauma Acute Care Surg.* 83, 635–642. doi: 10.1097/TA.0000000000001504

Contreras-Zentella, M. L., Sánchez-Sevilla, L., Suárez-Cuenca, J. A., Olguín-Martínez, M., Alatriste-Contreras, M. G., García-García, N., et al. (2019). The role of oxidant stress and gender in the erythrocyte arginine metabolism and ammonia management in patients with type 2 diabetes. *PLoS One* 14:e0219481. doi: 10.1371/journal.pone.0219481

Cox, L. A., Comuzzie, A. G., Havill, L. M., Karere, G. M., Spradling, K. D., Mahaney, M. C., et al. (2013). Baboons as a model to study genetics and epigenetics of human disease. *ILAR J.* 54, 106–121. doi: 10.1093/ilar/ilt038

D'Alessandro, A., Giardina, B., Gevi, F., Timperio, A. M., and Zolla, L. (2012). Clinical metabolomics: the next stage of clinical biochemistry. *Blood Transfus.* 10, s19–s24. doi: 10.2450/2012.005S

D'Alessandro, A., Nemkov, T., Reisz, J., Dzieciatkowska, M., Wither, M. J., and Hansen, K. C. (2017a). Omics markers of the red cell storage lesion and metabolic linkage. *Blood Transfus.* 15, 137–144. doi: 10.2450/2017.0341-16

D'Alessandro, A., Nemkov, T., Yoshida, T., Bordbar, A., Palsson, B. O., and Hansen, K. C. (2017b). Citrate metabolism in red blood cells stored in additive solution-3. *Transfusion* 57, 325–336. doi: 10.1111/trf.13892

D'Alessandro, A., Nemkov, T., Sun, K., Liu, H., Song, A., Monte, A. A., et al. (2016). AltitudeOmics: red blood cell metabolic adaptation to high altitude hypoxia. *J. Proteome Res.* 15, 3883–3895. doi: 10.1021/acs.jproteome.6b00733

D'Alessandro, A., Reisz, J. A., Culp-Hill, R., Korsten, H., van Bruggen, R., and de Korte, D. (2018). Metabolic effect of alkaline additives and guanosine/gluconate in storage solutions for red blood cells. *Transfusion* 58, 1992–2002. doi: 10.1111/trf.14620

D'Alessandro, A., Reisz, J. A., Zhang, Y., Gehrke, S., Alexander, K., Kanias, T., et al. (2019a). Effects of aged stored autologous red blood cells on human plasma metabolome. *Blood Adv.* 3, 884–896. doi: 10.1182/bloodadvances.2018029629

D'Alessandro, A., Zimring, J. C., and Busch, M. (2019b). Chronological storage age and metabolic age of stored red blood cells: are they the same? *Transfusion* 59, 1620–1623. doi: 10.1111/trf.15248

Doctor, A., and Stamler, J. S. (2011). Nitric oxide transport in blood: a third gas in the respiratory cycle. *Compr. Physiol.* 1, 541–568. doi: 10.1002/cphy.c090009

Fonseca, L. L., Alezi, H. S., Moreno, A., Barnwell, J. W., Galinski, M. R., and Voit, E. O. (2016). Quantifying the removal of red blood cells in *Macaca mulatta* during a Plasmodium coatneyi infection. *Malar. J.* 15:410. doi: 10.1186/s12936-016-1465-5

Fonseca, L. L., Joyner, C. J., Saney, C. L., Moreno, A., Barnwell, J. W., Galinski, M. R., et al. (2018). Analysis of erythrocyte dynamics in Rhesus macaque monkeys during infection with Plasmodium cynomolgi. *Malar. J.* 17:410. doi: 10.1186/s12936-018-2560-6

Fu, X., Felcyn, J. R., Odem-Davis, K., and Zimring, J. C. (2016). Bioactive lipids accumulate in stored red blood cells despite leukoreduction: a targeted metabolomics study. *Transfusion* 56, 2560–2570. doi: 10.1111/trf.13748

Gehrke, S., Shah, N., Gamboni, F., Kamyszek, R., Srinivasan, A. J., Gray, A., et al. (2019). Metabolic impact of red blood cell exchange with rejuvenated red blood cells in sickle cell patients. *Transfusion* 59, 3102–3112. doi: 10.1111/trf.15467

Harewood, W. J., Gillin, A., Hennessy, A., Armistead, J., Horvath, J. S., and Tiller, D. J. (1999). Biochemistry and haematology values for the baboon (Papio

hamadryas): the effects of sex, growth, development and age. *J. Med. Primatol.* 28, 19–31. doi: 10.1111/j.1600-0684.1999.tb00085.x

Havel, P. J., Kievit, P., Comuzzie, A. G., and Bremer, A. A. (2017). Use and importance of nonhuman primates in metabolic disease research: current state of the field. *ILAR J.* 58, 251–268. doi: 10.1093/ilar/ilx031

Herman, C. M., Rodkey, F. L., Valeri, C. R., and Fortier, N. L. (1971). Changes in the oxyhemoglobin dissociation curve and peripheral blood after acute red cell mass depletion and subsequent red cell mass restoration in baboons. *Ann. Surg.* 174, 734–743. doi: 10.1097/00000658-197111000-00002

Howie, H. L., Hay, A. M., de Wolski, K., Waterman, H., Lebedev, J., Fu, X., et al. (2019). Differences in Steap3 expression are a mechanism of genetic variation of RBC storage and oxidative damage in mice. *Blood Adv.* 3, 2272–2285. doi: 10.1182/bloodadvances.2019000605

Kaestner, L., and Minetti, G. (2017). The potential of erythrocytes as cellular aging models. *Cell Death Differ.* 24, 1475–1477. doi: 10.1038/cdd.2017.100

Kanias, T., Lanteri, M. C., Page, G. P., Guo, Y., Endres, S. M., Stone, M., et al. (2017). Ethnicity, sex, and age are determinants of red blood cell storage and stress hemolysis: results of the REDS-III RBC-Omics study. *Blood Adv.* 1, 1132–1141. doi: 10.1182/bloodadvances.2017004820

Kanias, T., Wang, L., Lippert, A., Kim-Shapiro, D. B., and Gladwin, M. T. (2013). Red blood cell endothelial nitric oxide synthase does not modulate red blood cell storage hemolysis. *Transfusion* 53, 981–989. doi: 10.1111/j.1537-2995.2012.03850.x

Klein, H. G. (2017). The red cell storage lesion(s): of dogs and men. *Blood Transfus.* 15, 107–111. doi: 10.2450/2017.0306-16

Kleinbongard, P., Schulz, R., Rassaf, T., Lauer, T., Dejam, A., Jax, T., et al. (2006). Red blood cells express a functional endothelial nitric oxide synthase. *Blood* 107, 2943–2951. doi: 10.1182/blood-2005-10-3992

Mahaney, M. C., Brugnara, C., Lease, L. R., and Platt, O. S. (2005). Genetic influences on peripheral blood cell counts: a study in baboons. *Blood* 106, 1210–1214. doi: 10.1182/blood-2004-12-4863

Mahdi, A., Jiao, T., Yang, J., Kövamees, O., Alvarsson, M., von Heijne, M., et al. (2019). The effect of glycemic control on endothelial and cardiac dysfunction induced by red blood cells in type 2 diabetes. *Front. Pharmacol.* 10:861. doi: 10.3389/fphar.2019.00861

McKenney, J., Valeri, C. R., Mohandas, N., Fortier, N., Giorgio, A., and Snyder, L. M. (1990). Decreased in vivo survival of hydrogen peroxide-damaged baboon red blood cells. *Blood* 76, 206–211.

Misra, B. B., Bassey, E., Bishop, A. C., Kusel, D. T., Cox, L. A., and Olivier, M. (2018). High resolution GC/MS metabolomics of non-human primate serum. *Rapid Commun. Mass Spectrom.* 32, 1497–1506. doi: 10.1002/rcm.8197

Nemkov, T., Hansen, K. C., and D'Alessandro, A. (2017). A three-minute method for high-throughput quantitative metabolomics and quantitative tracing experiments of central carbon and nitrogen pathways. *Rapid Commun. Mass Spectrom.* 31, 663–673. doi: 10.1002/rcm.7834

Nemkov, T., Hansen, K. C., Dumont, L. J., and D'Alessandro, A. (2016). Metabolomics in transfusion medicine. *Transfusion* 56, 980–993. doi: 10.1111/trf.13442

Nemkov, T., Reisz, J. A., Xia, Y., Zimring, J. C., and D'Alessandro, A. (2018a). Red blood cells as an organ? How deep omics characterization of the most abundant cell in the human body highlights other systemic metabolic functions beyond oxygen transport. *Expert Rev. Proteomics* 15, 855–864. doi: 10.1080/14789450.2018.1531710

Nemkov, T., Sun, K., Reisz, J. A., Song, A., Yoshida, T., Dunham, A., et al. (2018b). Hypoxia modulates the purine salvage pathway and decreases red blood cell and supernatant levels of hypoxanthine during refrigerated storage. *Haematologica* 103, 361–372. doi: 10.3324/haematol.2017.178608

Olivier, E. N., Wang, K., Grossman, J., Mahmud, N., and Bouhassira, E. E. (2019). Differentiation of baboon (*Papio anubis*) induced-pluripotent stem cells into enucleated red blood cells. *Cells* 8:1282. doi: 10.3390/cells8101282

Pernow, J., Mahdi, A., Yang, J., and Zhou, Z. (2019). Red blood cell dysfunction: a new player in cardiovascular disease. *Cardiovasc. Res.* 115, 1596–1605. doi: 10.1093/cvr/cvz156

Reckelhoff, J. F., Hennington, B. S., Moore, A. G., Blanchard, E. J., and Cameron, J. (1998). Gender differences in the renal nitric oxide (NO) system. *Am. J. Hypertens.* 11, 97–104. doi: 10.1016/S0895-7061(97)00360-9

Reisz, J. A., Nemkov, T., Dzieciatkowska, M., Culp-Hill, R., Stefanoni, D., Hill, R. C., et al. (2018). Methylation of protein aspartates and deamidated asparagines as a function of blood bank storage and oxidative stress in human red blood cells. *Transfusion* 58, 2978–2991. doi: 10.1111/trf.14936

Reisz, J. A., Slaughter, A. L., Culp-Hill, R., Moore, E. E., Silliman, C. C., Fragoso, M., et al. (2017). Red blood cells in hemorrhagic shock: a critical role for glutaminolysis in fueling alanine transamination in rats. *Blood Adv.* 1, 1296–1305. doi: 10.1182/bloodadvances.2017007187

Reisz, J. A., Zheng, C., D'Alessandro, A., and Nemkov, T. (2019). Untargeted and semi-targeted lipid analysis of biological samples using mass spectrometry-based metabolomics. *Methods Mol. Biol. Clifton N. J.* 1978, 121–135. doi: 10.1007/978-1-4939-9236-2_8

Reynolds, J. D., Jenkins, T., Matto, F., Nazemian, R., Farhan, O., Morris, N., et al. (2018). Pharmacologic targeting of red blood cells to improve tissue oxygenation. *Clin. Pharmacol. Ther.* 104, 553–563. doi: 10.1002/cpt.979

Rhesus Macaque Genome Sequencing and Analysis Consortium, Gibbs, R. A., Rogers, J., Katze, M. G., Bumgarner, R., Weinstock, G. M., et al. (2007). Evolutionary and biomedical insights from the rhesus macaque genome. *Science* 316, 222–234. doi: 10.1126/science.1139247

Ripps, H., and Shen, W. (2012). Review: taurine: a "very essential" amino acid. *Mol. Vis.* 18, 2673–2686.

Rolfsson, Ó, Sigurjonsson, ÓE., Magnusdottir, M., Johannsson, F., Paglia, G., Guðmundsson, S., et al. (2017). Metabolomics comparison of red cells stored in four additive solutions reveals differences in citrate anticoagulant permeability and metabolism. *Vox Sang.* 112, 326–335. doi: 10.1111/vox.12506

Seim, G. L., Britt, E. C., and Fan, J. (2019). Analysis of arginine metabolism using LC-MS and isotopic labeling. *Methods Mol. Biol. Clifton N. J.* 1978, 199–217. doi: 10.1007/978-1-4939-9236-2_13

Siepel, A. (2009). Phylogenomics of primates and their ancestral populations. *Genome Res.* 19, 1929–1941. doi: 10.1101/gr.084228.108

Spector, E. B., Rice, S. C., Hendrickson, R., and Cederbaum, S. D. (1985). Comparison of arginase activity in red blood cells of lower mammals, primates, and man: evolution to high activity in primates. *Am. J. Hum. Genet.* 7, 138–145.

Stefanoni, D., Shin, H. K. H., Baek, J. H., Champagne, D. P., Nemkov, T., Thomas, T., et al. (2020). Red blood cell metabolism in Rhesus macaques and humans: comparative biology of blood storage. *Haematologica.* 105, 2174–2186. doi: 10.3324/haematol.2019.229930

Tuo, W.-W., Wang, D., Liang, W.-J., and Huang, Y.-X. (2014). How cell number and cellular properties of blood-banked red blood cells of different cell ages decline during storage. *PLoS One* 9:e105692. doi: 10.1371/journal.pone.0105692

Tzounakas, V. L., Georgatzakou, H. T., Kriebardis, A. G., Papageorgiou, E. G., Stamoulis, K. E., Foudoulaki-Paparizos, L. E., et al. (2015). Uric acid variation among regular blood donors is indicative of red blood cell susceptibility to storage lesion markers: a new hypothesis tested. *Transfusion* 55, 2659–2671. doi: 10.1111/trf.13211

Valeri, C. R., Ellis, A., Donahue, K., Curran, T., and Pivacek, L. (1985). The viability of young and old baboon red cells stored in the liquid state at 4 C. *Prog. Clin. Biol. Res.* 195, 429–441.

Valeri, C. R., Lindberg, J. R., Contreras, T. J., Pivacek, L. E., Austin, R. M., Valeri, D. A., et al. (1981a). Liquid preservation of baboon red blood cells in acid-citrate-dextrose or citrate-phosphate-dextrose anticoagulant: effects of washing liquid-stored red blood cells. *Am. J. Vet. Res.* 42, 1011–1013.

Valeri, C. R., Lindberg, J. R., Contreras, T. J., Pivacek, L. E., Austin, R. M., Valeri, D. A., et al. (1981b). Measurement of red blood cell volume, plasma volume, and total blood volume in baboons. *Am. J. Vet. Res.* 42, 1025–1029.

Valeri, C. R., Pivacek, L. E., Cassidy, G. P., and Ragno, G. (2002). Volume of RBCs, 24- and 48-hour posttransfusion survivals, and the lifespan of (51)Cr and biotin-X-N-hydroxysuccinimide (NHS)-labeled autologous baboon RBCs: effect of the anticoagulant and blood pH on (51)Cr and biotin-X-NHS elution in vivo. *Transfusion* 42, 343–348. doi: 10.1046/j.1537-2995.2002.00071.x

Valeri, C. R., and Ragno, G. (2005). The 24-hour posttransfusion survival of baboon red blood cells preserved in citrate phosphate dextrose/ ADSOL (CPD/AS-1) for 49 days. *Contemp. Top. Lab. Anim. Sci.* 44, 38–40.

Valeri, C. R., and Ragno, G. (2006). The survival and function of baboon red blood cells, platelets, and plasma proteins: a review of the experience from 1972 to 2002 at the naval blood research Laboratory, Boston, Massachusetts. *Transfusion* 46, 1–42. doi: 10.1111/j.1537-2995.2006.00922.x

VandeBerg, J. L., Williams-Blangero, S., and Tardif, S. D. (eds) (2009). *The Baboon in Biomedical Research.* New York, NY: Springer-Verlag, doi: 10.1007/978-0-387-75991-3

Williams, A. T., Jani, V. P., Nemkov, T., Lucas, A., Yoshida, T., Dunham, A., et al. (2019). Transfusion of anaerobically or conventionally stored blood after hemorrhagic shock. *Shock Augusta Ga.* 53, 352–362. doi: 10.1097/SHK.0000000000001386

Wu, H., Bogdanov, M., Zhang, Y., Sun, K., Zhao, S., Song, A., et al. (2016). Hypoxia-mediated impaired erythrocyte Lands' Cycle is pathogenic for sickle cell disease. *Sci. Rep.* 6:29637. doi: 10.1038/srep29637

Yang, J., Zheng, X., Mahdi, A., Zhou, Z., Tratsiakovich, Y., Jiao, T., et al. (2018). Red blood cells in type 2 diabetes impair cardiac post-ischemic recovery through an arginase-dependent modulation of nitric oxide synthase and reactive oxygen species. *JACC Basic Transl. Sci.* 3, 450–463. doi: 10.1016/j.jacbts.2018.03.006

Yoshida, T., Prudent, M., and D'alessandro, A. (2019). Red blood cell storage lesion: causes and potential clinical consequences. *Blood Transfus.* 17, 27–52. doi: 10.2450/2019.0217-18

2

The Many Facets of Erythropoietin Physiologic and Metabolic Response

Sukanya Suresh[†], Praveen Kumar Rajvanshi and Constance T. Noguchi[*]

Molecular Medicine Branch, National Institute of Diabetes and Digestive and Kidney Diseases, National Institutes of Health, Bethesda, MD, United States

Correspondence:
Constance T. Noguchi
connien@niddk.nih.gov

[†] Present address:
Sukanya Suresh,
Division of Endocrinology, Department of Medicine, Indiana University School of Medicine, Indianapolis, IN, United States

In mammals, erythropoietin (EPO), produced in the kidney, is essential for bone marrow erythropoiesis, and hypoxia induction of EPO production provides for the important erythropoietic response to ischemic stress, such as during blood loss and at high altitude. Erythropoietin acts by binding to its cell surface receptor which is expressed at the highest level on erythroid progenitor cells to promote cell survival, proliferation, and differentiation in production of mature red blood cells. In addition to bone marrow erythropoiesis, EPO causes multi-tissue responses associated with erythropoietin receptor (EPOR) expression in non-erythroid cells such neural cells, endothelial cells, and skeletal muscle myoblasts. Animal and cell models of ischemic stress have been useful in elucidating the potential benefit of EPO affecting maintenance and repair of several non-hematopoietic organs including brain, heart and skeletal muscle. Metabolic and glucose homeostasis are affected by endogenous EPO and erythropoietin administration affect, in part via EPOR expression in white adipose tissue. In diet-induced obese mice, EPO is protective for white adipose tissue inflammation and gives rise to a gender specific response in weight control associated with white fat mass accumulation. Erythropoietin regulation of fat mass is masked in female mice due to estrogen production. EPOR is also expressed in bone marrow stromal cells (BMSC) and EPO administration in mice results in reduced bone independent of the increase in hematocrit. Concomitant reduction in bone marrow adipocytes and bone morphogenic protein suggests that high EPO inhibits adipogenesis and osteogenesis. These multi-tissue responses underscore the pleiotropic potential of the EPO response and may contribute to various physiological manifestations accompanying anemia or ischemic response and pharmacological uses of EPO.

Keywords: erythropoietin, erythropoietin receptor, nitric oxide, gender-specific, obesity, inflammation, bone

---

Abbreviations: ACTH, Adrenocorticotropic hormone; AKT, Protein kinase B (PKB); ARNT, Aryl hydrocarbon nuclear translocator (HIF-1β); BMP, Bone morphogenetic protein; BMSC, Bone marrow stromal cells; C2C12, Mouse myoblast cell line; EGLN1, Egl-9 Family Hypoxia Inducible Factor 1 (PHD2); eNOS, Endothelial nitric oxide synthase (NOS3); EPAS1, Endothelial PAS Domain Protein 1 (HIF-2α); EPO, Erythropoietin; EPOR, Erythropoietin receptor; ERK, Extracellular signal-regulated kinase; ET-1, Endothelin-1; FGF, Fibroblast growth factor; FIH-1, Factor inhibiting HIF-1; GATA, GATA binding protein; HIF, Hypoxia-inducible factor; JAK, Janus kinase; MDS, Myelodysplastic syndrome; mEPOR−/−, Erythropoietin receptor knockout mouse; Myf-5, Myogenic factor 5; MYF6, Myogenic factor 6; MyoD, Myoblast determination protein 1; NAD, Nicotinamide adenine dinucleotide; NF-κB, Nuclear factor kappa-light-chain-enhancer of activated B cells; NO, Nitric oxide; PAX7, Paired box 7; PGC-1, Peroxisome proliferator-activated receptor gamma coactivator 1-alpha; PHD, Prolyl hydroxylase; PI3K, Phosphoinositide 3-kinase; POMC, Proopiomelanocortin; PPARα, Peroxisome proliferator-activated receptor alpha; PRDM16, PR-domain-containing 16; RAW264.7, Mouse macrophage cell line; Sirt1, Sirtuin 1; STAT, Signal transducers and activators of transcription; TAL1, T-cell acute lymphocytic leukemia protein 1; U/kg, Units per kilogram (EPO dose); UCP1, Uncoupling protein 1; VHL, Von Hippel-Lindau protein; WSXWS, Amino acid motif (tryptophan-serine-X-tryptophan-serine); ΔEpoR$_E$, Mice with EPOR restricted to erythroid tissue; α-MSH, Alpha-melanocyte stimulating hormone.

INTRODUCTION

Erythropoietin is the hormone that regulates the daily production of 200 billion new red blood cells in the human body. Red blood cells are a renewable resource with a limited lifespan of about 120 days, have an intracellular protein content of about 95% hemoglobin, a tetrameric globular protein that binds oxygen cooperatively, and function primarily to transport oxygen from the lungs to the tissues. EPO binding to erythroid progenitor cells promotes their survival, proliferation, and differentiation to mature erythrocytes. EPO production is hypoxia inducible and is made in the interstitial cells in the adult kidney (Kobayashi et al., 2017; Anusornvongchai et al., 2018). In response to anemia, ischemic stress or high altitude, EPO production is induced (Pugh and Ratcliffe, 2017) and stimulates erythroid progenitor cells in the bone marrow to expand the erythroid lineage thus markedly increasing erythropoiesis and mature red blood cell production. EPO and EPOR on the surface of erythroid progenitor cells are required for red blood cell production and mice with targeted deletion of EPO or EPOR die during embryonic development of severe anemia (Wu et al., 1995; Lin et al., 1996). However, EPOR expression is not restricted to erythroid tissue. This review will address EPO activity in erythroid cells and in select non-hematopoietic tissues expressing EPOR and lessons gleaned from studies in animal models on the protective effects of EPO in ischemic injury and wound healing, regulation of metabolic homeostasis including gender specific EPO response and bone remodeling (**Figure 1**).

ERYTHROPOIETIN ACTION IN ERYTHROID CELLS

Erythropoietin is a glycoprotein produced primarily in the fetal liver and adult kidney to regulate red blood cell production. Human EPO is encoded as a 193 amino acid polypeptide with a NH_2-terminal 27 amino acid signal peptide and three potential N-linked glycosylation sites; posttranslational cleavage of the carboxyl terminal arginine gives rise to a 165 amino acid mature polypeptide for human EPO (Jacobs et al., 1985; Lin et al., 1985; Recny et al., 1987). Recombinant EPO produced in Chinese hamster ovary cells yields a glycoprotein with apparent molecular weight about 34,000 that is biologically active (Recny et al., 1987). EPO received approval in 1989 from the U.S. Food and Drug Administration for clinical treatment of anemia associated with chronic renal failure due to insufficient EPO production, which has markedly improved treatment of this disease (Paoletti and Cannella, 2006; Jelkmann, 2013; Wright et al., 2015).

Sites of EPO Production

During mammalian development, observations of EPO production in mice suggest that EPO is first expressed transiently in neural crest cells during mid-gestation to stimulate yolk sac primitive erythropoiesis for oxygen transport in mid-stage embryos (Malik et al., 2013; Suzuki et al., 2013; Hirano and Suzuki, 2019). As development progresses, the liver becomes the site of EPO production and definitive erythropoiesis

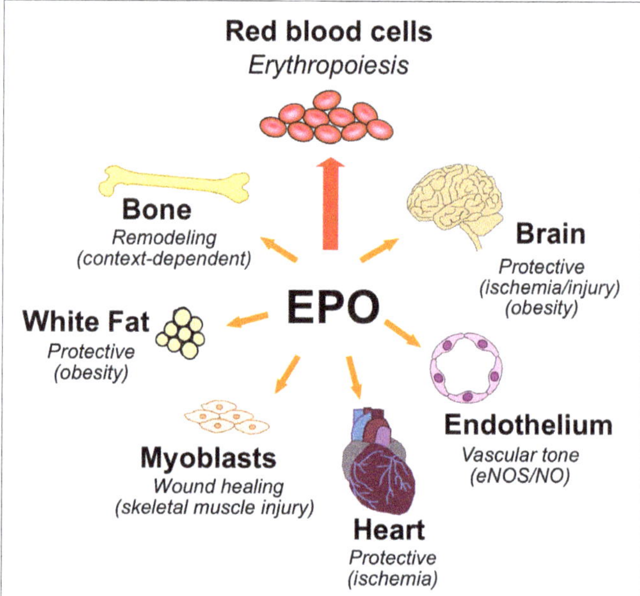

FIGURE 1 | Pleiotropic effects of erythropoietin. High level of EPOR on erythroid progenitor cells accounts for the sensitive erythropoietic response in the bone marrow to hypoxic induction of EPO. EPOR expression determines EPO response and expression beyond erythroid tissue provides for EPO response in non-hematopoietic tissues that include the following: brain for a neuroprotective and metabolic response; cardiovascular system for regulating vascular tone and oxygen delivery in endothelium and protection in heart against ischemic injury; skeletal muscle for muscle maintenance and repair; white adipose tissue for protection for inflammation associated with diet-induced obesity and fat mass accumulation, particularly in males; and bone remodeling.

(Palis, 2014; Palis and Koniski, 2018). EPO is required for definitive erythropoiesis and knockout of EPO or EPOR in mice results in death *in utero* around day 13.5 due to disruption of erythropoiesis in the fetal liver resulting in severe anemia (Wu et al., 1995; Lin et al., 1996). By the last third of gestation in mammalian development, the site of EPO production gradually switches to the kidney which becomes the major site of EPO production in the adult (Zanjani et al., 1981; Dame et al., 1998) and red blood cell production switches from the fetal liver to bone marrow, the site of adult hematopoiesis (Ho et al., 2015). Interstitial peritubular cells of the kidney are the EPO-producing cells and hypoxia induction of EPO production results mainly from the increase in the number of cells producing EPO (Tan et al., 1991; Eckardt et al., 1993; Juul et al., 1998; Obara et al., 2008). EPO expression is detected in tissues beyond liver and kidney including brain and neural cells, spleen, lung, and bone marrow (Fandrey and Bunn, 1993; Masuda et al., 1994; Marti et al., 1996; Dame et al., 1998; Juul et al., 1998), but does not substitute for the required erythropoietic regulation provided by the kidney. Interestingly, genetically over-stabilizing the hypoxic response in osteoblasts in mice resulted in selective expansion of the erythroid lineage leading to development of severe polycythemia due to high level of EPO expression in osteoblasts compared to relatively lower levels of EPO induced by hypoxia in control animals (Rankin et al., 2012).

EPO Is Hypoxia Inducible

Erythropoietin production is hypoxia responsive mediated via binding of HIF to the hypoxic responsive element located downstream of the coding region under hypoxic conditions (Semenza, 2009). HIF is a heterodimer between HIF-α (HIF-1α, HIF-2α or HIF-3α) and HIF-1β (or ARNT), and HIF-2α (or EPAS1) is particularly associated with EPO regulation (Rankin et al., 2007; Suzuki et al., 2017). Oxygen dependent hydroxylases, PHD and FIH-1 down regulate HIF-α stability/activity and provide oxygen sensitivity for HIF regulation of EPO expression (Jaakkola et al., 2001; Mahon et al., 2001; Lando et al., 2002). At normoxia, HIF-α is marked for degradation by proline hydroxylation, primarily by PHD2 (Jaakkola et al., 2001), providing a binding site for VHL which targets HIF-α for ubiquitination and proteasome degradation (Ohh et al., 2000; Ang et al., 2002; Minamishima et al., 2008; Takeda et al., 2008; Kobayashi et al., 2016). With reduced oxygen, HIF-α is stabilized and increased HIF-2α in renal EPO-producing cells up regulates EPO gene expression (Semenza, 2009). The asparaginyl hydroxylase, FIH-1 binds the HIF-α transactivation domain at normoxia and inhibits HIF-α transactivation by hydroxylating asparagine residue in the carboxy-terminal transactivation domain and blocks interactions with coactivator proteins (Mahon et al., 2001; Lando et al., 2002). Mutations in VHL, PHD2, and HIF-2α have been identified in patients with familial erythrocytosis. The Chuvash population of the Russian Federation is associated with a high prevalence of polycythemia due to VHL gene mutation that reduces oxygen dependent HIF-2α degradation and increases EPO production (Ang et al., 2002; Pastore et al., 2003). Mutations in HIF-2α or the EGLN1 gene that encodes PHD2 also give rise to erythrocytosis associated with increased EPO production (Percy et al., 2006, 2008).

EPOR Gene Regulation in the Erythroid Lineage

Erythropoietin receptor is expressed at the highest level on erythroid progenitor cells at the colony forming unit-erythroid (CFU-E) stage that becomes the most responsive to changes in EPO level (Broudy et al., 1991). EPO is required for erythroid progenitor cell survival as cells differentiate from early erythroid progenitors or burst forming unit-erythroid (BFU-E) to CFU-E. Mice that lack EPO or its receptor die *in utero* at day 13.5 due to severe anemia (Wu et al., 1995; Lin et al., 1996). EPO binding to its receptor on erythroid progenitor cells increases expression of erythroid transcription factors, GATA1 and the basic-helix-loop-helix protein, TAL1, that in turn transactivate EPOR expression; hence, EPO regulates expression of its own receptor (Zon et al., 1991; Kassouf et al., 2010; Rogers et al., 2012). The EPOR promoter region contains conserved binding sites for ubiquitous Sp1 transcription factor and GATA1 (AGATAA) and in the 5′ untranslated transcribed region 3 E-box (CAGCTG) TAL1 binding sites (**Figure 2A**). EPO binding at the early erythroid progenitor BFU-E stage with low level EPOR induces GATA1 and TAL1 to activate the erythroid program including EPOR. EPOR is down regulated with progression of erythroid differentiation and is not detected on reticulocytes (Broudy et al., 1991).

EPOR Structure and Signaling

Erythropoietin receptor is a member of the class I cytokine receptor superfamily that contains a WSXWS motif in the extracellular domain, a single transmembrane domain, a cytoplasmic domain that lacks tyrosine kinase activity and associates with JAK kinase, and form complexes that are homodimeric, heterodimeric or heterotrimeric (Bazan, 1990; Liongue et al., 2016). EPO binding to its homodimeric receptor complex brings the cytoplasmic associated JAK2 kinases in close proximity allowing for JAK transphosphorylation, phosphorylation of the receptor, and phosphorylation and activation of STAT and other downstream signaling pathways including AKT and ERK1/2 (Witthuhn et al., 1993; Miura et al., 1994; Watowich, 2011; Kuhrt and Wojchowski, 2015). Phosphorylated STAT5 subunits, STAT5A, and STAT5B, dimerize and translocate to the nucleus to activate select gene expression (Socolovsky et al., 2001).

EPO RESPONSE BEYOND ERYTHROID CELLS

Availability of recombinant human EPO facilitated discovery of EPO response in non-hematopoietic tissue. These include endothelial cells and the cardiovascular system, neural cells and the brain, myoblasts and skeletal muscle, adipocytes and fat depots, and bone (; McGee et al., 2012; Zhang et al., 2014; Hiram-Bab et al., 2017; **Figure 1**).

Endothelial EPO Response

Erythropoietin activity beyond erythropoiesis was first observed in endothelial cells. Cultures of primary endothelial cells from human umbilical vein and bovine adrenal capillary exhibited EPO binding and dose dependent proliferative and chemoattractant response to EPO, and EPOR is expressed in endothelial cells (Anagnostou et al., 1990, 1994). Prior to death *in utero*, mice that lack EPO or EPOR exhibit angiogenic defects including decreased vessel networks (Kertesz et al., 2004).

In endothelial cells, endothelial nitric oxide (NO) synthase (eNOS or NOS3) produces NO to regulate vascular tone and blood pressure. EPO stimulation of endothelial cells activates eNOS and NO production, particularly at reduced oxygen (Beleslin-Cokic et al., 2004, 2011). Transgenic mice expressing high level of human EPO with hematocrit of about 80% also exhibit markedly increased eNOS level and NO production (Ruschitzka et al., 2000). EPO induction by hypoxia or ischemic stress stimulates production of red blood cells to improve oxygen delivery and increase transport of oxygen from the lungs to the tissues. However, increased red blood cell production requires increased proliferation and differentiation of erythroid progenitor cells in bone marrow (and spleen in mice) to expand the erythroid lineage. EPO stimulated induction of NO provides the potential for an acute response regulating vascular tone and improving oxygen delivery.

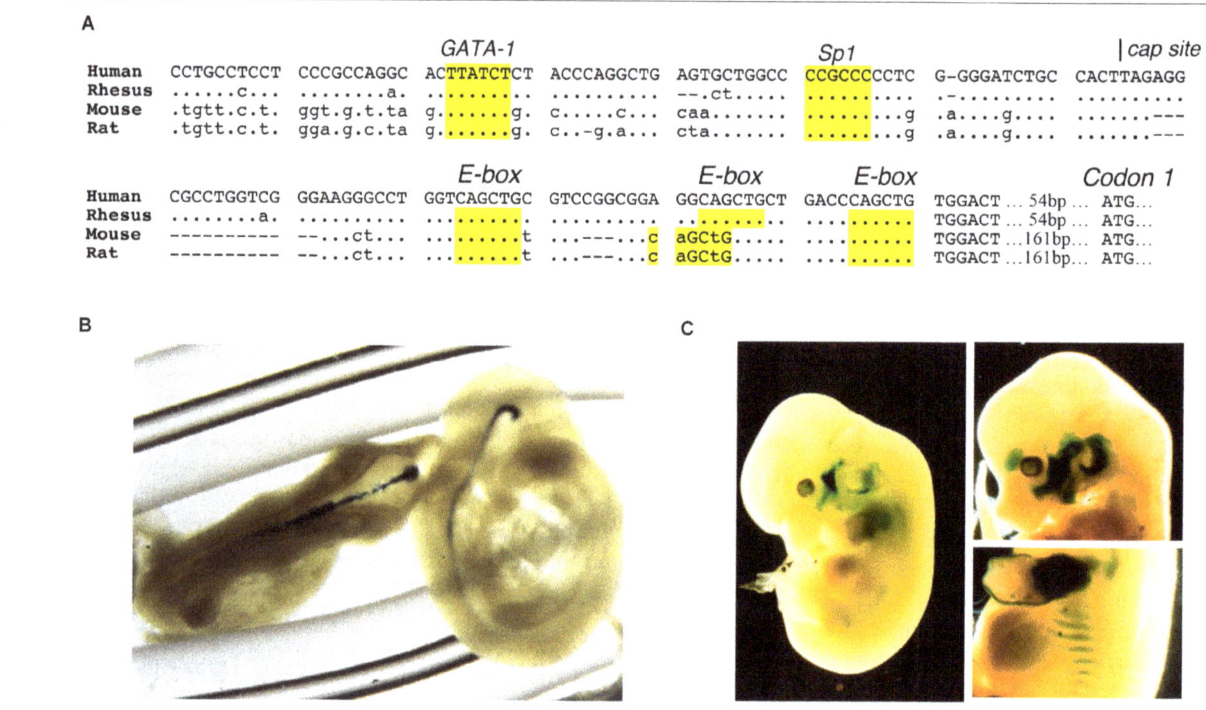

FIGURE 2 | Expression of EPOR reporter gene in transgenic mice. **(A)** The proximal promoter region of the human EPOR gene extending to the translation start site with ATG at +135 in the human EPOR gene, contains conserved regulatory binding sites for GATA proteins (AGATAA) and Sp1 (CCGCCC), and 3-E-boxes (CAGCTG) in the 5′ untranslated transcribed region that can bind basic-helix-loop-helix transcription factors such as erythroid TAL1 and skeletal muscle transcription factors Myf5 and MyoD. **(B)** Transgenic mice containing the human EPOR proximal promoter region extending 1778 bp 5′ of the transcription start site driving the β-galactosidase reporter gene shows EPOR expression in the embryonic brain at embryonic day E9.5 (from Liu et al., 1997, with permission). **(C)** Reporter gene expression at embryonic day 12.5 (left) and embryonic day 13.5 (right) in the visceral arches, base of limbs, intercostal rib regions, and fetal liver (from Ogilvie et al., 2000, with permission).

Erythropoietin stimulation of endothelial cells increases ET-1 secretion that may exacerbate adverse vascular effects (Carlini et al., 1993; Barhoumi et al., 2014). Mice expressing high level of transgenic human EPO with elevated hematocrit and increased NO production also have elevated ET-1 levels, but do not exhibit hypertension (Ruschitzka et al., 2000). Exposure to NO synthase inhibitor (N-nitro-L-arginine methyl ester) decreases eNOS production of NO, increases vasoconstriction and hypertension, and decreases survival of these mice, while pretreatment with ET(A) receptor antagonist (darusentan) improves survival, indicating that increased NO production in these mice suppresses the adverse effects associated with increased ET-1 (Quaschning et al., 2003).

EPO and Cardioprotection

During embryonic development, mice that lack EPO signaling exhibit ventricular hyperplasia indicating defect in proliferation and expansion of the myocardium (Wu et al., 1999; Yu et al., 2001). In studies of isolated adult hearts from rodents, EPO promoted cardioprotection in ischemia reperfusion injury (Cai et al., 2003; Parsa et al., 2003) and EPO administration in rabbits was cardioprotective from myocardial infarction (Cai and Semenza, 2004). In adult rats, EPO administration immediately after myocardial infarction reduced infarct size and improved cardiac function linked to neovascularization

(Van Der Meer et al., 2005). Furthermore, EPO treatment in a rat model for chronic heart failure suggested that long term EPO treatment stimulated homing of endothelial progenitor cells to induce neovascularization (Westenbrink et al., 2006). Delayed weekly EPO treatment beginning 7 days after coronary occlusion in rats resulted in reduced injury and improved cardiac function associated with mobilized endothelial progenitor cells (Prunier et al., 2007). Cardioprotective effects of EPO is mediated via induction of coronary endothelial production of NO and require activation of eNOS (Mihov et al., 2009). In genetically modified mice with EPOR restricted to hematopoietic and endothelial cells, EPO cardioprotection was observed in acute ischemia reperfusion injury in heart (Teng et al., 2011a). Cardioprotection was comparable to that observed in wild type mice, but not in eNOS knockout mice, providing evidence that EPOR expression in endothelial cells was sufficient for EPO protection in ischemia reperfusion injury in heart and that activation of eNOS and increased NO production is required (Teng et al., 2011a). In addition, high hematocrit associated with chronic EPO treatment could offset the EPO cardioprotective effect.

EPO and Neuroprotection

Erythropoietin receptor expression in neurons and select locations in brain (Masuda et al., 1993; Digicaylioglu et al., 1995; Morishita et al., 1997) combined with EPO production in astrocytes and neurons in a hypoxia inducible manner (Masuda et al., 1994; Marti et al., 1996) provide evidence for EPO signaling on the other side of the blood brain barrier. During mouse development, the brain expresses high level of EPOR midgestation, localized using reporter gene expression, and mice that lack EPOR exhibit thinning of the neuroepithelium, reduced neural progenitor cells with increased sensitivity to hypoxia, and increased brain apoptosis prior to embryonic death due to severe anemia (Liu et al., 1997; Yu et al., 2002; **Figure 2B**). Mice with EPOR expression restricted to hematopoietic and endothelial tissue and mice with targeted deletion of EPOR in neural cells show no gross morphological defects but do exhibit reduced neural cell proliferation and viability, and increased susceptibility to glutamate damage and stroke (Tsai et al., 2006; Chen et al., 2007). Conversely, EPO infusion into adult rodent brain increases the number of newly generated interneurons (Shingo et al., 2001). In animal models of brain injury, preconditioning with EPO infusion is protective for ischemia or middle cerebral artery occlusion – induced learning disability and neuron death (Sadamoto et al., 1998; Sakanaka et al., 1998; Bernaudin et al., 1999). In a rodent model of neonatal hypoxia/ischemia, EPO protection was associated with brain revascularization and neurogenesis (Iwai et al., 2007).

For clinical application of EPO for ischemic injury, a Phase I clinical trial for EPO treatment of ischemic stroke provided evidence for the safety and potential efficacy showing an association with improvement in clinical outcome at 1 month (Ehrenreich et al., 2002). A following Phase II/III clinical trial of EPO treatment with acute ischemic stroke treated 460 patients with EPO or placebo. Unexpectedly, 63% received recombinant tissue plasminogen activator for treating systemic thrombolysis that was not approved for use in ischemic stroke at the time of the Phase I trial. Favorable effects of EPO were not demonstrated, and overall death rate was 1.8 times higher in the EPO group than in the placebo group (Ehrenreich et al., 2009). Further subgroup analysis including post stroke biomarkers suggested that patients with ischemic stroke not receiving thrombolysis likely benefited from EPO treatment (Ehrenreich et al., 2009, 2011). This negative German Multicenter EPO Stroke Trial underscores the complexity posed by standard of care and difficulties in translating positive results from animal studies to the clinic.

Alternate EPO Receptors and EPO Derivatives

High hematocrit resulting from chronically administered EPO, particularly at high dose, is associated with adverse effects such as hypertension and thromboembolism and could counteract the neuroprotective and cardioprotective effects of EPO. The potential to activate EPO/EPOR protective response in non-hematopoietic tissue via an alternate EPO receptor or an EPO mimetic without increasing erythropoietic activity and hematocrit is of particular interest. An alternate EPO receptor has been proposed for non-hematopoietic tissues such as brain and heart, consisting of a heterodimer between EPOR and the beta common (βc) receptor, also a member of the class I cytokine

receptor superfamily and shared by receptors for granulocyte-macrophage colony stimulating factor and interleukins 3 and 5 (Jubinsky et al., 1997; Brines et al., 2004). EPO protection in a model of experimental colitis in mice is proposed to be mediated via activation of the EPOR/βc receptor heterodimer (Nairz et al., 2017). The role of the βc receptor in EPO response is controversial and evidence of direct interaction between EPOR and the βc receptor is lacking. In EPO responsive neural SH-Sy5y and PC12 cells, βc receptor is below the level of detection, and in rat brain, βc receptor does not colocalize with either EPO or EPOR (Nadam et al., 2007; Um et al., 2007). Furthermore, cardioprotection in mice by darbepoetin, a long acting derivative of EPO, did not require the βc receptor (Kanellakis et al., 2010). Biophysical analyses show that the extracellular domains of EPOR and the βc receptor do not directly interact in the presence or absence of EPO (Cheung Tung Shing et al., 2018), and the role of the βc receptor in EPOR response to EPO remains uncertain. Other proposed EPO binding receptors include the Ephrin B4 receptor on cancer cells and the orphan cytokine receptor CRLF3 on insect neural cells (Pradeep et al., 2015; Hahn et al., 2017).

Erythropoietin derivatives that can promote non-hematopoietic tissue protective effects, especially neuroprotection and cardioprotection, without stimulating erythropoiesis, have been proposed to bind alternate EPO receptors such as the EPOR/βc receptor heterodimer. These include asialoerythropoietin, carbamylated EPO, ARA 290 [cibinetide; helix B surface peptide (11 amino acid peptide derived from the EPO sequence)], and recombinant EV-3 (an EPO derived a spliced variant with exon 3 deleted) (Erbayraktar et al., 2003; Brines et al., 2004, 2008; Fiordaliso et al., 2005; Robertson et al., 2013; Bonnas et al., 2017). Clinical studies explored the safety and use of EPO and carbamylated EPO to increase frataxin levels for treatment of Friedreich's Ataxia (Boesch et al., 2014; Egger et al., 2014; Santner et al., 2014). ARA 290 treatment in phase 2 trials for Sarcoidosis-associated small nerve fiber loss showed improved abundance of corneal nerve fiber and neuropathic pain following 28 day treatment (Dahan et al., 2013; Culver et al., 2017). Another phase 2 trial focused on the potential benefit of ARA 290 treatment in subjects with type 2 diabetes for neuropathy as well as metabolic control (Brines et al., 2015). These EPO derivatives suggest the potential to therapeutically activate the tissue-protective activity associated with EPO while minimizing the risk attributed to increases in erythropoiesis and hematocrit.

Skeletal Muscle EPO Response

Reporter gene expression in transgenic mice during development revealed EPOR expression shared resemblance with the expression pattern in developing muscle associated with E-box binding basic-helix-loop-helix muscle transcription factors, MyoD and Myf-5, localizing in the visceral arches, proximal forelimb and intercostal area (Sadamoto et al., 1998; **Figure 2C**). Primary satellite cells isolated from mouse and human skeletal muscle express EPOR (Ogilvie et al., 2000; Rundqvist et al., 2009). Like erythroid progenitor cells, EPO increases proliferation of C2C12 myoblasts and induces EPOR expression that decreases with cell differentiation (Ogilvie et al., 2000). EPOR expression is regulated by GATA3, GATA4, and TAL1, and can also be transactivated by MyoD and Myf-5 (Ogilvie et al., 2000; Wang et al., 2012). EPO stimulates myoblast proliferation and survival, increases GATA4 and TAL1 that retard myogenic differentiation, mediated in part by Sirt1 activity, and inhibits expression of myogenin associated with differentiating myoblasts and myotube formation (Wang et al., 2012). Improved survival was demonstrated by transplantation of myoblasts over-expressing EPOR into skeletal muscle (Jia et al., 2009). In skeletal muscle, Pax-7$^+$ satellite cells can self-renew or differentiate to Myf5$^+$ committed muscle progenitor cells that contribute to growth, maintenance and repair of skeletal muscle (Kuang et al., 2007). Mice that express high transgenic human EPO have increased skeletal muscle Pax-7$^+$ satellite cells and isolated primary myoblasts had enhanced proliferation in culture compared to wild type cultures (Jia et al., 2012). These mice subjected to skeletal muscle injury exhibited improved muscle repair and recovery and increased maximum load tolerated by isolated muscle. In contrast, ΔEpoR$_E$ have fewer Pax-7$^+$ satellite cells and isolated primary myoblasts from ΔEpoR$_E$ mice do not proliferate in culture, and in skeletal muscle injury, these mice show delayed muscle repair and recovery, and reduced maximum load tolerated by isolated muscle (Jia et al., 2012). Furthermore, EPO treatment increases Pax-7$^+$ satellite cells and promotes repair and recovery from skeletal muscle injury (Jia et al., 2012). Skeletal muscle myoblasts produced endogenous EPO that increased at low oxygen, and transgenic mice with high transgenic human EPO exhibit mouse EPO and elevated human EPO expression in primary myoblasts, raising the possibility of an autocrine EPO response in skeletal muscle (Jia et al., 2012). Transgenic knockdown of circulating EPO levels did not show any change in EPOR gene expression in mouse skeletal muscle (Hagström et al., 2010; Mille-Hamard et al., 2012).

In humans, a single injection of EPO increased myogenic regulatory factor *MYF6* mRNA (Lundby et al., 2008) and EPO treatment increased PAX7 and MYOD1 content in human satellite cells (Hoedt et al., 2016), suggesting a role for EPO and its receptor in muscle development or remodeling. EPO administration also enhanced muscle mitochondrial oxidative phosphorylation and electron chain transport capacity following 8 weeks of treatment (Plenge et al., 2012). EPO stimulation in C2C12 myoblasts increased JAK2, STAT5 (Sadamoto et al., 1998) and AKT phosphorylation (Jia et al., 2009), and mice expressing elevated EPO in skeletal muscle by gene transfer also exhibited increased AKT phosphorylation (Hojman et al., 2009). In humans, a single EPO injection followed by exercise was not sufficient to activate the AKT pathway in skeletal muscle (Lamon et al., 2016).

EPO and Skeletal Muscle Fiber Type

Skeletal muscles of vertebrates contain mainly two types of muscle myofibers, type I (slow twitch) and type II (fast twitch) that differ in their function, mitochondrial density, and metabolic properties (Zierath and Hawley, 2004). Type I muscle fibers contain a high concentration of mitochondria and high oxidative capacity, and exhibit fatigue resistance and prolonged duration of muscle activity (Zierath and Hawley, 2004). Endogenous

EPO contributes to muscle myofiber type. In ΔEpoRE mice with EPO activity restricted to erythroid tissue, skeletal muscles exhibit fewer type I muscle fibers and reduced mitochondrial activity (Wang et al., 2013a). In contrast, skeletal muscles from transgenic mice with high EPO production with high transgenic EPO production show an increase in the proportion of type I muscle fibers and increased mitochondrial activity (Wang et al., 2013a). Furthermore, mice with high EPO production show prolonged position holding time to an inverted wire grid, suggesting that elevated EPO results in improved muscle response to fatigue (Jia et al., 2012). PGC-1α is expressed mainly and preferentially in type I muscle fibers and activates mitochondrial biogenesis and oxidative metabolism in mice (Lin et al., 2002). EPO treatment of primary skeletal myoblast cultures increases mitochondrial biogenesis gene expression including PGC-1α, increased cytochrome C and oxygen consumption rate that can contribute to skeletal muscle fiber programming and development of type I muscle fibers (Wang et al., 2013a).

EPO PROTECTION IN DIET INDUCED OBESITY

Erythropoietin regulation of metabolism extends beyond oxygen delivery and contributes to maintenance of white adipose tissue and metabolic homeostasis. Diet-induced obesity gives rise to glucose intolerance and insulin resistance, leading to type 2 diabetes. Animal studies suggest that EPO may be protective in diet-induced obesity, improves glucose tolerance, reduces insulin resistance and regulates fat mass accumulation, particularly in male mice (Wang et al., 2014; Zhang et al., 2014; Alnaeeli and Noguchi, 2015).

EPO and Inflammation in Obese White Adipose Tissue

The anti-apoptotic and protective effects of EPO in select tissues such as adult and preterm brain contribute to an anti-inflammatory response, inhibiting expression of proinflammatory cytokines and reducing macrophage infiltration (Villa et al., 2003; Wassink et al., 2017). The immune-modulatory activity of EPO as observed in the gut is mediated by JAK2 activation and inhibition of macrophage NF-kB response (Nairz et al., 2011). In diet-induced obesity, EPO modulates the proinflammatory response of macrophage infiltration in white adipose tissue and promotes an anti-inflammatory phenotype (Alnaeeli et al., 2014).

In white adipose tissue, macrophages are found in the stromal vascular fraction. High fat diet feeding in C57BL/6 mice results in inflammation and macrophage infiltration of white adipose tissue, resulting in crown-like structures of macrophages surrounding necrotic adipocytes. Among non-hematopoietic tissues, EPOR is highly expressed in white adipose tissue and specifically on adipocytes and macrophages in the stromal vascular fraction, and EPO treatment exhibits anti-inflammatory activity in white adipose tissue of obese mice (Alnaeeli and Noguchi, 2015). In obese C57BL/6 male mice, short term EPO treatment (2 weeks) increases hematocrit without affecting body mass, improves glucose tolerance and insulin resistance, and shifts the inflammation of white adipose tissue associated with high fat diet feeding toward an anti-inflammatory phenotype (Alnaeeli et al., 2014). EPO treatment reduces inflammation and crown-like structures in white adipose tissue, decreases pro-inflammatory cytokine/chemokine expression, and increases anti-inflammatory cytokine interleukin 10 expression. EPO reduces the total number of macrophages and shifts the remaining macrophage population toward the anti-inflammatory macrophage subtype. EPO promotes STAT3 phosphorylation in white adipose tissue macrophages and the EPO stimulated increase in the anti-inflammatory macrophage subtype is dependent on interleukin 4/STAT6 signaling. Furthermore, ΔEpoR$_E$ lack EPOR expression in white adipose tissue and exhibit higher circulating inflammatory monocytes on high fat diet feeding compared with wild type mice, suggesting a role for immune regulation by endogenous EPO/EPOR signaling (Alnaeeli and Noguchi, 2015). ΔEpoR$_E$ mice on high fat diet show even greater inflammation and crown-like structures in white adipose tissue, elevated cytokine/chemokine expression in perigonadal white adipose tissue stromal vascular fraction, glucose intolerance and insulin resistance. When treated with EPO, ΔEpoR$_E$ mice exhibit the expected increase in hematocrit without significant difference in glucose tolerance or inflammation in white adipose tissue (Alnaeeli et al., 2014).

Insulin resistance associated with diet induced obesity has been linked to inflammation of white adipose tissue (Lumeng and Saltiel, 2011; Han and Levings, 2013; Chen et al., 2019). This suggests that the activity of EPO to reduce white adipose tissue inflammation in diet induced obesity may contribute to EPO stimulated improvement in insulin resistance. Other EPO associated metabolic activity can also affect insulin resistance. Adipocyte response to EPO contributes to insulin sensitivity and C57BL/6 mice with adipocyte-specific deletion of EPOR on high fat diet exhibit decreased glucose tolerance and insulin sensitivity, an effect that may depend on mouse background strain (Luk et al., 2013; Wang et al., 2013b). In pancreatic β-cells EPO exerts JAK2 dependent protective effects and induces proliferative, anti-inflammatory and angiogenic activity within the islets in mouse diabetic models (Choi et al., 2010). EPO enhances AKT activation in liver, inhibits gluconeogenesis in high fat diet fed mice and reduces liver inflammation associated with diet induced obesity (Meng et al., 2013). EPO activity in brain, particularly the hypothalamus, also influences metabolic homeostasis (Teng et al., 2011b; Dey et al., 2016).

Endogenous EPO Is Required for Erythropoiesis and Regulates Fat Mass Accumulation

Expression of EPOR beyond erythroid tissue and the protective effects of EPO administration in animal models of ischemia and traumatic injury in non-hematopoietic tissue such as the cardiovascular system, brain and skeletal muscle raise questions about the requisite role of EPO beyond regulation of red blood cell production. ΔEpoR$_E$ mice with EPOR expression restricted to erythroid tissue were created using an erythroid specific transgene expressing EPOR cDNA driven by the erythroid transcription regulatory regions of the GATA1 erythroid transcription factor to rescue the EPOR knockout mouse (mEPOR-/-) (Suzuki et al., 2002). These mice survive through adulthood, providing evidence that the primary and necessary function of EPO is to regulate red blood cell production, and that the non-hematopoietic EPO activity is dispensable for embryonic development.

While ΔEpoR$_E$ mice created on a C56BL/6 background exhibit no gross morphologic defects, they exhibit a disproportionate accumulation of fat mass with age (Teng et al., 2011b). By 4 months, the body mass is 60% greater in female ΔEpoR$_E$ mice than wild type control mice and 25% greater in male ΔEpoR$_E$ mice than wild type control mice (**Figures 3A,B**). The increase in body mass in ΔEpoR$_E$ mice is due to an increase in both the visceral and subcutaneous white fat. Fat mass continues to disproportionately increase in ΔEpoR$_E$ mice and by 8 months, fat mass is more than doubled in female ΔEpoR$_E$ mice compared to control. ΔEpoR$_E$ mice are glucose intolerant and with increasing accumulation of fat mass become insulin resistant by 4 months in female and by 6 months in male mice. While there is no difference in food intake by ΔEpoR$_E$ mice on normal chow, ΔEpoR$_E$ mice exhibited greater body weight gain normalized to food intake, consistent with decreased energy expenditure. Even before overt obesity, young ΔEpoR$_E$ mice exhibited decreased locomotor activity and decreased metabolic rate assessed by indirect calorimetry. On high fat diet, ΔEpoR$_E$ mice behaved similarly with food intake comparable to wild type control mice and had decreased locomotor activity and metabolic rate.

C57BL/6 mice with targeted deletion of EPOR in adipocytes fed normal chow exhibited a modest increase in body weight and fat mass with lower total activity and oxygen consumption (Wang et al., 2013b). High fat diet feeding further accentuated these differences, although food intake was comparable with wild type control mice. Young mice that lack EPOR in adipocytes fed high fat diet for 6 weeks showed a 1.4-fold increase in fat mass compared with control mice and higher blood glucose and serum insulin levels, glucose intolerance and insulin resistance, suggesting that endogenous EPOR expression on adipocytes contributes significantly to metabolic regulation.

EPO Activity in Adipocytes and Regulation of Fat Mass Accumulation

Initial studies in mice suggesting EPO activity in metabolic homeostasis involved skeletal muscle gene transfer to over-express EPO that resulted in weight reduction in obese mice due to reduction in fat mass accompanied by increased muscle oxidation and normalization of glucose sensitivity (Hojman et al., 2009; Teng et al., 2011b). Further studies of exogenous EPO treatment in mice including genetic mouse models of obesity and transgenic mice constitutively overexpressing human EPO showed that elevated serum EPO levels decreased blood glucose and decreased body weight, especially body weight gain in obese mice (Katz et al., 2010). Interestingly, hemodialysis patients on short-term EPO treatment showed improved glucose metabolism

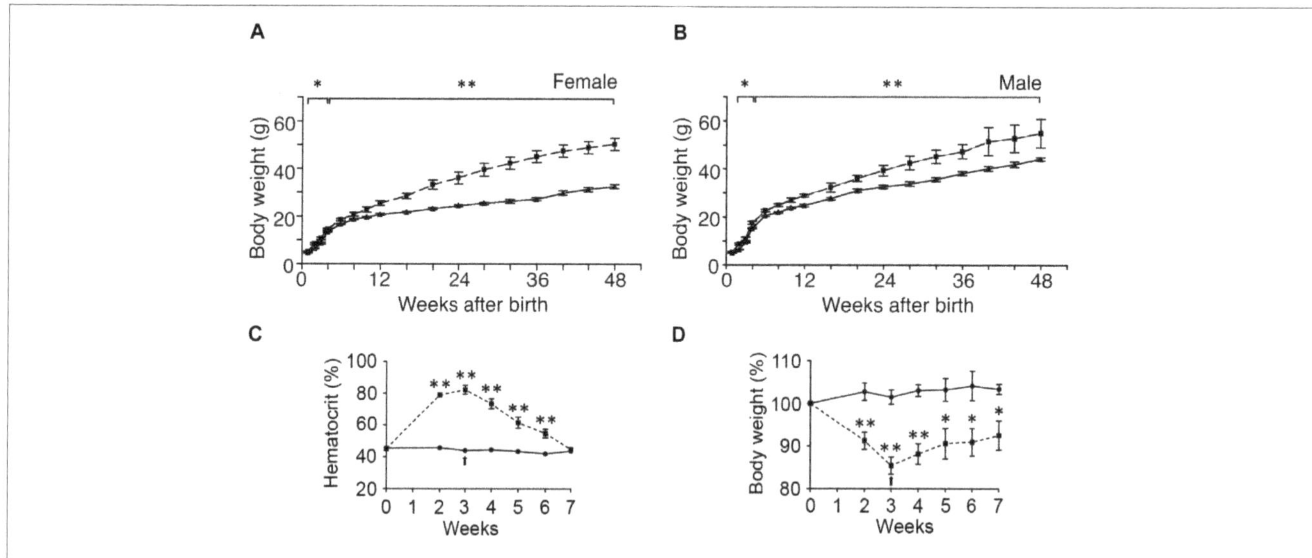

FIGURE 3 | Endogenous and exogenous EPO signaling regulates white fat mass accumulation. **(A,B)** Body weight to 48 weeks of age are indicated for wild type mice (solid line) and mice with EPOR restricted to erythroid tissue (Suzuki et al., 2002) (ΔEpoR$_E$) (dashed line) for females **(A)** and males **(B)**. **(C,D)** Hematocrit **(C)** and % body weight normalized to starting body weight **(D)** for male wild type mice subjected to 3 weeks of EPO treatment at 3000 U/kg three times weekly (dashed line) or saline (solid line). Arrow indicates end of EPO treatment. *$p < 0.05$; **$p < 0.01$ (from Teng et al., 2011b, with permission).

indicated by a reduction in insulin resistance and decreased glycated hemoglobin and hyperleptinemia (Osman et al., 2017). Accompanying the increase in hematocrit with EPO treatment in C57BL/6 male mice is a decrease of body weight in mice fed normal chow and a reduction in weight gain and fat mass accumulation in mice on high fat diet feeding (Teng et al., 2011b; **Figures 3C,D**). However, ΔEpoR$_E$ treated with EPO show the expected increase in hematocrit without change in body weight indicating that EPO regulation of fat mass is independent of EPO stimulated red blood cell production (Teng et al., 2011b). Mice with selective deletion of EPOR on adipocytes exhibit increased hematocrit with EPO treatment, but only a non-significant decreasing trend in body weight, exemplifying the direct role of EPO response in adipocytes to EPO regulation of body weight and fat mass accumulation in addition to glucose metabolism and insulin sensitivity (Wang et al., 2013b).

In white adipose tissue, EPOR is expressed at high level (about 60% of spleen, a mouse hematopoietic tissue), and in culture, EPO treatment decreases preadipocyte differentiation and induces ERK activation in primary mouse embryonic fibroblasts but not in embryonic fibroblasts generated from ΔEpoR$_E$ mice (Teng et al., 2011b). Despite the increase in fat mass in ΔEpoR$_E$ mice, analysis of adipocyte size distribution in gonadal fat pads showed a shift to smaller cells in ΔEpoR$_E$ mice indicating a marked increase in adipocyte number with loss of EPOR in non-hematopoietic tissue, providing further evidence that endogenous EPO contributes to regulation of adipocyte number in addition to fat mass accumulation (Teng et al., 2011b).

Insulin stimulates AKT activation and analysis of white adipose tissue from male C57BL/6 mice revealed that EPO treatment also increased AKT phosphorylation in white adipose tissue, but not in mice with targeted deletion of EPOR in adipocytes, suggesting that EPO modulates AKT activation in

white adipose tissue and potentially affects insulin signaling (Wang et al., 2013b). Further analysis of white adipose tissue shows that EPO treatment promotes a brown fat-like program, increases mitochondrial biogenesis independent of changes in body weight, and increases cellular respiration rate, and decreases the white fat-like program (Wang et al., 2013b). Conversely, white adipose tissue from mice with targeted deletion of EPOR in adipocytes shows decreased mitochondrial biogenesis, decreased cellular respiration rate and suppression of brown fat-like program and increase in white fat-like program, and no response to EPO stimulation (Wang et al., 2013b). EPO activity in white adipocytes is mediated via increased PPARa that cooperates with increased activity of metabolic sensor Sirt1, an NAD-dependent class III histone deacetylase sirtuin.

In addition to EPO regulation of fat mass accumulation in white adipose tissue, EPO decreased lipid accumulation in the liver while stimulating STAT3/STAT5 activation and promoting lipolysis in white adipose tissue, suggesting benefit in non-alcoholic fatty liver disease (Tsuma et al., 2019). In brown adipose tissue of young C57BL/6 mice, EPO treatment stimulated STAT3 and upregulated transcription factor PRDM16 that controls brown adipocyte differentiation and total UCP1 that is essential for brown adipose tissue thermogenesis (Kodo et al., 2017).

EPO Regulation of Proopiomelanocortin (POMC) and Food Intake

Erythropoietin treatment in male mice fed high fat diet showed both increase in activity and decrease in food intake, pointing to potential EPO regulation in the central nervous system to regulate food intake and energy expenditure (Teng et al., 2011b). The hypothalamus regulates appetite by production in the arcuate nucleus of the orexigenic or appetite-increasing neuropeptide

Y and agouti-related protein, and the anorexigenic or appetite-suppressing precursor protein, POMC. EPOR is expressed in brain, and EPOR expression level in hypothalamus is comparable to white adipose tissue and EPOR in the hypothalamus colocalizes to POMC expressing neurons in the arcuate nucleus (Teng et al., 2011b). EPO treatment in C57BL6 male mice increases hypothalamus expression of POMC and α-MSH, a POMC cleavage product, but not expression of neuropeptide Y or agouti-related protein (Teng et al., 2011b; Dey et al., 2016). Leptin stimulates STAT3 activation in the hypothalamus resulting in production of several neuropeptides including POMC to regulate appetite; hypothalamus neural cultures show that EPO also activates STAT3 (Dey et al., 2016). Conversely, $\Delta EpoR_E$ mice show a decrease in STAT3 activation in hypothalamus and reduced POMC levels with and without EPO treatment, suggesting that EPO regulates appetite, and potentiates leptin response (Dey et al., 2016).

In the pituitary, POMC production gives rise to ACTH. Although EPO treatment in wild type mice increases hypothalamus production of POMC, EPO treatment decreases plasma ACTH level (Dey et al., 2015). On the other hand, plasma concentration of ACTH is high in $\Delta EpoR_E$ mice with reduced POMC production in the hypothalamus, suggesting that both endogenous and exogenous EPO contributes to regulation of plasma ACTH. EPO treatment in cultures of mouse corticotroph pituitary cell line shows decrease in basal intracellular calcium levels, no change in POMC mRNA and increased intracellular ACTH, indicating disruption of post-translational processing of POMC and inhibition of ACTH secretion. EPO regulation of pituitary derived ACTH plasma levels suggests a wider role for EPO regulation of metabolism and obesity via the neuroendocrine hypothalamic-pituitary axis (Dey and Noguchi, 2017).

Gender Specificity of EPO Action and Regulation of Fat Mass

Estrogen can affect EPO action via direct regulation of EPO production or modulation of tissue specific EPO response. For example, in adult female mice, estrogen dependent induction of EPO in the mouse uterus contributes to angiogenic activity and blood vessel formation in the uterine endometrium, contributing to the cyclic remodeling in the estrus cycle transition from diestrus to proestrus (Yasuda et al., 1998). With regards to environmental hypoxia, EPO affects the hypoxic ventilatory response via EPOR expression in brain and carotid body, and this response is increased in women and female mice compared with men and male mice (Soliz et al., 2012). This sexual dimorphism of EPO stimulated hypoxic ventilatory response is attributed, in part, to carotid body sensitivity to sex hormones, particularly estrogen.

Erythropoietin regulation of metabolism also appears to be sex-dependent. In $\Delta EpoR_E$ mice the age dependent accumulation of excess body fat is greater in female that develop obesity and insulin resistance by 4 months of age compared with male mice that exhibit a slower rate of body fat accumulation, becoming obese and insulin resistance at 6 months of age

(Teng et al., 2011b; **Figures 3A,B**). In contrast, fat mass is not altered by EPO treatment in female C57BL/6 mice on normal chow or high fat diet, while EPO treatment in male mice on normal chow decreases fat mass and EPO treatment in male mice on high fat diet reduces the accumulation of fat mass (Zhang et al., 2017; **Figures 4A,B**). Gender-specific response was also observed in gene expression in white adipose tissue where EPO treatment in mice increased expression in select oxidative genes in male mice on normal chow and on high fat diet, but not in female mice. The greater increase in fat mass in male mice on high fat diet compared with female mice is evidence of the protective effect of female hormones against diet induced obesity (**Figures 4A,B**) and raises the possibility that female hormones interfere with the anti-obesity effect of EPO observed in male mice. The anti-obesity effect of EPO on mice fed high fat diet was restored in ovariectomized mice. Like male mice, ovariectomized mice on high fat diet treated with EPO showed the decrease in fat mass (**Figure 4C**). The protective effect of estrogen to diet induced obesity was demonstrated in ovariectomized mice supplemented with estradiol which was more effective in reducing fat mass than EPO, and fat mass was not further enhanced with estrogen combined with EPO (**Figure 4C**). EPO stimulated increase in hematocrit was comparable in male in female mice indicating that the sex-dependent regulation of fat mass is independent of EPO regulated response in erythroid tissue.

To further examine the relationship of EPO level with body weight in human, endogenous plasma EPO concentration was assessed in a subset of full-heritage Southwestern Native Americans studied to understand the high prevalence of obesity and type 2 diabetes (Smith et al., 1996; Pavkov et al., 2007). As expected, endogenous plasma EPO level negatively associated with hemoglobin ($p = 0.005$) and no association was found for EPO and percent weight change per year in the study group of 79 individuals (Reinhardt et al., 2016). However, when segregated by sex, males exhibited an association between higher EPO concentrations and higher 24-h energy expenditure and an inverse association of endogenous EPO level with percent weight change per year ($p = 0.02$) (**Figure 5A**). In contrast, females exhibited a positive association of EPO plasma level with weight change per year ($p = 0.02$) (**Figure 5B**). Hence, endogenous EPO association with weight loss in men and with weight gain in women is distinct from EPO regulation of erythropoiesis that is comparable in both men and women and provides additional evidence for non-hematopoietic and gender-specific endogenous EPO action on regulation of body weight.

Elevated EPO in Human and Decreasing Obesity Prevalence at Increasing Altitude

In humans, as one ascends to high altitude increased EPO production via HIF regulation induces an EPO response that increases iron utilization for hemoglobin synthesis with expansion of red blood cell production with elevated blood hemoglobin and hematocrit (Gassmann and Muckenthaler, 1985; Smith et al., 2008). The relationship between elevated EPO and reduction in fat mass suggested by metabolic studies in

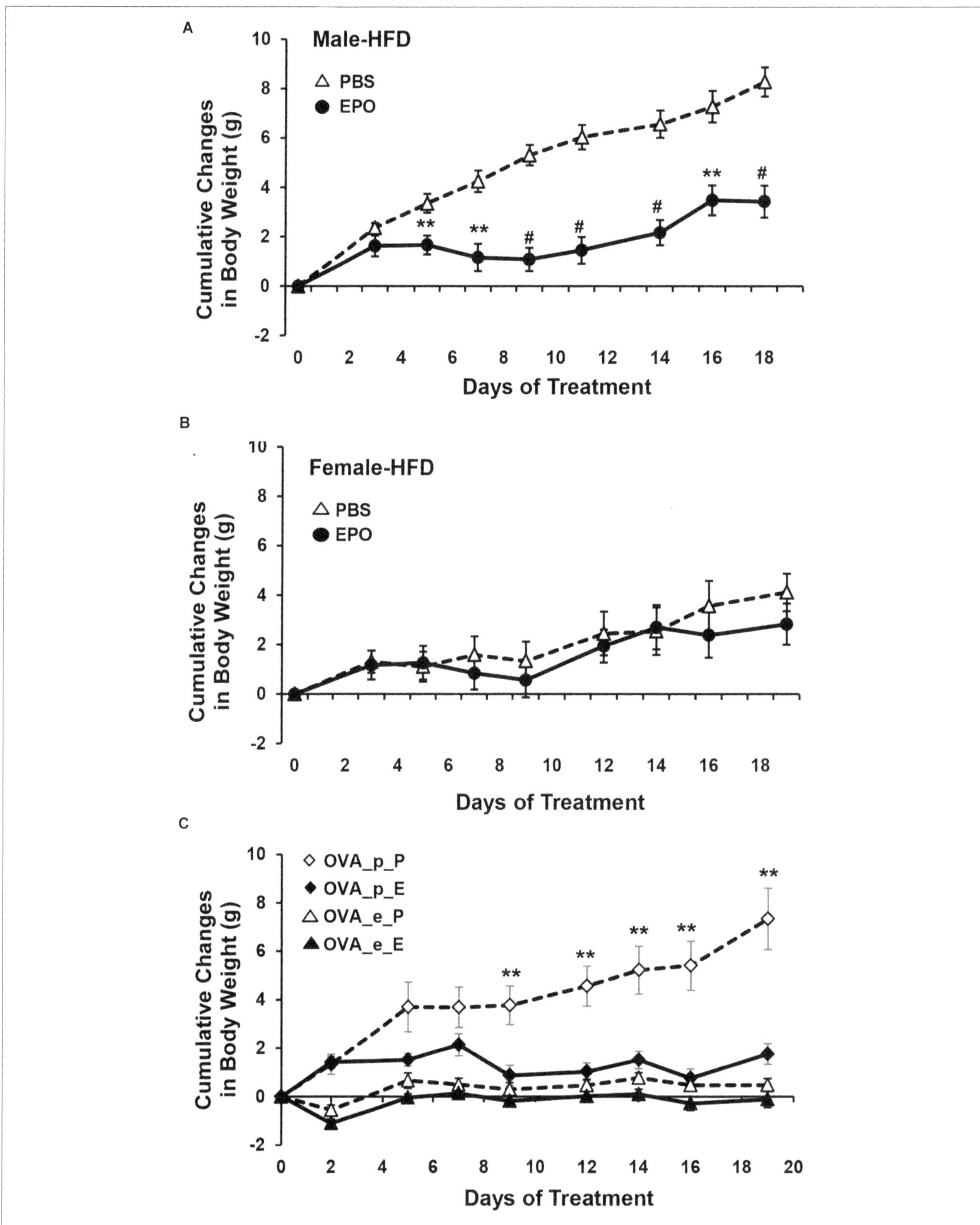

FIGURE 4 | EPO regulation of fat mass is gender specific. **(A,B)** Cumulative body weight change was monitored in male **(A)** and female **(B)** mice (16 weeks) fed high fat diet and treated with EPO at 3000 U/kg three times weekly (solid symbol) or saline (open symbol) for 3 weeks. **(C)** Cumulative body weight change in female ovariectomized (OVA) mice with placebo (p) or estradiol pellet-supplement and treated with EPO (E) or phosphate-buffered saline (P). $**p < 0.01$; $^{\#}p < 0.001$ (from Zhang et al., 2017, with permission).

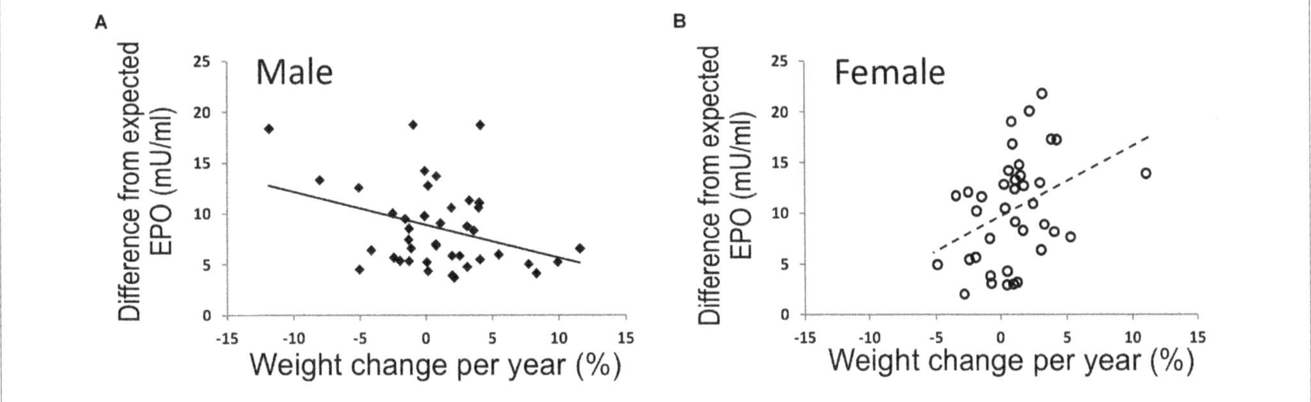

FIGURE 5 | EPO and % weight change per year were associated in opposing directions in males and females. **(A,B)** In full-heritage Southwestern Native Americans, the association of plasma EPO concentrations (adjusted for creatinine, hemoglobin, and storage time) and % weight change per year was negative in men **(A)** (closed squares, $N = 41$; $r = -0.35$, $p = 0.02$). Plasma EPO concentration was positively associated with % weight change per year in females **(B)** (open circles, $N = 38$; $r = 0.37$, $p = 0.02$) (modified from Reinhardt et al., 2016, with permission).

male mice (Teng et al., 2011b; Zhang et al., 2017), raises the possibility that increased EPO levels may explain in part the lower obesity rate reported for humans associated with residence at high altitude (Voss et al., 2013, 2014). In the United States, obesity prevalence is associated inversely with elevation and urbanization after adjusting for diet, ambient temperature, physical activity, smoking and demographic factors, and among overweight service members (proportion male of 93% at low altitude and 94% at high altitude), those stationed at high altitude were associated with lower incidence of obesity (Voss et al., 2013, 2014).

Note that high EPO levels can increase the risks of adverse effects such as cardiovascular complications and thromboembolic events, seizures, pulmonary embolism and death. For individuals in residence above 2500 m above sea level, about 5–10% can develop excessive erythrocytosis and associated Chronic Mountain Sickness (Villafuerte and Corante, 2016). However, Tibetan high-altitude natives are the exception and exhibit a different adaptive response from the hypoxia induction of EPO, allowing more efficient use of oxygen in their tissues (Horscroft et al., 2017). Even without high altitude, increase risk of adverse effects are evident with high endogenous EPO. In the elderly, elevated EPO levels were associated with increased risk of death, and in renal transplant recipients higher endogenous plasma EPO was associated with cardiovascular mortality and all-cause mortality (den Elzen et al., 2010; Sinkeler et al., 2012). Elevated EPO levels may be indicative of EPO resistance resulting from underlying co-morbidities including inflammation, iron status, malnutrition and disrupted NO metabolism (Ganz and Nemeth, 2016; Yokoro et al., 2017). The increase risk of adverse effects with high exogenous EPO administration is exemplified in extreme athletes who use EPO as a performance-enhancing drug. Around the time that EPO received FDA approval in 1989 for treatment of anemia in chronic renal disease, death associated with EPO doping was suspected in eighteen young professional cyclists who died from unknown causes (Noakes, 2004). While patients with renal failure clearly benefit from EPO treatment to correct anemia (Eschbach et al., 1987;

Us Renal Data System, 2007; Hasegawa et al., 2018) several clinical trials of EPO treatment to normalize hemoglobin values in kidney disease showed increase cardiovascular morbidity and mortality associated with high dose EPO treatment (Drueke et al., 2006; Singh et al., 2006). With increased concerns about safety of high dose EPO, the FDA issued a "Black Box" warning in 2007 indicating reduced EPO dose to achieve lower target hemoglobin levels. Labeling for erythropoiesis-stimulating agents was further modified by the FDA in 2011 to lower dosing recommendations. Hence, potential benefits and potential risks must be considered with EPO treatment.

EPO PROMOTES CONTEXT-DEPENDENT BONE REMODELING

In bone, functional EPO receptors are expressed by osteoblasts and osteoclasts that remodel the bone and by BMSCs that differentiate into osteoblasts, bone marrow adipocytes and chondrocytes. Studies on the effects of EPO on these specific cell types and overall bone homeostasis have utilized *in vitro* cultures (Kim et al., 2012; Rolfing et al., 2014a; Li et al., 2015), *in vivo* BMSC osteogenic assays using ectopic ossification models (Suresh et al., 2019), bone fracture healing models (Holstein et al., 2007; Mihmanli et al., 2009; Rolfing et al., 2014b; Omlor et al., 2016), transgenic mouse model with high human EPO (model for chronic EPO exposure) (Hiram-Bab et al., 2015; Suresh et al., 2019), administration of exogenous EPO in healthy mice (Shiozawa et al., 2010; Singbrant et al., 2011; Hiram-Bab et al., 2015; Suresh et al., 2019) (model for acute EPO exposure) and, recently, $\Delta EpoR_E$ mice that do not express EPOR in non-erythroid cells (Suresh et al., 2019) (model for endogenous EPO signaling).

Osteoclasts and EPO Response

Several reports including conflicting results suggest EPO regulating osteoblasts, osteoclasts and BMSCs in *in vitro* culture conditions. Osteoclast differentiation assays using marrow mononuclear cells (Shiozawa et al., 2010), non-adherent bone marrow cells (Hiram-Bab et al., 2015) and RAW264.7 mouse monocyte/macrophage cell line show EPO increasing osteoclast numbers (Li et al., 2015), but not osteoclast activity (Shiozawa et al., 2010; Li et al., 2015). However, increase in osteoclast numbers was not reported in other studies using mouse bone marrow cultures (Suresh et al., 2019) and in osteoblast-osteoclast cocultures treated with EPO (Singbrant et al., 2011). In cultures of preosteoclasts, EPO activates JAK2/PI3K pathways without affecting proliferation and in osteoclasts, EPOR expression decreases with differentiation (Hiram-Bab et al., 2015). Transgenic mice with chronic exposure to high levels of human EPO had more osteoclasts lining the bone surface (Hiram-Bab et al., 2015). Osteoclasts from these mice produced EPO and corresponding BMSC derived cultures from these mice exhibited a greater number of giant multinucleated osteoclasts (Suresh et al., 2019).

Osteoblasts and EPO Response

For osteoblast differentiation assays, studies have utilized mouse primary calvarial osteogenic cells as well as adherent BMSCs. EPO treatment of primary mouse calvarial osteogenic cells did not affect differentiation (Hiram-Bab et al., 2015; Suresh et al., 2019). However, in transgenic mice expressing high human EPO, calvarial osteoblasts produced human EPO and showed increased ALP expression and mineralization (Suresh et al., 2019). Conversely, primary calvarial osteoblasts lacking endogenous EPO signaling had reduced ALP expression and mineralization (Suresh et al., 2019). Other osteoblast assays using human and mouse BMSCs treated with EPO reported increased osteoblast differentiation with the activation of EphrinB2/EphrinB4 (Li et al., 2015), mTOR (Kim et al., 2012), JAK2/PI3K pathways (Rolfing et al., 2014a). BMSC cultures with low EPO doses less than 5 U/mL reduced osteoblast mineralization whereas high EPO doses (50 U/mL–250 U/mL) increased the BMSC proliferation and differentiation into osteoblasts (Rauner et al., 2016). EPO stimulation of HSCs to produce BMP and thereby stimulating osteoblasts are also reported (Shiozawa et al., 2010). In a recent study using mesenchymal stem cells from young and old healthy individuals and patients with MDS, EPO treatment increased mineralization only in cells from young healthy individuals and not in cells from older healthy donors and cells from MDS patients (Balaian et al., 2018), suggesting that EPO bone remodeling activity may be age dependent. EPO treatment of mesenchymal stem cells from MDS patients inhibited Wnt signaling and reactivation of Wnt signaling combined with EPO treatment promoted their differentiation to osteoblasts (Balaian et al., 2018).

EPO and Bone Fracture Repair

A role for EPO in bone regulation was initially reported in mouse models of fracture repair, where 5000 U/kg EPO doses for 6 days stimulated early endochondral ossification and bone mineralization along with reduced EPOR in differentiating chondrocytes (Holstein et al., 2007). A follow up study showed accelerated bone healing with much reduced dose of EPO at 500 U/kg for five-week treatment where EPO promoted

endosteal vascularization and reduced NF-KB expression in the fracture callus suggesting anti-inflammatory role for EPO (Garcia et al., 2011). In alveolar bone regeneration studies in rats, EPO administration into the tooth sockets, promoted new bone formation while inhibiting bone resorption (Li et al., 2015). In rabbits with mandibular distraction osteogenesis, 150 IU/kg EPO treatment for 30 days resulted in new bone formation with increased osteoblasts, blood vessels and reduced osteoclasts (Mihmanli et al., 2009). In rabbits, implantation of gelatin sponges soaked with EPO near the bone defects followed by a single high EPO dose of 4900 IU/kg accelerated bone healing and vascularization in the callus (Omlor et al., 2016). In porcine models with calvarial defects, a single EPO dose of 900 IU/ml in collagen carrier moderately increased bone healing without any increase in vasculature (Rolfing et al., 2014b). However, in a separate study in porcine models, EPO enhanced bone healing only in combination with bone marrow concentrate containing mesenchymal stem cells (Betsch et al., 2014).

Studies have also utilized the potential of EPO in recruiting BMSCs to sites of bone healing. In rats with intratibial fracture, tail vein injection of BMSCs along with intramuscular injection of EPO was shown to mobilize BMSCs to bone defects and enhanced bone regeneration while increasing bone strength (Li et al., 2019). EPO loaded scaffolds in BALB/c mice showed increased recruitment of multipotent stem cells, this study also demonstrated that in mouse calvarial defect model, implantation of scaffolds with EPO resulted in accelerated bone healing compared to BMP2 scaffolds (Nair et al., 2013). Two-week EPO treatment of murine cranial defect models with BMP2 scaffolds implanted at the defect site also accelerated healing while promoting stem cell recruitment to the scaffolds (Sun et al., 2012). In patients with tibiofibular fracture, administration of 4000 IU of EPO in the fractures site 2 weeks after surgery had 2 week faster union and fewer non-union rates compared to the placebo group (Bakhshi et al., 2013; Nemati and Fallahi, 2013), suggesting possibility of clinical EPO use in accelerating fracture healing. Thus, using multiple species and various EPO doses and mode of administration, EPO treatment has been demonstrated to exert a beneficial role in bone healing.

Bone Loss Concomitant With EPO Stimulated Erythropoiesis in Mice

In contrast to the bone formation potential of EPO in fracture healing studies, EPO administration in healthy mice results in significant reduction of trabecular bone volume. EPO administration in 9 week old male C57BL/6 mice for 10 days at a physiological dose of 300 U/kg resulted in significant trabecular bone loss in tibia along with increase in the number of osteoclasts and osteoblasts, indicating increased bone turnover induced by EPO treatment (Singbrant et al., 2011). However, dynamic morphometry analysis revealed that EPO administration did not affect mineral apposition rate or double labeled surface, but significantly reduced single labeled surface, suggesting a reduction in osteoblast activity

(Singbrant et al., 2011). Reduction in trabecular bone coupled with increased osteoclast numbers in the femurs of 12-week-old female mouse models was also observed in chronic and acute EPO exposure (Hiram-Bab et al., 2015). Since addition of EPO increased only osteoclast differentiation *in vitro* and did not affect osteoblast differentiation in cultures, the reduced bone formation *in vivo* with EPO treatment was attributed to an indirect effect of EPO (Hiram-Bab et al., 2015). However in subsequent studies, a very low dose of EPO administration in mice was shown to significantly reduce osteoblast activity without affecting bone resorption (Rauner et al., 2016). In contrast, in mice with chronically elevated high EPO, both osteoblast and osteoclast differentiation was increased suggesting increased remodeling contributing to their reduced bone mass (Suresh et al., 2019). Thus, the dose of EPO is important in determining the effect on osteoblasts and subsequently on its impact in bone health.

Studies in transgenic mice with conditional deletion of PHD2 have shown the importance of PHD2-HIF2α-EPO signaling in bone remodeling. Low bone density due to reduced osteoblast activity without affecting osteoclasts was observed in transgenic mice with conditional deletion of PHD2 in hematopoietic lineage, renal and neural cells. In these mice PHD2 deletion resulted in excessive HIF-2α induced EPO production increasing the hematocrit to 86%. Surprisingly, deletion of PHD2 in osteoblastic lineage increased bone density and reduced osteoclast numbers *in vivo* with no change in hematocrits. While deletion of PHD2 in osteoclast lineage did not result in any bone phenotype (Rauner et al., 2016).

Reduced trabecular bone was also observed in the femurs of C57BL/6 mice with 10 day EPO administration at 1200 U/kg (Suresh et al., 2019). In contrast to these studies, increased bone formation with EPO treatment was reported in the vertebrae of young and older mice receiving supraphysiologic doses of EPO up to 6000 U/kg for three times a week for 28 days (Shiozawa et al., 2010). Of note, the reported modest increase in hematocrits with such high doses of EPO treatment has not been explained (Shiozawa et al., 2010). These changes in hematocrit are in marked contrast to the expected increase in hematocrit associated with EPO dose. For example, low dose of 150 U/kg of EPO elevated hematocrits to 60% (Foskett et al., 2011) whilst higher doses of 3000 U/kg of EPO increased hematocrits to 70% in C57BL/6 mice (Zhang et al., 2017).

Osteoblasts exhibit the potential for EPO production. In genetically modified mice, targeted deletion of VHL in osteoblasts over-stabilizes the hypoxic response and gives rise to high EPO production in osteoblasts that leads to severe polycythemia (Rankin et al., 2012). Conversely, inactivation of HIF-2 in osteoblasts resulted in decreased EPO expression in bones of neonatal mice and remained at the limit of detection in adult mice. This novel finding that osteoblasts could produce EPO raises the potential for autocrine regulation of EPO response in osteoblasts. In other studies, EPO stimulated FGF23 production in HSCs was associated with an increase in serum FGF23 and reduced serum phosphate suggesting a possible mechanism of EPO induced bone reduction due to disrupted mineralization (Clinkenbeard et al., 2017).

Endogenous EPO Contributes to Bone Formation and Maintenance

Understanding of the effect of EPO in bone is derived predominantly from animal models with overexpression of EPO or acute EPO administration. Recent studies using $\Delta EpoR_E$ mice demonstrated the importance of endogenous EPO signaling in bone formation and maintenance. These mice had significant reduction in trabecular bone in both male and female mice of mature skeletal age (Suresh et al., 2019; **Figure 6**). Osteoblasts and BMSCs from $\Delta EpoR_E$ mice did not express EPOR, while osteoclasts expressed EPOR because the erythroid transgene in this model was GATA1 dependent and preosteoclasts express GATA1, suggesting that reduction in bone formation in these mice is related to loss of EPO response in osteoblasts rather than osteoclast activity. Administration of EPO in $\Delta EpoR_E$ mice increased hematocrits but did not reduce trabecular bone (**Figure 6**) indicating that EPO induced bone reduction is independent of erythropoiesis.

In addition to reduced trabecular bone, lack of endogenous EPO-EPOR signaling increased bone marrow adipocytes leading to fatty marrow. This raised the possibility of endogenous EPO signaling regulating BMSC differentiation to either osteoblastic or adipogenic lineage.

EPO Regulates Bone Marrow Stromal Cell Differentiation and Bone Formation

The potential for BMSCs to differentiate to osteoblasts or adipocytes can be assessed using transplantation into

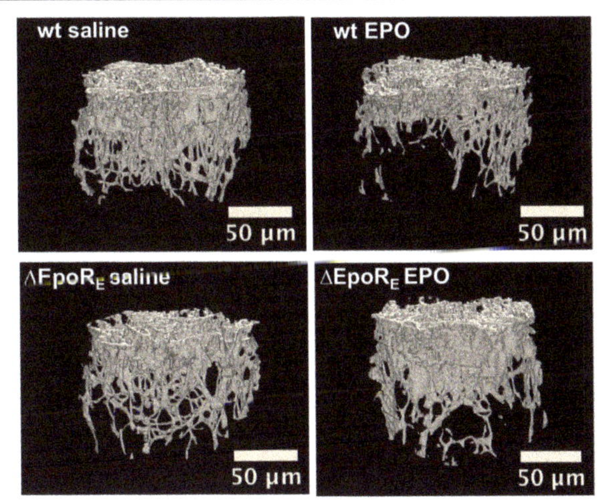

FIGURE 6 | Endogenous and exogenous EPO signaling regulates bone formation. Micro-CT 3D images of trabecular bone from wild type (wt) mice and mice with EPOR restricted to erythroid tissue ($\Delta EpoR_E$) (Suzuki et al., 2002). Mice (age 8 weeks) were treated with EPO at 1200 U/kg for 10 days (EPO) or saline. Images from saline treatment **(left)** show reduction in bone formation in $\Delta EpoR_E$ mice that lack EPOR in non-hematopoietic tissue. Furthermore, the reduction in bone parameters with EPO treatment in wild type mice **(top)** is not seen in $\Delta EpoR_E$ mice **(bottom)**, indicating that bone loss with EPO treatment is mediated by non-erythroid response (from Suresh et al., 2019, with permission).

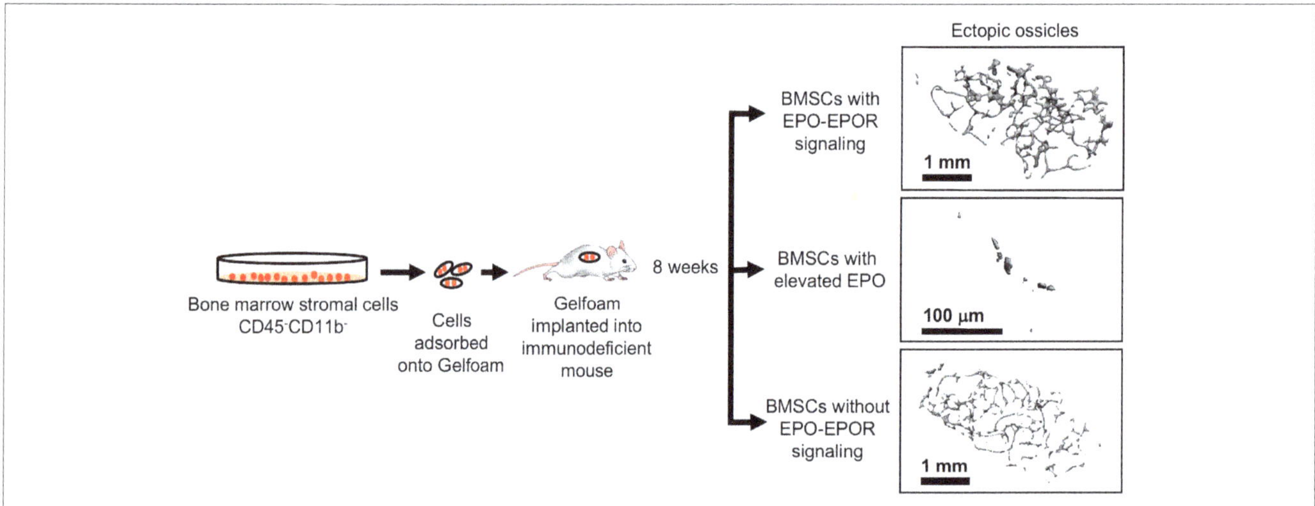

FIGURE 7 | Ectopic bone formation assay with BMSC. Adherent BMSC were selected from cultures of bone marrow cells harvested from wild type and genetically altered mice. Cells were absorbed into Gelfoam and surgically transplanted into immunocompromised NSG mice (Jackson Laboratory). After 8 weeks, ectopic bone ossicles formed and were analyzed by micro-CT. Ossicles formed in transplants from $\Delta EpoR_E$ mice that lack EPOR in non-hematopoietic tissue **(bottom)** showed similar structure to control mice **(top)** with a trend toward fewer trabeculae, consistent with a reduction in endogenous bone formation in $\Delta EpoR_E$ mice. Bone formation was markedly reduced with hardly any trabecular bone in ossicles in transplants from transgenic mice expressing high level of human EPO (Ruschitzka et al., 2000) **(middle)** compared with control mice **(top)**, reflecting the even greater reduction of endogenous bone formation in these transgenic mice compared with $\Delta EpoR_E$ mice and control mice (modified from Suresh et al., 2019, with permissions).

immunodeficient mice and monitoring bone ossicle formation. *In vivo* implantation of gelatin sponge cubes (Gelfoam) carrying BMSCs in immunodeficient mice model showed that BMSCs from $\Delta EpoR_E$ mice that lack of endogenous EPO signaling results in reduced ectopic bone formation and increased marrow adipogenesis (Suresh et al., 2019; **Figure 7**). In contrast, transplanting BMSCs from transgenic mice expressing elevated EPO significantly attenuated both ectopic bone formation and marrow adiposity (**Figure 7**). Furthermore, systemic elevated EPO inhibited BMP2 induced ectopic bone formation without affecting osteoclasts (Suresh et al., 2019). Thus, EPO is a critical regulator of bone. In the absence of endogenous EPO signaling there is reduced bone and increased marrow adiposity. Elevated EPO is also detrimental to bone formation demonstrated by several animal models of EPO administration in healthy animals resulting in significant bone reduction. On the other hand, other animal studies point to EPO potential in accelerating bone healing. Future studies focusing on optimizing the EPO dose and duration of treatment for clinical use along with the observation of any adverse events are warranted.

CONCLUSION

Hypoxia induction of EPO is an important response to ischemic stress resulting in increased red blood cell production to increase tissue oxygen delivery. Animal models have been useful in demonstrating EPOR expression beyond erythroid tissue and EPO response in non-hematopoietic tissue. For example, EPOR expression in vascular endothelium provides the potential for direct response via increase eNOS activation and NO production to regulate vascular tone

and oxygen delivery (Beleslin-Cokic et al., 2004, 2011). This acute response is in contrast to the time required to stimulate survival, proliferation and differentiation of erythroid progenitor cells to produce mature red blood cells. Requirement for eNOS for EPO protective activity in heart ischemic reperfusion injury also points to the potential benefit of this acute endothelial response. EPO stimulation of improved oxygen delivery may also contribute to protective action in other non-hematopoietic tissue such as brain ischemia and skeletal muscle injury.

Erythropoietin receptor expression is not for erythroid cells only and expression in non-erythroid tissues provides for tissue specific EPO protective response during ischemic challenge or injury. EPOR expression and EPO activity include the following tissues: endothelium to regulate vascular tone, improve oxygen delivery and provide cardioprotection to ischemic injury; brain (particularly neurons) to provide protection to ischemic stress or injury; skeletal muscle (myoblasts) for tissue maintenance or repair; white adipose tissue (adipocytes and macrophages) to protect from inflammation and from increase in fat mass in male mice during diet-induced obesity; and BMSC and osteoblasts to maintain normal bone development and bone remodeling accompanying exogenous EPO stimulated erythropoiesis (**Figure 1**). Sites of EPO production also extend beyond fetal liver and adult kidney and include the other side of the blood-brain barrier (astrocytes and neurons) (Masuda et al., 1994; Marti et al., 1996), skeletal muscle myoblasts (Kuang et al., 2007), and osteoblasts (Rankin et al., 2012), and estrogen stimulated induction in the uterus contributing to angiogenesis (Yasuda et al., 1998), raising the possibility of EPO/EPOR autocrine response even with low EPOR expression in select non-hematopoietic cells.

In regulation of skeletal bone formation, endogenous EPO contributes to development and maintenance of skeletal bone and bone marrow adipocytes, and loss of non-hematopoietic EPO activity results in decreased bone formation and increased marrow adiposity (Suresh et al., 2019). EPO administration has also demonstrated the capacity to affect bone remodeling, but in two disparate ways: accelerate bone healing in animal models of bone fracture, and cause bone loss in healthy animals responding with increased erythropoiesis. More studies in understanding the role of EPO in bone in specific pathological conditions are warranted as patients with diseases associated with high circulating EPO such as thalassemia (Vichinsky, 1998), sickle cell disease (Sarrai et al., 2007), and polycythemia vera (Farmer et al., 2013) have debilitating bone conditions.

Sex-specific differences in plasma concentration of EPO are not detected (Jelkmann and Wiedemann, 1989). However, estrogen can affect EPO response and confer gender specific EPO action. In ventilatory response in mice, hypoxia induction of EPO modulates ventilatory response, which exhibits a sex-dimorphic behavior mediated via interaction with carotid body cells that is also sensitive to ovarian steroids (Soliz et al., 2012). The sex-dependent EPO regulation of fat mass is demonstrated with EPO treatment during high fat diet feeding that reduces fat mass gain in male mice and in female ovariectomized mice but not in female non-ovariectomized mice and not in ovariectomized mice supplemented with estradiol pellets (Zhang et al., 2017). Endogenous EPO signaling may cooperate with estrogen regulation of fat mass in female mice but mask reduction of fat mass by exogenous EPO treatment. In human, sexual dimorphic response is observed in the association between endogenous plasma EPO concentration and percent weight change per year. A negative relationship is observed in males where the higher EPO level is related to lower percent weight change, while a positive relationship is evident in females

(Reinhardt et al., 2016). Elucidation of the potential cross-talk between estrogen and EPO will further the understanding of gender-specific EPO response in non-hematopoietic tissue.

Translation of animal studies that demonstrate protective effects of EPO treatment in non-hematopoietic tissues to human disease manifestations can be problematic. Neuroprotection observed with EPO treatment in animal models of ischemic stroke and brain injury and the encouraging results in the Phase I clinical trial of EPO administration for ischemic stroke (Ehrenreich et al., 2002) did not predict the adverse events associated with the combination of EPO and systemic thrombolysis therapy (Ehrenreich et al., 2009), although some benefit with EPO treatment was suggested in further subgroup analysis of patients not receiving thrombolysis therapy (Ehrenreich et al., 2011). EPO shows promise in treatment of extremely low birth weight premature infants where thrombolysis therapy is not used. EPO treatment in extremely low birth weight infants appears to be well tolerated with no excess morbidity or mortality, and, in addition to improving anemia, is associated with overall benefit in developmental assessment and cognitive development (Fauchere et al., 2008; Juul et al., 2008; Neubauer et al., 2010; Fischer et al., 2017). The potential long-term neurodevelopmental outcomes and whether benefit is a consequence of EPO stimulated erythropoiesis to improved oxygen delivery or a direct consequence of EPO response in the premature brain await further study (Fischer et al., 2017).

AUTHOR CONTRIBUTIONS

All authors listed have made a substantial, direct and intellectual contribution to the work, and approved it for publication.

REFERENCES

Alnaeeli, M., and Noguchi, C. T. (2015). Erythropoietin and obesity-induced white adipose tissue inflammation: redefining the boundaries of the immunometabolism territory. *Adipocyte* 4, 153–157. doi: 10.4161/21623945.2014.978654

Alnaeeli, M., Raaka, B. M., Gavrilova, O., Teng, R., Chanturiya, T., and Noguchi, C. T. (2014). Erythropoietin signaling: a novel regulator of white adipose tissue inflammation during diet-induced obesity. *Diabetes Metab. Res. Rev.* 63, 2415–2431. doi: 10.2337/db13-0883

Anagnostou, A., Lee, E. S., Kessimian, N., Levinson, R., and Steiner, M. (1990). Erythropoietin has a mitogenic and positive chemotactic effect on endothelial cells. *Proc. Natl. Acad. Sci. U.S.A.* 87, 5978–5982. doi: 10.1073/pnas.87.15.5978

Anagnostou, A., Liu, Z., Steiner, M., Chin, K., Lee, E. S., Kessimian, N., et al. (1994). Erythropoietin receptor mRNA expression in human endothelial cells. *Proc. Natl. Acad. Sci. U.S.A.* 91, 3974–3978. doi: 10.1073/pnas.91.9.3974

Ang, S. O., Chen, H., Hirota, K., Gordeuk, V. R., Jelinek, J., Guan, Y., et al. (2002). Disruption of oxygen homeostasis underlies congenital Chuvash polycythemia. *Nat. Genet.* 32, 614–621. doi: 10.1038/ng1019

Anusornvongchai, T., Nangaku, M., Jao, T. M., Wu, C. H., Ishimoto, Y., Maekawa, H., et al. (2018). Palmitate deranges erythropoietin production via transcription factor ATF4 activation of unfolded protein response. *Kidney Int.* 94, 536–550. doi: 10.1016/j.kint.2018.03.011

Bakhshi, H., Kazemian, G., Emami, M., Nemati, A., Karimi Yarandi, H., and Safdari, F. (2013). Local erythropoietin injection in tibiofibular fracture healing. *Trauma Mon.* 17, 386–388. doi: 10.5812/traumamon.7099

Balaian, E., Wobus, M., Weidner, H., Baschant, U., Stiehler, M., Ehninger, G., et al. (2018). Erythropoietin inhibits osteoblast function in myelodysplastic syndromes via the canonical Wnt pathway. *Haematologica* 103, 61–68. doi: 10.3324/haematol.2017.172726

Barhoumi, T., Briet, M., Kasal, D. A., Fraulob-Aquino, J. C., Idris-Khodja, N., Laurant, P., et al. (2014). Erythropoietin-induced hypertension and vascular injury in mice overexpressing human endothelin-1: exercise attenuated hypertension, oxidative stress, inflammation and immune response. *J. Hypertens.* 32, 784–794. doi: 10.1097/HJH.0000000000000101

Bazan, J. F. (1990). Structural design and molecular evolution of a cytokine receptor superfamily. *Proc. Natl. Acad. Sci. U.S.A.* 87, 6934–6938. doi: 10.1073/pnas.87.18.6934

Beleslin-Cokic, B. B., Cokic, V. P., Wang, L., Piknova, B., Teng, R., Schechter, A. N., et al. (2011). Erythropoietin and hypoxia increase erythropoietin receptor and nitric oxide levels in lung microvascular endothelial cells. *Cytokine* 54, 129–135. doi: 10.1016/j.cyto.2011.01.015

Beleslin-Cokic, B. B., Cokic, V. P., Yu, X., Weksler, B. B., Schechter, A. N., and Noguchi, C. T. (2004). Erythropoietin and hypoxia stimulate erythropoietin receptor and nitric oxide production by endothelial cells. *Blood* 104, 2073–2080. doi: 10.1182/blood-2004-02-0744

Bernaudin, M., Marti, H. H., Roussel, S., Divoux, D., Nouvelot, A., MacKenzie, E. T., et al. (1999). A potential role for erythropoietin in focal permanent cerebral ischemia in mice. *J. Cereb. Blood Flow Metab.* 19, 643–651. doi: 10.1097/00004647-199906000-00007

Betsch, M., Thelen, S., Santak, L., Herten, M., Jungbluth, P., Miersch, D., et al. (2014). The role of erythropoietin and bone marrow concentrate in the treatment of osteochondral defects in mini-pigs. *PLoS One* 9:e92766. doi: 10.1371/journal.pone.0092766

Boesch, S., Nachbauer, W., Mariotti, C., Sacca, F., Filla, A., Klockgether, T., et al. (2014). Safety and tolerability of carbamylated erythropoietin in Friedreich's ataxia. *Mov. Disord.* 29, 935–939. doi: 10.1002/mds.25836

Bonnas, C., Wustefeld, L., Winkler, D., Kronstein-Wiedemann, R., Dere, E., Specht, K., et al. (2017). EV-3, an endogenous human erythropoietin isoform with distinct functional relevance. *Sci. Rep.* 7:3684. doi: 10.1038/s41598-017-03167-0

Brines, M., Dunne, A. N., van Velzen, M., Proto, P. L., Ostenson, C. G., Kirk, R. I., et al. (2015). ARA 290, a nonerythropoietic peptide engineered from erythropoietin, improves metabolic control and neuropathic symptoms in patients with type 2 diabetes. *Mol. Med.* 20, 658–666. doi: 10.2119/molmed.2014.00215

Brines, M., Grasso, G., Fiordaliso, F., Sfacteria, A., Ghezzi, P., Fratelli, M., et al. (2004). Erythropoietin mediates tissue protection through an erythropoietin and common beta-subunit heteroreceptor. *Proc. Natl. Acad. Sci. U.S.A.* 101, 14907–14912. doi: 10.1073/pnas.0406491101

Brines, M., Patel, N. S., Villa, P., Brines, C., Mennini, T., De Paola, M., et al. (2008). Nonerythropoietic, tissue-protective peptides derived from the tertiary structure of erythropoietin. *Proc. Natl. Acad. Sci. U.S.A.* 105, 10925–10930. doi: 10.1073/pnas.0805594105

Broudy, V. C., Lin, N., Brice, M., Nakamoto, B., and Papayannopoulou, T. (1991). Erythropoietin receptor characteristics on primary human erythroid cells. *Blood* 77, 2583–2590. doi: 10.1182/blood.v77.12.2583.bloodjournal77122583

Cai, Z., Manalo, D. J., Wei, G., Rodriguez, E. R., Fox-Talbot, K., Lu, H., et al. (2003). Hearts from rodents exposed to intermittent hypoxia or erythropoietin are protected against ischemia-reperfusion injury. *Circulation* 108, 79–85. doi: 10.1161/01.cir.0000078635.89229.8a

Cai, Z., and Semenza, G. L. (2004). Phosphatidylinositol-3-kinase signaling is required for erythropoietin-mediated acute protection against myocardial ischemia/reperfusion injury. *Circulation* 109, 2050–2053. doi: 10.1161/01.cir.0000127954.98131.23

Carlini, R. G., Dusso, A. S., Obialo, C. I., Alvarez, U. M., and Rothstein, M. (1993). Recombinant human erythropoietin (rHuEPO). increases endothelin-1 release by endothelial cells. *Kidney Int.* 43, 1010–1014. doi: 10.1038/ki.1993.142

Chen, X., Zhuo, S., Zhu, T., Yao, P., Yang, M., Mei, H., et al. (2019). Fpr2 deficiency alleviates diet-induced insulin resistance through reducing body weight gain and inhibiting inflammation mediated by macrophage chemotaxis and M1 polarization. *Diabetes Metab. Res. Rev.* 68, 1130–1142. doi: 10.2337/db18-0469

Chen, Z. Y., Asavaritikrai, P., Prchal, J. T., and Noguchi, C. T. (2007). Endogenous erythropoietin signaling is required for normal neural progenitor cell proliferation. *J. Biol. Chem.* 282, 25875–25883. doi: 10.1074/jbc.m701988200

Cheung Tung Shing, K. S., Broughton, S. E., Nero, T. L., Gillinder, K., Ilsley, M. D., Ramshaw, H., et al. (2018). EPO does not promote interaction between the erythropoietin and beta-common receptors. *Sci. Rep.* 8:12457.

Choi, D., Schroer, S. A., Lu, S. Y., Wang, L., Wu, X., Liu, Y., et al. (2010). Erythropoietin protects against diabetes through direct effects on pancreatic beta cells. *J. Exp. Med.* 207, 2831–2842. doi: 10.1084/jem.20100665

Clinkenbeard, E. L., Hanudel, M. R., Stayrook, K. R., Appaiah, H. N., Farrow, E. G., Cass, T. A., et al. (2017). Erythropoictin stimulates murine and human fibroblast growth factor-23, revealing novel roles for bone and bone marrow. *Haematologica* 102, e427–e430. doi: 10.3324/haematol.2017.167882

Culver, D. A., Dahan, A., Bajorunas, D., Jeziorska, M., van Velzen, M., Aarts, L., et al. (2017). Cibinetide improves corneal nerve fiber abundance in patients with sarcoidosis-associated small nerve fiber loss and neuropathic pain. *Invest. Ophthalmol. Vis. Sci.* 58, BIO52–BIO60. doi: 10.1167/iovs.16-21291

Dahan, A., Dunne, A., Swartjes, M., Proto, P. L., Heij, L., Vogels, O., et al. (2013). ARA 290 improves symptoms in patients with sarcoidosis-associated small nerve fiber loss and increases corneal nerve fiber density. *Mol. Med.* 19, 334–345. doi: 10.2119/molmed.2013.00122

Dame, C., Fahnenstich, H., Freitag, P., Hofmann, D., Abdul-Nour, T., Bartmann, P., et al. (1998). Erythropoietin mRNA expression in human fetal and neonatal tissue. *Blood* 92, 3218–3225. doi: 10.1182/blood.v92.9.3218.421k11_3218_3225

den Elzen, W. P., Willems, J. M., Westendorp, R. G., de Craen, A. J., Blauw, G. J., Ferrucci, L., et al. (2010). Effect of erythropoietin levels on mortality in old age: the Leiden 85-plus Study. *CMAJ* 182, 1953–1958. doi: 10.1503/cmaj.100374

Dey, S., Li, X., Teng, R., Alnaeeli, M., Chen, Z., Rogers, H., et al. (2016). Erythropoietin regulates POMC expression via STAT3 and potentiates leptin response. *J. Mol. Endocrinol.* 56, 55–67. doi: 10.1530/JME-15-0171

Dey, S., and Noguchi, C. T. (2017). Erythropoietin and hypothalamic-pituitary axis. *Vitam. Horm.* 105, 101–120. doi: 10.1016/bs.vh.2017.02.007

Dey, S., Scullen, T., and Noguchi, C. T. (2015). Erythropoietin negatively regulates pituitary ACTH secretion. *Brain Res.* 1608, 14–20. doi: 10.1016/j.brainres.2015.02.052

Digicaylioglu, M., Bichet, S., Marti, H. H., Wenger, R. H., Rivas, L. A., Bauer, C., et al. (1995). Localization of specific erythropoietin binding sites in defined areas of the mouse brain. *Proc. Natl. Acad. Sci. U.S.A.* 92, 3717–3720. doi: 10.1073/pnas.92.9.3717

Drueke, T. B., Locatelli, F., Clyne, N., Eckardt, K. U., Macdougall, I. C., Tsakiris, D., et al. (2006). Normalization of hemoglobin level in patients with chronic kidney disease and anemia. *N. Engl. J. Med.* 355, 2071–2084. doi: 10.1056/nejmoa062276

Eckardt, K. U., Koury, S. T., Tan, C. C., Schuster, S. J., Kaissling, B., Ratcliffe, P. J., et al. (1993). Distribution of erythropoietin producing cells in rat kidneys during hypoxic hypoxia. *Kidney Int.* 43, 815–823. doi: 10.1038/ki.1993.115

Egger, K., Clemm von Hohenberg, C., Schocke, M. F., Guttmann, C. R., Wassermann, D., Wigand, M. C., et al. (2014). White matter changes in patients with friedreich ataxia after treatment with erythropoietin. *J. Neuroimaging* 24, 504–508. doi: 10.1111/jon.12050

Ehrenreich, H., Hasselblatt, M., Dembowski, C., Cepek, L., Lewczuk, P., Stiefel, M., et al. (2002). Erythropoietin therapy for acute stroke is both safe and beneficial. *Mol. Med.* 8, 495–505. doi: 10.1007/bf03402029

Ehrenreich, H., Kastner, A., Weissenborn, K., Streeter, J., Sperling, S., Wang, K. K., et al. (2011). Circulating damage marker profiles support a neuroprotective effect of erythropoietin in ischemic stroke patients. *Mol. Med.* 17, 1306–1310. doi: 10.2119/molmed.2011.00259

Ehrenreich, H., Weissenborn, K., Prange, H., Schneider, D., Weimar, C., Wartenberg, K., et al. (2009). Recombinant human erythropoietin in the treatment of acute ischemic stroke. *Stroke* 40, e647–e656. doi: 10.1161/STROKEAHA.109.564872

Erbayraktar, S., Grasso, G., Sfacteria, A., Xie, Q. W., Coleman, T., Kreilgaard, M., et al. (2003). Asialoerythropoietin is a nonerythropoietic cytokine with broad neuroprotective activity in vivo. *Proc. Natl. Acad. Sci. U.S.A.* 100, 6741–6746. doi: 10.1073/pnas.1031753100

Eschbach, J. W., Egrie, J. C., Downing, M. R., Browne, J. K., and Adamson, J. W. (1987). Correction of the anemia of end-stage renal disease with recombinant human erythropoietin. Results of a combined phase I and II clinical trial. *N. Engl. J. Med.* 316, 73–78.

Fandrey, J., and Bunn, H. F. (1993). In vivo and in vitro regulation of erythropoietin mRNA: measurement by competitive polymerase chain reaction. *Blood* 81, 617–623. doi: 10.1182/blood.v81.3.617.bloodjournal813617

Farmer, S., Horvath-Puho, E., Vestergaard, H., Hermann, A. P., and Frederiksen, H. (2013). Chronic myeloproliferative neoplasms and risk of osteoporotic fractures; a nationwide population-based cohort study. *Br. J. Haematol.* 163, 603–610. doi: 10.1111/bjh.12581

Fauchere, J. C., Dame, C., Vonthein, R., Koller, B., Arri, S., Wolf, M., et al. (2008). An approach to using recombinant erythropoietin for neuroprotection in very preterm infants. *Pediatrics* 122, 375–382. doi: 10.1542/peds.2007-2591

Fiordaliso, F., Chimenti, S., Staszewsky, L., Bai, A., Carlo, E., Cuccovillo, I., et al. (2005). A nonerythropoietic derivative of erythropoietin protects the myocardium from ischemia-reperfusion injury. *Proc. Natl. Acad. Sci. U.S.A.* 102, 2046–2051. doi: 10.1073/pnas.0409329102

Fischer, H. S., Reibel, N. J., Buhrer, C., and Dame, C. (2017). Prophylactic early erythropoietin for neuroprotection in preterm infants: a meta-analysis. *Pediatrics* 139:e20164317. doi: 10.1542/peds.2016-4317

Foskett, A., Alnaeeli, M., Wang, L., Teng, R., and Noguchi, C. T. (2011). The effects of erythropoietin dose titration during high-fat diet-induced obesity. *J. Biomed. Biotechnol.* 2011, 373781. doi: 10.1155/2011/373781

Ganz, T., and Nemeth, E. (2016). Iron balance and the role of hepcidin in chronic kidney disease. *Semin. Nephrol.* 36, 87–93. doi: 10.1016/j.semnephrol.2016.02.001

Garcia, P., Speidel, V., Scheuer, C., Laschke, M. W., Holstein, J. H., Histing, T., et al. (2011). Low dose erythropoietin stimulates bone healing in mice. *J. Orthopaed. Res.* 29, 165–172. doi: 10.1002/jor.21219

Gassmann, M., and Muckenthaler, M. U. (1985). Adaptation of iron requirement to hypoxic conditions at high altitude. *J. Appl. Physiol.* 119, 1432–1440. doi: 10.1152/japplphysiol.00248.2015

Hagström, L., Agbulut, O., El-Hasnaoui-Saadani, R., Marchant, D., Favret, F., Richalet, J.-P., et al. (2010). Epo is relevant neither for microvascular formation nor for the new formation and maintenance of mice skeletal muscle fibres in both normoxia and hypoxia. *BioMed. Res. Int.* 2010:137817. doi: 10.1155/2010/137817

Hahn, N., Knorr, D. Y., Liebig, J., Wustefeld, L., Peters, K., Buscher, M., et al. (2017). The insect ortholog of the human orphan cytokine receptor CRLF3 is a neuroprotective erythropoietin receptor. *Front. Mol. Neurosci.* 10:223. doi: 10.3389/fnmol.2017.00223

Han, J. M., and Levings, M. K. (2013). Immune regulation in obesity-associated adipose inflammation. *J. Immunol.* 191, 527–532. doi: 10.4049/jimmunol.1301035

Hasegawa, T., Koiwa, F., and Akizawa, T. (2018). Anemia in conventional hemodialysis: finding the optimal treatment balance. *Semin. Dial.* 31, 599–606. doi: 10.1111/sdi.12719

Hiram-Bab, S., Liron, T., Deshet-Unger, N., Mittelman, M., Gassmann, M., Rauner, M., et al. (2015). Erythropoietin directly stimulates osteoclast precursors and induces bone loss. *FASEB J.* 29, 1890–1900. doi: 10.1096/fj.14-259085

Hiram-Bab, S., Neumann, D., and Gabet, Y. (2017). Context-dependent skeletal effects of erythropoietin. *Vitam Horm* 105, 161–179. doi: 10.1016/bs.vh.2017.02.003

Hirano, I., and Suzuki, N. (2019). The neural crest as the first production site of the erythroid growth factor erythropoietin. *Front. Cell Dev. Biol.* 7:105. doi: 10.3389/fcell.2019.00105

Ho, M. S., Medcalf, R. L., Livesey, S. A., and Traianedes, K. (2015). The dynamics of adult haematopoiesis in the bone and bone marrow environment. *Br. J. Haematol.* 170, 472–486. doi: 10.1111/bjh.13445

Hoedt, A., Christensen, B., Nellemann, B., Mikkelsen, U. R., Hansen, M., Schjerling, P., et al. (2016). Satellite cell response to erythropoietin treatment and endurance training in healthy young men. *J. Physiol.* 594, 727–743. doi: 10.1113/JP271333

Hojman, P., Brolin, C., Gissel, H., Brandt, C., Zerahn, B., Pedersen, B. K., et al. (2009). Erythropoietin over-expression protects against diet-induced obesity in mice through increased fat oxidation in muscles. *PLoS One* 4:e5894. doi: 10.1371/journal.pone.0005894

Holstein, J. H., Menger, M. D., Scheuer, C., Meier, C., Culemann, U., Wirbel, R. J., et al. (2007). Erythropoietin (EPO): EPO-receptor signaling improves early endochondral ossification and mechanical strength in fracture healing. *Life Sci.* 80, 893–900. doi: 10.1016/j.lfs.2006.11.023

Horscroft, J. A., Kotwica, A. O., Laner, V., West, J. A., Hennis, P. J., Levett, D. Z. H., et al. (2017). Metabolic basis to Sherpa altitude adaptation. *Proc. Natl. Acad. Sci. U.S.A.* 114, 6382–6387. doi: 10.1073/pnas.1700527114

Iwai, M., Cao, G., Yin, W., Stetler, R. A., Liu, J., and Chen, J. (2007). Erythropoietin promotes neuronal replacement through revascularization and neurogenesis after neonatal hypoxia/ischemia in rats. *Stroke* 38, 2795–2803. doi: 10.1161/strokeaha.107.483008

Jaakkola, P., Mole, D. R., Tian, Y. M., Wilson, M. I., Gielbert, J., Gaskell, S. J., et al. (2001). Targeting of HIF-alpha to the von Hippel-Lindau ubiquitylation complex by O2-regulated prolyl hydroxylation. *Science* 292, 468–472. doi: 10.1126/science.1059796

Jacobs, K., Shoemaker, C., Rudersdorf, R., Neill, S. D., Kaufman, R. J., Mufson, A., et al. (1985). Isolation and characterization of genomic and cDNA clones of human erythropoietin. *Nature* 313, 806–810. doi: 10.1038/313806a0

Jelkmann, W. (2013). Physiology and pharmacology of erythropoietin. *Transfus. Med. Hemother.* 40, 302–309. doi: 10.1159/000356193

Jelkmann, W., and Wiedemann, G. (1989). Lack of sex dependence of the serum level of immunoreactive erythropoietin in chronic anemia. *Klin. Wochenschr.* 67, 1218. doi: 10.1007/bf01716210

Jia, Y., Suzuki, N., Yamamoto, M., Gassmann, M., and Noguchi, C. T. (2012). Endogenous erythropoietin signaling facilitates skeletal muscle repair and recovery following pharmacologically induced damage. *FASEB J.* 26, 2847–2858. doi: 10.1096/fj.11-196618

Jia, Y., Warin, R., Yu, X., Epstein, R., and Noguchi, C. T. (2009). Erythropoietin signaling promotes transplanted progenitor cell survival. *FASEB J.* 23, 3089–3099. doi: 10.1096/fj.09-130237

Jubinsky, P. T., Krijanovski, O. I., Nathan, D. G., Tavernier, J., and Sieff, C. A. (1997). The beta chain of the interleukin-3 receptor functionally associates with the erythropoietin receptor. *Blood* 90, 1867–1873. doi: 10.1182/blood.v90.5.1867

Juul, S. E., McPherson, R. J., Bauer, L. A., Ledbetter, K. J., Gleason, C. A., and Mayock, D. E. (2008). A phase I/II trial of high-dose erythropoietin in extremely low birth weight infants: pharmacokinetics and safety. *Pediatrics* 122, 383–391. doi: 10.1542/peds.2007-2711

Juul, S. E., Yachnis, A. T., and Christensen, R. D. (1998). Tissue distribution of erythropoietin and erythropoietin receptor in the developing human fetus. *Early Hum. Dev.* 52, 235–249. doi: 10.1016/s0378-3782(98)00030-9

Kanellakis, P., Pomilio, G., Agrotis, A., Gao, X., Du, X. J., Curtis, D., et al. (2010). Darbepoetin-mediated cardioprotection after myocardial infarction involves multiple mechanisms independent of erythropoietin receptor-common beta-chain heteroreceptor. *Br. J. Pharmacol.* 160, 2085–2096. doi: 10.1111/j.1476-5381.2010.00876.x

Kassouf, M. T., Hughes, J. R., Taylor, S., McGowan, S. J., Soneji, S., Green, A. L., et al. (2010). Genome-wide identification of TAL1's functional targets: insights into its mechanisms of action in primary erythroid cells. *Genome Res.* 20, 1064–1083. doi: 10.1101/gr.104935.110

Katz, O., Stuible, M., Golishevski, N., Lifshitz, L., Tremblay, M. L., Gassmann, M., et al. (2010). Erythropoietin treatment leads to reduced blood glucose levels and body mass: insights from murine models. *J. Endocrinol.* 205, 87–95. doi: 10.1677/JOE-09-0425

Kertesz, N., Wu, J., Chen, T. H., Sucov, H. M., and Wu, H. (2004). The role of erythropoietin in regulating angiogenesis. *Dev. Biol.* 276, 101–110. doi: 10.1016/j.ydbio.2004.08.025

Kim, J., Jung, Y., Sun, H., Joseph, J., Mishra, A., Shiozawa, Y., et al. (2012). Erythropoietin mediated bone formation is regulated by mTOR signaling. *J. Cell. Biochem.* 113, 220–228. doi: 10.1002/jcb.23347

Kobayashi, H., Liu, J., Urrutia, A. A., Burmakin, M., Ishii, K., Rajan, M., et al. (2017). Hypoxia-inducible factor prolyl-4-hydroxylation in FOXD1 lineage cells is essential for normal kidney development. *Kidney Int.* 92, 1370–1383. doi: 10.1016/j.kint.2017.06.015

Kobayashi, H., Liu, Q., Binns, T. C., Urrutia, A. A., Davidoff, O., Kapitsinou, P. P., et al. (2016). Distinct subpopulations of FOXD1 stroma-derived cells regulate renal erythropoietin. *J. Clin. Invest.* 126, 1926–1938. doi: 10.1172/JCI83551

Kodo, K., Sugimoto, S., Nakajima, H., Mori, J., Itoh, I., Fukuhara, S., et al. (2017). Erythropoietin (EPO). ameliorates obesity and glucose homeostasis by promoting thermogenesis and endocrine function of classical brown adipose tissue (BAT). in diet-induced obese mice. *PLoS One* 12:e0173661. doi: 10.1371/journal.pone.0173661

Kuang, S., Kuroda, K., Le Grand, F., and Rudnicki, M. A. (2007). Asymmetric self-renewal and commitment of satellite stem cells in muscle. *Cell* 129, 999–1010. doi: 10.1016/j.cell.2007.03.044

Kuhrt, D., and Wojchowski, D. M. (2015). Emerging EPO and EPO receptor regulators and signal transducers. *Blood* 125, 3536–3541. doi: 10.1182/blood-2014-11-575357

Lamon, S., Zacharewicz, E., Arentson-Lantz, E., Gatta, P. A. D., Ghobrial, L., Gerlinger-Romero, F., et al. (2016). Erythropoietin does not enhance skeletal muscle protein synthesis following exercise in young and older adults. *Front. Physiol.* 7:292. doi: 10.3389/fphys.2016.00292

Lando, D., Peet, D. J., Gorman, J. J., Whelan, D. A., Whitelaw, M. L., and Bruick, R. K. (2002). FIH-1 is an asparaginyl hydroxylase enzyme that regulates the transcriptional activity of hypoxia-inducible factor. *Genes Dev.* 16, 1466–1471. doi: 10.1101/gad.991402

Li, C., Shi, C., Kim, J., Chen, Y., Ni, S., Jiang, L., et al. (2015). Erythropoietin promotes bone formation through EphrinB2/EphB4 signaling. *J. Dent. Res.* 94, 455–463. doi: 10.1177/0022034514566431

Li, J., Huang, Z., Li, B., Zhang, Z., and Liu, L. (2019). Mobilization of transplanted bone marrow mesenchymal stem cells by erythropoietin facilitates the reconstruction of segmental bone defect. *Stem Cells Int.* 2019:5750967. doi: 10.1155/2019/5750967

Lin, C. S., Lim, S. K., D'Agati, V., and Costantini, F. (1996). Differential effects of an erythropoietin receptor gene disruption on primitive and definitive erythropoiesis. *Genes Dev.* 10, 154–164. doi: 10.1101/gad.10.2.154

Lin, F. K., Suggs, S., Lin, C. H., Browne, J. K., Smalling, R., Egrie, J. C., et al. (1985). Cloning and expression of the human erythropoietin gene. *Proc. Natl. Acad. Sci. U.S.A.* 82, 7580–7584.

Lin, J., Wu, H., Tarr, P. T., Zhang, C. Y., Wu, Z., Boss, O., et al. (2002). Transcriptional co-activator PGC-1 alpha drives the formation of slow-twitch muscle fibres. *Nature* 418, 797–801. doi: 10.1038/nature00904

Liongue, C., Sertori, R., and Ward, A. C. (2016). Evolution of cytokine receptor signaling. *J. Immunol.* 197, 11–18. doi: 10.4049/jimmunol.1600372

Liu, C., Shen, K., Liu, Z., and Noguchi, C. T. (1997). Regulated human erythropoietin receptor expression in mouse brain. *J. Biol. Chem.* 272, 32395–32400. doi: 10.1074/jbc.272.51.32395

Luk, C. T., Shi, S. Y., Choi, D., Cai, E. P., Schroer, S. A., and Woo, M. (2013). In vivo knockdown of adipocyte erythropoietin receptor does not alter glucose or energy homeostasis. *Endocrinology* 154, 3652–3659. doi: 10.1210/en.2013-1113

Lumeng, C. N., and Saltiel, A. R. (2011). Inflammatory links between obesity and metabolic disease. *J. Clin. Invest.* 121, 2111–2117. doi: 10.1172/JCI57132

Lundby, C., Robach, P., Boushel, R., Thomsen, J. J., Rasmussen, P., Koskolou, M., et al. (2008). Does recombinant human Epo increase exercise capacity by means other than augmenting oxygen transport? *J. Appl. Physiol.* 105, 581–587. doi: 10.1152/japplphysiol.90484.2008

Mahon, P. C., Hirota, K., and Semenza, G. L. (2001). FIH-1: a novel protein that interacts with HIF-1alpha and VHL to mediate repression of HIF-1 transcriptional activity. *Genes Dev.* 15, 2675–2686. doi: 10.1101/gad.924501

Malik, J., Kim, A. R., Tyre, K. A., Cherukuri, A. R., and Palis, J. (2013). Erythropoietin critically regulates the terminal maturation of murine and human primitive erythroblasts. *Haematologica* 98, 1778–1787. doi: 10.3324/haematol.2013.087361

Marti, H. H., Wenger, R. H., Rivas, L. A., Straumann, U., Digicaylioglu, M., Henn, V., et al. (1996). Erythropoietin gene expression in human, monkey and murine brain. *Eur. J. Neurosci.* 8, 666–676. doi: 10.1111/j.1460-9568.1996.tb01252.x

Masuda, S., Nagao, M., Takahata, K., Konishi, Y., Gallyas, F. Jr., Tabira, T., et al. (1993). Functional erythropoietin receptor of the cells with neural characteristics. Comparison with receptor properties of erythroid cells. *J. Biol. Chem.* 268, 11208–11216.

Masuda, S., Okano, M., Yamagishi, K., Nagao, M., Ueda, M., and Sasaki, R. (1994). A novel site of erythropoietin production. Oxygen-dependent production in cultured rat astrocytes. *J. Biol. Chem.* 269, 19488–19493.

McGee, S. J., Havens, A. M., Shiozawa, Y., Jung, Y., and Taichman, R. S. (2012). Effects of erythropoietin on the bone microenvironment. *Growth Factors* 30, 22–28. doi: 10.3109/08977194.2011.637034

Meng, R., Zhu, D., Bi, Y., Yang, D., and Wang, Y. (2013). Erythropoietin inhibits gluconeogenesis and inflammation in the liver and improves glucose intolerance in high-fat diet-fed mice. *PLoS One* 8:e53557. doi: 10.1371/journal.pone.0053557

Mihmanli, A., Dolanmaz, D., Avunduk, M. C., and Erdemli, E. (2009). Effects of recombinant human erythropoietin on mandibular distraction osteogenesis. *J. Oral Maxil. Surg.* 67, 2337–2343. doi: 10.1016/j.joms.2008.06.082

Mihov, D., Bogdanov, N., Grenacher, B., Gassmann, M., Zund, G., Bogdanova, A., et al. (2009). Erythropoietin protects from reperfusion-induced myocardial injury by enhancing coronary endothelial nitric oxide production. *Eur. J. Cardiothorac. Surg.* 35, 839–46; discussion 846. doi: 10.1016/j.ejcts.2008.12.049

Mille-Hamard, L., Billat, V. L., Henry, E., Bonnamy, B., Joly, F., Benech, P., et al. (2012). Skeletal muscle alterations and exercise performance decrease in erythropoietin-deficient mice: a comparative study. *BMC Med. Genomics* 5:29. doi: 10.1186/1755-8794-5-29

Minamishima, Y. A., Moslehi, J., Bardeesy, N., Cullen, D., Bronson, R. T., and Kaelin, W. G. Jr. (2008). Somatic inactivation of the PHD2 prolyl hydroxylase causes polycythemia and congestive heart failure. *Blood* 111, 3236–3244. doi: 10.1182/blood-2007-10-117812

Miura, Y., Miura, O., Ihle, J. N., and Aoki, N. (1994). Activation of the mitogen-activated protein kinase pathway by the erythropoietin receptor. *J. Biol. Chem.* 269, 29962–29969.

Morishita, E., Masuda, S., Nagao, M., Yasuda, Y., and Sasaki, R. (1997). Erythropoietin receptor is expressed in rat hippocampal and cerebral cortical neurons, and erythropoietin prevents in vitro glutamate-induced neuronal death. *Neuroscience* 76, 105–116. doi: 10.1016/s0306-4522(96)00306-5

Nadam, J., Navarro, F., Sanchez, P., Moulin, C., Georges, B., Laglaine, A., et al. (2007). Neuroprotective effects of erythropoietin in the rat hippocampus after pilocarpine-induced status epilepticus. *Neurobiol. Dis.* 25, 412–426. doi: 10.1016/j.nbd.2006.10.009

Nair, A. M., Tsai, Y. T., Shah, K. M., Shen, J., Weng, H., Zhou, J., et al. (2013). The effect of erythropoietin on autologous stem cell-mediated bone regeneration. *Biomaterials* 34, 7364–7371. doi: 10.1016/j.biomaterials.2013.06.031

Nairz, M., Haschka, D., Dichtl, S., Sonnweber, T., Schroll, A., Asshoff, M., et al. (2017). Cibinetide dampens innate immune cell functions thus ameliorating the course of experimental colitis. *Sci. Rep.* 7:13012. doi: 10.1038/s41598-017-13046-3

Nairz, M., Schroll, A., Moschen, A. R., Sonnweber, T., Theurl, M., Theurl, I., et al. (2011). Erythropoietin contrastingly affects bacterial infection and experimental colitis by inhibiting nuclear factor-kappaB-inducible immune pathways. *Immunity* 34, 61–74. doi: 10.1016/j.immuni.2011.01.002

Nemati, A., and Fallahi, A. H. (2013). In reply to: queries regarding local erythropoietin injection in tibiofibular fracture healing. *Trauma Mon.* 18, 103–104. doi: 10.5812/traumamon.14208

Neubauer, A. P., Voss, W., Wachtendorf, M., and Jungmann, T. (2010). Erythropoietin improves neurodevelopmental outcome of extremely preterm infants. *Ann. Neurol.* 67, 657–666. doi: 10.1002/ana.21977

Noakes, T. D. (2004). Tainted glory–doping and athletic performance. *N. Engl. J. Med.* 351, 847–849. doi: 10.1056/nejmp048208

Obara, N., Suzuki, N., Kim, K., Nagasawa, T., Imagawa, S., and Yamamoto, M. (2008). Repression via the GATA box is essential for tissue-specific erythropoietin gene expression. *Blood* 111, 5223–5232. doi: 10.1182/blood-2007-10-115857

Ogilvie, M., Yu, X., Nicolas-Metral, V., Pulido, S. M., Liu, C., Ruegg, U. T., et al. (2000). Erythropoietin stimulates proliferation and interferes with differentiation of myoblasts. *J. Biol. Chem.* 275, 39754–39761. doi: 10.1074/jbc.m004999200

Ohh, M., Park, C. W., Ivan, M., Hoffman, M. A., Kim, T. Y., Huang, L. E., et al. (2000). Ubiquitination of hypoxia-inducible factor requires direct binding to the beta-domain of the von Hippel-Lindau protein. *Nat. Cell Biol.* 2, 423–427. doi: 10.1038/35017054

Omlor, G. W., Kleinschmidt, K., Gantz, S., Speicher, A., Guehring, T., and Richter, W. (2016). Increased bone formation in a rabbit long-bone defect model after single local and single systemic application of erythropoietin. *Acta Orthop.* 87, 425–431. doi: 10.1080/17453674.2016.1198200

Osman, H. M., Khamis, O. A., Elfeky, M. S., El Amin Ali, A. M., and Abdelwahed, M. Y. (2017). Effect of short-term erythropoietin therapy on insulin resistance and serum levels of leptin and neuropeptide Y in hemodialysis patients. *Indian J. Endocrinol. Metab.* 21, 724–730. doi: 10.4103/ijem.IJEM_462_16

Palis, J. (2014). Primitive and definitive erythropoiesis in mammals. *Front. Physiol.* 5:3. doi: 10.3389/fphys.2014.00003

Palis, J., and Koniski, A. (2018). Functional analysis of erythroid progenitors by colony-forming assays. *Methods Mol. Biol.* 1698, 117–132. doi: 10.1007/978-1-4939-7428-3_7

Paoletti, E., and Cannella, G. (2006). Update on erythropoietin treatment: should hemoglobin be normalized in patients with chronic kidney disease? *J. Am. Soc. Nephrol.* 17, S74–S77.

Parsa, C. J., Matsumoto, A., Kim, J., Riel, R. U., Pascal, L. S., Walton, G. B., et al. (2003). A novel protective effect of erythropoietin in the infarcted heart. *J. Clin. Invest.* 112, 999–1007. doi: 10.1172/jci200318200

Pastore, Y. D., Jelinek, J., Ang, S., Guan, Y., Liu, E., Jedlickova, K., et al. (2003). Mutations in the VHL gene in sporadic apparently congenital polycythemia. *Blood* 101, 1591–1595. doi: 10.1182/blood-2002-06-1843

Pavkov, M. E., Hanson, R. L., Knowler, W. C., Bennett, P. H., Krakoff, J., and Nelson, R. G. (2007). Changing patterns of type 2 diabetes incidence among Pima Indians. *Diabetes Care* 30, 1758–1763. doi: 10.2337/dc06-2010

Percy, M. J., Furlow, P. W., Lucas, G. S., Li, X., Lappin, T. R., McMullin, M. F., et al. (2008). A gain-of-function mutation in the HIF2A gene in familial erythrocytosis. *N. Engl. J. Med.* 358, 162–168. doi: 10.1056/NEJMoa073123

Percy, M. J., Zhao, Q., Flores, A., Harrison, C., Lappin, T. R., Maxwell, P. H., et al. (2006). A family with erythrocytosis establishes a role for prolyl hydroxylase domain protein 2 in oxygen homeostasis. *Proc. Natl. Acad. Sci. U.S.A.* 103, 654–659. doi: 10.1073/pnas.0508423103

Plenge, U., Belhage, B., Guadalupe-Grau, A., Andersen, P. R., Lundby, C., Dela, F., et al. (2012). Erythropoietin treatment enhances muscle mitochondrial capacity in humans. *Front. Physiol.* 3:50. doi: 10.3389/fphys.2012.00050

Pradeep, S., Huang, J., Mora, E. M., Nick, A. M., Cho, M. S., Wu, S. Y., et al. (2015). Erythropoietin stimulates tumor growth via EphB4. *Cancer Cell* 28, 610–622. doi: 10.1016/j.ccell.2015.09.008

Prunier, F., Pfister, O., Hadri, L., Liang, L., del Monte, F., Liao, R., et al. (2007). Delayed erythropoietin therapy reduces post-MI cardiac remodeling only at a dose that mobilizes endothelial progenitor cells. *Am. J. Physiol. Heart Circ. Physiol.* 292, H522–H529.

Pugh, C. W., and Ratcliffe, P. J. (2017). New horizons in hypoxia signaling pathways. *Exp. Cell Res.* 356, 116–121. doi: 10.1016/j.yexcr.2017.03.008

Quaschning, T., Ruschitzka, F., Stallmach, T., Shaw, S., Morawietz, H., Goettsch, W., et al. (2003). Erythropoietin-induced excessive erythrocytosis activates the tissue endothelin system in mice. *FASEB J.* 17, 259–261. doi: 10.1096/fj.02-0296fje

Rankin, E. B., Biju, M. P., Liu, Q., Unger, T. L., Rha, J., Johnson, R. S., et al. (2007). Hypoxia-inducible factor-2 (HIF-2). regulates hepatic erythropoietin in vivo. *J. Clin. Invest.* 117, 1068–1077. doi: 10.1172/jci30117

Rankin, E. B., Wu, C., Khatri, R., Wilson, T. L., Andersen, R., Araldi, E., et al. (2012). The HIF signaling pathway in osteoblasts directly modulates erythropoiesis through the production of EPO. *Cell* 149, 63–74. doi: 10.1016/j.cell.2012.01.051

Rauner, M., Franke, K., Murray, M., Singh, R. P., Hiram-Bab, S., Platzbecker, U., et al. (2016). Increased EPO levels are associated with bone loss in mice lacking PHD2 in EPO-Producing cells. *J. Bone Miner. Res.* 31, 1877–1887. doi: 10.1002/jbmr.2857

Recny, M. A., Scoble, H. A., and Kim, Y. (1987). Structural characterization of natural human urinary and recombinant DNA-derived erythropoietin. Identification of des-arginine 166 erythropoietin. *J. Biol. Chem.* 262, 17156–17163.

Reinhardt, M., Dey, S., Tom Noguchi, C., Zhang, Y., Krakoff, J., and Thearle, M. S. (2016). Non-hematopoietic effects of endogenous erythropoietin on lean mass and body weight regulation. *Obesity* 24, 1530–1536. doi: 10.1002/oby.21537

Robertson, C. S., Garcia, R., Gaddam, S. S., Grill, R. J., Cerami Hand, C., Tian, T. S., et al. (2013). Treatment of mild traumatic brain injury with an erythropoietin-mimetic peptide. *J. Neurotrauma* 30, 765–774. doi: 10.1089/neu.2012.2431

Rogers, H., Wang, L., Yu, X., Alnaeeli, M., Cui, K., Zhao, K., et al. (2012). T-cell acute leukemia 1 (TAL1). regulation of erythropoietin receptor and association with excessive erythrocytosis. *J. Biol. Chem.* 287, 36720–36731. doi: 10.1074/jbc.M112.378398

Rolfing, J. H., Baatrup, A., Stiehler, M., Jensen, J., Lysdahl, H., and Bunger, C. (2014a). The osteogenic effect of erythropoietin on human mesenchymal stromal cells is dose-dependent and involves non-hematopoietic receptors and multiple intracellular signaling pathways. *Stem Cell Rev.* 10, 69–78. doi: 10.1007/s12015-013-9476-x

Rolfing, J. H., Jensen, J., Jensen, J. N., Greve, A. S., Lysdahl, H., Chen, M., et al. (2014b). A single topical dose of erythropoietin applied on a collagen carrier enhances calvarial bone healing in pigs. *Acta Orthop.* 85, 201–209. doi: 10.3109/17453674.2014.889981

Rundqvist, H., Rullman, E., Sundberg, C. J., Fischer, H., Eisleitner, K., Stahlberg, M., et al. (2009). Activation of the erythropoietin receptor in human skeletal muscle. *Eur. J. Endocrinol.* 161, 427–434. doi: 10.1530/EJE-09-0342

Ruschitzka, F. T., Wenger, R. H., Stallmach, T., Quaschning, T., de Wit, C., Wagner, K., et al. (2000). Nitric oxide prevents cardiovascular disease and determines survival in polyglobulic mice overexpressing erythropoietin. *Proc. Natl. Acad. Sci. U.S.A.* 97, 11609–11613. doi: 10.1073/pnas.97.21.11609

Sadamoto, Y., Igase, K., Sakanaka, M., Sato, K., Otsuka, H., Sakaki, S., et al. (1998). Erythropoietin prevents place navigation disability and cortical infarction in rats with permanent occlusion of the middle cerebral artery. *Biochem. Biophys. Res. Commun.* 253, 26–32. doi: 10.1006/bbrc.1998.9748

Sakanaka, M., Wen, T. C., Matsuda, S., Masuda, S., Morishita, E., Nagao, M., et al. (1998). In vivo evidence that erythropoietin protects neurons from ischemic damage. *Proc. Natl. Acad. Sci. U.S.A.* 95, 4635–4640. doi: 10.1073/pnas.95.8.4635

Santner, W., Schocke, M., Boesch, S., Nachbauer, W., and Egger, K. (2014). A longitudinal VBM study monitoring treatment with erythropoietin in patients with Friedreich ataxia. *Acta Radiol. Short Rep.* 3:2047981614531573. doi: 10.1177/2047981614531573

Sarrai, M., Duroseau, H., D'Augustine, J., Moktan, S., and Bellevue, R. (2007). Bone mass density in adults with sickle cell disease. *Br. J. Haematol.* 136, 666–672. doi: 10.1111/j.1365-2141.2006.06487.x

Semenza, G. L. (2009). Involvement of oxygen-sensing pathways in physiologic and pathologic erythropoiesis. *Blood* 114, 2015–2019. doi: 10.1182/blood-2009-05-189985

Shingo, T., Sorokan, S. T., Shimazaki, T., and Weiss, S. (2001). Erythropoietin regulates the in vitro and in vivo production of neuronal progenitors by mammalian forebrain neural stem cells. *J. Neurosci.* 21, 9733–9743. doi: 10.1523/jneurosci.21-24-09733.2001

Shiozawa, Y., Jung, Y., Ziegler, A. M., Pedersen, E. A., Wang, J., Wang, Z., et al. (2010). Erythropoietin couples hematopoiesis with bone formation. *PLoS One* 5:e10853. doi: 10.1371/journal.pone.0010853

Singbrant, S., Russell, M. R., Jovic, T., Liddicoat, B., Izon, D. J., Purton, L. E., et al. (2011). Erythropoietin couples erythropoiesis, B-lymphopoiesis, and bone homeostasis within the bone marrow microenvironment. *Blood* 117, 5631–5642. doi: 10.1182/blood-2010-11-320564

Singh, A. K., Szczech, L., Tang, K. L., Barnhart, H., Sapp, S., Wolfson, M., et al. (2006). Correction of anemia with epoetin alfa in chronic kidney disease. *N. Engl. J. Med.* 355, 2085–2098.

Sinkeler, S. J., Zelle, D. M., Homan van der Heide, J. J., Gans, R. O., Navis, G., and Bakker, S. J. (2012). Endogenous plasma erythropoietin, cardiovascular mortality and all-cause mortality in renal transplant recipients. *Am. J. Transplant.* 12, 485–491. doi: 10.1111/j.1600-6143.2011.03825.x

Smith, C. J., Nelson, R. G., Hardy, S. A., Manahan, E. M., Bennett, P. H., and Knowler, W. C. (1996). Survey of the diet of Pima Indians using quantitative food frequency assessment and 24-hour recall. Diabetic Renal Disease Study. *J. Am. Diet. Assoc.* 96, 778–784. doi: 10.1016/s0002-8223(96)00216-7

Smith, T. G., Robbins, P. A., and Ratcliffe, P. J. (2008). The human side of hypoxia-inducible factor. *Br. J. Haematol.* 141, 325–334. doi: 10.1111/j.1365-2141.2008.07029.x

Socolovsky, M., Nam, H., Fleming, M. D., Haase, V. H., Brugnara, C., and Lodish, H. F. (2001). Ineffective erythropoiesis in Stat5a(-/-)5b(-/-). mice due to decreased survival of early erythroblasts. *Blood* 98, 3261–3273. doi: 10.1182/blood.v98.12.3261

Soliz, J., Khemiri, H., Caravagna, C., and Seaborn, T. (2012). Erythropoietin and the sex-dimorphic chemoreflex pathway. *Adv. Exp. Med. Biol.* 758, 55–62. doi: 10.1007/978-94-007-4584-1_8

Sun, H., Jung, Y., Shiozawa, Y., Taichman, R. S., and Krebsbach, P. H. (2012). Erythropoietin modulates the structure of bone morphogenetic protein 2-engineered cranial bone. *Tissue Eng. Part A* 18, 2095–2105. doi: 10.1089/ten.TEA.2011.0742

Suresh, S., de Castro, L. F., Dey, S., Robey, P. G., and Noguchi, C. T. (2019). Erythropoietin modulates bone marrow stromal cell differentiation. *Bone Res.* 7, 21. doi: 10.1038/s41413-019-0060-0

Suzuki, N., Gradin, K., Poellinger, L., and Yamamoto, M. (2017). Regulation of hypoxia-inducible gene expression after HIF activation. *Exp. Cell Res.* 356, 182–186. doi: 10.1016/j.yexcr.2017.03.013

Suzuki, N., Hirano, I., Pan, X., Minegishi, N., and Yamamoto, M. (2013). Erythropoietin production in neuroepithelial and neural crest cells during primitive erythropoiesis. *Nat. Commun.* 4:2902. doi: 10.1038/ncomms3902

Suzuki, N., Ohneda, O., Takahashi, S., Higuchi, M., Mukai, H. Y., Nakahata, T., et al. (2002). Erythroid-specific expression of the erythropoietin receptor rescued its null mutant mice from lethality. *Blood* 100, 2279–2288. doi: 10.1182/blood-2002-01-0124

Takeda, K., Aguila, H. L., Parikh, N. S., Li, X., Lamothe, K., Duan, L. J., et al. (2008). Regulation of adult erythropoiesis by prolyl hydroxylase domain proteins. *Blood* 111, 3229–3235. doi: 10.1182/blood-2007-09-114561

Tan, C. C., Eckardt, K. U., and Ratcliffe, P. J. (1991). Organ distribution of erythropoietin messenger RNA in normal and uremic rats. *Kidney Int.* 40, 69–76. doi: 10.1038/ki.1991.181

Teng, R., Calvert, J. W., Sibmooh, N., Piknova, B., Suzuki, N., Sun, J., et al. (2011a). Acute erythropoietin cardioprotection is mediated by endothelial response. *Basic Res. Cardiol.* 106, 343–354. doi: 10.1007/s00395-011-0158-z

Teng, R., Gavrilova, O., Suzuki, N., Chanturiya, T., Schimel, D., Hugendubler, L., et al. (2011b). Disrupted erythropoietin signalling promotes obesity and alters hypothalamus proopiomelanocortin production. *Nat. Commun.* 2:520. doi: 10.1038/ncomms1526

Tsai, P. T., Ohab, J. J., Kertesz, N., Groszer, M., Matter, C., Gao, J., et al. (2006). A critical role of erythropoietin receptor in neurogenesis and post-stroke recovery. *J. Neurosci.* 26, 1269–1274. doi: 10.1523/jneurosci.4480-05.2006

Tsuma, Y., Mori, J., Ota, T., Kawabe, Y., Morimoto, H., Fukuhara, S., et al. (2019). Erythropoietin and long-acting erythropoiesis stimulating agent ameliorate non-alcoholic fatty liver disease by increasing lipolysis and decreasing lipogenesis via EPOR/STAT pathway. *Biochem. Biophys. Res. Commun.* 509, 306–313. doi: 10.1016/j.bbrc.2018.12.131

Um, M., Gross, A. W., and Lodish, H. F. (2007). A "classical" homodimeric erythropoietin receptor is essential for the antiapoptotic effects of erythropoietin on differentiated neuroblastoma SH-SY5Y and pheochromocytoma PC-12 cells. *Cell. Signal.* 19, 634–645. doi: 10.1016/j.cellsig.2006.08.014

Us Renal Data System (2007). *USRDS 2007 Annual Data Report 2007.* Minneapolis, MN: US Renal Data System.

Van Der Meer, P., Lipsic, E., Henning, R. H., Boddeus, K., Van Der Velden, J., Voors, A. A., et al. (2005). Erythropoietin induces neovascularization and improves cardiac function in rats with heart failure after myocardial infarction. *J. Am. Coll. Cardiol.* 46, 125–133. doi: 10.1016/j.jacc.2005.03.044

Vichinsky, E. P. (1998). The morbidity of bone disease in thalassemia. *Ann. N. Y. Acad. Sci.* 850, 344–348. doi: 10.1111/j.1749-6632.1998.tb10491.x

Villa, P., Bigini, P., Mennini, T., Agnello, D., Laragione, T., Cagnotto, A., et al. (2003). Erythropoietin selectively attenuates cytokine production and inflammation in cerebral ischemia by targeting neuronal apoptosis. *J. Exp. Med.* 198, 971–975. doi: 10.1084/jem.20021067

Villafuerte, F. C., and Corante, N. (2016). Chronic mountain sickness: clinical aspects, etiology, management, and treatment. *High Alt. Med. Biol.* 17, 61–69. doi: 10.1089/ham.2016.0031

Voss, J. D., Allison, D. B., Webber, B. J., Otto, J. L., and Clark, L. L. (2014). Lower obesity rate during residence at high altitude among a military population with frequent migration: a quasi experimental model for investigating spatial causation. *PLoS One* 9:e93493. doi: 10.1371/journal.pone.0093493

Voss, J. D., Masuoka, P., Webber, B. J., Scher, A. I., and Atkinson, R. L. (2013). Association of elevation, urbanization and ambient temperature with obesity prevalence in the United States. *Int. J. Obes.* 37, 1407–1412. doi: 10.1038/ijo.2013.5

Wang, L., Di, L., and Noguchi, C. T. (2014). Erythropoietin, a novel versatile player regulating energy metabolism beyond the erythroid system. *Int. J. Biol. Sci.* 10, 921–939. doi: 10.7150/ijbs.9518

Wang, L., Jia, Y., Rogers, H., Suzuki, N., Gassmann, M., Wang, Q., et al. (2013a). Erythropoietin contributes to slow oxidative muscle fiber specification via PGC-1alpha and AMPK activation. *Int. J. Biochem. Cell Biol.* 45, 1155–1164. doi: 10.1016/j.biocel.2013.03.007

Wang, L., Teng, R., Di, L., Rogers, H., Wu, H., Kopp, J. B., et al. (2013b). PPARalpha and Sirt1 mediate erythropoietin action in increasing metabolic activity and browning of white adipocytes to protect against obesity and metabolic disorders. *Diabetes Metab. Res. Rev.* 62, 4122–4131. doi: 10.2337/db13-0518

Wang, L., Jia, Y., Rogers, H., Wu, Y. P., Huang, S., and Noguchi, C. T. (2012). GATA-binding protein 4 (GATA-4). and T-cell acute leukemia 1 (TAL1). regulate myogenic differentiation and erythropoietin response via cross-talk with Sirtuin1 (Sirt1). *J. Biol. Chem.* 287, 30157–30169. doi: 10.1074/jbc.M112.376640

Wassink, G., Davidson, J. O., Dhillon, S. K., Fraser, M., Galinsky, R., Bennet, L., et al. (2017). Partial white and grey matter protection with prolonged infusion of recombinant human erythropoietin after asphyxia in preterm fetal sheep. *J. Cereb. Blood Flow Metab.* 37, 1080–1094. doi: 10.1177/0271678X16650455

Watowich, S. S. (2011). The erythropoietin receptor: molecular structure and hematopoietic signaling pathways. *J. Invest. Med.* 59, 1067–1072. doi: 10.2310/jim.0b013e31820fb28c

Westenbrink, B. D., Visser, F. W., Voors, A. A., Smilde, T. D., Lipsic, E., Navis, G., et al. (2006). Anaemia in chronic heart failure is not only related to impaired renal perfusion and blunted erythropoietin production, but to fluid retention as well. *Eur. Heart J.* 28, 166–171. doi: 10.1093/eurheartj/ehl419

Witthuhn, B. A., Quelle, F. W., Silvennoinen, O., Yi, T., Tang, B., Miura, O., et al. (1993). JAK2 associates with the erythropoietin receptor and is tyrosine phosphorylated and activated following stimulation with erythropoietin. *Cell* 74, 227–236. doi: 10.1016/0092-8674(93)90414-l

Wright, D. G., Wright, E. C., Narva, A. S., Noguchi, C. T., and Eggers, P. W. (2015). Association of erythropoietin dose and route of administration with clinical outcomes for patients on hemodialysis in the United States. *Clin. J. Am. Soc. Nephrol.* 10, 1822–1830. doi: 10.2215/CJN.01590215

Wu, H., Lee, S. H., Gao, J., Liu, X., and Iruela-Arispe, M. L. (1999). Inactivation of erythropoietin leads to defects in cardiac morphogenesis. *Development* 126, 3597–3605.

Wu, H., Liu, X., Jaenisch, R., and Lodish, H. F. (1995). Generation of committed erythroid BFU-E and CFU-E progenitors does not require erythropoietin or the erythropoietin receptor. *Cell* 83, 59–67. doi: 10.1016/0092-8674(95)90234-1

Yasuda, Y., Masuda, S., Chikuma, M., Inoue, K., Nagao, M., and Sasaki, R. (1998). Estrogen-dependent production of erythropoietin in uterus and its implication in uterine angiogenesis. *J. Biol. Chem.* 273, 25381–25387. doi: 10.1074/jbc.273.39.25381

Yokoro, M., Nakayama, Y., Yamagishi, S. I., Ando, R., Sugiyama, M., Ito, S., et al. (2017). Asymmetric dimethylarginine contributes to the impaired response to erythropoietin in CKD-Anemia. *J. Am. Soc. Nephrol.* 28, 2670–2680. doi: 10.1681/ASN.2016111184

Yu, X., Lin, C. S., Costantini, F., and Noguchi, C. T. (2001). The human erythropoietin receptor gene rescues erythropoiesis and developmental defects in the erythropoietin receptor null mouse. *Blood* 98, 475–477. doi: 10.1182/blood.v98.2.475

Yu, X., Shacka, J. J., Eells, J. B., Suarez-Quian, C., Przygodzki, R. M., Beleslin-Cokic, B., et al. (2002). Erythropoietin receptor signalling is required for normal brain development. *Development* 129, 505–516.

Zanjani, E. D., Ascensao, J. L., McGlave, P. B., Banisadre, M., and Ash, R. C. (1981). Studies on the liver to kidney switch of erythropoietin production. *J. Clin. Invest.* 67, 1183–1188. doi: 10.1172/jci110133

Zhang, Y., Rogers, H. M., Zhang, X., and Noguchi, C. T. (2017). Sex difference in mouse metabolic response to erythropoietin. *FASEB J.* 31, 2661–2673. doi: 10.1096/fj.201601223RRR

Zhang, Y., Wang, L., Dey, S., Alnaeeli, M., Suresh, S., Rogers, H., et al. (2014). Erythropoietin action in stress response, tissue maintenance and metabolism. *Int. J. Mol. Sci.* 15, 10296–10333. doi: 10.3390/ijms150610296

Zierath, J. R., and Hawley, J. A. (2004). Skeletal muscle fiber type: influence on contractile and metabolic properties. *PLoS Biol.* 2:e348. doi: 10.1371/journal.pbio.0020348

Zon, L. I., Youssoufian, H., Mather, C., Lodish, H. F., and Orkin, S. H. (1991). Activation of the erythropoietin receptor promoter by transcription factor GATA-1. *Proc. Natl. Acad. Sci. U.S.A.* 88, 10638–10641. doi: 10.1073/pnas.88.23.10638

3

Impaired Cytoskeletal and Membrane Biophysical Properties of Acanthocytes in Hypobetalipoproteinemia

Anne-Sophie Cloos[1†], Laura G. M. Daenen[2†], Mauriane Maja[1], Amaury Stommen[1], Juliette Vanderroost[1], Patrick Van Der Smissen[1], Minke Rab[3], Jan Westerink[4], Eric Mignolet[5], Yvan Larondelle[5], Romano Terrasi[6], Giulio G. Muccioli[6], Andra C. Dumitru[5], David Alsteens[5], Richard van Wijk[3] and Donatienne Tyteca[1]*

[1] CELL Unit & PICT Imaging Platform, de Duve Institute, UCLouvain, Brussels, Belgium, [2] Department of Hematology, University Medical Center Utrecht, Utrecht University, Utrecht, Netherlands, [3] Central Diagnostic Laboratory – Research, University Medical Center Utrecht, Utrecht University, Utrecht, Netherlands, [4] Department of Vascular Medicine, University Medical Center Utrecht, Utrecht University, Utrecht, Netherlands, [5] Louvain Institute of Biomolecular Science and Technology, UCLouvain, Ottignies-Louvain-la-Neuve, Belgium, [6] Bioanalysis and Pharmacology of Bioactive Lipids Research Group, Louvain Drug Research Institute, UCLouvain, Brussels, Belgium

Correspondence:
Donatienne Tyteca
donatienne.tyteca@uclouvain.be

[†] These authors share first authorship

Familial hypobetalipoproteinemia is a metabolic disorder mainly caused by mutations in the *apolipoprotein B* gene. In its homozygous form it can lead without treatment to severe ophthalmological and neurological manifestations. In contrast, the heterozygous form is generally asymptomatic but associated with a low risk of cardiovascular disease. Acanthocytes or thorny red blood cells (RBCs) are described for both forms of the disease. However, those morphological changes are poorly characterized and their potential consequences for RBC functionality are not understood. Thus, in the present study, we asked whether, to what extent and how acanthocytes from a patient with heterozygous familial hypobetalipoproteinemia could exhibit altered RBC functionality. Acanthocytes represented 50% of the total RBC population and contained mitoTracker-positive surface patches, indicating the presence of mitochondrial fragments. While RBC osmotic fragility, calcium content and ATP homeostasis were preserved, a slight decrease of RBC deformability combined with an increase of intracellular free reactive oxygen species were observed. The spectrin cytoskeleton was altered, showing a lower density and an enrichment in patches. At the membrane level, no obvious modification of the RBC membrane fatty acids nor of the cholesterol content were detected but the ceramide species were all increased. Membrane stiffness and curvature were also increased whereas transversal asymmetry was preserved. In contrast, lateral asymmetry was highly impaired showing: (i) increased abundance and decreased functionality of sphingomyelin-enriched domains; (ii) cholesterol enrichment in spicules;

and (iii) ceramide enrichment in patches. We propose that oxidative stress induces cytoskeletal alterations, leading to increased membrane stiffness and curvature and impaired lipid lateral distribution in domains and spicules. In addition, ceramide- and spectrin-enriched patches could result from a RBC maturation defect. Altogether, the data indicate that acanthocytes are associated with cytoskeletal and membrane lipid lateral asymmetry alterations, while deformability is only mildly impaired. In addition, familial hypobetalipoproteinemia might also affect RBC precursors leading to disturbed RBC maturation. This study paves the way for the potential use of membrane biophysics and lipid vital imaging as new methods for diagnosis of RBC disorders.

Keywords: acanthocytosis, lipidomics, lipid domains, membrane biophysical properties, reactive oxygen species, ceramide, mitochondria, erythropoiesis

INTRODUCTION

Abetalipoproteinemia and familial homozygous hypobetalipoproteinemia are two very rare metabolic disorders which are clinically indistinguishable (Biemer, 1980). Abetalipoproteinemia is provoked by mutations of both alleles of the *microsomal triglyceride transfer protein (MTP)* encoding gene (Calzada et al., 2013). MTP allows the transfer of triglycerides, cholesteryl ester and phospholipids onto the apolipoprotein B (ApoB). Familial hypobetalipoproteinemia is considered as autosomal codominant and most frequently caused by mutations affecting directly the *ApoB* gene leading to truncated forms of the protein (Clarke et al., 2006). In both abetalipoproteinemia and familial homozygous hypobetalipoproteinemia the consequence of gene mutations is failure of release of ApoB-containing lipoproteins, especially chylomicrons and very low density lipoproteins (VLDL) (Burnett and Hooper, 2015). As a result, lipid and lipid soluble vitamin absorption is disturbed, clinically manifesting by extremely low plasma cholesterol and triacylglycerol levels, vitamin A, E, and K levels as well as an almost complete lack of circulating chylomicrons, VLDL and low density lipoproteins (LDL) (Granot and Kohen, 2004). Patients suffering from these diseases present early in life a series of gastroenterological complications including fat malabsorption, steatorrhea, vomiting, abdominal distension and failure to thrive. Later in life, neurological and ophthalmological manifestations, due to lipid soluble vitamin deficiency, as well as non-alcoholic fatty liver disease (NAFLD) become part of the clinical picture (Burnett and Hooper, 2015). For this reason, therapy implies, besides of a strict low fat diet, a life-long supplementation with vitamins A and E (Granot and Kohen, 2004).

Besides abetalipoproteinemia and familial homozygous hypobetalipoproteinemia, a heterozygous form of familial hypobetalipoproteinemia has also been described. Patients are often asymptomatic but frequently associated with NAFLD and in rare cases with neurological disorders (Musialik et al., 2020). Despite the frequently asymptomatic appearance of the heterozygous form of familial hypobetalipoproteinemia, strongly reduced plasma ApoB and LDL-cholesterol levels combined eventually with vitamin E levels below or at the limit of reference values are typical clinical features (Clarke et al.,

2006). This heterozygous form is also associated with a low risk of cardiovascular disease (Whitfield et al., 2004).

More surprisingly, acanthocytes (or thorny red blood cells, RBCs) are usually observed on blood smears from patients with abetalipoproteinemia and familial homozygous hypobetalipoproteinemia and to a lesser extent in the heterozygous form (Whitfield et al., 2004; Cuerq et al., 2018). Acanthocytes are RBCs with few membrane projections varying in length and presenting a non-uniform distribution along the RBC surface, in contrast to echinocytes which exhibit more regular membrane projections that are evenly distributed on the RBC membrane. Acanthocytes are, however, not specific to the above mentioned metabolic disorders but are also associated with a series of other pathologies and disorders including chorea-acanthocytosis, McLeod phenotype, In(Lu) phenotype, hereditary spherocytosis with a β-spectrin deficiency, alcoholic cirrhosis, uremia, vitamin E deficiency, anorexia nervosa and hypothyroidism (Wong, 2004).

Such acanthocyte shape could lead to alteration of RBC deformability, which mainly results from the RBC typical biconcave shape as determined by a membrane surface area to cytoplasmatic volume excess. Besides, RBC deformability is also determined by a finely regulated cytoplasmic viscosity controlled by hemoglobin (Hb) concentration and a cytoskeleton composed of a meshwork of spectrin tetramers linked to the membrane by the 4.1R- and ankyrin-based anchorage complexes (Salomao et al., 2008; Baines, 2010). RBC deformation also depends on the intracellular ATP content, the antioxidant defense, the ion balance and subsequent volume control regulated by ion channels, symporters, antiporters and pumps. Among ion channels, the mechanosensitive non-selective cation channel PIEZO1 has been recently identified as the link between mechanical forces, calcium influx and RBC volume homeostasis. The calcium-activated K^+ channel (named Gardos), the Cl^-/HCO_3^- antiporter Band3 and the plasma membrane calcium ATPase pump (PMCA) are also essential for the RBC homeostasis. For instance, the transient increase in intracellular calcium upon deformation activates Gardos channels, leading to cell dehydration and favoring local membrane:cytoskeleton uncoupling (Bogdanova et al., 2013; Cahalan et al., 2015; Lew and Tiffert, 2017; Kaestner et al., 2020). In addition, RBC membrane lipid composition and distribution are also important for RBC deformation. Actually, the RBC plasma membrane exhibits a high cholesterol content as compared to other cells and shows lipid clustering in domains (Carquin et al., 2016). Three types of lipid domains coexist at the RBC outer membrane leaflet (D'Auria et al., 2013; Carquin et al., 2014, 2015; Leonard et al., 2017b; Conrard et al., 2018). The first, associated with high-curvature membrane areas, is mainly enriched in cholesterol and gathers and/or stabilizes high curvature membrane areas upon RBC deformation. The second and third domains are associated with low-curvature membrane areas. They are, respectively, enriched in ganglioside GM1, phosphatidylcholine and cholesterol (hereafter named GM1-enriched domains) and in sphingomyelin, phosphatidylcholine and cholesterol (hereafter named sphingomyelin-enriched domains). Both GM1- and sphingomyelin-enriched domains seem to participate in calcium exchanges during RBC deformation as GM1-enriched domain abundance increases upon calcium influx after PIEZO1 activation while sphingomyelin-enriched domain abundance increases during calcium efflux (Leonard et al., 2017b; Conrard et al., 2018). A fourth type of lipid domain, mainly enriched in ceramide and associated to the membrane inner leaflet, has been recently identified (Cloos et al., 2020).

In the present study, we asked whether, to what extent and how acanthocytes from a patient with heterozygous familial hypobetalipoproteinemia (pHypoβ) could exhibit alteration of RBC deformability. For this purpose, the RBCs of the patient, who presented a high number of acanthocytes on blood smears, were analyzed for the main features associated with RBC deformability, i.e., RBC morphology, the spectrin cytoskeleton integrity, the intracellular ATP content, the antioxidant defense, the ion balance and the RBC membrane lipid composition and biophysical properties.

The experimental approaches were chosen based on our previous work (Carquin et al., 2014, 2015; Leonard et al., 2017a; Pollet et al., 2018, 2020; Cloos et al., 2020): (i) for RBC morphology: blood smears, optical microscopy of living RBCs and scanning electron microscopy (SEM) of fixed RBCs; (ii) for RBC deformability: osmotic fragility through Hb release and deformability upon shear stress by ektacytometry; (iii) for cytoskeleton integrity: spectrin confocal imaging; (iv) for intracellular ATP content: a biochemical assay; (v) for the antioxidant defense: extent of oxidative stress through measurement of free intracellular reactive oxygen species (ROS) content, lipid peroxidation and Hb oxidation; (vi) for the ion balance: intracellular calcium content evaluated by flow cytometry; (vii) for membrane lipid composition: determination of fatty acids (FA), cholesterol, sphingolipids and phospholipids content; and (viii) for membrane biophysical properties: evaluation of elastic modulus by atomic force microscopy (AFM), curvature by optical microscopy of living RBCs in suspension, transversal asymmetry by phosphatidylserine (PS) surface exposure and lateral asymmetry by vital confocal imaging using validated fluorescent lipid analogs or toxin fragments specific to endogenous lipids. We finally explored whether impairments resulted from defects during RBC maturation or from acceleration of aging. Our findings improve our

understanding of hypobetalipoproteinemia and might be extended to other RBC diseases.

MATERIALS AND METHODS

Blood Collection and Preparation

The study was approved by the Medical Ethics Committee of UCLouvain and University Medical Center Utrecht (Study 17–450). After informed consent, blood from the patient and 4 healthy volunteers was collected by venipuncture into K^+/EDTA-coated tubes at the University Medical Center Utrecht. Whenever possible, healthy donors were selected to be age- and gender-matched. After collection, the tubes were transferred to the research laboratory at UCLouvain. Two splenectomised healthy donors were included in the study. Before experiments, RBCs were collected through 10-fold blood dilution in a glucose- and HEPES-containing medium [Dulbecco's modified eagle medium (DMEM), Invitrogen]. Diluted blood was centrifuged at 200 g for 2 min, the supernatant removed and RBCs suspended in medium. RBCs were washed a second time by centrifugation at 200 g for 2 min and resuspended, as in Cloos et al. (2020). The number of RBCs used for experiments were as follows: (i) Hb release, 12.5×10^6; (ii) intracellular ATP content, 5×10^6; (iii) calcium and ROS contents, 15×10^6; (iv) lipidomic and FA analyses, $2.5–6 \times 10^8$; (v) PS surface exposure, 0.5×10^6; (vi) cholesterol assay, 37.5×10^6; and (vii) fluorescence imaging, 12.5×10^6.

RBC Chemical Treatments

Chemical treatments were performed on washed RBCs. To activate calcium entry through PIEZO1, RBCs were incubated with 0.1 μM Yoda1 (Bio-Techne) for 30 s at RT. To deplete the intracellular calcium content, RBCs were preincubated in a calcium-free medium containing 1mM EGTA (Sigma-Aldrich) for 10 min at RT and maintained during the experiment. To induce oxidative stress, RBCs were submitted to 100 mM H_2O_2 for 60 min at 37°C.

RBC Morphology Determination

RBC morphology was determined (i) on blood smears, (ii) upon suspension of living RBCs in μ-dish IBIDI chambers, (iii) by SEM on fixed RBCs, and (iv) after immobilization of living RBCs on poly-L-lysine (PLL)-coated coverslips. For the RBCs in suspension, washed RBCs were diluted 24-fold in DMEM, deposed in μ-dish IBIDI chambers and observed with a wide-field fluorescence microscope (Observer.Z1; plan-Apochromat 100× 1.4 oil Ph3 objective). For SEM microscopy experiments, RBCs were prepared and analyzed exactly as in Cloos et al. (2020). For RBC immobilization, PLL was deposed on coverslips for 30 min at 37°C and then washed. RBCs were then spread on the PLL-coated coverslip during 4 min and the coverslip was then placed upside down in a medium-filled LabTek chamber (Fisher Scientific). RBCs were observed with the fluorescence microscope Observer.Z1 as above.

RBC Hemoglobin Release Measurement

Washed RBCs were incubated for 10 min at RT in isotonic (320 mOsm, DMEM) or increasingly hypotonic media (264–0 mOsm, DMEM mixed with water to obtain the desired osmolarity) and then pelleted by centrifugation at 200 g for 2 min. Supernatants and pellets broken with 0.2% (w/v) Triton X-100 were both assessed for Hb at 450 nm in 96-well plates (SpectraCount^{TM}, Packard BioScience Co.). Hb release in the supernatant was expressed as percentage of the total Hb present in the sample.

Ektacytometry

Deformability of RBCs was measured with the Laser Optical Rotational Red Cell Analyzer (Lorrca, RR Mechatronics, Zwaag, Netherlands). In this ektacytometer, RBCs are exposed to shear stress in a viscous solution (Elon-Iso, RR Mechatronics), forcing the cells to elongate in an elliptical shape. The diffraction pattern, that is generated by a laser beam, is measured by a camera. The vertical axis (A) and the horizontal axis (B) of the ellipse are used to calculate the elongation index (EI) by the formula $(A - B)/(A + B)$. The EI reflects the deformability of the total population of RBCs. Two different forms of ektacytometry were used: osmotic gradient ektacytometry and cell membrane stability test (CMST). Osmotic gradient ektacytometry measurements of RBCs were obtained using the osmoscan module, according to the manufacturer's instructions and as described elsewhere (Da Costa et al., 2016; Lazarova et al., 2017). Briefly, 250 µl whole blood was standardized to a fixed RBC count of $1000*10^6$ and mixed with 5 ml of Elon-Iso. RBCs in polyvinylpyrrolidone (PVP) were exposed to an osmolarity gradient from approximately 60–600 mOsmol/L, whereas shear stress was kept constant (30 Pa). The CMST was performed using the CMST module on the ektacytometer. To perform a CMST, 50 µl of whole blood was standardized to a fixed RBC count of $200*10^6$ and mixed with 5 ml of Elon-Iso. In the CMST, RBCs are exposed to a shear stress of 100 Pa for 3,600 s (1 h) while the EI is continuously measured. The change in the elongation index (ΔEI) was calculated by determining the median of the first and the last 100 s of the CMST and subsequently calculating the difference between the medians. The ΔEI depicts the capacity of the RBCs to shed membrane and resist shear stress.

Intracellular ATP Measurement

The intracellular ATP level was determined by a chemiluminescence assay kit (Abcam) as described in Conrard et al. (2018) and Cloos et al. (2020). ATP levels were then reported to the corresponding Hb content measured by spectrophotometry.

Intracellular Calcium Determination

Intracellular calcium content was determined by RBC labeling with Fluo4-AM (Invitrogen) as in Cloos et al. (2020). Labeled RBCs were then either analyzed by flow cytometry (FACSVerse, BD Biosciences) or fluorimetry (GloMax; Promega; λ_{exc} of 490 nm and λ_{em} of 520 nm). Flow cytometry data, obtained by medium flow rate on 10.000 events, was analyzed with the

software FlowJo to determine the median fluorescence intensity (MFI) of analyzed RBCs. Fluorimetry data were normalized to the global Hb content determined spectrophotometrically as above.

RBC Fatty Acid Content

RBC FA composition was assessed exactly as described in Cloos et al. (2020).

RBC Lipid Composition

The cholesterol content of washed RBCs was determined with a fluorescent assay kit (Invitrogen) as in Cloos et al. (2020) and reported to the global Hb content as described above. The membrane composition in phospholipids, lysophospholipids, sphingolipids and oxysterols was analyzed by lipidomics according to Guillemot-Legris et al. (2016) and Mutemberezi et al. (2016), as described in Cloos et al. (2020).

Oxidative Stress Measurements

Intracellular free ROS were assessed by RBC labeling with 2,7-dichlorodihydrofluoresceindiacetate (H_2DCFDA) in Krebs–Ringer–Hepes (KRH) solution for 30 min at 37°C. MFI of the whole RBC population was acquired as explained above for Fluo4-AM. Lipid peroxidation was evaluated by measurement of malondialdehyde (MDA) with a lipid peroxidation assay kit (Abcam) according to the high sensitivity protocol as in Cloos et al. (2020). Methemoglobin (metHb) was analyzed in RBC lysates, produced through repeated freeze–thaw cycles, with a sandwich Elisa kit (Lifespan Biosciences) as described in Cloos et al. (2020). Both MDA levels and metHb contents were reported to the global Hb content.

Membrane Lipid Imaging on Living RBCs

Endogenous cholesterol was evidenced through living RBC labeling with mCherry-Theta toxin fragment followed by immobilization on PLL-coated coverslips (Cloos et al., 2020). To visualize GM1 ganglioside, sphingomyelin and ceramide, living RBCs were spread onto PLL-coated coverslips and then labeled by trace insertion in the plasma membrane of BODIPY fluorescent analogs of those lipids (Invitrogen) as described in Conrard et al. (2018). Coverslips with living immobilized RBCs were placed in medium-filled LabTek chambers and observed with the wide-field fluorescence microscope (Observer.Z1; plan-Apochromat 100× 1.4 oil Ph3 objective).

Mitochondria Fragment Imaging on Living RBCs

To determine the presence of mitochondria fragments in RBCs, washed RBCs were labeled with 100 nM mitoTracker (Invitrogen) at RT for 30 min, washed, placed on PLL-coated coverslips and observed with the wide-field fluorescence microscope Observer.Z1.

Membrane Curvature on Living RBCs

RBCs were placed into IBIDI chambers as described above for RBC morphology in suspension. Microscopy images were then analyzed with the last version of Shape Analysis by Fourier

Descriptors computation plugging for ImageJ as in Leonard et al. (2017b) and Pollet et al. (2020).

Membrane Transversal Asymmetry on Living RBCs

Exposure of PS was assessed by RBC labeling with Annexin-V coupled to FITC (Invitrogen) and analyzed by flow cytometry as in Cloos et al. (2020). The proportion of PS-exposing RBCs was then determined with the FlowJo software by positioning the cursor at the edge of the healthy RBC population.

Force Distance-Based Atomic Force Microscopy on Living RBCs

AFM experiments were performed with a Bioscope Resolve AFM (Bruker) at ∼25–30°C in DMEM. PeakForce QNM Live Cell probes (Bruker) with spring constants of 0.10 ± 0.02 N m^{-1} and tip radius of curvature of 65 nm were used. The spring constant of the cantilevers was calibrated with a vibrometer (OFV-551, Polytec, Waldbronn) by the manufacturer. The pre-calibrated spring constant was used to determine the deflection sensitivity (Schillers et al., 2017) using the thermal noise method (Hutter and Bechhoefer, 1993) before each experiment. In fast indentation experiments (Dumitru et al., 2018), the AFM was operated in PeakForce QNM mode and Force-distance (FD)-based multiparametric maps were acquired using a force setpoint of 300 pN. The AFM cantilever was oscillated vertically at 0.25 kHz with a peak-to-peak oscillation amplitude of 500 nm. Height and Young's modulus maps were recorded using a scan rate of 0.2 Hz and 256 pixels per line. Slow indentation experiments were performed in Force-Volume (FV) mode. Individual FD curves were recorded in contact mode on the RBC surface with a force setpoint of 300 pN, using ramp speeds of 2 μm s^{-1} for a 2 μm ramp size. Measurements were performed in the central region of RBCs to avoid substrate effects. Indentations ranged between 300 nm and 1 μm depending on cell type.

RBC Spectrin Immunofluorescence

Immunolabeling of spectrin was performed as in Cloos et al. (2020) and Pollet et al. (2020). All coverslips were mounted with Dako and examined with a Zeiss Cell Observer Spinning Disk (COSD) confocal microscope using a plan-Apochromat 100× NA 1.4 oil immersion objective and the same settings for illumination.

Image Analysis, Data Quantification and Statistical Analyses

RBC membrane area, RBC curvature and spectrin intensity/occupation were determined using the Fiji software. Lipid domain abundance was determined by manual counting and expressed by reference to the hemi-RBC projected area. The proportion of spiculated RBCs and of RBCs presenting lipid- or mitoTracker-enriched vesicles, patches or spicules was also assessed by manual counting on fluorescence images. For AFM, at least 64 FD curves were recorded per cell and analyzed using the Nanoscope Analysis v9.1 software (Bruker). Hertz model was used to extract Young's modulus values from individual FD curves. Whole RBC elasticity was analyzed from Slow indentation experiments in FD-AFM mode, by fitting the repulsive part of FD curves with the Hertz model. To analyze the contribution of the plasma membrane and cytoskeleton to the cellular mechanical behavior, two different fit ranges were defined in the repulsive part of FD curves obtained in Fast indentation experiments (PeakForce QNM mode). Plasma membrane elastic contribution was estimated for indentations δ < 50 nm and cytoskeleton elasticity was extracted from δ between 50 and 100 nm. Data are expressed as means ± SEM when the number of independent experiments was $n \geq 3$ or means ± SD if $n \leq 2$. For all experiments, statistical tests were performed only when $n \geq 3$. Tests were non-parametrical Mann–Whitney test or Kruskal–Wallis followed by Dunn's comparison test, except for AFM and membrane curvature data for which two sample t-tests were applied as RBCs were analyzed individually and data obtained for all RBCs are presented. To evaluate the effect of chemical agents, the paired data were analyzed by paired non-parametrical tests (Wilcoxon matched-pairs signed rank tests). ns, not significant; $^*p < 0.05$, $^{**}p < 0.01$, $^{****}p < 0.0001$.

RESULTS

Case Presentation

The patient pHypoβ is a 49 year-old adult. Due to idiopathic thrombocytopenic purpura, his spleen has been removed at the age of 17 as a curative treatment. He has no additional treatment. The transaminases are increased (**Supplementary Table 1**) but he has no anemia and is otherwise healthy. He exhibits since more than 30 years a high number of acanthocytes on blood smears (**Figure 1A**), which was suspected to result from hypobetalipoproteinemia. The patient also presented a vitamin E level close to the lower reference limit (**Supplementary Table 1**). Despite of acanthocytosis, reticulocyte count, RBC mean corpuscular volume and Hb levels are within the normal range (**Supplementary Table 1**).

pHypoβ Presents a Nucleotide Deletion in *APOB* Resulting in a Truncated ApoB

Next-Generation Sequencing of pHypoβ revealed a heterozygous single nucleotide deletion in *APOB*, the gene encoding ApoB. It concerns the deletion of adenosine at nucleotide 2,534 in exon 17, which leads to a shift in the reading frame and consequent premature end of translation at residue 862 (Human Gene Mutation Database, CD051293). Nonsense-mediated mRNA decay likely will prevent translation of such mutant *APOB* mRNA into a severely truncated protein. This mutation has already been identified (Fouchier et al., 2005), but consequences for RBC morphology, cytoskeletal and biophysical properties as well as functionality were not evaluated.

pHypoβ Exhibits a High Proportion of Acanthocytes

We started by confirming the presence of spiculated RBCs detected on a blood smear (**Figure 1A**) by optical microscopy

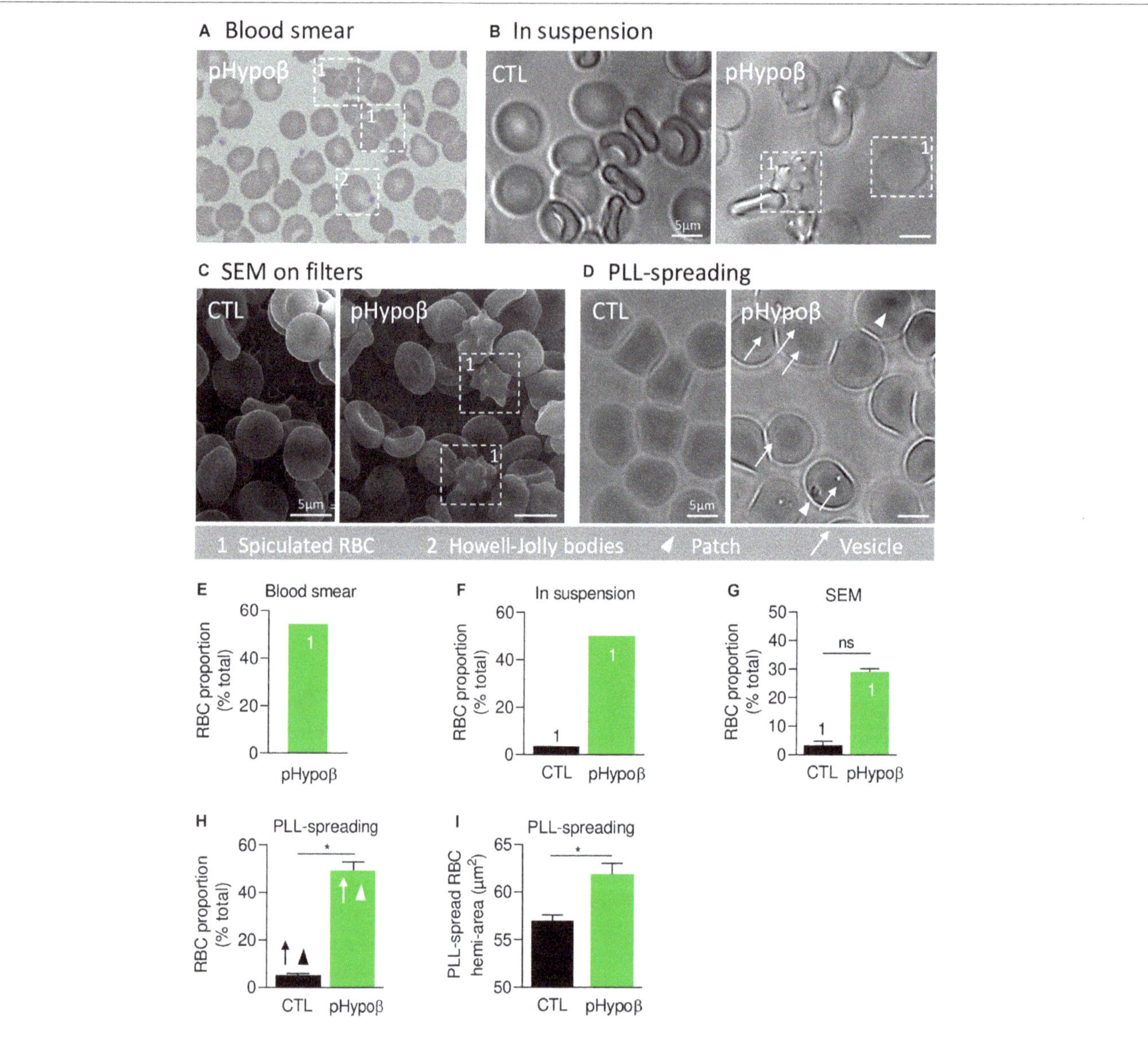

FIGURE 1 | A high proportion of acanthocytes is visible on pHypoβ blood smears and RBCs in suspension while patches and small vesicles are evidenced upon RBC spreading. The morphology of RBCs of the patient with hypobetalipoproteinemia (pHypoβ; green columns) was compared to those of healthy donors (CTL; black columns), either on blood smear **(A)**, on RBCs in suspension **(B,C)**, or upon spreading on poly-L-lysine (PLL)-coated coverslips **(D)**. Then, the proportion of RBCs with spicules (indicated with 1; **E–G**), patches (arrowheads, **H**) and/or vesicles (arrows, **H**) and the RBC membrane area were determined **(I)**.
(A) May–Grünwald Giemsa-stained blood smear. 1, spiculated RBCs; 2, Howell-Jolly bodies. **(B)** Light microscopy of living RBCs in suspension in IBIDI chambers. Images are representative of 2 independent experiments. **(C)** Scanning electron microscopy of glutaraldehyde-fixed RBCs on filters. Images are representative of 2 experiments with 3 filters each. **(D)** Light microscopy of living RBCs spread on PLL-coated coverslips. Images are representative of at least 10 experiments. **(E–G)** Quantification of the abundance of spiculated RBCs, expressed as % of total RBCs. Data are means from 1 experiment for **(E,F)** and means ± SEM of 3–4 filters in **(G)**. Mann–Whitney test; ns, not significant. **(H)** Quantification of the proportion of PLL-spread RBCs presenting patches and/or vesicles. Data are means ± SEM of 3–6 independent experiments. Mann–Whitney test; *$p < 0.05$. **(I)** Quantification of the membrane surface area of PLL-spread RBCs. Data are means ± SEM of 6–7 independent experiments. Mann–Whitney test; *$p < 0.05$.

of living RBCs in suspension and by SEM of fixed RBCs on filters (**Figures 1B,C**). Quantification indicated that spiculated RBCs represented ∼50% of all pHypoβ RBCs in blood smear and optical microscopy images and ∼30% in electron microscopy images while they accounted only for a minority of RBCs from healthy donors (**Figures 1E–G**). Discrepancies regarding the proportion of spiculated RBCs detected by electron and

optical microscopy could result from the shear stress induced by filtration during RBC preparation for SEM, eventually leading to hemolysis of spiculated RBCs.

When RBCs were spread on PLL-coated coverslips, spiculated RBCs could still be distinguished from other RBCs as they presented big dark patches and smaller clearer vesicles at their surface (arrowheads and arrows at **Figure 1D**). The proportion of such RBCs was significantly increased in pHypoβ (**Figure 1H**) and corresponded exactly to the proportion determined above for non-spread RBCs. RBC morphological alterations were accompanied by an increased membrane area of spread RBCs (**Figure 1I**) and by the presence of Howell-Jolly bodies on blood smears (**Figure 1A**), compatible with splenectomy. In contrast, the increase of membrane area did not result from splenic absence since a healthy splenectomised donor did not show such a membrane surface area increase (**Supplementary Figure 1A**).

pHypoβ RBCs Show no Evidence for Increased Osmotic Fragility nor Modifications of ATP and Calcium Homeostasis but a Reduced Ability to Shed Membrane Upon Shear Stress

Despite the high proportion of pHypoβ RBCs with a modified morphology, their resistance to hemolysis was similar to healthy controls (splenectomised or not), as shown by Hb release after RBC incubation in isotonic and increasingly hypotonic media (**Figures 2A,B** and **Supplementary Figure 1B**). Nevertheless, the more sensitive osmotic gradient ektacytometry test revealed that the O_{min} value was slightly increased in pHypoβ (153 vs. 125–148 in healthy donors), indicative of a slight decrease in surface area-to-volume ratio and a slight increase in osmotic fragility (**Figure 2C**). Moreover, the EI_{max}, and therefore the surface area, were slightly decreased in pHypoβ RBCs (0.566 vs. 0.585–0.607). This decrease *a priori* contrasted with the increased surface area determined by optical microscopy on spread RBCs which could result from the increase of the outer leaflet membrane area specifically (**Figure 1I**). Those changes did not result from splenectomy since all osmotic gradient ektacytometry-derived parameters were normal in the healthy splenectomized control (**Figure 2C**). To further investigate pHypoβ RBC deformability, the CMST was performed to assess the RBC ability to shed membrane when exposed to prolonged supraphysiological shear stress force. This is a physiological property of healthy RBCs and the decrease in EI that occurs under these circumstances is proposed to be a measure of membrane health. pHypoβ RBCs showed a ΔEI of −0.144, which was considerably less than the ΔEI of −0.187 from the healthy splenectomized control, and even more less than the ΔEI of −0.208 seen in healthy donors (**Figure 2D**). These results indicated that splenectomy on itself was accompanied with loss of the ability to shed membrane upon shear stress, but that this loss was more pronounced in pHypoβ RBCs. We also evaluated the intracellular contents in ATP and calcium, two key parameters for RBC functionality and deformation (Bogdanova et al., 2013; Conrard et al., 2018). The intracellular ATP content was slightly increased but the calcium content remained unchanged (**Figures 2E,F**).

pHypoβ RBCs Exhibit Normal Membrane Fatty Acid Profile and Cholesterol Content but Increased Ceramide Species

As acanthocytotic RBCs in abetalipoproteinemia and homozygous familial hypobetalipoproteinemia are associated with profound alterations of membrane lipid composition (Biemer, 1980; Barenholz et al., 1981; Gheeraert et al., 1988), we then determined whether RBC morphology alterations in pHypoβ could be associated with modified membrane composition in FAs (**Figures 3A–E**), cholesterol (**Figure 3F**), phospholipids and/or sphingolipids (**Figure 4**). The relative proportion of saturated (SFA), monounsaturated (MUFA) and polyunsaturated (PUFA) FAs was unchanged in pHypoβ RBCs (**Figure 3A**). Closer examination of major SFAs and MUFAs did not reveal more changes (**Figures 3B,C**). For PUFAs, however, the proportion of long chain C22 PUFAs appeared to decrease in favor of PUFAs with shorter C18 or C20 chains (**Figure 3D**). The increase of C18 and C20 PUFAs appeared to result from the higher relative contents in linoleic (C18:2) and arachidonic (C20:4) acids (**Figure 3E**).

Since the plasmatic cholesterol content was decreased in pHypoβ (**Supplementary Table 1**), we then evaluated the RBC membrane cholesterol content. Surprisingly, it was not modified in the patient as compared to the healthy donors (**Figure 3F**). Sphingomyelin species were also largely maintained (**Figure 4A**) whereas all ceramide and dihydroceramide species, whatever their fatty acid length and unsaturation number, were increased by 1.5- to 2-fold in pHypoβ (**Figure 4B**). Among phospholipids, no obvious change was detected for PS and phosphatidylinositol (PI) but phosphatidylcholine (PC) species were decreased and phosphatidylethanolamine (PE) species were very slightly increased (**Figures 4C,E,G,I**). In agreement with the heightened proportion of PE, four lysoPE species were also increased. The other lysophospholipids remained unaffected in pHypoβ (**Figures 4D,F,H,J**).

pHypoβ RBCs Have Increased Free Reactive Oxygen Species but No Modification of Phospholipid, Cholesterol and Hemoglobin Oxidation

As the proportion of long chain PUFAs was decreased, we investigated whether pHypoβ RBCs could suffer from oxidative stress due to low circulating vitamin E levels (**Supplementary Table1**), eventually leading to lipid and protein oxidation. We started by measuring ROS using H_2DCFDA. A significant twofold increase of free ROS was observed in pHypoβ RBCs (**Figure 5A**) but not in a healthy donor without spleen (**Supplementary Figure 1C**), indicating that the ROS increase in pHypoβ was not due to splenectomy.

The extent of lipid peroxidation, determined through the level of MDA, was not modified in pHypoβ RBCs (**Figure 5B**). This might seem surprising because increased lipid peroxidation could be detected in HDL and platelets of patients with abetalipoproteinemia (Calzada et al., 2013). Nevertheless, only a slight increase was observed even for healthy RBCs treated with H_2O_2 (**Figure 5B**) and might be explained by the fact that MDA is not the only end product of lipid peroxidation. Nevertheless, oxysterol species were

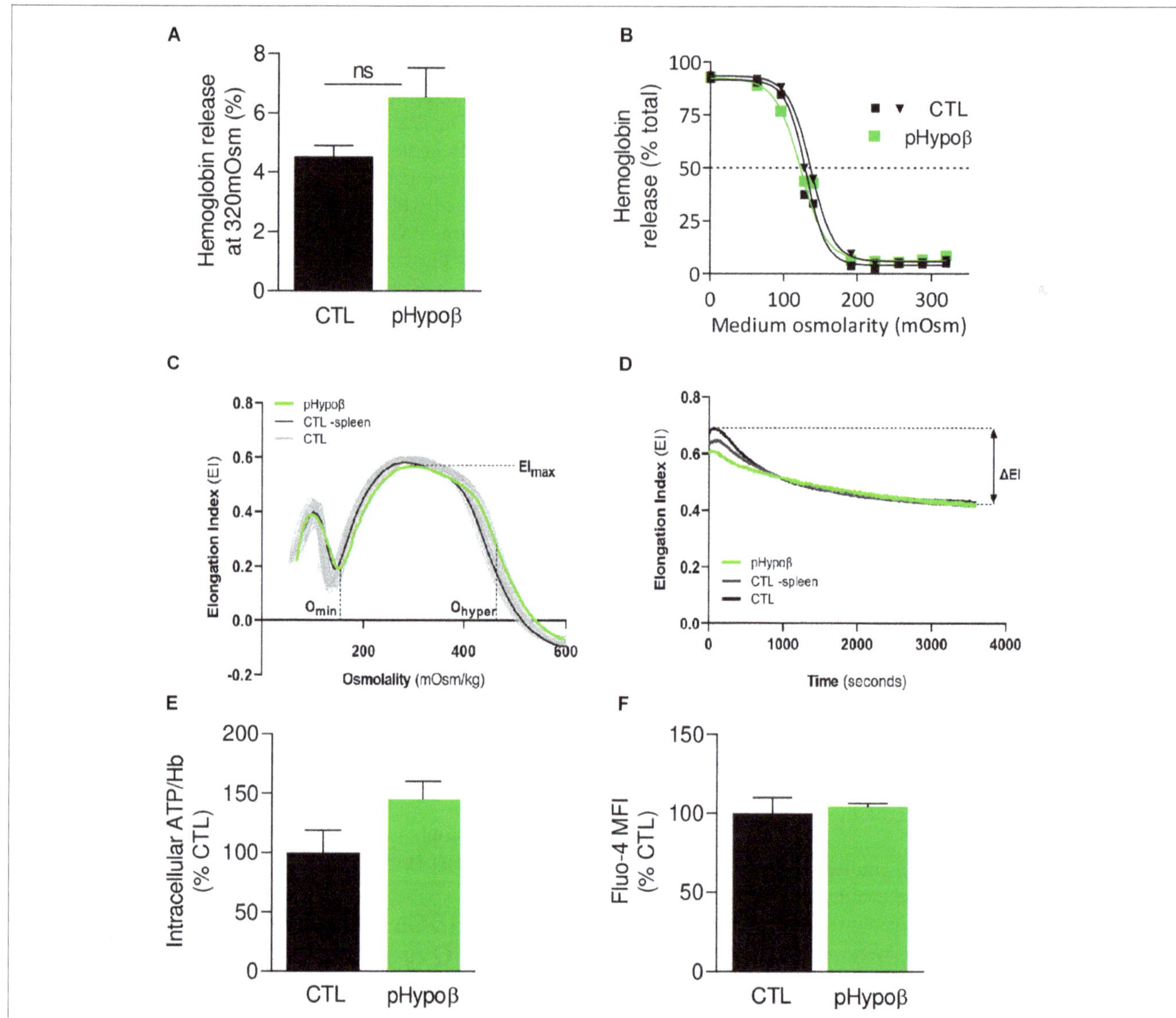

FIGURE 2 | The extent of hemoglobin release and the intracellular calcium and ATP levels are preserved in pHypoβ RBCs while deformability upon shear stress is slightly decreased. RBCs from healthy donors (black and gray in panel **C**) or pHypoβ (green) were evaluated for osmotic fragility **(A,B)**, deformability **(C,D)**, intracellular ATP **(E)** and calcium content **(F)**. **(A,B)** RBC osmotic fragility. RBCs were incubated in isotonic **(A)** or increasingly hypotonic **(B)** media and then centrifugated. Hemoglobin (Hb) in supernatants and in RBC pellets was assessed spectrophotometrically to determine hemolysis. The horizontal dotted line in **(B)** indicates the medium osmolarity at which 50% of the RBCs were lysed. Data are means ± SEM of 4 independent experiments for **(A)** and are representative for 2 independent experiments in **(B)**. Mann–Whitney test; ns, not significant. **(C,D)** RBC deformability. **(C)** Osmotic gradient ektacytometry curve and derived EI_{max}, O_{min}, and O_{hyper} parameters, which, respectively, reflect membrane surface area, surface area-to-volume ratio and cellular hydration. pHypoβ RBCs (green curve) were compared to healthy controls (gray curves obtained from 25 healthy subjects) and a healthy splenectomized control (black curve). **(D)** Cell membrane stability test (CMST) curve and the derived ΔEI parameter which depicts the capacity of the RBCs to shed membrane and resist shear stress. Data are representative of 2 experiments in **(C)** and 1 experiment in **(D)**. **(E)** Intracellular ATP. ATP levels were determined with a kit based on the activity of the firefly luciferase in presence of ATP and the consequent light emission in presence of luciferin. Intracellular ATP levels were normalized to Hb and expressed as percentage of the CTL RBCs. Data are means ± SD of triplicates from 1 experiment. **(F)** RBC calcium content. RBCs were labeled with the non-fluorescent Fluo4-AM which is transformed in RBCs into the fluorescent Fluo4 after de-esterification and interaction with calcium ions. Labeled RBCs were analyzed by flow cytometry for median fluorescence intensity (MFI) and then expressed as percentage of CTLs. Data are means ± SD of triplicates from 1 experiment.

also maintained at levels similar to those found in healthy RBCs (**Figure 5C**).

Besides lipids, Hb is also a major target of oxidative stress in RBCs, generating metHb and eventually hemichromes (Mohanty et al., 2014). By ELISA we showed that pHypoβ RBCs exhibited a very slight metHb increase, comparable to the one obtained upon treatment of healthy RBCs with H_2O_2 (**Figure 5D**). All those data suggested that, despite major morphological changes

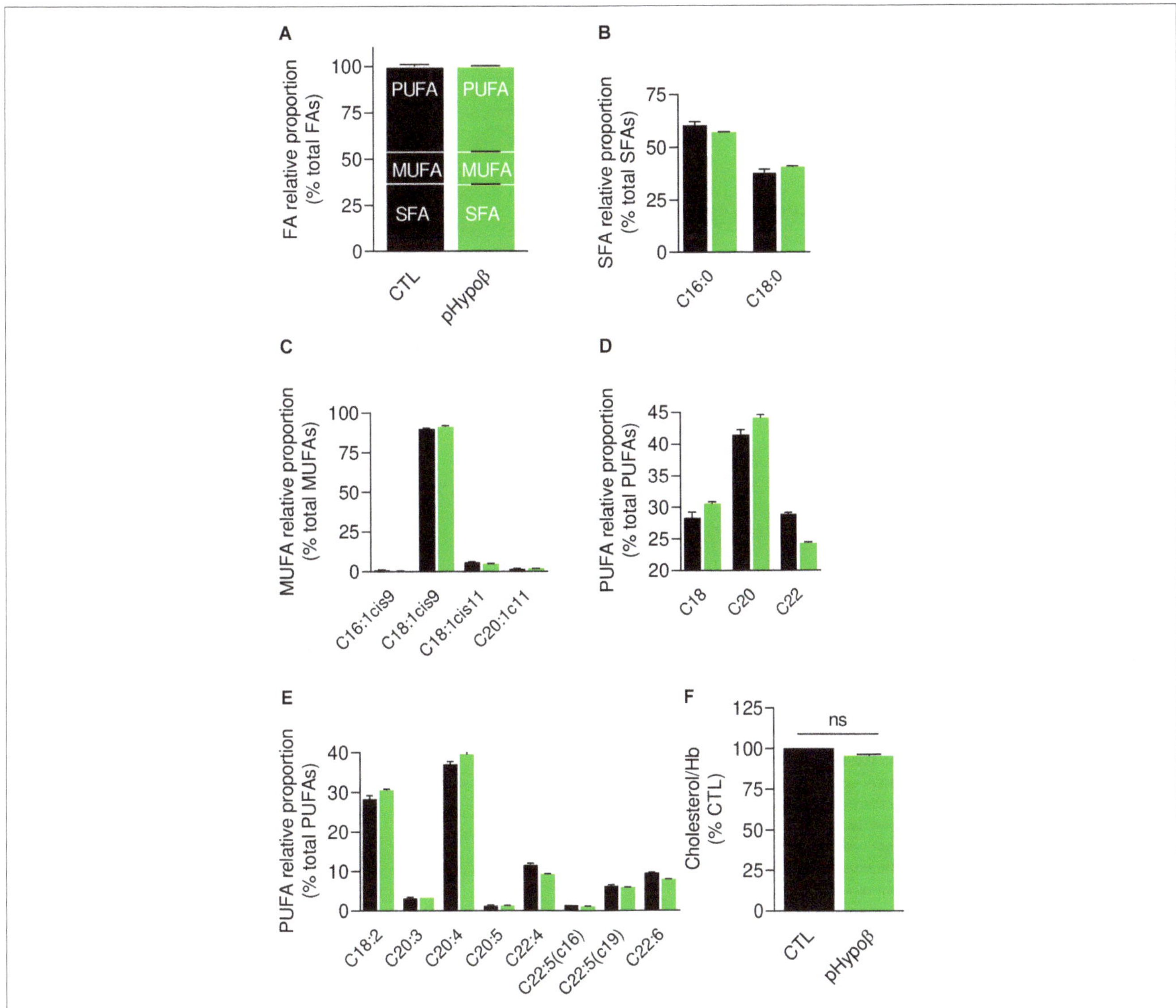

FIGURE 3 | Membrane fatty acid and cholesterol contents are largely preserved in pHypoβ RBCs. RBCs from healthy donors (black columns) or pHypoβ (green columns) were evaluated for membrane fatty acid (FA) content **(A–E)** and cholesterol level **(F)**. **(A–E)** RBC FA composition. Lipids were extracted from isolated RBCs and prepared for gas chromatography to analyze FA content. **(A)** Relative proportion of saturated (SFA), monounsaturated (MUFA) and polyunsaturated (PUFA) FAs expressed as percentage of total FAs. **(B)** Relative proportion of the two major SFAs (C16:0 and C18:0). **(C)** Relative proportion of the four major MUFAs (C16 and C18 with one double bond on position 9 and C18 and C20 with one double bound on position 11). **(D)** Relative proportion of PUFA according to the carbon chain length (chains of 18, 20, or 22C). **(E)** Relative proportion of the major PUFAs (chains of 18–22C and 2–6 double bonds). All data are means ± SD of triplicates from 1 experiment. **(F)** Cholesterol content. Lysed RBCs were evaluated for their cholesterol content through a fluorescent assay kit which uses several enzymatic reactions starting with cholesterol and ending with the transformation of Amplex Red into fluorescent resofurin. Cholesterol content was normalized to Hb content and expressed as percentage of the CTL RBCs. Data are means ± SEM of 3 independent experiments. Mann–Whitney test; ns, not significant.

and increase of free ROS, no obvious alterations in lipid and Hb oxidation appeared to occur in pHypoβ RBCs.

pHypoβ RBCs Are Altered for Their Spectrin Network, Showing Either a Lower Density or a Patchy or Vesiculated Pattern

We next evaluated by immunofluorescence the distribution of spectrin, another major target of oxidative stress (Voskou et al., 2015). Although the spectrin network was homogeneous in ~99% of healthy RBCs, this proportion was significantly reduced to ~75% in pHypoβ (population 1 at **Figures 6A,B**). Closer examination of this population indicated a tendency to decrease of the spectrin occupancy per RBC surface combined with a lower variance of spectrin occupancy (**Figures 6C,D**), suggesting that the spectrin network at the surface of pHypoβ RBCs was less dense than in healthy RBCs. Besides, the patient exhibited two additional populations with differential spectrin patterns, namely spectrin-enriched patches and vesicles (populations 2 and 3 at

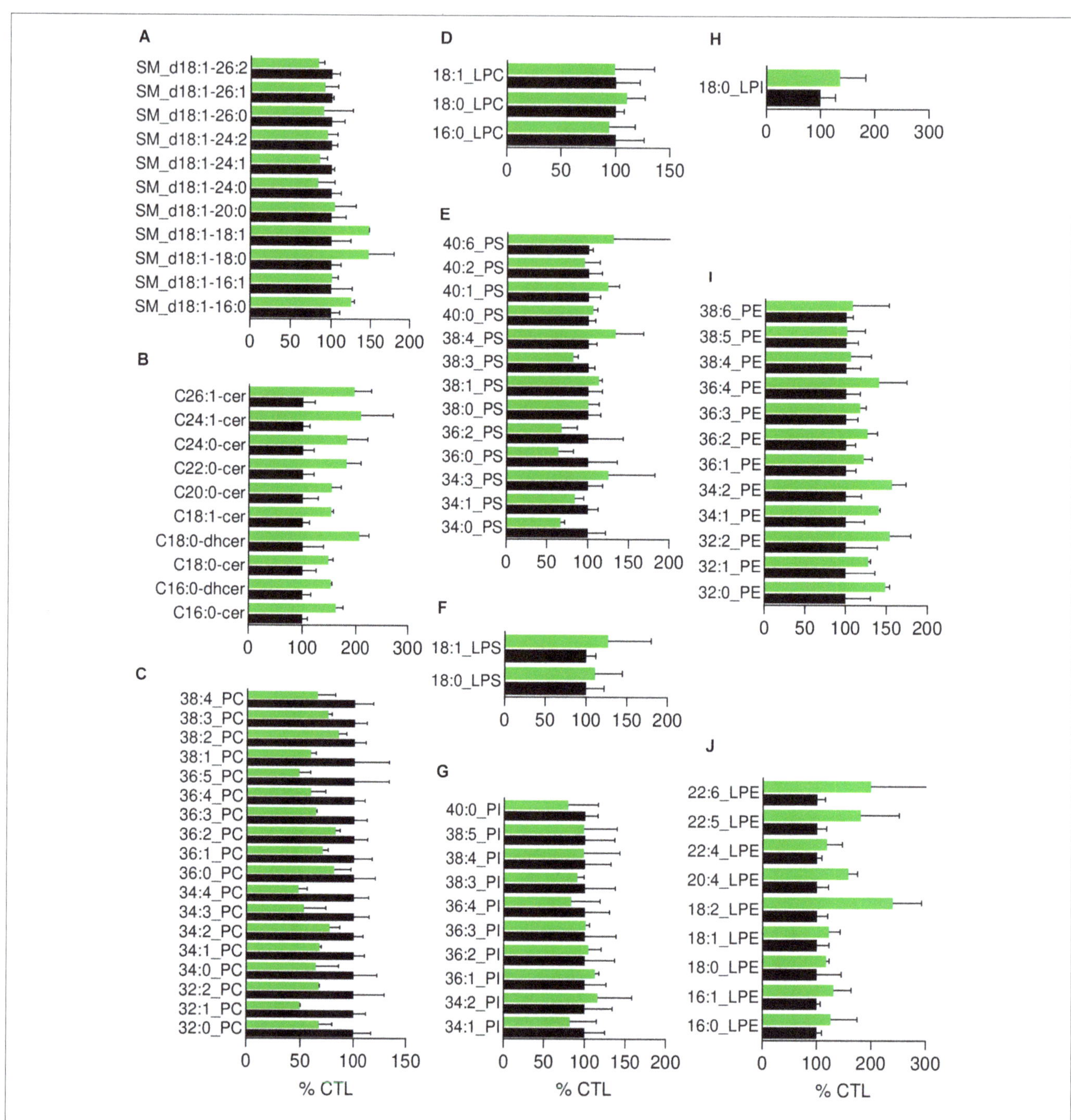

FIGURE 4 | The most obvious membrane lipid changes in pHypoβ RBCs are an increase of ceramide species and a decrease of phosphatidylcholine species. Lipid species were assessed by HPLC-MS on washed, lysed and lipid-extracted RBCs. Content in sphingomyelin (d18:1 D-erythro-sphingosine backbone, SM; **A**), ceramide (Cer) and dihydroceramide (dhcer; **B**), phosphatidylcholine (PC; **C**), lysophosphatidylcholine (LPC; **D**), phosphatidylserine (PS; **E**), lysophosphatidylserine (LPS; **F**), phosphatidylinositol (PI; **G**), lysophosphatidylinositol (LPI; **H**), phosphatidylethanolamnine (PE; **I**) and lysophosphatidylethanolamine (LPE; **J**). Results are expressed as percentage of controls (CTL, mean of 4 donors). Data are means ± SD of 2 independent experiments.

Figure 6A). The proportion of those two populations together increased by ~25-fold as compared to healthy RBCs (populations 2 and 3 at **Figure 6B**). In conclusion, the spectrin cytoskeleton was altered in pHypoβ RBCs, ~25% of them showing a patchy and vesiculated pattern and the remaining ~75% exhibiting a less dense network.

RBC Whole Stiffness and Curvature in Low and High Curvature Areas Are Increased in pHypoβ

To determine whether spectrin cytoskeleton modifications were accompanied by altered membrane biophysical properties, we

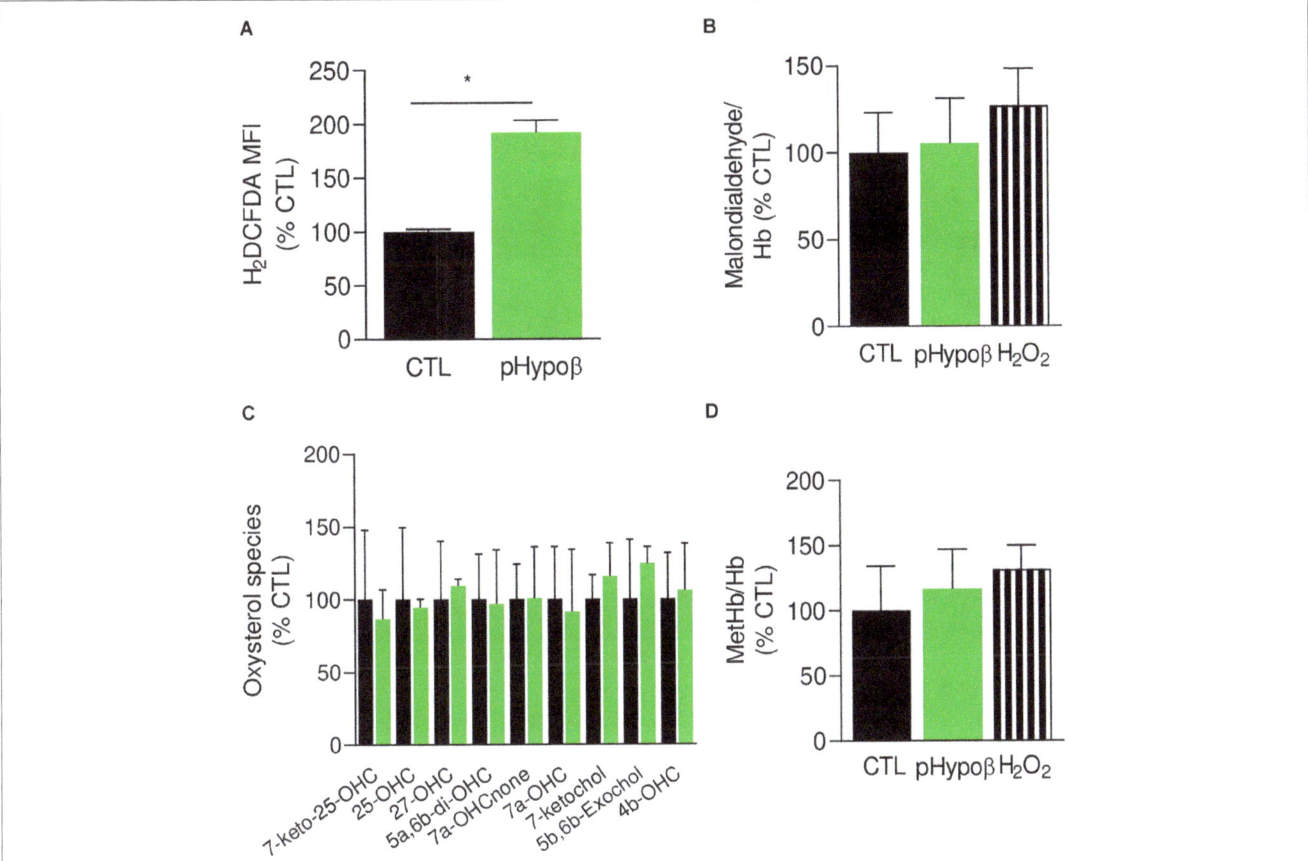

FIGURE 5 | Accumulation of free reactive oxygen species in pHypoβ RBCs is not accompanied by lipid or hemoglobin oxidation. RBCs from healthy donors (black columns) or pHypoβ (green columns) were evaluated for free reactive oxygen species (ROS) accumulation **(A)**, lipid peroxidation **(B)**, oxysterol content **(C)**, and methemoglobin (MetHb) accumulation **(D)**. Hydrogen peroxide was used as positive control in **(B,D)** (hatched columns). All data are expressed as percentage of healthy untreated RBCs. **(A)** Intracellular ROS accumulation. RBCs were labeled with the non-fluorescent H_2DCFDA which is transformed into fluorescent DCF inside the RBCs after de-esterification and interaction with ROS. Flow cytometry analysis allowed to determine the MFI of the global RBC population. Data are means ± SEM of 3–6 independent experiments. Mann–Whitney test; *$p < 0.05$. **(B)** Lipid peroxidation. Malondialdehyde (MDA), one final product of lipid peroxidation, was detected through interaction with thiobarbituric acid forming a fluorescent adduct. MDA levels were normalized to Hb content and data are means ± SD of triplicates from 2 independent experiments. **(C)** Membrane content in oxysterols. RBCs were washed, lysed, extracted for lipids and determined for 7-Keto-25-hydroxycholesterol (7-keto-25-OHC), 25-hydroxycholesterol (25-OHC), 27-hydroxycholesterol (27-OHC), 5α,6β-dihydroxycholesterol (5α,6β-diOHC), 7α-hydroxycholestenone (7α-OHCnone), 7α-hydroxycholesterol (7α-OHC), 7-ketocholesterol (7-ketochol), 5β,6β-epoxycholesterol (5β,6β-exochol) and 4β-hydroxycholesterol (4β-OHC). Results are expressed as percentage of control RBCs and are means ± SD of 2 independent experiments. **(D)** MetHb content. MetHb was determined using a sandwich Elisa and reported to the global Hb content. Data are means ± SD of triplicates from 1 experiment.

analyzed RBC membrane stiffness and curvature. AFM at SLOW indentation revealed that pHypoβ RBCs were stiffer than healthy RBCs (**Figure 7A**). Since our previous experiments on RBCs treated with cytoskeleton-depolymerizing drugs using the same indentation approach revealed changes in the Young's modulus, we inferred that the main contribution in the whole Young's modulus originated from the cytoskeleton even if some contribution from the cytoplasmic viscosity cannot be discarded. To further test the contribution of the cytoskeleton in the whole RBC Young's modulus vs. the plasma membrane stiffness, RBCs were analyzed at Fast indentation. Plasma membrane elasticity was similar in CTL and pHypoβ RBCs, while a stiffening of the cytoskeleton was observed for pHypoβ RBCs (**Figure 7B**). Hence, a higher variability of measurements was observed for the patient than healthy donors, which might reflect

the presence of both acanthocytes and discocytes that were difficult to discriminate after RBC spreading. Interestingly, for both control and pHypoβ RBCs, the highest plasma membrane elastic modulus corresponded to the highest cytoskeleton stiffness (**Supplementary Figure 2**), suggesting that in diseased RBCs both the plasma membrane and the cytoskeleton were altered.

Because it was not possible to discriminate between spiculated and non-spiculated RBCs in AFM due to spreading we then used chambers compatible with RBC morphology preservation and quantified the membrane curvature of the non-spiculated RBCs (**Figure 7C**, population 1). Although the differential curvature between high (HC) and low curvature (LC) areas was preserved in the patient, curvature in both HC and LC areas was increased (**Figures 7C,D**). This latter increase was in good agreement with the lower spectrin occupancy per RBC surface in pHypoβ. Of

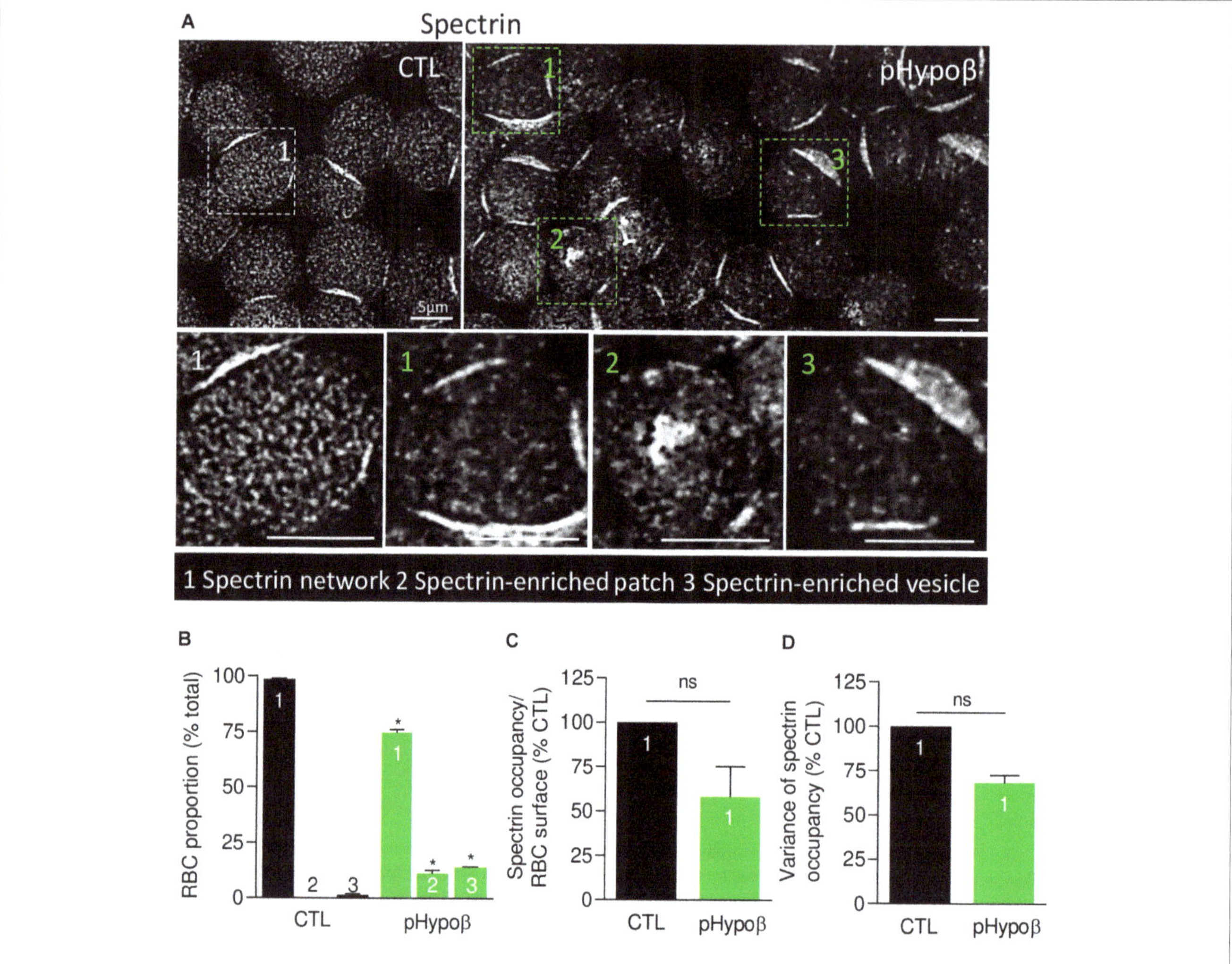

FIGURE 6 | The spectrin network is altered in pHypoβ RBCs, showing either a lower density or a patchy or vesiculated pattern. RBCs from healthy donors (CTL, black) or pHypoβ (green) were spread onto PLL-coated coverslips, fixed/permeabilized, immunolabelled for spectrin and visualized by confocal fluorescence microscopy using the same settings for sample illumination. **(A)** Representative general views and zooms of RBCs with homogenous spectrin network (1), spectrin-enriched patches (2) and spectrin-enriched vesicles (3). **(B)** Quantification of the relative proportion of RBCs with homogenous spectrin network (1), spectrin-enriched patches (2) or spectrin-enriched vesicles (3). Data are means ± SEM of 3–6 independent experiments. Mann–Whitney tests to compare each RBC population in healthy vs. pHypoβ RBCs. *$p < 0.05$. **(C,D)** Quantification of spectrin occupancy normalized to the RBC surface **(C)** and variance of the spectrin labeling **(D)** in RBCs with homogenous spectrin network (population 1). Data are expressed as percentage of CTLs and are means ± SEM of 3 independent experiments. Mann–Whitney tests; ns, not significant.

note, no modification of membrane curvature was detected for RBCs from a healthy splenectomised donor (**Supplementary Figure 1D**). Altogether our data indicated that the cytoskeleton alteration in pHypoβ RBCs was accompanied by increased stiffness and curvature. To further analyze the potential alteration of the plasma membrane in the disease, we determined the membrane asymmetry both at the transversal and lateral levels.

The Proportion of RBCs Exhibiting Phosphatidylserine Surface Exposure Is Preserved in pHypoβ

RBC PS exposure, a measure of membrane transversal asymmetry integrity loss, was not increased in pHypoβ RBCs as compared to healthy RBCs. As internal control, we used blood stored for 2 weeks in K⁺/EDTA tubes at 4°C, which we previously validated as a model of accelerated RBC aging *in vitro* based on morphological, biophysical and biochemical storage lesions, including PS surface exposure (Cloos et al., 2020). Thus, according to our previous work, a strong increase of PS surface exposure was observed in this condition (**Figure 7E**).

Sphingomyelin-Enriched Domains at the pHypoβ RBC Surface Increase in Abundance and Do Not Respond to Intracellular Calcium Depletion

We then took benefit from our expertise in submicrometric lipid domain organization at the RBC outer plasma membrane

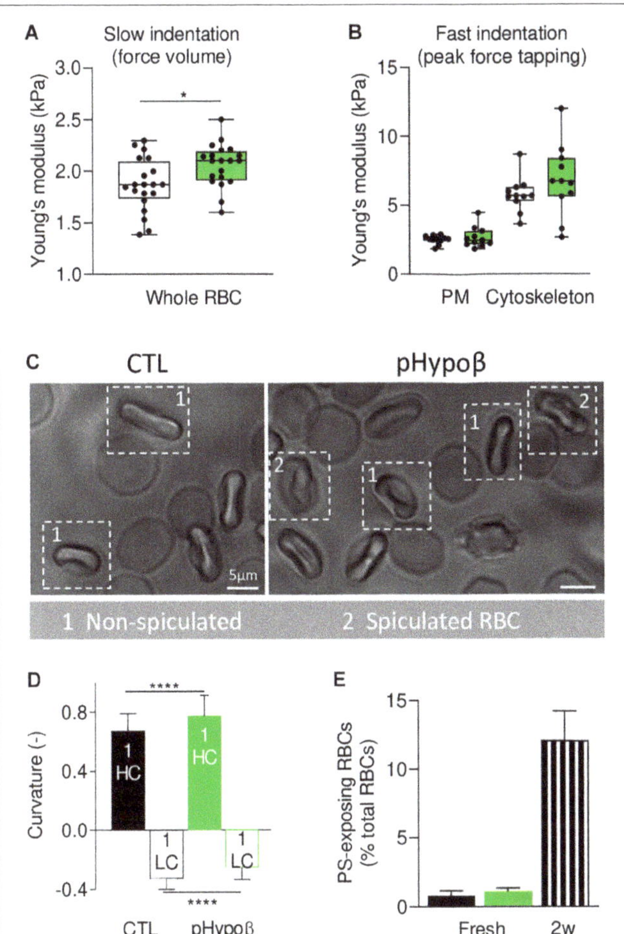

FIGURE 7 | Membrane stiffness and curvature are increased in pHypoβ RBCs whereas membrane transversal asymmetry is not modified. RBCs from healthy donors (CTL, black) or pHypoβ (green) were evaluated for membrane stiffness (A,B), curvature (C,D) and transversal asymmetry (E). (A,B) Membrane stiffness. (A) Young's modulus values extracted for CTL and pHypoβ RBCs in Slow indentation experiments, where the whole RBC mechanical behavior was analyzed. (B) Young's modulus of CTL and pHypoβ RBCs obtained in Fast indentation experiments, where the elastic contribution of the plasma membrane (PM) and cytoskeleton were evaluated separately. Each data point represents the mean Young's modulus value calculated for one RBC. Box plots depict 25th–75th percentiles, horizontal lines and cantered squares show mean values and error bars indicate SD. 20 RBCs were analyzed in (A) and 11 RBCs in (B). Two-sample t-test. *$p < 0.05$. (C,D) Membrane curvature. RBCs were diluted, dropped off in IBIDI chambers and immediately observed by microscopy. Spiculated RBCs (indicated by 2) were distinguished from non-spiculated RBCs (indicated by 1) and the latter were then quantified for curvature in high curvature maxima (HC) and low curvature maxima (LC). 110 RBCs were analyzed for healthy donors and 50 RBCs for pHypoβ. Two-sample t-test. ****$p < 0.0001$. (E) Membrane transversal asymmetry. RBCs were labeled with fluorescent Annexin-V and then analyzed with FlowJo to determine the proportion of phosphatidylserine (PS)-exposing cells by positioning the cursor at the edge of the healthy fresh RBC population. RBCs from blood stored for 2 weeks at 4°C (2w) were used as positive control. Data are means ± SD of triplicates from 1 experiment.

to evaluate the respective abundance of cholesterol-, GM1- and sphingomyelin-enriched domains, which contribute to the RBC deformation process (Leonard et al., 2017b;

Conrard et al., 2018). Indeed, global quantitative analyses at the whole cell level may not reveal more localized alterations of acanthocyte membranes. Moreover, lipid domain abundance depends on membrane:cytoskeleton anchorage (Conrard et al., 2018). To visualize lipid domains, RBCs were labeled with a mCherry-toxin fragment specific to endogenous cholesterol or BODIPY fluorescent analogs of GM1 and sphingomyelin. Although the three types of lipid domains showed a tendency to increase in pHypoβ, only the abundance of sphingomyelin-enriched domains per hemi-RBC was significantly increased by 2.5-fold (**Figures 8A,B**). In contrast, no modifications of lipid domain abundance could be observed for RBCs from healthy splenectomised donors (**Supplementary Figure 1E**).

As GM1- and sphingomyelin-enriched domains were proposed to be implicated in calcium exchanges necessary for RBC deformation, we investigated their functionality in pHypoβ by stimulation of RBCs with Yoda1 and EGTA in a calcium-free medium to induce calcium entry and intracellular depletion, respectively. RBC incubation with Yoda1 led to an increased abundance of GM1-enriched domains in healthy RBCs, as previously shown (Conrard et al., 2018), and the same amplitude of response could be observed for pHypoβ RBCs (**Figure 8C**). In contrast, after intracellular calcium depletion, a heightened abundance of sphingomyelin-enriched domains was detected for healthy RBCs, as expected (Conrard et al., 2018), but not for pHypoβ RBCs (**Figure 8D**).

pHypoβ RBCs Show Low Curvature-Associated Ceramide-Enriched Patches and High Curvature-Associated Cholesterol-Enriched Spicules

Besides well-defined submicrometric lipid domains, larger lipid-enriched patches (arrowheads at **Figure 8A**) and peripheral spicules (open arrows at **Figure 8A**) were seen in pHypoβ RBCs. We quantified the proportion of those lipid-enriched patches and spicules and their potential relationship with the patches and vesicles observed by contrast phase microscopy (see **Figure 1D**). Although none of the patches were enriched in cholesterol or GM1 ganglioside (white arrowheads at **Figure 9A**), some sphingomyelin-enriched patches were observed (red arrowheads at **Figure 9A**). However, their abundance was non-significantly different from the one in healthy RBCs (**Figure 9B**) and represented a low proportion of the patches evidenced at **Figure 1D**.

Intrigued by the increased content of ceramide and dihydroceramide species in pHypoβ (**Figure 4B**), we wondered if the patches could be enriched in ceramide. A ~30-fold increase of the number of RBCs presenting ceramide-enriched patches was observed (red arrowheads at **Figure 9A** and quantification at **Figure 9C**). A heightened proportion of RBCs with ceramide-enriched patches was

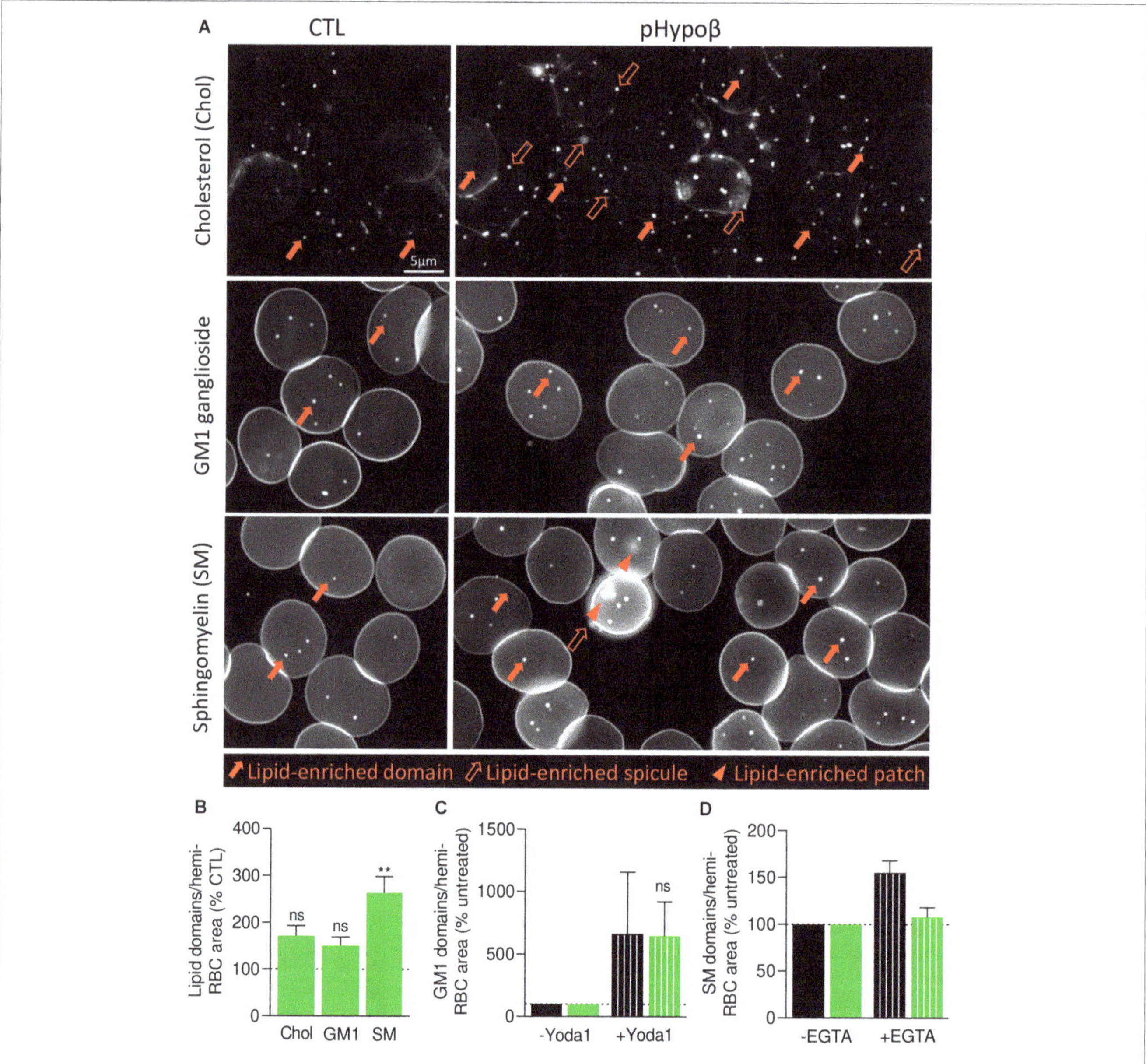

FIGURE 8 | In contrast to cholesterol- and GM1-enriched domains, those enriched in sphingomyelin are modified in abundance and functionality at the pHypoβ RBC surface. RBCs from healthy donors (CTL, black columns) or pHypoβ (green columns) were either left untreated (not-hatched columns) or incubated with Yoda1 for 30 s (hatched columns in **C**) or EGTA for 10 min in a calcium-free medium (hatched columns in **D**). All RBCs were then either labeled with the fluorescent Theta toxin fragment specific to endogenous cholesterol and then immobilized on PLL-coated coverslips (Chol; **A,B**); or immobilized on PLL-coated coverslips and then labeled with fluorescent BODIPY analogs of GM1 ganglioside (GM1; **A–C**) or sphingomyelin (SM; **A,B,D**). All coverslips were then directly observed by vital fluorescence microscopy. **(A)** Representative images of Chol-, GM1-, and SM-enriched domains, spicules or patches in untreated RBCs. Large filled arrows, lipid-enriched domains; large open arrows, lipid-enriched spicules; arrowheads, lipid-enriched patches. **(B–D)** Quantification of lipid domain abundance in RBCs either untreated or treated with Yoda1 or EGTA to, respectively, activate PIEZO1 **(C)** or induce intracellular calcium depletion **(D)**. In **(D)** the calcium-free medium containing EGTA is maintained all along the experiment. Data are normalized to the hemi-RBC area and are means $\pm$ SEM of 4–5 independent experiments **(B)** or means $\pm$ SD/SEM of 2–3 independent experiments **(C,D)**. Kruskal–Wallis test followed by Dunn's comparison test **(B)** and Wilcoxon matched-pairs signed rank tests **(C)**. ns, not significant; **$p < 0.01$.

also observed for a healthy splenectomised donor but those RBCs accounted only for ~10% of all RBCs compared to ~35% for pHypoβ (**Figure 9C** and **Supplementary Figure 1F**). This could suggest that those patches are normally eliminated by the spleen.

Besides patches, ceramide-enriched vesicles can also be found but they did not significantly increase in abundance as compared to healthy RBCs (**Figure 9C**). The same was true for sphingomyelin-enriched vesicles (**Figure 9B**). In contrast, cholesterol-enriched spicules were observed at the edges of

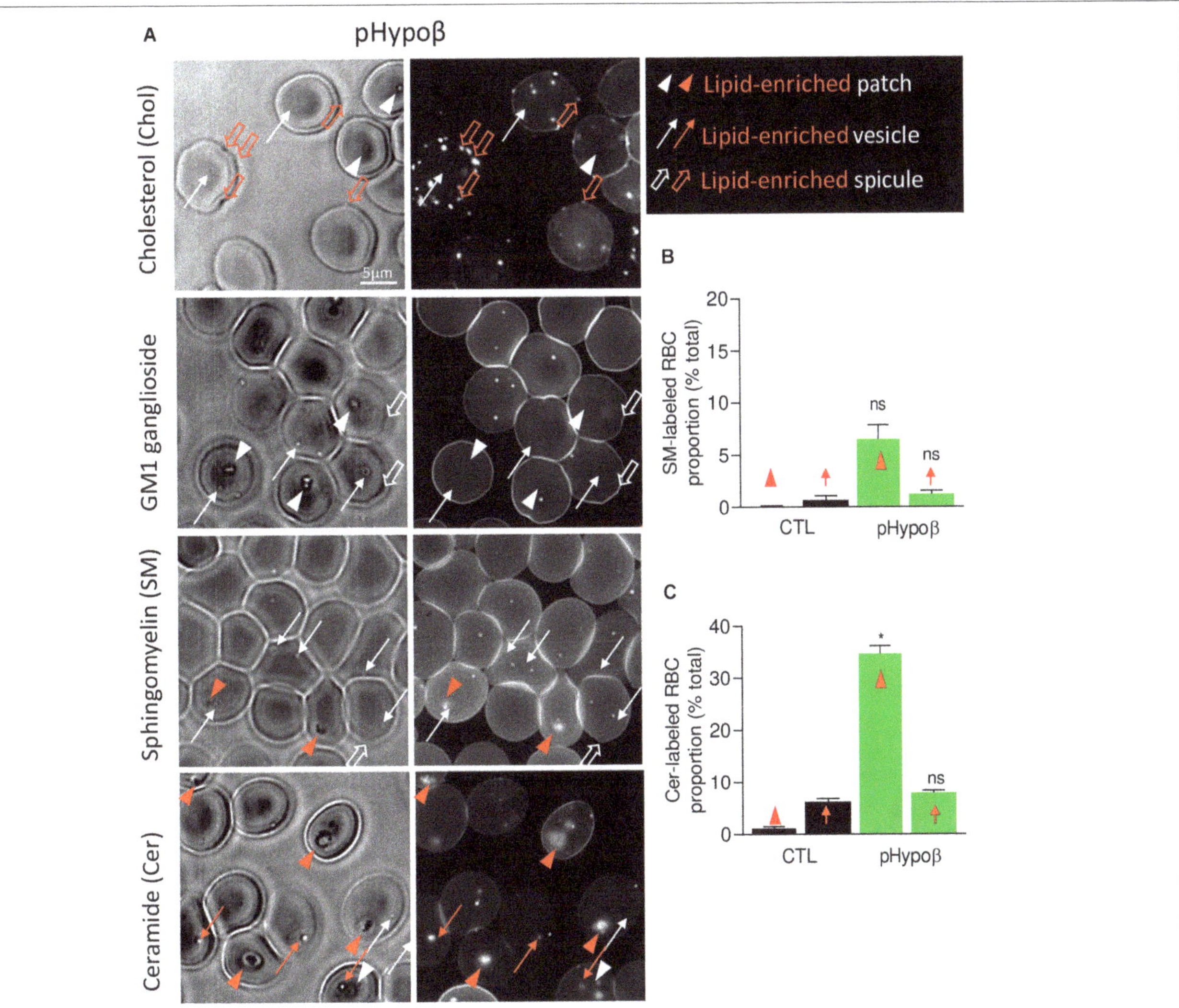

FIGURE 9 | The spicules at the pHypoβ surface are mainly enriched in cholesterol whereas the patches and vesicles are mainly enriched in ceramide. RBCs from healthy donors (black columns) or pHypoβ (green columns) were either labeled with the fluorescent Theta toxin fragment specific to endogenous cholesterol and then immobilized on PLL-coated coverslips (Chol; **A**) or immobilized on PLL-coated coverslips and then labeled with fluorescent BODIPY analogs of GM1 ganglioside (GM1; **A**), sphingomyelin (SM; **A,B**); or ceramide (Cer; **A,C**). All coverslips were then directly observed by vital fluorescence microscopy. **(A)** Representative images of pHypoβ RBCs. Left, transmission microscopy image; right, fluorescence microscopy image. Lipid-enriched (red) or non-enriched (white) patches (arrowheads), vesicles (arrows), spicules (large open arrows) detected on the RBC surface on transmission images. **(B,C)** Quantification of the proportion of RBCs presenting SM- **(B)** or cer- **(C)** enriched patches or vesicles. Data are expressed as means ± SEM of 3–6 independent experiments. Mann–Whitney tests to compare enriched structures at the surface of healthy vs. pHypoβ RBCs; ns, not significant; *$p < 0.05$.

pHypoβ RBCs. As for patches, GM1 was enriched neither in vesicles nor in spicules (**Figure 9A**, white thin and thick arrows).

pHypoβ RBCs Do Not Represent an Accelerated Model of RBC Aging

Altogether, our data indicated that pHypoβ RBCs were spiculated and showed enhanced ROS content, altered spectrin cytoskeleton integrity as well as increased membrane stiffness and curvature and abundance of sphingomyelin-enriched domains. These elements are consistent with the RBC morphological and biochemical storage lesions we previously revealed upon blood storage in K^+/EDTA tubes for up to 4 weeks at 4°C as a model of accelerated RBC aging *in vitro* (Cloos et al., 2020). On the other hand, pHypoβ had increased membrane surface area and proportion of ceramide-enriched patches, rather consistent with reticulocyte properties (Ney, 2011). To therefore evaluate whether morphological, biochemical and biophysical alterations of pHypoβ RBCs could result from acceleration of RBC aging, RBCs from pHypoβ and healthy donors were stored for up to 3 weeks at 4°C and compared for morphology, fragility, functionality and biophysical properties. After 1 week of storage, the proportion of pHypoβ RBCs with patches did not increase any more (**Figure 10A**) and the differences in RBC membrane area and Hb release between pHypoβ and healthy donors seen in fresh RBCs were both abrogated (**Figures 10B,C**). The increase of

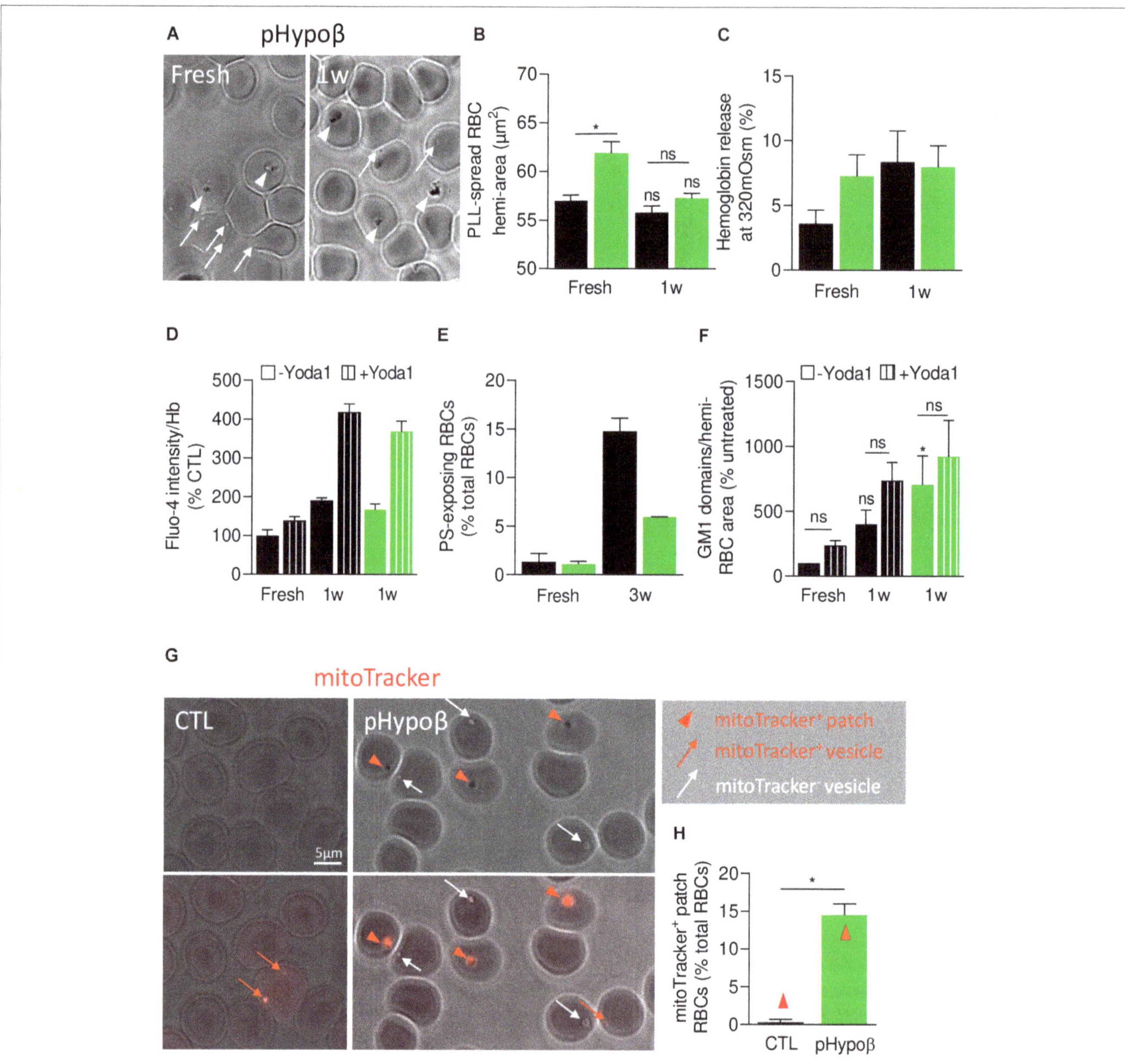

FIGURE 10 | pHypoβ RBCs partly resist to RBC aging upon blood storage in K^+/EDTA tubes at 4°C but exhibit a high abundance of residual mitochondrial fragments. RBCs from healthy donors (black columns) or pHypoβ (green columns) were either stored for 1–3 weeks at 4°C (1-3w) and assessed for RBC morphology and surface area **(A,B)**, osmotic fragility **(C)**, calcium content **(D)**, PS surface exposure **(E)** and GM1-enriched domain abundance **(F)**; or freshly analyzed for the presence of mitochondria remnants **(G,H)**. **(A–F)** RBC morphology and functionality upon storage at 4°C. **(A)** Morphology of fresh and 1 week-stored pHypoβ RBCs immobilized on PLL-coated coverslips as in **Figure 1D**. Images are representative of 3 experiments. **(B)** RBC surface determined as in **Figure 1I**. Data are means ± SEM of 3–7 independent experiments. Wilcoxon matched-pairs signed rank tests and Mann–Whitney test; ns, not significant; *$p < 0.05$. **(C)** RBC fragility determined as in **Figure 2A**. Data are means ± SD of triplicates from 1 experiment. **(D)** RBC calcium content. RBCs were labeled with Fluo4-AM as in **Figure 2F**, incubated with Yoda1 (hatched columns) and analyzed by fluorimetry. Data are normalized to Hb content and are means ± SD of triplicates from 1 experiment. **(E)** RBC PS exposure assessed as in **Figure 7E**. Data are means ± SD of triplicates from 1 experiment. **(F)** GM1-enriched domain abundance on RBCs at resting state and upon stimulation with Yoda1 (hatched columns), determined as in **Figure 8C**. Data are means ± SEM of 3 independent experiments. Kruskal–Wallis test followed by Dunn's comparison test and Wilcoxon matched-pairs signed rank tests for the effect of Yoda1. ns, not significant; *$p < 0.05$. **(G,H)** RBC evaluation for the presence of mitochondrial remnants. RBCs were incubated for 30 min with a fluorescent mitoTracker, washed, dropped on PLL-coated coverslips and immediately observed by fluorescence microscopy. **(G)** Representative images. Upper panels, transmission; lower panels, transmission with fluorescence. Red arrowheads, mitoTracker-positive patches; red arrows, mitoTracker-positive vesicles; white arrows, mitotracker-negative vesicles. **(H)** Relative proportion of RBCs presenting mitoTracker-labeled patches as percentage of total RBCs. Data are means ± SEM of 3–6 independent experiments. Mann–Whitney test. *$p < 0.05$.

intracellular calcium content measured after 1 week of storage in untreated and Yoda1-treated conditions were similar in pHypoβ and healthy donors (**Figure 10D**). Quite surprisingly, the extent of PS surface exposure measured after 3 weeks of storage increased three times less in pHypoβ than in healthy donors (**Figure 10E**). Regarding GM1-enriched domains, they increased in healthy RBCs after 1 week as expected from our previous data (Cloos et al., 2020), but this increase was even more pronounced and significant in pHypoβ RBCs. However, the increase of GM1-enriched domain abundance resulting from Yoda1 stimulation was lower in pHypoβ than healthy RBCs which might be due to the fact that GM1-enriched domain abundance was already higher at the surface of 1 week-stored pHypoβ RBCs than healthy RBCs stored for the same time (**Figure 10F**). We conclude that pHypoβ RBCs did not appear to represent an accelerated model of RBC aging.

pHypoβ Shows a Very High Proportion of RBCs With Mitochondrial Fragments

We therefore evaluated the alternative possibility of a maturation defect of pHypoβ RBCs by labeling with a fluorescent mitoTracker. The proportion of pHypoβ RBCs presenting mitoTracker-positive patches at their surface increased by ∼40-fold as compared to healthy RBCs (**Figures 10G,H**). Importantly, the presence of those patches cannot be explained by the absence of spleen (**Supplementary Figure 1G**).

DISCUSSION

We here describe a case of familial hypobetalipoproteinemia, resulting from a heterozygosity for the pathogenic Gln845Argfs*18 mutation in the *ApoB* gene. Similar to most patients with familial hypobetalipoproteinemia, pHypoβ presented reduced plasma cholesterol and ApoB levels, heightened plasmatic liver enzymes as well as a vitamin E level close to the lower reference limit (Clarke et al., 2006; Musialik et al., 2020). More surprisingly, a high proportion of acanthocytes was detected in this patient, most likely corresponding to the RBCs with surface patches and/or vesicles upon spreading on PLL-coated coverslips. Although those morphology changes were not accompanied by altered resistance to osmotic stress and cellular homeostasis, RBC deformability, oxidative stress, and membrane cytoskeletal and biophysical properties (curvature and stiffness) were impaired. Hence, a deeper imaging analysis revealed (i) heterogeneous spectrin cytoskeleton distribution in patches and vesicles; (ii) ceramide-enrichment and mitoTracker-partitioning in patches, suggesting the presence of mitochondria fragments resulting from RBC maturation defect; (iii) cholesterol-enrichment in spicules; and (iv) sphingomyelin-enriched domain increased abundance and decreased functionality.

Extent of Acanthocytosis in Heterozygous Hypobetalipoproteinemia

Acanthocytes provide a typical clinical feature of abetalipoproteinemia, representing 50–90% of total RBCs (Biemer, 1980). In contrast, the proportion of acanthocytes and their existence are more debated for homozygous and heterozygous familial hypobetalipoproteinemia. In homozygous familial hypobetalipoproteinemia, while it has been stated that acanthocytes are as numerous as in abetalipoproteinemia Biemer (1980) and Tamura et al. (1988) reported two patients suffering from this disease with acanthocytes representing only 10–15% of all RBCs. For the heterozygous form of the disease, their presence is less frequently detected but not excluded (Biemer, 1980; Hardie, 1989; Clarke et al., 2006). Actually, splenectomy could be partly responsible for the unusual high proportion of spiculated RBCs detected for pHypoβ as (i) in some cases, acanthocytes can be observed post-splenectomy (Shah and Hamad, 2020); and (ii) the absence of the spleen in this patient could prevent hemolysis of acanthocytes normally provoked through splenic sequestration. Even if the post-splenectomy hypothesis could be supported by Howell–Jolly bodies observed on pHypoβ blood smears, it seems more likely that reduced RBC elimination as consequence of the absence of a spleen contributes to the abundance of acanthocytes in pHypoβ as no spiculated RBCs could be detected for a heathy splenectomised donor and acanthocytosis is part of the usual clinical picture of lipoprotein disorders and vitamin E deficiencies (Wong, 2004).

Oxidative Stress, Cytoskeleton Defects and Membrane Stiffness Contribute to the RBC Phenotype in Heterozygous Hypobetalipoproteinemia

Vitamins E and A are important antioxidants and their supplementation is part of therapy for patients with homozygous familial hypobetalipoproteinemia and abetalipoproteinemia in order to prevent neurological and ophthalmological disorders (Granot and Kohen, 2004). This is not the case for heterozygous familial hypobetalipoproteinemia even if vitamin E levels below the normal range were detected (Burnett and Hooper, 2015). Regarding pHypoβ, although vitamin A was within the normal range, vitamin E was at the lower level. Clarke et al. actually detected reduced plasma vitamin E levels combined with reduced alpha-tocopherol content in RBCs of 9 patients suffering from heterozygous hypobetalipoproteinemia (Clarke et al., 2006). For patients suffering from abetalipoproteinemia and supplemented with liposoluble vitamins, observations differ. According to Granot and Kohen (2004), no signs for oxidative stress can be detected in the plasma of those patients while Calzada et al. (2013) reported decreased alpha-tocopherol levels in both platelets and HDLs of these patients combined with increased lipid peroxidation, consistent with the fact that vitamin E is a potent inhibitor of lipid peroxidation (Burnett and Hooper, 2015).

A high amount of free ROS was measured in pHypoβ RBCs but was not accompanied by increase in lipid peroxidation and metHb. However, it remains unclear where these ROS are coming from. Actually, vitamin K is not the typical antioxidant but it has nevertheless been shown to prevent oxidative stress in neurons (Li et al., 2003). Plasma levels of the latter were very low in pHypoβ. Taken together with low vitamin E levels, oxidative balance could be disturbed in pHypoβ RBCs because of reduced amount in circulating antioxidant vitamins eventually sufficient to avoid important lipid peroxidation but not to totally prevent RBC oxidative stress.

Besides membrane lipids and Hb, spectrin and several membrane:cytoskeleton anchorage proteins such as Band3 also represent targets of oxidative stress (Voskou et al., 2015). In pHypoβ RBCs, we provided several lines of evidence supporting the role of the altered cytoskeleton in RBC morphology alteration. First, the abundance of RBCs with a homogenously dense spectrin network was decreased at the benefit of RBCs presenting spectrin-enriched patches and/or vesicles. Second, in the RBCs with an homogeneous spectrin network, the spectrin membrane occupancy was decreased, which could reflect a reduced membrane:cytoskeleton anchorage and lead to increased stiffness. Third, the latter increase in pHypoβ RBCs was confirmed by AFM.

The cytoskeleton alterations found in pHypoβ were in agreement with the literature. First, spectrin enrichment in the thorny projections of acanthocytes was already described by electron microscopy (Siegl et al., 2013). Second, spectrin modification through treatment with urea combined with lysophosphatidylcholine membrane insertion has been shown to induce the transformation of discocytes into acanthocytes (Khodadad et al., 1996). Third, cytoskeletal alterations and abnormalities of Band3 immunolabeling have been revealed in neuro-acanthocytosis (Wong, 2004; Adjobo-Hermans et al., 2015). Acanthocytes could be actually formed by alterations in Band3 conformation similar to echinocytes in which the ratio of outward and inward facing Band3 is responsible for the morphological alteration (Wong, 2004). Thus, spectrin but also Band3 could be the target of ROS in pHypoβ RBCs, leading to decreased membrane:cytoskeleton anchorage and to acanthocytes with spectrin-enriched projections. However, the role of Band3 in the disease remains to be explored.

Alteration of Membrane Lipid Lateral Distribution Represents a New Contributor to the Acanthocytosis Seen in Heterozygous Hypobetalipoproteinemia

Analysis of the whole pHypoβ RBC population revealed no modification in the total cholesterol content in agreement with findings from other groups obtained on RBCs from patients with heterozygous hypobetalipoproteinemia (Biemer, 1980; Gheeraert et al., 1988). In abetalipoproteinemia and homozygous hypobetalipoproteinemia, observations are less consistent. Some groups reported increased cholesterol-to-phospholipid ratios (McBride and Jacob, 1970; Biemer, 1980; Barenholz et al., 1981) while others detected increased total cholesterol content without any increase in the cholesterol-to-phospholipid ratio (Iida et al., 1984) and still others described no alterations at all of the total cholesterol amount (Simon and Ways, 1964). Fatty acid composition was also largely maintained in pHypoβ RBCs, except for the slight increase of the linoleic and arachidonic acids, which could be related to specific dietary intake (Takkunen et al., 2013; Ford et al., 2016). To the best of

our knowledge, the fatty acid content has not been studied in heterozygous hypobetalipoproteinemia. Abetalipoproteinemia and homozygous hypobetalipoproteinemia are instead generally associated with decreased linoleic acid content together with lower membrane fluidity. Besides, an increased sphingomyelin/lecithin or sphingomyelin/phosphatidylcholine ratio is also seen in RBCs of those patients (Ways et al., 1963; Simon and Ways, 1964; Cooper et al., 1977; Biemer, 1980; Barenholz et al., 1981; Iida et al., 1984). An increased sphingomyelin/phosphatidylcholine ratio resulting from a reduced abundance of phosphatidylcholine species was also evidenced in pHypoβ. A higher (dihydro)ceramide content was also seen in pHypoβ RBCs in agreement with the increased ceramide-enriched patches. More surprisingly, a specific increase of lysophosphatidylethanolamine species was revealed while lysophosphatidylcholine is generally described to be associated with RBC morphological transformation into acanthocytes or echinocytes (Fujii and Tamura, 1984; Khodadad et al., 1996).

Lipid distribution at the RBC surface was analyzed in parallel to evaluate whether local changes in lipid domains could be revealed. We showed three types of membrane lipid distribution alterations. First, the abundance of sphingomyelin-enriched domains increased by 2.5-fold at the surface of pHypoβ RBCs. As membrane sphingomyelin content was not altered, the decreased membrane to cytoskeletal anchorage in pHypoβ RBCs could be responsible for this increase. This hypothesis is based on the observation by Conrard et al. that impairment of the membrane:cytoskeleton anchorage by PKC activation leads to an increase of sphingomyelin-enriched domains (Conrard et al., 2018). Second, ~35 and ~7% of pHypoβ RBCs presented ceramide- and sphingomyelin-enriched patches, respectively. Third, a high number of RBCs presented cholesterol-enriched spicules. The evident changes in membrane lipid organization stressed the importance of combining quantitative lipidomic analyses with qualitative lipid imaging to better understand diseases characterized by a variable- not systematically inventoried- amount of acanthocytes.

RBC Functionality Is Poorly Affected in Heterozygous Hypobetalipoproteinemia

Quite surprisingly at first glance, only slight modifications of pHypoβ RBC functionality were observed. First, although no modification was seen in the poorly sensitive osmotic fragility test, osmotic gradient ektacytometry indicated a slight increase in surface area-to-volume ratio and therefore of osmotic fragility. Those observations were in accordance with the literature as for other models of acanthocytic RBCs no modification in osmotic fragility is detected (Cooper, 1969; McBride and Jacob, 1970). Second, the intracellular calcium and ATP contents were normal. Third, the response of GM1-enriched domains to PIEZO1 activation through Yoda1 in fresh RBCs was similar in the patient and healthy donors.

Nevertheless, a deeper analysis indicated that the number of GM1-enriched domains was increased- although not significantly- in fresh RBCs and significantly in 1 week-old RBCs. Likewise, the abundance of sphingomyelin-enriched domains was enhanced and their response to intracellular calcium depletion through EGTA was abrogated. Thus, a fine analysis of the RBC surface both upon resting state and upon stimulation of calcium exchanges revealed functionality impairments in pHypoβ. Those alterations were not accompanied by altered calcium content or large increases in fragility probably because those measurements were made on the whole RBC population without any distinction between acanthocytes and discocytes. Interestingly, Ohsaka et al. reported on a patient with acute myelodysplasia with myelofibrosis who presented acanthocytes a heightened calcium uptake in RBCs combined with normal calcium levels (Ohsaka et al., 1989). In agreement with slight but detectable functionality impairment in pHypoβ, a loss of deformability after prolonged exposure to high shear stress was observed. Since this loss could only be partly explained by splenectomy, this observation revealed a slight reduction of membrane health of pHypoβ RBCs. This conclusion was further supported by alterations of RBC membrane biophysical properties.

Defective RBC Maturation May Be a Novel Feature of Heterozygous Hypobetalipoproteinemia

In this study, we showed that the proportion of pHypoβ RBCs presenting mitoTracker-enriched patches, and thus mitochondrial fragments, increased by ∼40-fold as compared to healthy RBCs and represented ∼15% of total RBCs in pHypoβ. Moreover, ∼35% of pHypoβ RBCs presented ceramide-enriched patches. We propose that those ceramide-enriched patches corresponded to mitochondrial fragments, since (i) the increased abundance of those two types of patches in pHypoβ was similar; (ii) ceramides are known to be enriched in the outer mitochondrial membrane (Siskind, 2005); and (iii) mitoTracker and BODIPY-ceramide double-labeling in RBCs from patients with hereditary spherocytosis showed high colocalization (our unpublished data).

Besides the high number of mitochondria-positive RBCs, the RBC maturation defect in pHypoβ is further supported by (i) the increased membrane area measured after RBC spreading, normally reduced by vesiculation upon reticulocyte maturation (Trakarnsanga et al., 2017); (ii) the higher oxidative stress, as lipid peroxidation and activity of antioxidant enzymes are increased in reticulocytes as compared to mature RBCs (Sailaja et al., 2003); (iii) the increased abundance of GM1-enriched domains, as those domains are related to calcium influx in RBCs and calcium uptake is heightened in reticulocytosis (Ohsaka et al., 1989); and (iv) the threefold reduction in stored RBCs of surface exposure of PS, a lipid normally found in autophagic vesicles upon reticulocyte maturation (Trakarnsanga et al., 2017). Actually, the resistance to PS surface exposure has also been reported by Siegl et al. (2013) in chorea-acanthocytosis after RBC stimulation with lysophosphatidic acid. Thus, pHypoβ has a high proportion of acanthocytes which could result from a maturation defect during the R1 reticulocyte stage known to undergo significant rearrangements in reticulocyte membrane

and intracellular components via several mechanisms including exosome release and mitophagy (Minetti et al., 2018, 2020).

Experimental Strategy Strengths and Weaknesses

In this study, a wide range of research methods was used to evaluate morphological, biochemical and biophysical changes and their potential consequences for RBC functionality in hypobetalipoproteinemia. Moreover, thanks to a deep imaging analysis, we were able to reveal local changes in lipid composition and cytoskeleton distribution in acanthocytes. Nevertheless, our experimental strategy also presents some drawbacks that are inherent to the low prevalence of heterozygous hypobetalipoproteinemia with a high proportion of acanthocytes on one hand and to the study of membrane lipid lateral organization on the other hand. However, those limitations were minimized, based on the following evidences. First, although generalization of observations based on only one patient is difficult, our findings are in agreement with literature data on diseases associated with acanthocytosis. In fact, altered membrane composition and biophysical properties have also been demonstrated in a patient suffering from hereditary elliptocytosis and a series of patients suffering from hereditary spherocytosis, two diseases associated with increased membrane fragility due to cytoskeleton defects (Pollet et al., 2020; unpublished data), thereby supporting the importance of membrane:cytoskeleton interplay in RBC morphology and functionality. Second, although fluorescent lipid analogs were extensively used and validated in previous researches, their validity as *bona fide* surrogates of endogenous lipid counterparts is debated in view of the bulk fluorophore which can deeply modify biophysical properties. To minimize difficulties, we extensively validated all the lipid probes we used. For information regarding BODIPY-sphingomyelin and -GM1, we refer the reader to our previous papers (Tyteca et al., 2010; D'Auria et al., 2011, 2013; Carquin et al., 2016; Leonard et al., 2017b). For BODIPY-ceramide, the following lines of evidences supported a distribution as endogenous ceramides: (i) the fluorescent BODIPY-ceramide labeling correlated with the endogenous ceramide content measured by lipidomics both in pHypoβ and in RBCs upon blood storage at 4°C (Cloos et al., 2020); (ii) BODIPY-ceramide accumulates in the Golgi complex in nucleated cells (Tyteca et al., 2010) and in mitochondria remnants in RBCs from patients suffering from familial hypobetalipoproteinemia and from hereditary spherocytosis (data not shown), indicating that the lipid probe is able to join the intracellular compartments enriched in endogenous ceramide (Pagano et al., 1991); (iii) in the Golgi complex, BODIPY-ceramide can be efficiently converted into BODIPY-sphingomyelin, demonstrating the "one molecule at a time" conversion by the selectivity of the catalytic site of sphingomyelin synthase (Tyteca et al., 2010); (iv) BODIPY-ceramide accumulates in the inner plasma membrane leaflet, as revealed by resistance to back-exchange by BSA (Pollet et al., 2020) and demonstrating ability of the probe incorporated into the outer leaflet to flip to the inner one, as natural ceramide

(Contreras et al., 2010); (v) BODIPY-ceramide-enriched domain abundance correlates with the ceramide content, as revealed by the decrease of ceramide-enriched domain abundance at the RBC membrane and the concomitant enrichment of ceramide species in RBC-derived extracellular vesicles at the end of the blood storage period in K^+/EDTA tubes at 4°C (Cloos et al., 2020). Besides extensive validation of fluorescent lipid analogs, we developed the use of fluorescent Lysenin and Theta toxin fragments specific to endogenous sphingomyelin and cholesterol. However, it should be noted that, as fluorescent lipid analogs, toxin fragments also present drawbacks (Carquin et al., 2014, 2015, 2016) and it is therefore crucial to use complementary unrelated lipid probes to target one lipid, once possible. Third, regarding the imaging method used, although confocal microscopy offers several advantages such as vital imaging, multiple labeling and 3D-reconstruction, the size of observed lipid domains could have been overestimated.

Conclusion and Overall Significance

In a case of familial hypobetalipoproteinemia, resulting from heterozygosity for the pathogenic Gln845Argfs*18 mutation in the *ApoB* gene, we showed that the resulting acanthocytes exhibited impaired cytoskeleton and membrane biophysical properties without significant loss of RBC functionality as assessed by functional tests applied on the total RBC population, without any distinction between acanthocytes and discocytes (e.g., osmotic fragility test, intracellular calcium measurement). In contrast, more sensitive tests (e.g., ektacytometry) and assays aimed at investigating specific RBC populations (e.g., lipid domains, membrane curvature) did demonstrate functional impairments in pHypoβ.

Although our findings were generated from only one patient, the observed cytoskeleton and membrane alterations were consistent with literature data on diseases associated with acanthocytosis. Indeed, cytoskeletal alterations and abnormalities of Band3 immunolabeling have been shown in neuro-acanthocytosis (Wong, 2004; Adjobo-Hermans et al., 2015). The resistance to PS surface exposure has also been reported in chorea-acanthocytosis after RBC stimulation with lysophosphatidic acid (Siegl et al., 2013). The modest alterations of pHypoβ RBC functionality agreed with the absence of modification in osmotic fragility in other models of acanthocytotic RBCs (Cooper, 1969; McBride and Jacob, 1970).

From a more global point-of-view, altered membrane composition and biophysical properties have also been demonstrated in two RBC disorders associated with altered cytoskeleton function and a large diversity of RBC morphological changes and deformability alterations (Pollet et al., 2020; unpublished data), indicating the close membrane:cytoskeleton interplay and its role in RBC morphology and functionality. More specifically, a parallelism could be established with the patients with hereditary spherocytosis exhibiting a high proportion of spiculated RBCs (our unpublished data).

Our study demonstrates that evaluation of membrane biophysical properties and membrane lipid distribution could be of benefit in the diagnosis and better understanding of RBC disorders.

AUTHOR CONTRIBUTIONS

A-SC and DT designed the experiments, analyzed and interpreted the data, and wrote the manuscript. LGMD and JW identified the patient and established the diagnosis. MM, AS, and JV collected, analyzed, and quantified lipid imaging data. PVDS did the electron microscopy experiments. MR and RW performed ektacytometry measurements. EM and YL were responsible for fatty acid analysis while RT and GM performed lipidomics. ACD and DA generated and analyzed AFM data. All authors reviewed the final version of the manuscript.

ACKNOWLEDGMENTS

We thank Drs. A. Miyawaki, M. Abe, T. Kobayashi (Riken Brain Science Institute, Saitama, Japan & University of Strasbourg, France), and H. Mizuno (KU Leuven, Belgium) for the Dronpa-theta-D4 plasmid. The MASSMET platform (UCLouvain, Belgium) is acknowledged for the access to the LC-MS for lipid analysis.

SUPPLEMENTARY MATERIAL

Supplementary Figure 1 | Except for a slight increase in ceramide-enriched patches, RBCs from a healthy splenectomised donor exhibit similar morphology, biophysical properties, functionality and maturation than healthy non-splenectomised donors. RBCs from healthy non-splenectomised (+spleen; black columns) donors and a healthy splenectomised (−spleen; gray columns) donor were compared for RBC membrane area **(A)**, osmotic fragility **(B)**, ROS intracellular content **(C)**, curvature **(D)**, lipid domain abundance **(E)**, and proportion of RBCs with ceramide (cer)- or mitoTracker-enriched patches and vesicles **(F,G)**. **(A)** RBC membrane area assessed as in **Figure 1I**. Data are means ± SEM of 5 independent experiments. Mann–Whitney test. Ns, not significant. **(B)** RBC osmotic fragility determined as in **Figure 2B**. Data from 1 experiment. **(C)** ROS content evaluated as in **Figure 5A**. Data are means ± SD of triplicates from 1 experiment. **(D)** RBC curvature determined on RBCs in suspension as in **Figure 2B**. Representative images of 2 independent experiments.

(E) Lipid domain abundance determined as in **Figure 8B**. Data are means ± SEM of 4–5 independent experiments. Kruskal–Wallis test followed by Dunn's comparison test. Ns, not significant. **(F)** Proportion of RBCs with cer-enriched patches and vesicles determined as in **Figure 9C**. Upper panel, representative image of RBCs from the splenectomised donor; lower panel, quantification of cer-enriched patches (red arrowheads) and vesicles (red arrows) in RBCs from non-splenectomised (black columns) and splenectomised (gray columns) donors. Data are means ± SEM of 3–6 independent experiments. Mann–Whitney tests to compare non-splenectomised vs. splenectomised RBCs; ns, not significant; *p < 0.05. **(G)** MitoTracker labeling as in **Figure 10G**. Upper panels, transmission; lower panels, transmission combined with fluorescence. MitoTracker-positive vesicles are indicated with red arrows. Images are representative of 3 experiments.

Supplementary Figure 2 | Relation between RBC plasma membrane and cytoskeleton elastic modulus in the fast indentation experiments. RBCs from healthy donors or pHypoβ were evaluated for both plasma membrane and cytoskeleton Young's modulus. Each dot depicts one RBC measured for plasma membrane and cytoskeleton Young's modulus. Experimental data are from **Figure 7B**.

Supplementary Table 1 | pHypoβ clinical data. Clinical parameters were assessed for pHypoβ after blood collection. MCV, mean corpuscular volume; MCH, mean corpuscular hemoglobin; Gamma-GT, Gamma-glutamyltransferase; ASAT, aspartate transaminase; ALAT, alanine transaminase; LD, lactate dehydrogenase. Values upon or under the reference values are indicated in red or blue, respectively.

REFERENCES

Adjobo-Hermans, M. J., Cluitmans, J. C., and Bosman, G. J. (2015). Neuroacanthocytosis: Observations, Theories and Perspectives on the Origin and Significance of Acanthocytes. *Tremor. Other. Hyperkinet. Mov.* 5:328. doi: 10.5334/tohm.271

Baines, A. J. (2010). The spectrin-ankyrin-4.1-adducin membrane skeleton: adapting eukaryotic cells to the demands of animal life. *Protoplasma* 244, 99–131. doi: 10.1007/s00709-010-0181-1

Barenholz, Y., Yechiel, E., Cohen, R., and Deckelbaum, R. J. (1981). Importance of cholesterol-phospholipid interaction in determining dynamics of normal and abetalipoproteinemia red blood cell membrane. *Cell Biophys.* 3, 115–126. doi: 10.1007/bf02788128

Biemer, J. J. (1980). Acanthocytosis–biochemical and physiological considerations. *Ann. Clin. Lab. Sci.* 10, 238–249.

Bogdanova, A., Makhro, A., Wang, J., Lipp, P., and Kaestner, L. (2013). Calcium in red blood cells-a perilous balance. *Int. J. Mol. Sci.* 14, 9848–9872. doi: 10.3390/ijms14059848

Burnett, J. R., and Hooper, A. J. (2015). Vitamin E and oxidative stress in abetalipoproteinemia and familial hypobetalipoproteinemia. *Free Radic. Biol. Med.* 88, 59–62. doi: 10.1016/j.freeradbiomed.2015.05.044

Cahalan, S. M., Lukacs, V., Ranade, S. S., Chien, S., Bandell, M., and Patapoutian, A. (2015). Piezo1 links mechanical forces to red blood cell volume. Elife, e07370. doi: 10.7554/eLife.07370

Calzada, C., Vericel, E., Colas, R., Guillot, N., El Khoury, G., Drai, J., et al. (2013). Inhibitory effects of in vivo oxidized high-density lipoproteins on platelet aggregation: evidence from patients with abetalipoproteinemia. *FASEB J.* 27, 2855–2861. doi: 10.1096/fj.12-225169

Carquin, M., Conrard, L., Pollet, H., Van Der Smissen, P., Cominelli, A., Veiga-Da-Cunha, M., et al. (2015). Cholesterol segregates into submicrometric domains at the living erythrocyte membrane: evidence and regulation. *Cell. Mol. Life Sci.* 72, 4633–4651. doi: 10.1007/s00018-015-1951-x

Carquin, M., D'auria, L., Pollet, H., Bongarzone, E. R., and Tyteca, D. (2016). Recent progress on lipid lateral heterogeneity in plasma membranes: From rafts to submicrometric domains. *Prog. Lipid. Res.* 62, 1–24. doi: 10.1016/j.plipres.2015.12.004

Carquin, M., Pollet, H., Veiga-Da-Cunha, M., Cominelli, A., Van Der Smissen, P., N'kuli, F., et al. (2014). Endogenous sphingomyelin segregates into submicrometric domains in the living erythrocyte membrane. *J. Lipid Res.* 55, 1331–1342. doi: 10.1194/jlr.m018538

Clarke, M. W., Hooper, A. J., Headlam, H. A., Wu, J. H., Croft, K. D., and Burnett, J. R. (2006). Assessment of tocopherol metabolism and oxidative stress in familial hypobetalipoproteinemia. *Clin. Chem.* 52, 1339–1345. doi: 10.1373/clinchem.2006.068692

Cloos, A. S., Ghodsi, M., Stommen, A., Vanderroost, J., Dauguet, N., Pollet, H., et al. (2020). Interplay Between Plasma Membrane Lipid Alteration, Oxidative Stress and Calcium-Based Mechanism for Extracellular Vesicle Biogenesis From Erythrocytes During Blood Storage. *Front. Physiol.* 11:712. doi: 10.3389/fphys.2020.00712

Conrard, L., Stommen, A., Cloos, A. S., Steinkuhler, J., Dimova, R., Pollet, H., et al. (2018). Spatial Relationship and Functional Relevance of Three Lipid Domain Populations at the Erythrocyte Surface. *Cell Physiol. Biochem.* 51, 1544–1565. doi: 10.1159/000495645

Conrard, L., and Tyteca, D. (2019). Regulation of Membrane Calcium Transport Proteins by the Surrounding Lipid Environment. *Biomolecules* 9:513. doi: 10.3390/biom9100513

Contreras, F. X., Sanchez-Magraner, L., Alonso, A., and Goni, F. M. (2010). Transbilayer (flip-flop) lipid motion and lipid scrambling in membranes. *FEBS Lett.* 584, 1779–1786. doi: 10.1016/j.febslet.2009.12.049

Cooper, R. A. (1969). Anemia with spur cells: a red cell defect acquired in serum and modified in the circulation. *J. Clin. Invest.* 48, 1820–1831. doi: 10.1172/jci106148

Cooper, R. A., Durocher, J. R., and Leslie, M. H. (1977). Decreased fluidity of red cell membrane lipids in abetalipoproteinemia. *J. Clin. Invest.* 60, 115–121. doi: 10.1172/jci108747

Cuerq, C., Henin, E., Restier, L., Blond, E., Drai, J., Marcais, C., et al. (2018). Efficacy of two vitamin E formulations in patients with abetalipoproteinemia and chylomicron retention disease. *J. Lipid Res.* 59, 1640–1648. doi: 10.1194/jlr.m085043

Da Costa, L., Suner, L., Galimand, J., Bonnel, A., Pascreau, T., Couque, N., et al. (2016). Diagnostic tool for red blood cell membrane disorders: Assessment of a new generation ektacytometer. *Blood Cells Mol. Dis.* 56, 9–22. doi: 10.1016/j.bcmd.2015.09.001

D'Auria, L., Fenaux, M., Aleksandrowicz, P., Van Der Smissen, P., Chantrain, C., Vermylen, C., et al. (2013). Micrometric segregation of fluorescent membrane lipids: relevance for endogenous lipids and biogenesis in erythrocytes. *J. Lipid. Res.* 54, 1066–1076. doi: 10.1194/jlr.m034314

D'Auria, L., Van Der Smissen, P., Bruyneel, F., Courtoy, P. J., and Tyteca, D. (2011). Segregation of fluorescent membrane lipids into distinct micrometric domains: evidence for phase compartmentation of natural lipids? *PLoS One* 6:e17021. doi: 10.1371/journal.pone.0017021

Dumitru, A. C., Poncin, M. A., Conrard, L., Dufrêne, Y. F., Tyteca, D., and Alsteens, D. (2018). Nanoscale membrane architecture of healthy and pathological red blood cells. *Nanoscale Horiz.* 3, 293–304. doi: 10.1039/c7nh00187h

Ford, R., Faber, M., Kunneke, E., and Smuts, C. M. (2016). Dietary fat intake and red blood cell fatty acid composition of children and women from three different geographical areas in South Africa. *Prostagl. Leukot Essent Fatty Acids* 109, 13–21. doi: 10.1016/j.plefa.2016.04.003

Fouchier, S. W., Sankatsing, R. R., Peter, J., Castillo, S., Pocovi, M., Alonso, R., et al. (2005). High frequency of APOB gene mutations causing familial hypobetalipoproteinaemia in patients of Dutch and Spanish descent. *J. Med. Genet.* 42:e23. doi: 10.1136/jmg.2004.029454

Fujii, T., and Tamura, A. (1984). Shape change of human erythrocytes induced by phosphatidylcholine and lysophosphatidylcholine species with various acyl chain lengths. *Cell. Biochem. Funct.* 2, 171–176. doi: 10.1002/cbf.290020311

Gheeraert, P., De Buyzere, M., Delanghe, J., De Scheerder, I., Bury, J., and Rosseneu, M. (1988). Plasma and erythrocyte lipids in two families with heterozygous hypobetalipoproteinemia. *Clin. Biochem.* 21, 371–377. doi: 10.1016/s0009-9120(88)80020-1

Granot, E., and Kohen, R. (2004). Oxidative stress in abetalipoproteinemia patients receiving long-term vitamin E and vitamin A supplementation. *Am. J. Clin. Nutr.* 79, 226–230. doi: 10.1093/ajcn/79.2.226

Guillemot-Legris, O., Masquelier, J., Everard, A., Cani, P. D., Alhouayek, M., and Muccioli, G. G. (2016). High-fat diet feeding differentially affects the development of inflammation in the central nervous system. *J. Neuroinfl.* 13:206.

Hardie, R. J. (1989). Acanthocytosis and neurological impairment–a review. *Q. J. Med.* 71, 291–306.

Hutter, J. L., and Bechhoefer, J. (1993). Calibration of atomic–force microscope tips. *Rev. Scient. Instrum.* 64, 1868–1873. doi: 10.1063/1.1143970

Iida, H., Takashima, Y., Maeda, S., Sekiya, T., Kawade, M., Kawamura, M., et al. (1984). Alterations in erythrocyte membrane lipids in abetalipoproteinemia: phospholipid and fatty acyl composition. *Biochem. Med.* 32, 79–87. doi: 10.1016/0006-2944(84)90010-3

Kaestner, L., Bogdanova, A., and Egee, S. (2020). Calcium Channels and Calcium-Regulated Channels in Human Red Blood Cells. *Adv. Exp. Med. Biol.* 1131, 625–648. doi: 10.1007/978-3-030-12457-1_25

Khodadad, J. K., Waugh, R. E., Podolski, J. L., Josephs, R., and Steck, T. L. (1996). Remodeling the shape of the skeleton in the intact red cell. *Biophys. J.* 70, 1036–1044. doi: 10.1016/s0006-3495(96)79649-2

Lazarova, E., Gulbis, B., Oirschot, B. V., and Van Wijk, R. (2017). Next-generation osmotic gradient ektacytometry for the diagnosis of hereditary spherocytosis: interlaboratory method validation and experience. *Clin. Chem. Lab. Med.* 55, 394–402.

Leonard, C., Alsteens, D., Dumitru, A., Mingeot-Leclercq, M., and Tyteca, D. (2017a). "Lipid domains and membrane (re)shaping : from biophysics to biology," in *The role of the physical properties of membranes in influencing biological phenomena*, eds J. Ruysschaert and R. Epand (Netherland: Springer), 121–175. doi: 10.1007/978-981-10-6244-5_5

Leonard, C., Conrard, L., Guthmann, M., Pollet, H., Carquin, M., Vermylen, C., et al. (2017b). Contribution of plasma membrane lipid domains to red blood cell (re)shaping. *Sci. Rep.* 7:4264.

Lew, V. L., and Tiffert, T. (2017). On the Mechanism of Human Red Blood Cell Longevity: Roles of Calcium, the Sodium Pump, PIEZO1, and Gardos Channels. *Front. Physiol.* 8:977. doi: 10.3389/fphys.2017.00977

Li, J., Lin, J. C., Wang, H., Peterson, J. W., Furie, B. C., Furie, B., et al. (2003). Novel role of vitamin k in preventing oxidative injury to developing oligodendrocytes and neurons. *J. Neurosci.* 23, 5816–5826. doi: 10.1523/jneurosci.23-13-05816.2003

McBride, J. A., and Jacob, H. S. (1970). Abnormal kinetics of red cell membrane cholesterol in acanthocytes: studies in genetic and experimental abetalipoproteinaemia and in spur cell anaemia. *Br. J. Haematol.* 18, 383–397. doi: 10.1111/j.1365-2141.1970.tb01452.x

Minetti, G., Achilli, C., Perotti, C., and Ciana, A. (2018). Continuous Change in Membrane and Membrane-Skeleton Organization During Development From Proerythroblast to Senescent Red Blood Cell. *Front. Physiol.* 9:286. doi: 10.3389/fphys.2018.00286

Minetti, G., Bernecker, C., Dorn, I., Achilli, C., Bernuzzi, S., Perotti, C., et al. (2020). Membrane Rearrangements in the Maturation of Circulating Human Reticulocytes. *Front. Physiol.* 11:215. doi: 10.3389/fphys.2020.00215

Mohanty, J. G., Nagababu, E., and Rifkind, J. M. (2014). Red blood cell oxidative stress impairs oxygen delivery and induces red blood cell aging. *Front. Physiol.* 5:84. doi: 10.3389/fphys.2014.00084

Musialik, J., Boguszewska-Chachulska, A., Pojda-Wilczek, D., Gorzkowska, A., Szymanczak, R., Kania, M., et al. (2020). A Rare Mutation in The APOB Gene Associated with Neurological Manifestations in Familial Hypobetalipoproteinemia. *Int. J. Mol. Sci.* 21:1439. doi: 10.3390/ijms21041439

Mutemberezi, V., Masquelier, J., Guillemot-Legris, O., and Muccioli, G. G. (2016). Development and validation of an HPLC-MS method for the simultaneous quantification of key oxysterols, endocannabinoids, and ceramides: variations in metabolic syndrome. *Anal. Bioanal. Chem.* 408, 733–745. doi: 10.1007/s00216-015-9150-z

Ney, P. A. (2011). Normal and disordered reticulocyte maturation. *Curr. Opin. Hematol.* 18, 152–157. doi: 10.1097/moh.0b013e328345213e

Ohsaka, A., Yawata, Y., Enomoto, Y., Takahashi, A., Sato, Y., Sakamoto, S., et al. (1989). Abnormal calcium transport of acanthocytes in acute myelodysplasia with myelofibrosis. *Br. J. Haematol.* 73, 568–570. doi: 10.1111/j.1365-2141.1989.tb00301.x

Pagano, R. E., Martin, O. C., Kang, H. C., and Haugland, R. P. (1991). A novel fluorescent ceramide analogue for studying membrane traffic in animal cells: accumulation at the Golgi apparatus results in altered spectral properties of the sphingolipid precursor. *J. Cell. Biol.* 113, 1267–1279. doi: 10.1083/jcb.113.6.1267

Pollet, H., Cloos, A. S., Stommen, A., Vanderroost, J., Conrard, L., and Paquot, A. (2020). Aberrant Membrane Composition and Biophysical Properties Impair Erythrocyte Morphology and Functionality in Elliptocytosis. *Biomolecules* 10:1120. doi: 10.3390/biom10081120

Pollet, H., Conrard, L., Cloos, A. S., and Tyteca, D. (2018). Plasma Membrane Lipid Domains as Platforms for Vesicle Biogenesis and Shedding? *Biomolecules* 8:94. doi: 10.3390/biom8030094

Sailaja, Y. R., Baskar, R., and Saralakumari, D. (2003). The antioxidant status during maturation of reticulocytes to erythrocytes in type 2 diabetics. *Free Radic. Biol. Med.* 35, 133–139. doi: 10.1016/s0891-5849(03)00071-6

Salomao, M., Zhang, X., Yang, Y., Lee, S., Hartwig, J. H., Chasis, J. A., et al. (2008). Protein 4.1R-dependent multiprotein complex: new insights into the structural organization of the red blood cell membrane. *Proc. Natl. Acad. Sci. U. S. A.* 105, 8026–8031. doi: 10.1073/pnas.0803225105

Schillers, H., Rianna, C., Schäpe, J., Luque, T., Doschke, H., Wälte, M., et al. (2017). Standardized Nanomechanical Atomic Force Microscopy Procedure (SNAP) for Measuring Soft and Biological Samples. *Sci. Rep.* 7:5117.

Shah, P. R., and Hamad, H. (2020). *Acanthocytosis," in StatPearls*. Treasure Island, FL: StatPearls.

Siegl, C., Hamminger, P., Jank, H., Ahting, U., Bader, B., Danek, A., et al. (2013). Alterations of red cell membrane properties in neuroacanthocytosis. *PLoS One* 8:e76715. doi: 10.1371/journal.pone.0076715

Simon, E. R., and Ways, P. (1964). Incubation Hemolysis and Red Cell Metabolism in Acanthocytosis. *J. Clin. Invest.* 43, 1311–1321. doi: 10.1172/jci105006

Siskind, L. J. (2005). Mitochondrial ceramide and the induction of apoptosis. *J. Bioenerg. Biomembr.* 37, 143–153. doi: 10.1007/s10863-005-6567-7

Takkunen, M., Agren, J., Kuusisto, J., Laakso, M., Uusitupa, M., and Schwab, U. (2013). Dietary fat in relation to erythrocyte fatty acid composition in men. *Lipids* 48, 1093–1102. doi: 10.1007/s11745-013-3832-0

Tamura, A., Keiji, K., Fiujii, T., Harano, Y., Harada, M., Nakano, T., Hidaka, H., et al. (1988). Abnormalities in the Membrane of Erythrocytes from Two Patients Homozygous for Familial Hypobetalipoproteinemia. *J. Clin. Biochem. Nutr.* 5, 103–108. doi: 10.3164/jcbn.5.103

Trakarnsanga, K., Griffiths, R. E., Wilson, M. C., Blair, A., Satchwell, T. J., Meinders, M., et al. (2017). An immortalized adult human erythroid line facilitates sustainable and scalable generation of functional red cells. *Nat. Commun.* 8:14750.

Tyteca, D., D'auria, L., Der Smissen, P. V., Medts, T., Carpentier, S., Monbaliu, J. C., et al. (2010). Three unrelated sphingomyelin analogs spontaneously cluster into plasma membrane micrometric domains. *Biochim. Biophys. Acta* 1798, 909–927. doi: 10.1016/j.bbamem.2010.01.021

Voskou, S., Aslan, M., Fanis, P., Phylactides, M., and Kleanthous, M. (2015). Oxidative stress in beta-thalassaemia and sickle cell disease. *Redox. Biol.* 6, 226–239. doi: 10.1016/j.redox.2015.07.018

Ways, P., Reed, C. F., and Hanahan, D. J. (1963). Red-Cell and Plasma Lipids in Acanthocytosis. *J. Clin. Invest.* 42, 1248–1260. doi: 10.1172/jci104810

Whitfield, A. J., Barrett, P. H., Van Bockxmeer, F. M., and Burnett, J. R. (2004). Lipid disorders and mutations in the APOB gene. *Clin. Chem.* 50, 1725–1732. doi: 10.1373/clinchem.2004.038026

Wong, P. (2004). A basis of the acanthocytosis in inherited and acquired disorders. *Med. Hypotheses* 62, 966–969. doi: 10.1016/j.mehy.2003.12.032

Red Blood Cell Morphodynamics: A New Potential Marker in High-Risk Patients

Benedetta Porro[1†], Edoardo Conte[1†], Anna Zaninoni[2], Paola Bianchi[2], Fabrizio Veglia[1], Simone Barbieri[1], Susanna Fiorelli[1], Sonia Eligini[1], Alessandro Di Minno[1‡], Saima Mushtaq[1], Elena Tremoli[1], Viviana Cavalca[1*] and Daniele Andreini[1,3]

[1]Centro Cardiologico Monzino, Istituto di Ricovero e Cura a Carattere Scientifico (IRCCS), Milan, Italy, [2]Fondazione IRCCS Ca' Granda Ospedale Maggiore Policlinico Milano, Unità Operativa Complessa (UOC) Ematologia, Unità Operativa Semplice (UOS) Fisiopatologia delle Anemie, Milan, Italy, [3]Department of Clinical Sciences and Community Health, Cardiovascular Section, University of Milan, Milan, Italy

*Correspondence:
Viviana Cavalca
viviana.cavalca@ccfm.it

†These authors have contributed equally to this work

‡Present address:
Alessandro Di Minno,
Department of Pharmacy,
Federico II University Naples,
Naples, Italy

In the last years, a substantial contribution of red blood cells (RBCs) in cardiovascular homeostasis has been evidenced, as these cells are able to regulate cardiovascular function by the export of adenosine triphosphate and nitric oxide as well as to maintain redox balance through a well-developed antioxidant system. Recently a link between high-risk plaque (HRP) features and myocardial ischemia, in the absence of severe lumen stenosis, has been evidenced. Nonobstructive coronary artery disease (nonob CAD) has been associated in fact with a greater 1-year risk of myocardial infarction and all-cause mortality compared with no apparent CAD. This new evidence increases interest in searching new triggers to identify these high-risk patients, in the absence/or on top of traditional hazard markers. In this study, we investigated the existence of any association between RBC morphodynamics and HRP features in individuals with different grades of coronary stenosis detected by coronary computed tomography angiography (CCTA). Ninety-one consecutive individuals who underwent CCTA [33 no CAD; 26 nonobstructive (nonob), and 32 obstructive (ob) CAD] were enrolled. RBC morphodynamic features, i.e., RBC aggregability and deformability, were analyzed by means of Laser Assisted Optical Rotation Cell Analyzer (LoRRca MaxSis). The putative global RBC morphodynamic (RMD) score and the related risk chart, associating the extent of HRP (e.g., the non-calcified plaque volume) with both the RMD score and the max % stenosis were computed. In nonob CAD group only positive correlations between RBC rigidity, osmotic fragility or aggregability and HRP features (plaque necrotic core, fibro-fatty and fibro-fatty plus necrotic core plaque volumes) were highlighted. Interestingly, in this patient cohort three of these RBC morphodynamic features result to be independent predictors of the presence of non-calcified plaque volume in this patients group. The risk chart created shows that only in nonob CAD plaque vulnerability increases according to the score quartile. Findings of this work, by evidencing the association between erythrocyte morphodynamic characteristics assessed by LoRRca and plaque instability in a high-risk cohort of nonob CAD, suggest the use of these blood cell features in the identification of high-risk patients, in the absence of severe coronary stenosis.

Keywords: red blood cell morphodynamics, non-obstructive coronary artery disease, high-risk plaque, coronary computed tomography angiography, risk chart

INTRODUCTION

Coronary artery disease (CAD) is one of the major causes of morbidity and mortality in the western countries (Montalescot et al., 2013). Specific anatomic plaque features have been established as fundamental to the process leading to acute coronary thrombosis and, among them, plaque burden, thin-cap fibroatheroma, positive arterial remodeling, necrotic cores, spotty calcifications, and macrophage infiltration play a central role (Virmani et al., 2006).

Although the diagnosis of obstructive (ob) CAD is the milestone for risk stratification in cardiac disease, non-obstructive (nonob) CAD is a relatively common feature, occurring in 10–25% of patients undergoing coronary angiography (Bugiardini and Bairey Merz, 2005), and its presence has been defined as "insignificant" or "no significant CAD" in the medical literature so far (Hung and Cherng, 2003; Patel et al., 2006). Nevertheless, a high number of atheromatous plaques that are not flow limiting are responsible for acute coronary syndromes (ACS; Finn et al., 2010).

The need for improved methods and new markers beyond stenosis for high-risk plaque (HRP) identification follows from these premises.

In the last years, coronary computed tomography angiography (CCTA) has emerged as a non-invasive method for accurate detection and/or exclusion of the presence of CAD (Al-Mallah et al., 2015). Specifically, a high extent of literature evidenced the ability of CCTA to identify not only ob CAD but also early atherosclerotic lesions (Min et al., 2011; Andreini et al., 2012). As before the "CCTA era" patients with nonob CAD and without signs of inducible ischemia were included in the same group of those without evident disease, now with the aid of this imaging technique, we are able, in this patient group, to discriminate between individuals with low-risk plaque morphology and subjects in whom plaque characteristics are associated with an increased risk of future events (Libby, 2013; Conte et al., 2017).

These methodological improvements enhanced a new interest in the evaluation of atherosclerosis determinants, from lumen stenosis to myocardial ischemia. In this context, a new and emerging factor is red blood cell (RBC), not only the transporter of oxygen to tissues but also a cell able to modulate blood flow behavior. Several factors related to RBCs are associated with CAD including erythrocyte sedimentation rate, hemoglobin levels, hematocrit (Hct), and red blood cell distribution width (RDW; Danesh et al., 2000). In particular, it has been demonstrated that men who had the erythrocyte sedimentation rate in the upper quintile had more than twice the risk of coronary heart disease (CHD) death (Gillum et al., 1995). As regards hemoglobin, an observational study conducted on 2,059 patients undergoing coronary artery bypass surgery revealed that individuals with a preoperative hemoglobin concentration of 100 g/L or less had a five-fold higher in-hospital mortality rate after surgery than those with a higher hemoglobin concentration (Zindrou et al., 2002). Opposite results were published on the correlation between Hct and the risk of CHD, with high Hct level associated with an increased risk of myocardial infarction, coronary insufficiency, or CHD death (Sorlie et al., 1981). Recently, the role of RDW in identifying

mortality and cardiovascular risk among patients with CAD has been highlighted, being high RDW levels associated with increased risk of mortality and cardiovascular disease (CVD) events in patients with established CAD (Su et al., 2014). However, scarce knowledge is available nowadays regarding their functional profile in relation to morphodynamic features.

An increase in RBC aggregation has been described in patients with acute myocardial infarction (Lakshmi et al., 2011) and was associated with different cardiovascular risk factors, i.e., age (Vaya et al., 2013), obesity (Wiewiora et al., 2007), or diabetes mellitus (Martinez et al., 1998).

A reduced RBC deformability was reported in different vessel types (Baskurt and Meiselman, 2003; Keymel et al., 2011) and linked with pathological states related to microcirculatory disorders such as CAD (Pytel et al., 2013), hypertension (Odashiro et al., 2015), hypercholesterolemia (Kohno et al., 1997), and diabetes mellitus (Jain and Lim, 2000).

However, no data are available about the existence of a relation between RBC rheological properties and high-risk atherosclerotic plaque features.

In this study, we investigated the morphodynamic features of erythrocytes of individuals with different grades of coronary stenosis by means of Laser Assisted Optical Rotational Cell Analyzer (LoRRca MaxSis) in order to find any association between RBC features and plaque characteristics.

MATERIALS AND METHODS

Study Population and Blood Collection

In this study, we prospectively enrolled 91 consecutive patients who underwent CCTA between March 2016 and February 2018 for suspected but unknown stable CAD. In all patients, blood sample was obtained before CCTA and collected into EDTA tubes. Based on CCTA evaluation, patients were defined as having no apparent CAD in the absence of any plaque in the coronary tree (0% stenosis and no luminal irregularities, namely no CAD). Nonobstructive disease was defined as the presence of limited atherosclerotic disease demonstrated by a stenosis <50% (1–49%, named nonob CAD). When the atherosclerotic disease was associated with a stenosis ≥50%, patients were classified as ob CAD.

All patients were further evaluated for the presence of traditional cardiovascular risk factors such as diabetes mellitus (fasting glucose level of 126 mg/dl or higher and/or the need for insulin or oral hypoglycemic agents), hypercholesterolemia (total cholesterol level >200 mg/dl, or treatment with lipid-lowering drugs), hypertension, smoking attitude, and family history of CAD.

This observational study was carried out in accordance with the Declaration of Helsinki and approved by the local ethics research committee of Centro Cardiologico Monzino. Written informed consent to participate was obtained from all subjects.

Coronary Computed Tomography Angiography Scan Protocol, Images Reconstruction, and Analysis

Before CT scan, patients were treated with intravenous beta-blocker (Metoprolol up to 20 mg) to optimize heart rate and with a standard dose of sublingual nitrates. CCTAs were performed using a last generation 256-CT scanner (Revolution CT GE Healthcare, Milwaukee, WI, USA) with prospective ECG-triggering. A BMI-adapted scanning protocol was used: BMI < 20 Kg/m^2, tube voltage and tube current of 100 kVp and 500 mA, respectively; 20 $\leq$ BMI < 25 Kg/m^2, tube voltage and tube current of 100 kVp and 550 mA, respectively; 25 $\leq$ BMI < 30 Kg/m^2, tube voltage and tube current of 100 kVp and 600 mA, respectively; 30 $\leq$ BMI < 35 Kg/m^2, tube voltage and tube current of 120 kVp and 650 mA, respectively. Patients received a 50-ml (for BMI $\leq$ 25 Kg/m^2) or 60-ml (for BMI > 25 Kg/m^2) bolus of contrast medium (Iomeron 400 mg/ml, Bracco, Milan, Italy) through an antecubital vein at an infusion rate of 5 ml/s, followed by 50 ml of saline solution. The imaging was performed using bolus tracking technique.

Image CCTA datasets were evaluated using vessel analysis software (CardioQ3 Package - GE Healthcare).

Coronary plaques were defined as structures of at least 1 mm^2 area adjacent to coronary lumen, clearly distinguishable from the vessel lumen and surrounded by pericardial tissue; tissue with signal intensity below 40 HU was considered a pericardial fat and excluded from analysis. Normal coronary arteries were defined when no atherosclerotic plaque (including focal and eccentric calcified plaques) could be detected in any segment within the coronary artery wall or lumen.

Advanced coronary atherosclerosis evaluation were performed as follows: arterial remodeling index assessed using vessel area = lesion plaque area/reference area, plaque burden = (lesion plaque area-lesion lumen area)/lesion plaque area, napkin ring sign defined as the presence of a semi-circular thin enhancement around the plaque along the outer contour of the vessel, and small spotty calcifications as any discrete calcification $\leq$3 mm in length and occupying $\leq$90° arc when viewed in short axis, low-attenuation plaque defined as the presence of any plaque voxel <30 HU 11. Total plaque volume was evaluated and reported in mm^3 as previously described (Conte et al., 2020b). Non-calcified fibro-fatty plaque volume was expressed as the amount of plaque <150 HU, reported in mm^3.

Measurement of Morphodynamic RBC Characteristics

The analysis of RBC morphodynamic features was performed by means of Laser Assisted Optical Rotational Red Cell Analyzer (LoRRca MaxSis, Mechatronics, Hoorn, The Netherlands) according to the manufacturer's instructions and detailed below.

Osmotic gradient dependent RBC deformability: 250 μl of whole blood was suspended in 5 ml of polyvinylpyrrolidone buffer (Mechatronics, Hoorn, The Netherlands) and used for the analysis. The osmotic gradient curve generated by the instrument shows the variation in deformability as a continuous function of the osmolality of the solution in which RBCs are dissolved. The following parameters were evaluated: the elongation index (EI)max, corresponding to the maximal deformability or elongation obtained near the isotonic osmolality and is an expression of the membrane surface; the EImin, corresponding to the osmolality at which the deformability reaches its minimum and represents the 50% of the RBCs hemolysis in conventional

osmotic fragility assays, reflecting mean cellular surface-to-volume ratio; the osmolality (O)EImax or Omin, corresponding to the osmolality value at which the deformability reaches its EImax or EImin; the area under the curve (AUC, reported in the text as Area), defined in the provided software as the AUC beginning from a starting point in the hypo-osmolar region and an ending point in the hyper-osmolar region (instrument settings 500 mOsm/kg; Baskurt and Meiselman, 2004; Baskurt et al., 2009).

Red blood cells aggregation and disaggregation: 1 ml of oxygenated blood was placed into a preheated (37°C) Couette system consisting of two cylinders. A photo diode, integrated in the fixed inner cylinder, detects the intensity of the backscattered light during the RBC aggregation and disaggregation processes. A syllectogram curve was generated at the end of the assay and the following parameters were automatically calculated: the amplitude (Amp), showing the total extent of aggregation; the aggregation index (AI), calculated as integral of the total syllectogram curve; and the aggregation halftime (t1/2), reflecting the kinetics of RBC aggregation.

Statistical Analysis

Continuous variables were presented as mean $\pm$ SD or as median with interquartile range [IQR: 25°–75°], if more appropriate. Continuous variables normally distributed were compared using the Student's t-test for independent samples. When the variable distribution was not normal, Mann-Whitney U tests for independent samples were used. Variables with positively skewed distributions were log-transformed before analysis. The proportion of the categorical variables was compared using a χ^2 analysis or Fisher exact test, as appropriate. A value of $p < 0.05$ was considered statistically significant.

Logistic regression analysis was used in order to evaluate the relationship between biological variables and CCTA advanced coronary atherosclerosis characteristics (i.e., plaque volume and HRP features).

For every biological variable associated with CCTA finding with a $p < 0.10$ at logistic regression analysis, a receiver operating characteristic (ROC) curve and a correlation analysis were performed.

The association between RBC morphodynamics and HRP features was assessed by multivariable general linear models, after stratifying the population into the three mentioned groups. The hypothesis that RBC morphodynamics is predictive of a HRP in nonob CAD patients but not in ob ones was tested by computing the appropriate interaction terms. The different RBC morphodynamics variables were tested individually and by multiple regression. Statistical analysis and graphics were produced with MedCalc (version 11.6.1.0, Med-Calc Software; 1993–2011) and by SAS v. 9.4 statistical package (SAS Inc. Cary NC, USA).

Global RBC Morphodynamic (RMD) Score

In order to summarize the potential association of the different RBC morphodynamic variables with HRP in nonob CAD patients, we created a RMD score with the following procedure: first, we ran a multiple regression with non-calcified plaque volume as dependent variable and the RBC morphodynamic variables as predictors; then, the score was computed as the sum of the variables that were independent predictors of non-calcified plaque volume in the multivariable analysis, each weighted by the beta coefficient. The strength of the associations of the individual variables and of the score with HRP was quantified by the partial R-square.

RESULTS

Population Features

A total of 91 patients were consecutively enrolled. Male prevalence was 68.1% (62 out of 91) and mean age was 61.07 ± 10.9 years. The presence of plaque was excluded in 33 patients (36.3%), while 26 patients had nonob CAD (28.6%) and 32 patients had at least one obstructive coronary stenosis (35.2%).

All demographic and clinical characteristics of the study population were reported in **Table 1**.

Patients with ob CAD were older compared to individuals without CAD. This difference is reflected in a low number of platelets in ob CAD individuals. Interestingly, both ob and nonob CAD patients showed higher mean corpuscular hemoglobin (MCH) and mean corpuscular hemoglobin concentration (MCHC) values compared to no CAD individuals. Prevalence of traditional risk factors and medical therapy at the time of cardiac CCTA were similar throughout the entire enrolled population.

CCTA Characteristics

In **Table 2**, the main features of atherosclerotic lesions detected by CCTA have been reported.

Total coronary plaque volume was 88.5 ± 128.2 mm³ in the whole cohort of patients enrolled in the study and resulted to be significantly higher in patients with ob CAD when compared to those with nonob disease ($p < 0.0001$). Similarly, non-calcified plaque volume resulted to be significantly higher in patients with ob CAD than in nonob CAD ones. In 26 of

TABLE 1 | Demographic and clinical characteristics of the study population.

Variables	All patients N = 91	no CAD N = 33	nonob CAD N = 26	ob CAD N = 32	p
Demographic and clinical characteristics					
Age, years	61.07 ± 10.9	57.1 ± 11.5	60.68 ± 10	65.48 ± 9.7*	0.007
Male, n (%)	62 (68.1)	18 (54.5)	20 (76.9)	24 (75)	0.07
BMI, kg/m²	25.42 ± 3.8	25.5 ± 4.5	25.01 ± 3.6	25.66 ± 3.3	0.80
WBC, 10³/μl	7.41 ± 2.2	7.78 ± 2.5	7.3 ± 1.9	7.11 ± 2.2	0.46
Platelet, 10³/μl	221.34 ± 61.8	242.76 ± 70	213.69 ± 55.7	205.47 ± 52*	0.04
MPV, fl	10.36 ± 1.4	10.55 ± 0.8	10.23 ± 0.9	10.27 ± 2	0.60
RBC, 10⁶/μl	4.75 ± 0.6	4.69 ± 0.5	4.83 ± 0.5	4.75 ± 0.8	0.72
Hb, g/dl	14.3 [13.2–15.2]	13.7 [13–14.6]	14.5 [13.9–15.6]	14.4 [13.4–15.4]	0.05
Hct, %	41.48 ± 4.6	40.4 ± 3.2	42.33 ± 3.7	41.91 ± 6.2	0.22
MCV, fl	87.56 ± 4.5	86.39 ± 5.5	88 ± 4.5	88.41 ± 3.1	0.17
MCH, pg	30.22 ± 1.8	29.49 ± 2.1	30.5 ± 1.9*	30.75 ± 1.2*	0.01
MCHC, %	34.51 ± 0.9	34.13 ± 1	34.66 ± 0.9*	34.78 ± 0.8*	0.01
RDW-CV, %	13.17 ± 0.9	13.22 ± 0.9	13.12 ± 0.7	13.15 ± 0.9	0.90
RDW-SD, fl	41.32 ± 2.8	40.84 ± 2.1	41.58 ± 3	41.59 ± 3.2	0.47
Total cholesterol, mg/dl	197.8 ± 39.5	196.3 ± 35.2	187.9 ± 46.8	206.8 ± 36.1	0.22
LDL cholesterol, mg/dl	118.5 ± 35.1	116.9 ± 29.1	106.7 ± 42.5	129.3 ± 31.4	0.05
HDL cholesterol, mg/dl	57.4 ± 15.1	60.2 ± 14.8	57.2 ± 18.6	54.8 ± 12.3	0.36
Triglycerides, mg/dl	104.77 ± 49.1	97.61 ± 47.6	103.19 ± 50.8	113.44 ± 49.5	0.43
Basal glucose, mg/dl	99.5 ± 16.7	103.3 ± 19.7	93 ± 10.4	100.8 ± 16.2	0.05
Hypertension, n (%)	43 (47.3)	14 (42.4)	12 (46.2)	17 (53.1)	0.39
Family history of CVD, n (%)	36 (39.6)	10 (30.3)	15 (57.7)	11 (34.4)	0.65
Dyslipidaemia, n (%)	37 (40.6)	13 (39.4)	14 (53.8)	10 (31.2)	0.51
Diabetes, n (%)	5 (5.5)	2 (6.1)	1 (3.8)	2 (6.2)	0.98
Active smokers, n (%)	20 (21.9)	5 (15.2)	9 (34.6)	6 (18.7)	0.17
Past smokers, n (%)	16 (17.6)	7 (21.2)	2 (7.7)	7 (21.8)	0.28
Pharmacological treatments					
β-blockers, n (%)	55 (60.4)	16 (48.5)	19 (71.1)	20 (62.5)	0.24
ACE-inhibitors, n (%)	18 (19.8)	6 (18.2)	6 (23.1)	6 (18.7)	0.95
Angiotensin receptor blockers, n (%)	16 (17.6)	4 (12.1)	5 (19.2)	7 (21.8)	0.30
Diuretics, n (%)	13 (16.3)	4 (12.1)	1 (3.8)	8 (25)	0.14
Aspirin, n (%)	20 (21.9)	8 (24.4)	4 (15.4)	8 (25)	0.95
Statins, n (%)	30 (32.9)	7 (21.1)	13 (50)	10 (31.2)	0.38

BMI, body mass index; CVD, cardiovascular disease; Hb, hemoglobin; Hct, hematocrit; HDL, high-density lipoprotein; LDL, low-density lipoprotein; MCH, mean corpuscular hemoglobin; MCHC, mean corpuscular hemoglobin concentration; MCV, mean corpuscular volume; MPV, mean platelet volume; RBC, red blood cell; RDW-CV, red blood cell distribution width-coefficient of variation; RDW-SD, red blood cell distribution width-standard deviation; WBC, white blood cell.
$p < 0.05$ vs. no CAD; †$p < 0.05$ vs. nonob CAD.

TABLE 2 | CCTA characteristics of the study population.

Parameters	All patients N = 91	no CAD N = 33	nonob CAD N = 26	ob CAD N = 32	p
Total plaque volume, mm³	88.5 ± 128.2	-	99.6 ± 126.7[*]	170.7 ± 139.1[*,†]	<0.001
Non-calcified plaque volume, mm³	25.2 ± 40.7	-	25.9 ± 35.6[*]	50.7 ± 49.3[*,†]	<0.001
High risk plaque features >2, n (%)	26 (28.6)	0	6 (23.1)	20 (62.5)[*,†]	<0.001
Non-calcified plaque volume HQ, n (%)	23 (25.3)	0	6 (23.1)	17 (53.1)[*,†]	<0.001
Total plaque volume HQ, n (%)	23 (25.3)	0	5 (19.2)	18 (56.3)[*,†]	<0.001

CAD, coronary artery disease; CCTA, coronary computed tomography angiography; HQ, high quartile.
[*]p < 0.05 vs. no CAD; [†]p < 0.05 vs. nonob CAD.

TABLE 3 | RBC morphodynamic parameters in the study population.

Parameters	All patients N = 91	no CAD N = 33	nonob CAD N = 26	ob CAD N = 32	p
Deformability					
Elmax	0.63 ± 0.02	0.62 ± 0.02	0.63 ± 0.01	0.63 ± 0.02	0.54
Elmin	0.12 ± 0.01	0.12 ± 0.01	0.12 ± 0.01	0.12 ± 0.01	0.96
O Elmax, mOsm/kg	290.85 ± 19.5	290.96 ± 18.4	285.95 ± 17.1	294.84 ± 22.1	0.31
Omin, mOsm/kg	134.81 ± 8.4	134.23 ± 7.8	133.81 ± 8.5	136.24 ± 9.1	0.57
Area	149.93 ± 5.2	152.02 ± 5.3	149.45 ± 4.9	148.14 ± 4.7[*]	0.02
Aggregation					
AI, %	64.76 ± 10.4	66.69 ± 8.7	61.81 ± 11.2	65.54 ± 10.8	0.24
Amp, au	37.09 ± 9.5	37.02 ± 11.6	39.79 ± 5.2	34.94 ± 10	0.19
$t_{1/2}$, sec	1.8 [1.5–2.5]	1.8 [1.3–2.3]	2.2 [1.6–2.5]	1.8 [1.3–2.4]	0.17
RMD score	−47.2 [−47.6, −47.0]	−47.1 [−47.6, −46.9]	−47.2 [−47.5, −47.0]	−47.3 [−48.1, −46.7]	0.75

AI, aggregation index; Amp, amplitude of the aggregate; CAD, coronary artery disease; Elmax, maximal elongation index; Elmin, minimum elongation index; O Elmax, osmolality corresponding to maximal elongation index; Omin osmolality corresponding to minimum elongation index; RMD, RBC morphodynamic; t1/2, aggregation halftime.
[*]p < 0.05 vs. no CAD.

91 patients (28.6%) more than two HRP features were identified; as expected, an high prevalence of HRP features was recorded among patients with ob CAD [6 (23.1%) vs. 20 (62.5%), value of p < 0.0001 for nonob vs. ob CAD patients].

RBC Morphodynamics

In our study population, the analysis of RBC morphodynamic characteristics in relation to the degree of stenosis did not evidence any difference. The only parameter altered in ob CAD patients compared to no CAD is the Area (value of p = 0.024, Table 3), evidencing a reduced ability of erythrocytes to modify their shape in response to an osmotic gradient curve in patients with a severe stenosis.

RBC Morphodynamics and Atherosclerotic Plaque Features at CCTA

While no statistically significant association between RBC morphodynamic characteristics assessed by LoRRca and HRP features was found in patients with severe stenosis, in nonob CAD group, several relationships were evidenced.

In particular, we highlighted a positive correlation between RBC osmotic fragility (evaluated using the Omin value from the osmoscan curve) and two HRP features: fibro-fatty (**Figure 1A**) and fibro-fatty plus necrotic core plaque (**Figure 1B**) volumes. In parallel with increased fragility, we found correlations between RBC rigidity (assessed by a reduction in the Area of the osmoscan graph), defined by the inability of cell to change its shape under various osmotic conditions, and necrotic core (**Figure 1C**), fibro-fatty (**Figure 1D**) and fibro-fatty plus necrotic core (**Figure 1E**) plaque volumes.

In addition, RBC from nonob CAD patients with a high necrotic core plaque volume displayed an increased kinetics of aggregation (**Figure 1F,G**).

In accordance to these findings, we evidenced only in nonob CAD group the existence of an interaction among all these morphodynamic RBC characteristics and fibro-fatty plus necrotic core plaque volume (value of p = 0.005).

In this patient group, the non-calcified plaque volume resulted to be inversely associated with the degree of RBC resistance to lysis (Omin; **Figure 2A**) and positively correlated with the Area (**Figure 2B**), while no association was found with the total plaque volume (**Figures 2C,D**), an index of the extent of atherosclerosis.

Of interest, neither Omin nor Area or RBC aggregation parameters resulted to be significantly correlated to HRP features among patients with ob CAD (**Supplementary Figures S1, S2**).

To corroborate the existence of a relationship between RBC morphodynamic characteristics and HRP features specifically in nonob CAD population, in this patient group, we found an apparent linear relation between five RBC morphodynamic features and non-calcified plaque volume. Although the correlation was significant only for O Elmax, the pattern was similar for all features, with a concomitant totally flat relation in patients with severe stenosis (ob CAD; **Figures 3A–E**).

Most importantly, in a multivariable analysis, three morphodynamic RBC variables were independent predictors of non-calcified plaque volume in nonob CAD patients, i.e.,

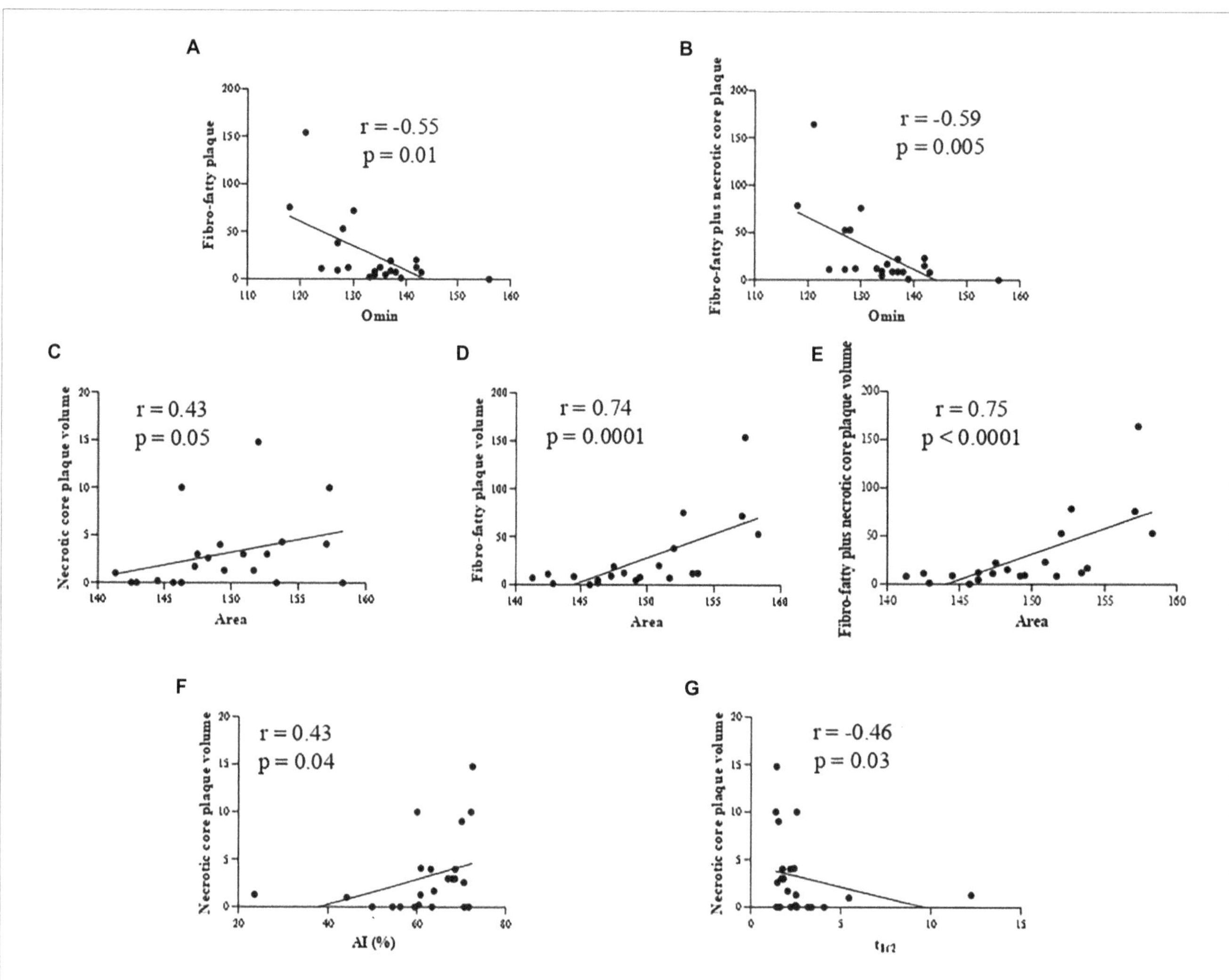

FIGURE 1 | Correlation analysis in nonob CAD patients revealing the association between: RBC osmotic fragility (evaluated using the Omin value from the osmoscan curve) and two HRP features (fibro-fatty **(A)** and fibro-fatty plus necrotic core plaque **(B)** volumes) and between RBC rigidity (assessed by a reduction in the Area of the osmoscan graph) and necrotic core **(C)**, fibro-fatty **(D)** and fibro-fatty plus necrotic core **(E)** plaque volumes. RBC from nonob CAD with a high necrotic core plaque volume displayed an increased kinetics of aggregation [displayed by AI **(F)** and $t_{1/2}$ **(G)**]. AI, aggregation index; Area, area under the elongation index-related osmolality curve; CAD, coronary artery disease; HRP, high-risk plaque; nonob CAD, non-obstructive coronary artery disease; Omin, minimal osmolality; RBC, red blood cell; $t_{1/2}$, aggregation half time.

EImax, O EImax and $t_{1/2}$ (**Table 4**). These variables were included in the putative global RMD score, which was computed using the following formula:

$$\text{RMD score} = 22.6 \times \log(\text{EImax}) + 17.2 \times \log(\text{O EImax}) + 1.6 \times \log(t1/2)$$

The RMD score was not associated with the presence or degree of CAD (**Table 3**), but it strongly correlated with the extent of HRP (such as the non-calcified plaque volume) in the nonob CAD group (**Figure 4A**). Therefore, the RMD score could be utilized in clinical practice to illustrate its potential use, we created a risk chart, relating three variables: the RMD score (in quartiles), the max % stenosis, and the non-calcified plaque volume (**Figure 4B**). This chart shows that only in nonob CAD patients the plaque instability increases according to RMD score quartile. Of note, a significant interaction between

the RMD score and patient group was found (value of $p = 0.001$), with an R-square (i.e., the proportion of the variability of non-calcified plaque volume explained by the score) in the nonob CAD group equal to 0.38, whereas in the ob CAD was 0.008.

DISCUSSION

In this study, we highlighted the presence of changes in morphodynamic features of RBCs from patients with nonob CAD assessed by means of LoRRca. In this clinical setting, RBC alterations correlate well with HRP features detected by CCTA, suggesting the existence of a possible link between erythrocyte morphodynamics and the increased risk of acute coronary events.

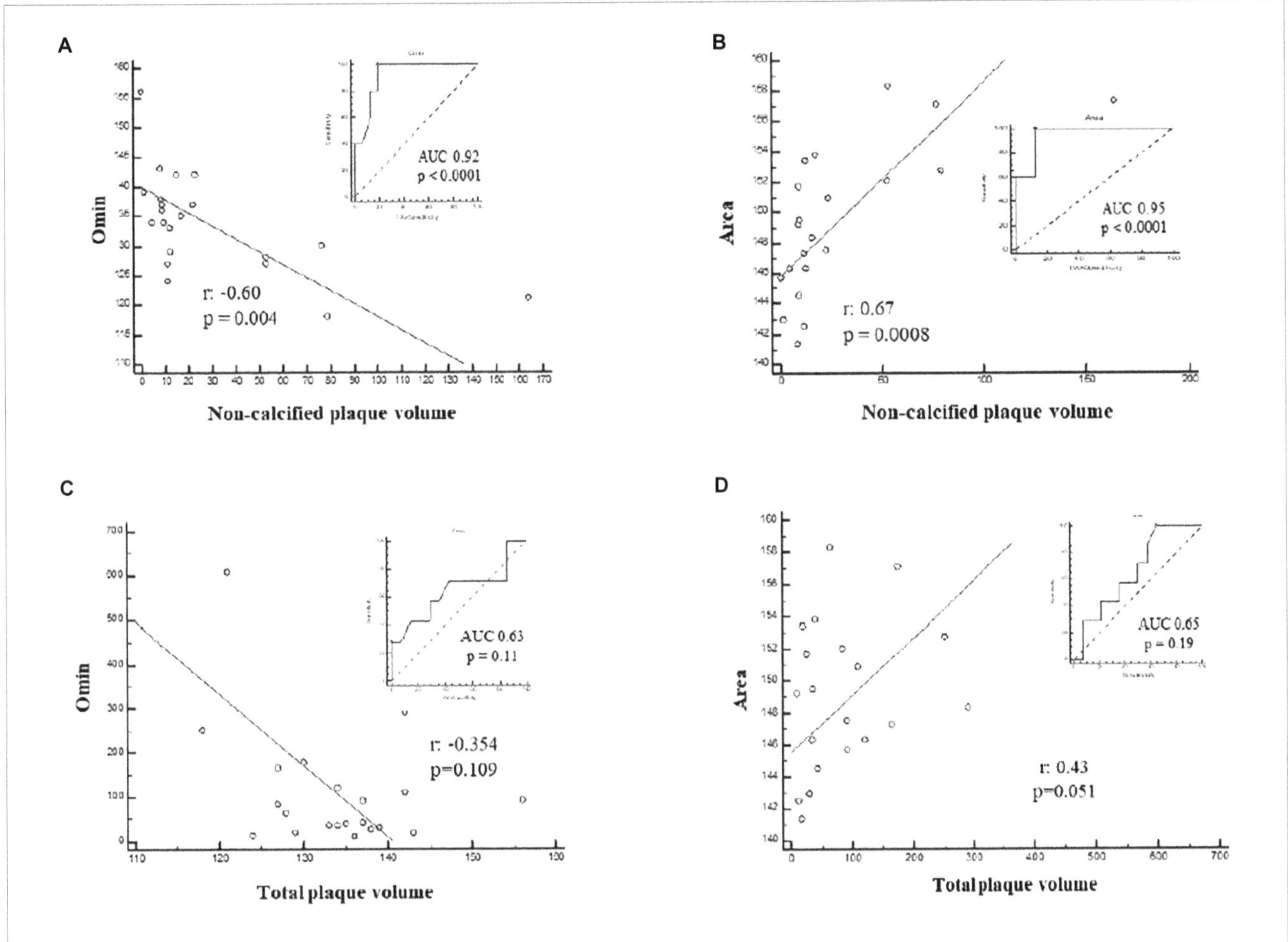

FIGURE 2 | ROC curves and correlation analysis in nonob CAD patients showing the association between non-calcified plaque volume and **(A)** Omin, corresponding to the degree of RBC resistance to lysis; **(B)** the Area and the absence of any correlation between total plaque volume and Omin **(C)** and the Area **(D)**. Area, area under the elongation index-related osmolality curve; CAD, coronary artery disease; nonob CAD, non-obstructive coronary artery disease; Omin, minimal osmolality; RBC, red blood cell; ROC, Receiver Operating Characteristic.

In the last decade, a high number of reports highlighted the presence of acute coronary syndrome (ACS) in the contest of nonob CAD. Up to now, the identification of HRPs is demanded to CCTA and, even if preliminary data on the CAPIRE study may suggest a potential association between HRP features and inflammatory markers (Conte et al., 2020a), the lack of circulating biomarkers associated with plaque instability constitutes a critical diagnostic issue.

Our findings, highlighting the existence of a connection between the erythrocyte morphodynamic behavior and plaque instability, turns on new lights on this circulating cell, moving the attention from the instable plaque to the instable patient.

During its long lifespan of 120 days, RBCs pass through the entire circulatory system where they play a key role in blood flow regulation and the consequent tissue perfusion (Yedgar et al., 2002). Even if up to now the precise mechanism underlining the involvement of this blood cell in the control of circulatory processes is not well described, some hypotheses can be formulated. In particular, it has been recognized the role of ATP release from RBC in modulating vasomotor tone in the microcirculation through a diffusive mechanism toward endothelial cells. This mechanical-dependent release is sensitive to several stimuli and, between them, low oxygen level, shear stress, and shape deformation and is believed to play an important role in the microcirculation, characterized by high levels of shear stress and shape deformation degree (Sprague et al., 1998; Dietrich et al., 2000; Wan et al., 2008; Forsyth et al., 2011). Another key mechanism through which RBCs regulate vascular function is NO production and release (Simmonds et al., 2014). Specifically, thanks to the internal compartmentalization of hemoglobin, RBCs are able to maintain hemostasis through the well-regulated delivery of oxygen and the balance of NO scavenging and production (Helms et al., 2018).

It becomes evident the importance of RBC regulatory role in vascular function in pathological states able to compromise the RBC membrane integrity leading to conditions of increased oxidative stress, hypertension, thrombosis, and vaso-occlusion.

In this regard, RBC aggregation and adherence to the endothelial wall were described to be influenced by both blood cellular and plasma factors, as this is a reversible process that depends on the concentration of high molecular weight proteins such as

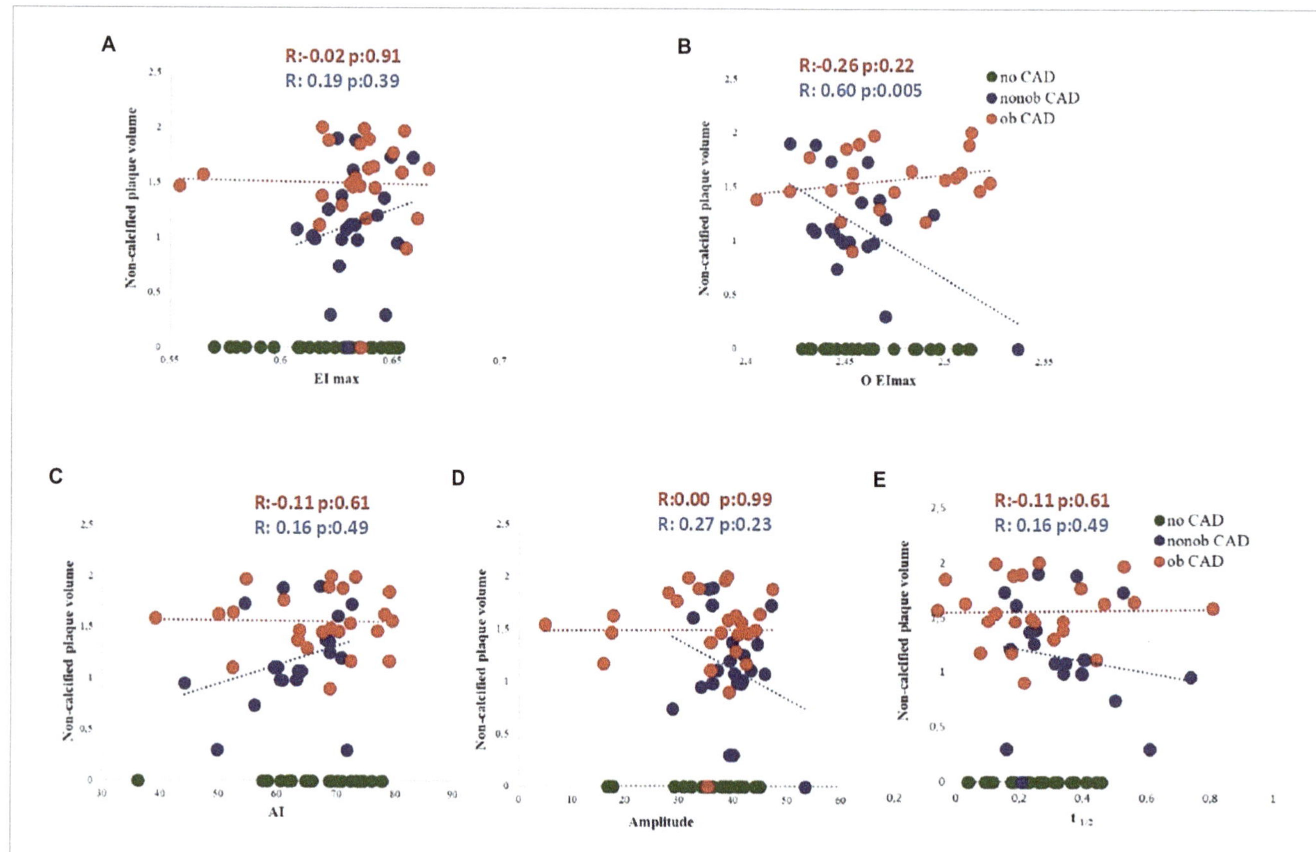

FIGURE 3 | Association between five RBC morphodynamic characteristics (AI, **A**; Amp, **B**; $t_{1/2}$, **C**; Elmax, **D**; O Elmax, **E**) and non-calcified plaque volume in nonob CAD (red dots), ob CAD (green dots) and no CAD (blue dots) patients. AI, aggregation index; Amp, amplitude of RBC aggregation; CAD, coronary artery disease; Elmax, maximal elongation index; nonob CAD, non-obstructive coronary artery disease; O Elmax, osmolality for Elmax; ob CAD, obstructive coronary artery disease; RBC, red blood cell; $t_{1/2}$, aggregation half time.

TABLE 4 | Morphodynamic RBC variables that are independent predictors of non-calcified plaque volume in nonob CAD patients.

Variable (log-transformed)	Beta coefficient	SE	p
Elmax	22.6	7.7	0.011
O Elmax	17.2	2.8	<0.0001
$t_{1/2}$	1.6	0.5	0.004

El max, maximal elongation index; O Elmax, osmolality corresponding to maximal elongation index; SE, standard error; t1/2, aggregation halftime.

fibrinogen (Baskurt, 2007). In this regard, it has been demonstrated that RBC aggregates strength directly correlates to the inflammatory state (Ami et al., 2001). Specifically, the RBC aggregation parameters considered in the study conducted by Ami and colleagues (i.e., the average aggregate size, the distribution of the RBC population into aggregate size ranges, the shear stress required to obtain 50% of RBC population in the small range of aggregate size, and the area under the curve (AUC) of the plot as a function of shear stress) correlated with the inflammatory indexes C reactive protein and fibrinogen in patients with cardiovascular and infectious conditions associated with an inflammatory response, regardless of the specific pathological state and despite the

variability of symptoms. In addition, RBC deformability was found to be able to control platelet-vessel interaction (Aarts et al., 1984). In this context, the morphodynamic characteristics we evidenced in RBCs of patients with nonob CAD could be the result of the interaction among RBC and cell components or humoral factors derived from plaque or are consequent to the alteration of the erythrocyte itself that negatively influences plaque progression (Aarts et al., 1984).

It could be expected HRP features to be associated to RBC dysfunctionality also in ob CAD patients. These data might be explained by the diffuse endothelial dysfunction that characterizes patients with ob CAD that could mask the interplay between atherosclerosis and RBC alterations. Of interest, when comparing individuals with or without ob stenosis, no difference in RBC morphodynamics was observed as total plaque volume was not associated with erythrocyte features, suggesting that overall atherosclerosis burden is a sub-optimal marker of atherosclerosis disease activity at least in the early stages.

Thus, it can be hypothesized that RBC could be mainly involved in the early phase of atherosclerosis process (nonob CAD), rather than in the more advanced stages (ob CAD).

Our data have also to be considered in view of the results of the CAPIRE study that specifically identified high non-calcified

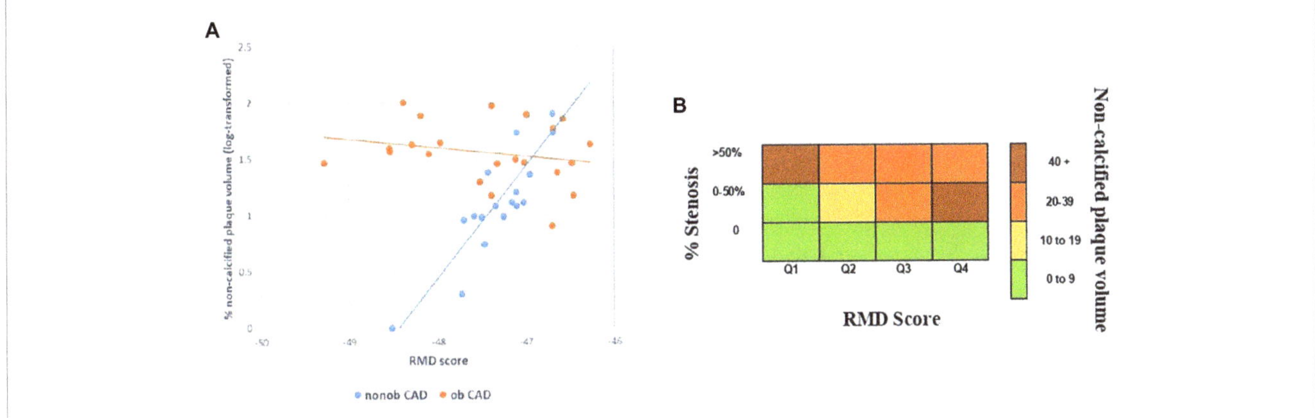

FIGURE 4 | Association between HRP (such as the non-calcified plaque volume) and the RMD score in nonob and ob CAD groups **(A)**. Risk chart, relating the RMD score and the maximal percentage of stenosis with the extent of HRP feature (i.e., the non-calcified plaque volume), showing the existence of a strong relationship in nonob CAD group. The strong association is depicted in dark red, while the weak one in green **(B)**. CAD, coronary artery disease; HRP, high-risk plaque; nonob CAD, non-obstructive coronary artery disease; ob CAD, obstructive coronary artery disease; RMD, RBC morphodynamic; RBC, red blood cell.

plaque volume as the most ACS-predictive parameter in CAD patients. The underlined association between RBC dysfunction indexes and adverse plaque features (i.e., non-calcified plaque volume) can be proposed as a new systemic non-invasive marker for the identification of high risk patients (Andreini et al., 2019).

In this context, one possible pathophysiological model to explain the correlation between elevated non-calcified plaque volume and RBC rigidity could be identified in the role of cholesterol in determining RBC deformability. More precisely, as previously described (Banerjee et al., 1998; Cicco and Pirrelli, 1999; Ercan et al., 2002), higher amount of cholesterol in RBC structure positively correlates with their hemorheological parameters and especially with rigidity. In our clinical setting, the decrease in the area under the curve (AUC) value in ob CAD vs. no CAD could be the result of increased cholesterol (Banerjee et al., 1998), as in our patient group, the levels of LDL cholesterol show a trend to increase, moving from nonob CAD to no CAD and to ob CAD. We cannot exclude that the absence of a difference in the EI max value can be attributable to the pharmacological treatment with statins, present in the 50% of nonob CAD and in the 31.2% of ob CAD.

Similarly, tissue with low CT attenuation are commonly considered as lipid rich and low-attenuation coronary plaque in CCTA could be considered as a diagnostic for lipid-core unstable plaque (Chen et al., 2010; Schlett et al., 2013). This pathophysiological hypothesis, even if of speculative nature, well fits with the widely recognized role of dyslipidemia as a major cardiovascular risk factor.

Another factor able to augment RBC rigidity is an increase in intracellular viscosity because of increased MCHC. Specifically, a variation in MCHC values as a result of a different hydrating states of erythrocytes is able to induce a modification in the osmotic deformability profile (and consequently in the area) without affecting the EImax (Clark et al., 1983). According to these findings, in our study population we registered a significant increase in MCHC values from no CAD to ob CAD that could explain in part this result.

As clinical tool, the morphodynamic characteristics combined into the RMD score could be used in a risk chart. This is a simple, cheap and non-invasive procedure that will allow identifying high risk patients in the early stage of atherosclerosis. The use of this chart in the clinical practice could help in better identifying patients who may merit CCTA, even if asymptomatic, and in the long-term follow-up of non-obstructive patients with a high probability of experiencing acute events.

Several limitations warrant discussion. This study was conducted on a small number of subjects. The RMD score and its relative risk chart here obtained have not to be considered as substitutes of CCTA. Moreover, further studies are needed before our findings could be used in clinical practice for early identification of asymptomatic patients who may merit screening CCTA, that nowadays remains not indicated.

CONCLUSION

In nonob CAD patients the HRP features correlated with an altered RBC morphodynamic behavior highlighted by means of LoRRca. Three of the erythrocyte flow-affecting properties analyzed are independent predictors of the extent of the non-calcified plaque volume in this population. The RMD score and the related risk chart clearly highlighted the rise of acute events probability with the increase of erythrocyte altered features.

Further studies are needed to validate this chart in a large cohort of subjects in order to define its potential use in patient identification and management, with the final goal of slowing the disease progression.

AUTHOR CONTRIBUTIONS

BP, EC, VC, and DA designed the study. BP, AZ, AM, and SF performed the ektacytometry assays. EC and SM enrolled patients. BP, EC, PB, DA, and VC interpreted the results. FV and SB performed statistical analysis. BP and EC wrote the manuscript. BP, EC, AZ, PB, FV, SB, SF, SE, AM, SM, ET, VC, and DA reviewed and approved the final version of the manuscript.

REFERENCES

Aarts, P. A., Heethaar, R. M., and Sixma, J. J. (1984). Red blood cell deformability influences platelets--vessel wall interaction in flowing blood. *Blood* 64, 1228–1233.

Al-Mallah, M. H., Aljizeeri, A., Villines, T. C., Srichai, M. B., and Alsaileek, A. (2015). Cardiac computed tomography in current cardiology guidelines. *J. Cardiovasc. Comput. Tomogr.* 9, 514–523. doi: 10.1016/j.jcct.2015.09.003

Ami, R. B., Barshtein, G., Zeltser, D., Goldberg, Y., Shapira, I., Roth, A., et al. (2001). Parameters of red blood cell aggregation as correlates of the inflammatory state. *Am. J. Physiol. Heart Circ. Physiol.* 280, H1982–H1988. doi: 10.1152/ajpheart.2001.280.5.H1982

Andreini, D., Magnoni, M., Conte, E., Masson, S., Mushtaq, S., Berti, S., et al. (2019). Coronary plaque features on CTA can identify patients at increased risk of cardiovascular events. *JACC Cardiovasc. Imaging* 13, 1704–1717. doi: 10.1016/j.jcmg.2019.06.019

Andreini, D., Pontone, G., Mushtaq, S., Bartorelli, A. L., Bertella, E., Antonioli, L., et al. (2012). A long-term prognostic value of coronary CT angiography in suspected coronary artery disease. *JACC Cardiovasc. Imaging* 5, 690–701. doi: 10.1016/j.jcmg.2012.03.009

Banerjee, R., Nageshwari, K., and Puniyani, R. R. (1998). The diagnostic relevance of red cell rigidity. *Clin. Hemorheol. Microcirc.* 19, 21–24.

Baskurt, O. K. (2007). *Handbook of Hemorheology and Hemodynamics.* IOS Press.

Baskurt, O. K., Hardeman, M. R., Uyuklu, M., Ulker, P., Cengiz, M., Nemeth, N., et al. (2009). Parameterization of red blood cell elongation index—shear stress curves obtained by ektacytometry. *Scand. J. Clin. Lab. Invest.* 69, 777–788. doi: 10.3109/00365510903266069

Baskurt, O. K., and Meiselman, H. J. (2003). Blood rheology and hemodynamics. *Semin. Thromb. Hemost.* 29, 435–450. doi: 10.1055/s-2003-44551

Baskurt, O. K., and Meiselman, H. J. (2004). Analyzing shear stress-elongation index curves: comparison of two approaches to simplify data presentation. *Clin. Hemorheol. Microcirc.* 31, 23–30.

Bugiardini, R., and Bairey Merz, C. N. (2005). Angina with "normal" coronary arteries: a changing philosophy. *J. Am. Med. Assoc.* 293, 477–484. doi: 10.1001/jama.293.4.477

Chen, Z., Ichetovkin, M., Kurtz, M., Zycband, E., Kawka, D., Woods, J., et al. (2010). Cholesterol in human atherosclerotic plaque is a marker for underlying disease state and plaque vulnerability. *Lipids Health Dis.* 9:61. doi: 10.1186/1476-511X-9-61

Cicco, G., and Pirrelli, A. (1999). Red blood cell (RBC) deformability, RBC aggregability and tissue oxygenation in hypertension. *Clin. Hemorheol. Microcirc.* 21, 169–177.

Clark, M. R., Mohandas, N., and Shohet, S. B. (1983). Osmotic gradient ektacytometry: comprehensive characterization of red cell volume and surface maintenance. *Blood* 61, 899–910.

Conte, E., Andreini, D., Magnoni, M., Masson, S., Mushtaq, S., Berti, S., et al. (2020a). Association of high-risk coronary atherosclerosis at CCTA with clinical and circulating biomarkers: insight from CAPIRE study. *J. Cardiovasc. Comput. Tomogr.* doi: 10.1016/j.jcct.2020.03.005 [Epub ahead of print]

Conte, E., Annoni, A., Pontone, G., Mushtaq, S., Guglielmo, M., Baggiano, A., et al. (2017). Evaluation of coronary plaque characteristics with coronary computed tomography angiography in patients with non-obstructive coronary artery disease: a long-term follow-up study. *Eur. Heart J. Cardiovasc. Imaging* 18, 1170–1178. doi: 10.1093/ehjci/jew200

Conte, E., Mushtaq, S., Pontone, G., Li Piani, L., Ravagnani, P., Galli, S., et al. (2020b). Plaque quantification by coronary computed tomography angiography using intravascular ultrasound as a reference standard: a comparison between standard and last generation computed tomography scanners. *Eur. Heart J. Cardiovasc. Imaging* 21, 191–201. doi: 10.1093/ehjci/jez089

Danesh, J., Collins, R., Peto, R., and Lowe, G. D. (2000). Haematocrit, viscosity, erythrocyte sedimentation rate: meta-analyses of prospective studies of coronary heart disease. *Eur. Heart J.* 21, 515–520. doi: 10.1053/euhj.1999.1699

Dietrich, H. H., Ellsworth, M. L., Sprague, R. S., and Dacey, R. G. Jr. (2000). Red blood cell regulation of microvascular tone through adenosine triphosphate. *Am. J. Physiol. Heart Circ. Physiol.* 278, H1294–H1298. doi: 10.1152/ajpheart.2000.278.4.H1294

Ercan, M., Konukoglu, D., Erdem, T., and Onen, S. (2002). The effects of cholesterol levels on hemorheological parameters in diabetic patients. *Clin. Hemorheol. Microcirc.* 26, 257–263.

Finn, A. V., Nakano, M., Narula, J., Kolodgie, F. D., and Virmani, R. (2010). Concept of vulnerable/unstable plaque. *Arterioscler. Thromb. Vasc. Biol.* 30, 1282–1292. doi: 10.1161/ATVBAHA.108.179739

Forsyth, A. M., Wan, J., Owrutsky, P. D., Abkarian, M., and Stone, H. A. (2011). Multiscale approach to link red blood cell dynamics, shear viscosity, and ATP release. *Proc. Natl. Acad. Sci. U. S. A.* 108, 10986–10991. doi: 10.1073/pnas.1101315108

Gillum, R. F., Mussolino, M. E., and Makuc, D. M. (1995). Erythrocyte sedimentation rate and coronary heart disease: the NHANES I epidemiologic follow-up study. *J. Clin. Epidemiol.* 48, 353–361. doi: 10.1016/0895-4356(94)00156-k

Helms, C. C., Gladwin, M. T., and Kim-Shapiro, D. B. (2018). Erythrocytes and vascular function: oxygen and nitric oxide. *Front. Physiol.* 9:125. doi: 10.3389/fphys.2018.00125

Hung, M. J., and Cherng, W. J. (2003). Comparison of white blood cell counts in acute myocardial infarction patients with significant versus insignificant coronary artery disease. *Am. J. Cardiol.* 91, 1339–1342. doi: 10.1016/s0002-9149(03)00325-4

Jain, S. K., and Lim, G. (2000). Lipoic acid decreases lipid peroxidation and protein glycosylation and increases $(Na^{(+)} + K^{(+)})$- and $Ca^{(++)}$-ATPase activities in high glucose-treated human erythrocytes. *Free Radic. Biol. Med.* 29, 1122–1128. doi: 10.1016/s0891-5849(00)00410-x

Keymel, S., Heiss, C., Kleinbongard, P., Kelm, M., and Lauer, T. (2011). Impaired red blood cell deformability in patients with coronary artery disease and diabetes mellitus. *Horm. Metab. Res.* 43, 760–765. doi: 10.1055/s-0031-1286325

Kohno, M., Murakawa, K., Yasunari, K., Yokokawa, K., Horio, T., Kano, H., et al. (1997). Improvement of erythrocyte deformability by cholesterol-lowering therapy with pravastatin in hypercholesterolemic patients. *Metabolism* 46, 287–291.

Lakshmi, A. B., Uma, P., Venkatachalam, C., and Nageswar Rao, G. S. (2011). A simple slide test to assess erythrocyte aggregation in acute ST-elevated myocardial infarction and acute ischemic stroke: its prognostic significance. *Indian J. Pathol. Microbiol.* 54, 63–69. doi: 10.4103/0377-4929.77327

Libby, P. (2013). Mechanisms of acute coronary syndromes and their implications for therapy. *N. Engl. J. Med.* 368, 2004–2013. doi: 10.1056/NEJMra1216063

Martinez, M., Vaya, A., Server, R., Gilsanz, A., and Aznar, J. (1998). Alterations in erythrocyte aggregability in diabetics: the influence of plasmatic fibrinogen and phospholipids of the red blood cell membrane. *Clin. Hemorheol. Microcirc.* 18, 253–258.

Min, J. K., Dunning, A., Lin, F. Y., Achenbach, S., Al-Mallah, M., Budoff, M. J., et al. (2011). Age- and sex-related differences in all-cause mortality risk based on coronary computed tomography angiography findings results from the International Multileft CONFIRM (Coronary CT angiography evaluation for clinical outcomes: an international Multileft registry) of 23,854 patients without known coronary artery disease. *J. Am. Coll. Cardiol.* 58, 849–860. doi: 10.1016/j.jacc.2011.02.074

Montalescot, G., Sechtem, U., Achenbach, S., Andreotti, F., Arden, C., Budaj, A., et al. (2013). 2013 ESC guidelines on the management of stable coronary artery disease: the task force on the management of stable coronary artery disease of the European Society of Cardiology. *Eur. Heart J.* 34, 2949–3003. doi: 10.1093/eurheartj/eht296

Odashiro, K., Saito, K., Arita, T., Maruyama, T., Fujino, T., and Akashi, K. (2015). Impaired deformability of circulating erythrocytes obtained from nondiabetic hypertensive patients: investigation by a nickel mesh filtration technique. *Clin. Hypertens.* 21:17. doi: 10.1186/s40885-015-0030-9

Patel, M. R., Chen, A. Y., Peterson, E. D., Newby, L. K., Pollack, C. V. Jr., Brindis, R. G., et al. (2006). Prevalence, predictors, and outcomes of patients with non-ST-segment elevation myocardial infarction and insignificant coronary artery disease: results from the Can Rapid risk stratification of Unstable angina patients Suppress ADverse outcomes with Early implementation of the ACC/AHA guidelines (CRUSADE) initiative. *Am. Heart J.* 152, 641–647. doi: 10.1016/j.ahj.2006.02.035

Pytel, E., Olszewska-Banaszczyk, M., Koter-Michalak, M., and Broncel, M. (2013). Increased oxidative stress and decreased membrane fluidity in erythrocytes of CAD patients. *Biochem. Cell Biol.* 91, 315–318. doi: 10.1139/bcb-2013-0027

Schlett, C. L., Ferencik, M., Celeng, C., Maurovich-Horvat, P., Scheffel, H., Stolzmann, P., et al. (2013). How to assess non-calcified plaque in CT angiography: delineation methods affect diagnostic accuracy of low-attenuation plaque by CT for lipid-core plaque in histology. *Eur. Heart J. Cardiovasc. Imaging* 14, 1099–1105. doi: 10.1093/ehjci/jet030

Simmonds, M. J., Detterich, J. A., and Connes, P. (2014). Nitric oxide, vasodilation and the red blood cell. *Biorheology* 51, 121–134. doi: 10.3233/BIR-140653

Sorlie, P. D., Garcia-Palmieri, M. R., Costas, R. Jr., and Havlik, R. J. (1981). Hematocrit and risk of coronary heart disease: the Puerto Rico Health Program. *Am. Heart J.* 101, 456–461. doi: 10.1016/0002-8703(81)90136-8

Sprague, R. S., Ellsworth, M. L., Stephenson, A. H., Kleinhenz, M. E., and Lonigro, A. J. (1998). Deformation-induced ATP release from red blood cells requires CFTR activity. *Am. J. Phys.* 275, H1726–H1732. doi: 10.1152/ajpheart.1998.275.5.H1726

Su, C., Liao, L. Z., Song, Y., Xu, Z. W., and Mei, W. Y. (2014). The role of red blood cell distribution width in mortality and cardiovascular risk among patients with coronary artery diseases: a systematic review and meta-analysis. *J. Thorac. Dis.* 6, 1429–1440. doi: 10.3978/j.issn.2072-1439.2014.09.10

Vaya, A., Alis, R., Romagnoli, M., Perez, R., Bautista, D., Alonso, R., et al. (2013). Rheological blood behavior is not only influenced by cardiovascular risk factors but also by aging itself. Research into 927 healthy Spanish Mediterranean subjects. *Clin. Hemorheol. Microcirc.* 54, 287–296. doi: 10.3233/CH-131734

Virmani, R., Burke, A. P., Farb, A., and Kolodgie, F. D. (2006). Pathology of the vulnerable plaque. *J. Am. Coll. Cardiol.* 47 (Suppl. 8), C13–C18. doi: 10.1016/j.jacc.2005.10.065

Wan, J., Ristenpart, W. D., and Stone, H. A. (2008). Dynamics of shear-induced ATP release from red blood cells. *Proc. Natl. Acad. Sci. U. S. A.* 105, 16432–16437. doi: 10.1073/pnas.0805779105

Wiewiora, M., Sosada, K., Wylezol, M., Slowinska, L., and Zurawinski, W. (2007). Red blood cell aggregation and deformability among patients qualified for bariatric surgery. *Obes. Surg.* 17, 365–371. doi: 10.1007/s11695-007-9066-6

Yedgar, S., Koshkaryev, A., and Barshtein, G. (2002). The red blood cell in vascular occlusion. *Pathophysiol. Haemost. Thromb.* 32, 263–268. doi: 10.1159/000073578

Zindrou, D., Taylor, K. M., and Bagger, J. P. (2002). Preoperative haemoglobin concentration and mortality rate after coronary artery bypass surgery. *Lancet* 359, 1747–1748. doi: 10.1016/S0140-6736(02)08614-2

5

Targeted Next Generation Sequencing and Diagnosis of Congenital Hemolytic Anemias: A Three Years Experience Monocentric Study

Elisa Fermo[1], Cristina Vercellati[1], Anna Paola Marcello[1], Ebru Yilmaz Keskin[2], Silverio Perrotta[3], Anna Zaninoni[1], Valentina Brancaleoni[4], Alberto Zanella[1], Juri A. Giannotta[1], Wilma Barcellini[1] and Paola Bianchi[1]*

[1] UOS Fisiopatologia delle Anemie, UOC Ematologia, Fondazione IRCCS Ca' Granda Ospedale Maggiore Policlinico, Milan, Italy, [2] Department of Pediatric Hematology and Oncology, Suleyman Demirel University, Isparta, Turkey, [3] Dipartimento della Donna, del Bambino e di Chirurgia Generale e Specialistica, Università degli Studi della Campania "Luigi Vanvitelli," Naples, Italy, [4] UOC Medicina Generale, Fondazione IRCCS Ca' Granda Ospedale Maggiore Policlinico, Milan, Italy

*Correspondence:
Paola Bianchi
paola.bianchi@policlinico.mi.it

Congenital hemolytic anemias (CHAs) are heterogeneous and rare disorders caused by alterations in structure, membrane transport, metabolism, or red blood cell production. The pathophysiology of these diseases, in particular the rarest, is often poorly understood, and easy-to-apply tools for diagnosis, clinical management, and patient stratification are still lacking. We report the 3-years monocentric experience with a 43 genes targeted Next Generation Sequencing (t-NGS) panel in diagnosis of CHAs; 122 patients from 105 unrelated families were investigated and the results compared with conventional laboratory pathway. Patients were divided in two groups: 1) cases diagnosed with hematologic investigations to be confirmed at molecular level, and 2) patients with unexplained anemia after extensive hematologic investigation. The overall sensitivity of t-NGS was 74 and 35% for families of groups 1 and 2, respectively. Inside this cohort of patients we identified 26 new pathogenic variants confirmed by functional evidence. The implementation of laboratory work-up with t-NGS increased the number of diagnoses in cases with unexplained anemia; cytoskeleton defects are well detected by conventional tools, deserving t-NGS to atypical cases; the diagnosis of Gardos channelopathy, some enzyme deficiencies, familial siterosterolemia, X-linked defects in females and other rare and ultra-rare diseases definitely benefits of t-NGS approaches.

Keywords: congenital hemolytic anemia, targeted-NGS, pathogenic variants, red blood cells, differential diagnosis

INTRODUCTION

Congenital hemolytic anemias (CHAs) comprise a group of very heterogeneous and rare disorders caused by alterations in structure, transport functions, metabolism, or defective production of red blood cells (RBCs). Since the pathophysiology of some rare forms is poorly understood, these disorders represent a group of diseases that still lack easy-to-apply tools for diagnosis, clinical management, and patient stratification.

The laboratory diagnostic pathway of CHAs was historically based on sequential steps using a panel of functional analyses that investigate the RBC membrane and metabolism. The first diagnostic step relies on hematological tests (complete blood count, red cell indices, hemolysis markers), peripheral blood smear examination, osmotic fragility tests and eosin-5-maleimide (EMA) binding test. The second includes biochemical tests such as quantitative assay of RBC enzymes activity, sodium dodecyl sulphate-polyacrylamide gel electrophoresis (SDS-PAGE) for the diagnosis of membranopathies, and specialized investigations requiring instruments which are available only in reference Centers, as ektacytometry (King et al., 2015; Fermo et al., 2021). The molecular characterization of the affected genes by traditional Sanger sequencing was usually, up to some years ago, the last step allowing the definitive confirmation of the diagnosis. Despite this, even after extensive and complete investigation, the differential diagnosis of these disorders may be difficult, and some patients remain undiagnosed with a consequent negative impact on clinical follow up, risk of inappropriate therapeutic decisions and lack of access to new specific treatments.

The advent and recent progresses on next generation sequencing (NGS) technologies has radically changed the diagnostic approach to CHAs, often placing the genetic analysis as a first line screening tool; different NGS strategies have been developed in the last years including targeted panels and whole exome sequencing (WES), with a progressive reduction of costs that allowed their routinely use (Steinberg-Shemer and Tamary, 2020; Russo et al., 2020). Targeted-NGS panels have been developed and applied by several groups, being currently the preferred approach for the molecular diagnosis of CHAs; custom panels include different numbers of genes and have been reported to have a wide range of diagnostic efficacy (38–90%) depending on the number of genes included and on the characteristics of the patients studied (Agarwal et al., 2016; Del Orbe Barreto et al., 2016; Niss et al., 2016; Roy et al., 2016; Russo et al., 2018; Shefer Averbuch et al., 2018; Choi et al., 2019; Kedar et al., 2019a,b; Svidnicki et al., 2020). In this paper we report the 3 years monocentric experience in targeted-NGS and diagnosis of CHAs, and we compare the efficacy of this methodological approach with the conventional laboratory diagnosis.

MATERIALS AND METHODS

Patients

Among the patients referred to our Institut (Pathophysiology of Anemias Unit, Fondazione IRCCS Ca' Granda Ospedale Maggiore of Milan–Italy) between 2017 and 2019 with a clinical suspect of CHAs we examined a cohort of 122 patients from 105 unrelated families (51% males and 49% females, median age 27 years, range 0–73 years). The inclusion criteria in the study were: a diagnostic indication for congenital hemoltic anemia; the specific request from clinicians to perform/complete diagnosis at molecular level; a specific scientific interest in cases with intra-family variability or atypical presentation; all cases with unexplained phenotype despite extensive investigations; signed informed consent to perform molecular investigations. Samples were collected from patients and controls during diagnostic procedures after obtaining informed consent and approval from the Institutional Human Research Committee. For patients under the age of 18, written informed consent was obtained from the parents. The procedures followed were in accordance with the Helsinki International ethical standards on human experimentation.

All patients but 18 (received after shipping from abroad) underwent extensive hematologic investigations, including: RBC morphology, complete blood count, hemolysis markers, screening for abnormal/unstable hemoglobins, direct antiglobulin test, osmotic fragility tests, EMA binding test (Bianchi et al., 2012), and RBC enzymes activities determination (hexokinase, glucosephosphate isomerase, phosphofructokinase, glyceraldehyde-P-dehydrogenase, phosphoglycerate kinase, pyruvate kinase, glucose-6-phosphate dehydrogenase, 6-phosphogluconate dehydrogenase, adenylate kinase) (Beutler et al., 1977).

In the vast majority of cases we also performed Osmoscan analysis by LoRRca MaxSis (Laser-Assisted Optical Rotational Cell Analyzer, Mechatronics, NL) (Zaninoni et al., 2018, see **Supplementary Materials** for more details), and RBC membrane protein analysis by SDS–PAGE performed according to Laemmli (1970) and Fairbanks et al. (1971) with minor modifications (Mariani et al., 2008).

In the remaining 18 cases (15 families), when possible, a minimum panel of investigation was performed (i.e., EMA binding test, RBC enzyme assay, Osmoscan) and evaluated together with clinical and laboratory information received from the local centers.

In 60 patients from 51 families a definite diagnostic orientation was obtained after first and second level hematological investigations. In particular, 6 were suspected of RBC enzyme defects, 28 of structural membrane defects (hereditary spherocytosis, elliptocytosis, pyropoikilocytosis, HS/HE/HPP), 10 of stomatocytosis and 7 were classified as congenital dyscrythropoietic anemia.

In the remaining 62 cases (from 54 families) the laboratory investigations did not allow a clear diagnosis.

NGS Sequencing and Analysis

Genomic DNA was extracted from peripheral blood, using standard manual methods and quantified by Nandrop One (Thermo Scientific, Italy). When available, relatives of affected cases were also enrolled to analyze allelic segregation and correctly assess the pathogenicity of each variant.

Using SureDesign software (Agilent) we created a NGS based panel containing 43 genes already described as disease causing for RBC membrane disorders ($n = 14$ genes), enzymopathies ($n = 18$), dyserythropoietic and sideroblastic anemias ($n = 11$) (**Supplementary Table 1**). For the probe design, coding regions, 5'UTR, 3'UTR, 50 bp flanking splice junctions were selected as regions of interest. Sequence length was set at 150×2 nucleotides and the average target coverage was 99.52%. The panel was validated on previously characterized patients affected by CHAs.

Libraries were obtained by HaloPlexHS Target Enrichment System Kit (Agilent) following the instruction's manufacturer. Samples were pooled (average 22 samples) and loaded at 7 pM on MiSeq platform using a v2-300 cycle reagent kit (Illumina). Sequencing reads were aligned against reference genome (UCSC hg19) and variants were called and annotated using the SureCall software (Agilent Technologies).

Filter settings were as follows: variant call quality threshold > 100; minimum number of reads supporting variant allele > 10; mutant allele frequency ≥ 25%; minor allele frequency (MAF) < 1% of the population. Common pathogenic modifiers of hereditary spherocytosis known to be present above the 1% cutoff, e.g., *SPTA1* LELY (Low Expression Lyon) allele (p.L1858V, found in association with the intronic variant c.6531-12 c > t), or the SPTA-Bughill allele (p.A970D) usually found in linkage with the LEPRA (Low Expression PRAgue) allele (SPTA c.4339-99 c > t) were also included.

Targeted filtering and annotation of protein-changing variants were performed using the wANNOVAR web tool[1].

Variants were assessed by mutation prediction and conservation programs including SIFT[2], polyphen-2 (Polymorphism Phenotyping v2)[3] and MutationTaster[4]; pathogenicity was evaluated according to the guidelines of American College of Medical Genetics and Genomics (ACMG) (Richards et al., 2015) using the online tool VarSome[5] (Kopanos et al., 2019).

Variants previously classified as pathogenic by databases such as ClinVar, HGMD, dbSNP, deleterious variants expected to produce truncated or abnormal protein, or splice site variants were considered as causatives. Variants of unknown significance (VUSs) were reported only if found in genes relevant to the primary indication for testing.

The mutations identified were confirmed by Sanger method (ABI PRISM 310 Genetic Analyzer, Applied Biosystems, Warrington, United Kingdom) using the Big Dye Terminator Cycle Sequencing Kit (Applied Biosystems, Warrington, United Kingdom).

Nucleotide numbering reflects cDNA numbering with + 1 corresponding to the A of ATG translation initiation codon in the reference sequence, according to the nomenclature for the description of sequence variants of Human Genome Variation Society[6]. The initiation codon is codon 1.

According with ACMG guidelines, well-established *in vitro* functional studies were performed to assess pathogenicity of VUSs, in particular: spectrophotometric enzyme assay in the suspect of RBC enzyme defects, SDS-PAGE electrophoresis analysis of RBC membrane proteins in the suspect of membrane defects and CDAII, ektacytometry and patch-clamp analysis in case of novel missense variants in *PIEZO1* and *KCNN4* mutations.

---

[1] http://wannovar.wglab.org/

[2] http://sift.jcvi.org/

[3] http://genetics.bwh.harvard.edu/pph2/

[4] http://www.mutationtaster.org/

[5] https://varsome.com/

[6] www.hgvs.org/mutnomen

RESULTS

The 122 patients examined have been divided in two groups: (1) patients who reached a diagnosis after first and second level hematologic investigations, to be confirmed at molecular level; (2) patients with unexplained chronic hemolytic anemia after extensive hematologic investigations.

Patients belonging to group 1 (60 patients, 51 families) were selected among the entire cohort of patients with CHAs studied between 2017 and 2019 basing on complexity of clinical presentation, intra-family clinical variability or incomplete molecular characterization by Sanger (e.g., enzyme defects or CDA cases). All the available patients who didn't reach a definitive diagnosis in the same period were investigated and included in group 2 (62 cases from 54 families, 39 of them studied in our Center, the other 15 received from abroad).

Group 1

We were able to identify pathogenic variants in 38/51 families belonging to Group 1, with an overall sensitivity of the t-NGS of 74% (**Figure 1A** and **Table 1**).

Enzyme Defects

Eight cases from six families had a biochemical diagnosis of enzyme defect; two were confirmed to have pyruvate kinase (PK) deficiency, three from two families glucose-6-phosphate dehydrogenase (G6PD) deficiency and one was found to have one mutation in *GPI* gene, in compound heterozygosity with the polymorphism c.489A > G (p.Gly163 = rs1801015), possibly associated to splicing alterations (Fermo et al., 2019); in two related cases no mutations in *PKLR* gene were found in spite of a slightly decreased PK activity detected by enzymatic assay.

Among these cases with enzyme defects, in four patients from three families a coinheritance of hereditary stomatocytosis was also suspected due to the presence of stomatocytes at peripheral blood smear and left-shifted Osmoscan curve; only in two of them (5-1 and 5-2) in addition to G6PD deficiency we found a mutation in *PIEZO1* gene causing hereditary stomatocytosis.

Congenital Dyserythropoietic Anaemias (CDAs)

Seven cases had a suspect of congenital dyserythropoietic anemia; five were classified as CDAII on the basis of band 3 deglycosylation at SDS-PAGE analysis, and two had an atypical CDA variant with distinct morphological abnormalities at bone marrow examination. Three out of five CDAII cases displayed biallelic mutations in *SEC23B* gene, whereas in two cases (9-1 and 10-1) we were able to find only one pathogenic variant; in the two cases with the atypical CDA we didn't find any pathological mutation in the genes associated to CDAs included in our panel.

Hereditary Stomatocytosis

Thirteen patients from 10 families had a suspect of hereditary stomatocytosis based on presence of stomatocytes at peripheral blood smear and/or abnormalities at ektacytometric curve; eight of them (from six families) were confirmed to have dehydrated stomatocytosis caused by *PIEZO1* gene pathogenic variants. In one case previously diagnosed as overhydrated stomatocytosis

(11-1) based on a clear right-shifted ektacytometric curve, we found mutations in *ABCG8* gene responsible for sitosterolemia. In the remaining case no pathogenic variants in genes associated with hereditary stomatocytosis were identified.

Membrane Defects

We included in this group chronic hemolytic anemias due to altered RBC membrane structural organization, as HS, HE, HPP; 32 cases from 28 families have received such diagnosis based on first and second level laboratory investigations. In 23 cases from 21 families we found pathogenic variants. One patient with a previous diagnosis of HS was found to be affected by dehydrated stomatocytosis (DHSt); in the remaining cases, mutations in genes responsible for HS and HE were identified. In particular among them we studied 3 families with a complex phenotype: case 38-2 (mixed HS/HE phenotype) carried two variants in *SPTA1* and *SLC4A1*, whereas the brother with typical HE (38-1) had only the *SPTA1* variant. Case 20-1, presenting severe HPP with combined 68% spectrin and 56% ankyrin deficiency, had two different mutations in *SPTA1* gene transmitted by the parents, one of them associated in *cis* with the alpha-LELY allele. Finally, case 19-2 (severe HS with 50% band3 deficiency) had 3 different variants in *SLC4A1* gene, the father (19-1) presenting only one missense mutation, thus justifying intra-family clinical variability (**Figure 2**).

Group 2

Inside this group we were able to identify pathogenic variants in 19/54 families (21/62 patients), with an overall sensitivity of the t-NGS of 35% (**Figure 1B** and **Table 1**).

Ten families were found to have pathogenic variants in RBC enzymes. In five of them (2 PK, 2 G6PD and 1 hexokinase (HK) deficiency, cases 39-1, 40-1, 43-1, 44-1, and 41-1) the enzyme activity was previously tested and found to be normal; in other four cases enzyme activity was not tested for sample unavailability. One case had known G6PD deficiency that was initially not considered during the laboratory screening based on chronic hemolysis. Triosephosphate isomerase (TPI) deficiency was diagnosed in a 7 months old child presenting at the time of the study only hemolytic anemia without neuromuscular manifestations.

Three unrelated cases were found to have pathogenic variants in *KCNN4* gene associated with Gardos channelopathy, two of them carrying the common variant p.R352H and one a private variant p.S314P resulting in channel activation (Fermo et al., 2020).

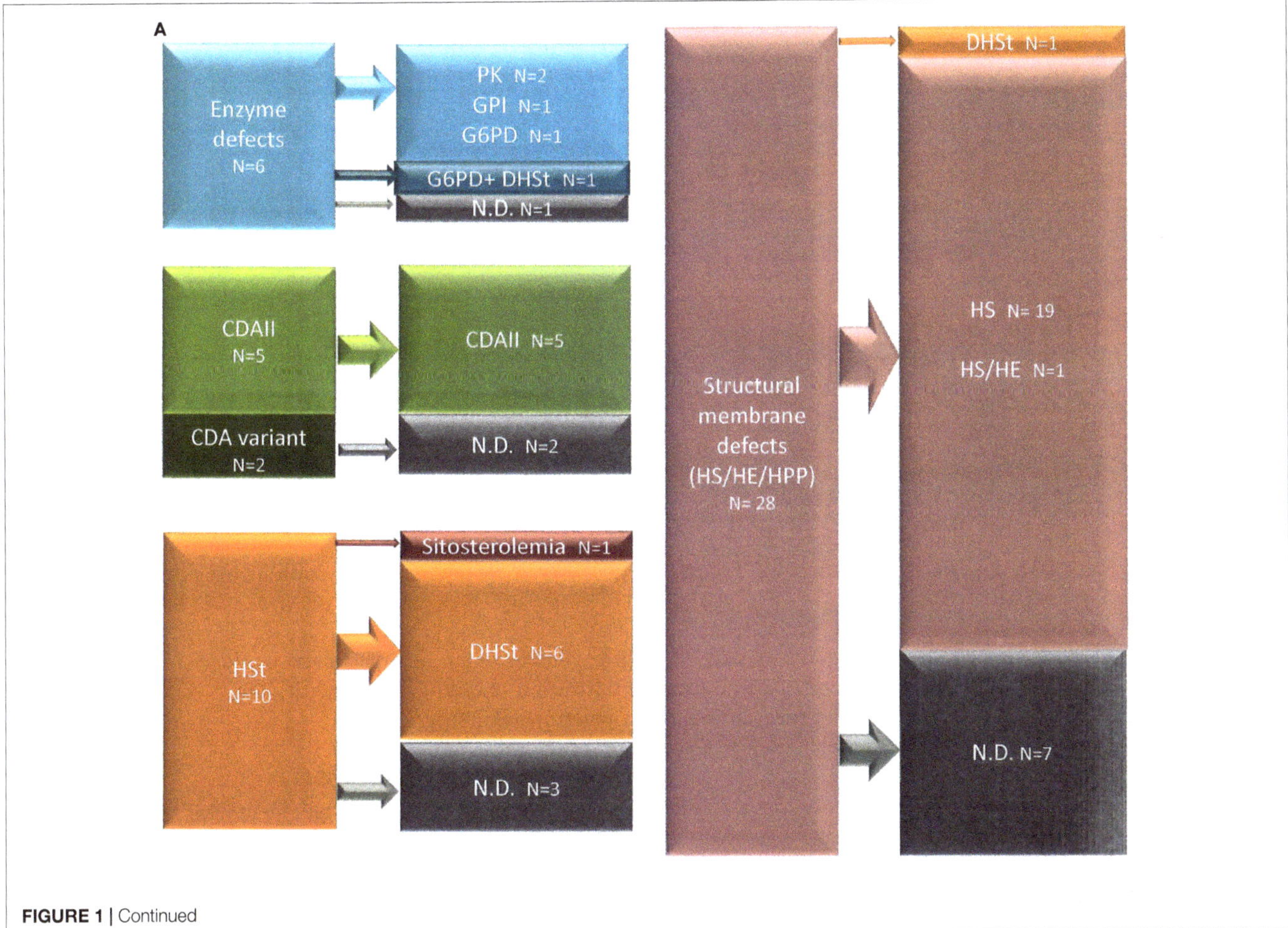

FIGURE 1 | Continued

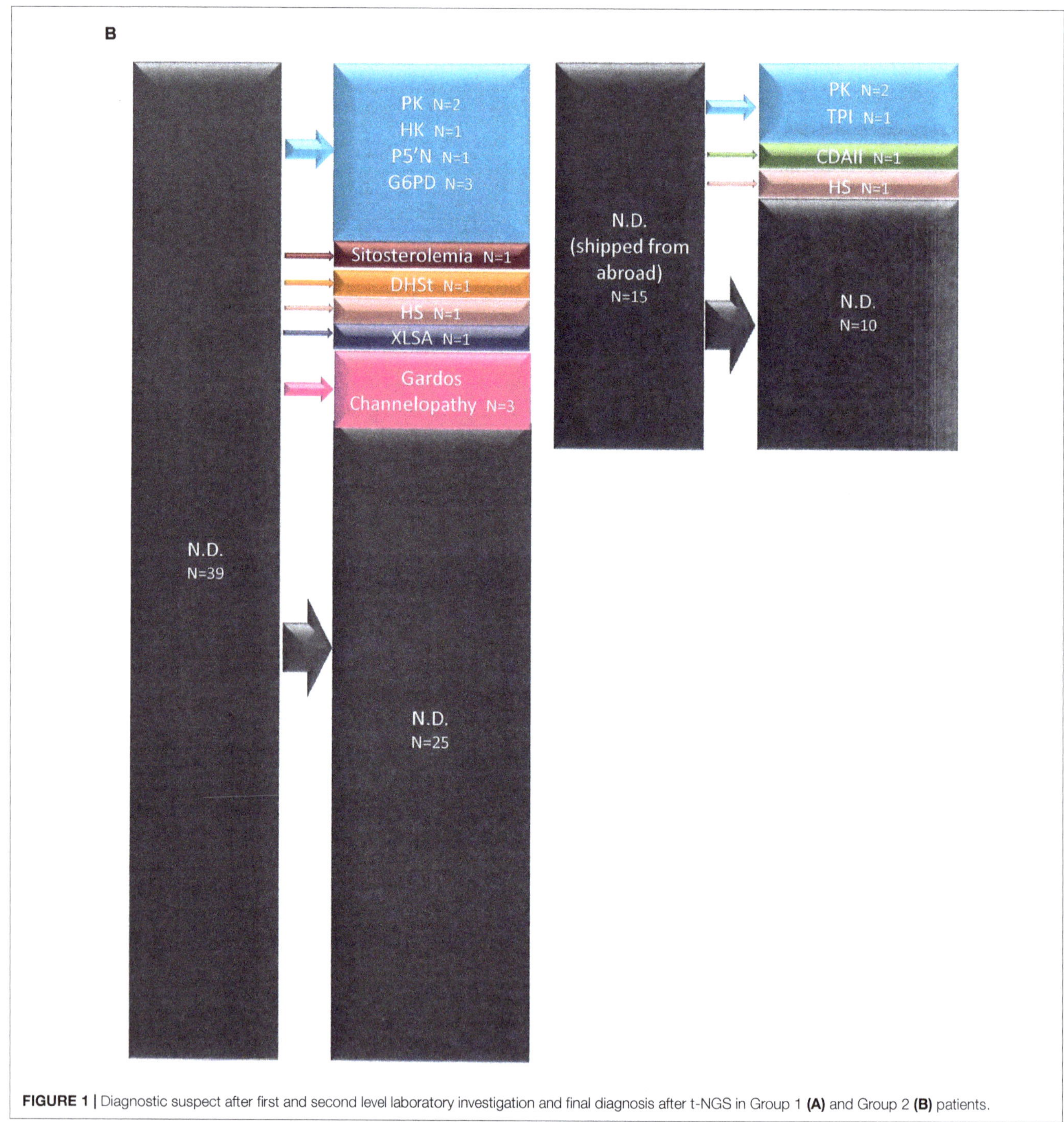

FIGURE 1 | Diagnostic suspect after first and second level laboratory investigation and final diagnosis after t-NGS in Group 1 **(A)** and Group 2 **(B)** patients.

Only one family with two affected siblings (47-1 and 47-2), addressed to our attention in the suspect of PK deficiency in absence of pathogenic variants in *PKLR* gene, carried a PIEZO1 pathogenic variant. Patient 46-1, who had 17% of stomatocytes at peripheral blood smear without any abnormality of the Osmoscan curve, was diagnosed with sitosterolemia due to *ABCG5* mutation p.Y458X. Finally, NGS analysis allowed the diagnosis of X-linked sideroblastic anemia (XLSA) in a woman affected by unexplained macrocytic anemia since birth (case 49-1, *ALAS2* mutation p.H524R). Human androgen receptor gene (HUMARA) analysis showed skewed X-Chromosome inactivation pattern; the mutation was transmitted by the asymptomatic mother.

New Variants Identified

Inside the analyzed cohort we identified 26 new variants, 20 of them associated with structural membrane defects (*SLC4A1*: p.QPLLPQ186Q, p.R870Q, p.T837K, p.K757X, p.G771V, p.S510R, c.1627-1g>t; *SPTB*: p.R52W, p.Q1036PfsX37, p.W1312X, p.Q342X, p.Q760X, p.L883WfsX15, c.4842+1g>c;

TABLE 1 | Results of NGS analysis in 64 patients with congenital hemolytic anemia.

Pt. ID	Laboratory diagnosis	Gene	Transcript	Mutation	Effect	Status	rs	Pathogenicity	Final diagnosis	Additional findings
Group 1										
1-1	Enzyme defect	*PKLR*	NM_000298.5	c.1618 + 2t>c	Abn. splicing	Comp het	rs983394596	5-P	PK deficiency	
		PKLR	NM_000298.5	c.1456C>T	p.R486W	Comp het	rs116100695	5-P		
2-1	Enzyme defect	*PKLR*	NM_000298.5	c.-73G>C	Promoter variant	Comp het		5-P	PK deficiency	*SPTA1* a - lely
		PKLR	NM_000298.5	c.1456C>T	p.R486W	Comp het	rs116100695	5-P		
3-1	Enzyme defect	*GPI*	NM_000175.3	c.1415G>A	p.R472H	Het	rs148811525	5-P	GPI deficiency	
4-1	Enzyme defect	*G6PD*	NM_001042351.1	c.202G>A	p.V68M	Hem	rs1050828	5-P	G6PD deficiency	*SPTA1* a - lely
5-1	Enzyme defect	*G6PD*	NM_001042351.1	c.563C>T	p.S188F	Hom	rs5030868	5-P	G6PD def./DHSt	
		PIEZO1	NM_001142864.3	c.7367G>A	p.R2456H	Het	rs587776988	5-P		
5-2	Enzyme defect	*G6PD*	NM_001042351.1	c.563C>T	p.S188F	Hom	rs5030868	5-P	G6PD def./DHSt	
		PIEZO1	NM_001142864.3	c.7367G>A	p.R2456H	Het	rs587776988	5-P		
6-1	CDAII	*SEC23B*	NM_006363.4	c.40C>T	p.R14W	Comp het	rs121918222	5-P	CDAII	*SPTA1* a - lely
		SEC23B	NM_006363.4	c.689 + 1 g>a	Abn. splicing	Comp het	rs398124226	5-P		
7-1	CDAII	*SEC23B*	NM_006363.4	c.325G>A	p.E109K	Comp het	rs121918221	5-P	CDAII	
		SEC23B	NM_006363.4	c.2270A>C	p.H757P	Comp het	rs1331894607	5-P		
8-1	CDAII	*SEC23B*	NM_006363.4	c.40C>T	p.R14W	Comp het	rs121918222	5-P	CDAII	
		SEC23B	NM_006363.4	c.1821delT	p.H608lfsX7	Comp het		5-P		
9-1	CDAII	*SEC23B*	NM_006363.4	c.235C>T	p.R79X	Het	rs150263014	5-P	CDAII	
10-1	CDAII	*SEC23B*	NM_006363.4	c.40C>T	p.R14W	Het	rs121918222	5-P	CDAII	*SPTA1* a - lely
11-1	HSt	*ABCG8*	NM_022437.2	c.788G>A	p.R263Q	Hom	rs137852990	5-P	Sitosterolemia	
12-1	HSt	*PIEZO1*	NM_001142864.3	c.6574C>A	p.L2192I	Het		4-LP	DHSt	
12-2	HSt	*PIEZO1*	NM_001142864.3	c.6574C>A	p.L2192I	Het		4-LP	DHSt	
13-1	HSt	*PIEZO1*	NM_001142864.3	c.7367G>A	p.R2456H	Het	rs587776988	5-P	DHSt	
14-1	HSt	*PIEZO1*	NM_001142864.3	c.6007G>A	p.A2003T	Het		4-LP	DHSt	
15-1	HSt	*PIEZO1*	NM_001142864.3	c.6328C>T	p.R2110W	Het	rs776531529	5-P	DHSt	*SPTA1* a - lely
16-1	HSt	*PIEZO1*	NM_001142864.3	c.1815G>A	p.M605I	Het		4-LP	DHSt	
16-2	HSt	*PIEZO1*	NM_001142864.3	c.1815G>A	p.M605I	Het		4-LP	DHSt	
17-1	HSt	*PIEZO1*	NM_001142864.3	c.1815G>A	p.M605I	Het		4-LP	DHSt	*SPTA1* a - lely
18-1	Membrane defect	*PIEZO1*	NM_001142864.3	c.1792G>A	p.V598M	Het		4-LP	DHSt	
19-1	Membrane defect	*SLC4A1*	NM_000342.3	c.1469G>A	p.R490H	Het		5-P	HS	
19-2	Membrane defect	*SLC4A1*	NM_000342.3	c.1469G>A	p.R490H	Comp het		5-P	HS	*SPTA1* a - lely
		SLC4A1	**NM_000342.3**	**c.558_572del15bp**	**p.QPLLPQ186Q**	Comp het		4-LP		
		SLC4A1	**NM_000342.3**	**c.1627-1g>t**	**Abn. splicing**	Comp het		5-P		
20-1	Membrane defect	**SPTA1**	**NM_003126.3**	**c.3477 + 1g>c**	**Abn. splicing**	Comp het		5-P	HS	*SPTA1* a - lely
		SPTA1	NM_003126.3	c.3139C>T	p.R1047X	Comp het		5-P		

(Continued)

TABLE 1 | Continued

Pt. ID	Laboratory diagnosis	Gene	Transcript	Mutation	Effect	Status	rs	Pathogenicity	Final diagnosis	Additional findings
21-1	Membrane defect	*SPTB*	NM_001355436.1	c.154C>T	p.R52W	Het		4-LP	HS	*SPTA1* a - lely
22-1	Membrane defect	*SPTB*	NM_001355436.1	c.3106_3107insC	p.Q1036Pfs*37	Het		5-P	HS	*SPTA1* a - lely
23-1	Membrane defect	*SPTB*	NM_001355436.1	c.3936G>A	p.W1312X	Het		5-P	HS	*PKLR* p.R486W
24-1	Membrane defect	*SLC4A1*	NM_000342.3	c.1462G>A	p.V488M	Het	rs28931584	5-P	HS	
25-1	Membrane defect	*SLC4A1*	NM_000342.3	c.2609G>A	p.R870Q	Het	rs746426065	4-LP	HS	*SPTA1* a - lely
26-1	Membrane defect	*SLC4A1*	NM_000342.3	c.2279G>A	p.R760Q	Het	rs121912755	5-P	HS	
27-1	Membrane defect	*SLC4A1*	NM_000342.3	c.2510C>A	p.T837K	Het		4-LP	HS	
28-1	Membrane defect	*SLC4A1*	NM_000342.3	c.2269A>T	p.K757X	Het		5-P	HS	*SEC23B* c.689 + 1G>A *SPTA1* a - lely
29-1	Membrane defect	*ANK1*	NM_000037.3	c.4541delA	p.Y1514SfsX33	Het		5-P	HS	*SPTA1* a - lely
30-1	Membrane defect	*ANK1*	NM_000037.3	c.1430_1431insGTGC	p.A478CfsX17	Het		5-P	HS	
31-1	Membrane defect	*ANK1*	NM_000037.3	c.4057C>T	p.Q1353X	Het		5-P	HS	*SPTA1* a - lely
32-1	Membrane defect	*SPTB*	NM_001355436.1	c.4842 + 1g>c	Abn. splicing	Het		5-P	HS	*SPTA1* a - lely
33-1	Membrane defect	*EPB42*	NM_000119.2	c.922G>C	p.V308L/Abn. splicing	Comp het	rs772330879	4-LP	HS	
		EPB42	NM_000119.2	c.413C>T	p.T138I	Comp het		4-LP		
34-1	Membrane defect	*SLC4A1*	NM_000342.3	c.2312G>T	p.G771V/Abn. splicing	Het		4-LP	HS	
35-1	Membrane defect	*SLC4A1*	NM_000342.3	c.163delC	p.H55TfsX11	Het		5-P	HS	
36-1	Membrane defect	*SPTB*	NM_001355436.1	c.1024C>T	p.Q342X	Het		5-P	HS	
37-1	Membrane defect	*SLC4A1*	NM_000342.3	c.2423G>A	p.R808H	Het	rs866727908	4-LP	HS	
38-1	Membrane defect	*SPTA1*	NM_003126.3	c.460_462dupTTG	p.L155dup	Het	rs757679761	4-LP	HE	
38-2	Membrane defect	*SPTA1*	NM_003126.3	c.460_462dupTTG	p.L155dup	Het	rs757679761	4-LP	HS/HE	
		SLC4A1	NM_000342.3	c.1530C>G	p.S510R	Het		3-VUS		
Group 2										
39-1	n.d	*PKLR*	NM_000298.5	c.1456C>T	p.R486W	Comp het	rs116100695	5-P	PK deficiency	
		PKLR	NM_000298.5	c.1151C>T	p.T384M	Comp het	rs74315362	5-P		
40-1	n.d	*PKLR*	NM_000298.5	c.1456C>T	p.R486W	Comp het	rs116100695	5-P	PK deficiency	*SPTA1* a - lely
		PKLR	NM_000298.5	c.958G>A	p.V320M	Comp het		5-P		
41-1	n.d	*HK1*	NM_033496.2	c.34C>T	p.R12X	Comp het	rs756166032	5-P	HK deficiency	
		HK1	NM_033496.2	c.1351G>C	p.G451R	Comp het		4-LP		
42-1	n.d	*NT5C3A*	NM_016489.12	c.64C>T	p.R22X	Hom	rs753346459	5-P	P5'N deficiency	*SPTA1* a - lely
43-1	n.d	*G6PD*	NM_001042351.1	c.563C>T	p.S188F	Het	rs5030868	5-P	G6PD deficiency	
44-1	n.d	*G6PD*	NM_001042351.1	c.1160G>A	p.R387H	Hem	rs137852321	5-P	G6PD deficiency	
45-1	n.d	*G6PD*	NM_001042351.1	c.1180G>C	p.V394L	Hem	rs137852335	5-P	G6PD deficiency	*NT5C3A* p.N178H
46-1	n.d	*ABCG5*	NM_022436.2	c.1374C>G	p.Y458X	Hom		5-P	Sitosterolemia	

(Continued)

TABLE 1 | Continued

Pt. ID	Laboratory diagnosis	Gene	Transcript	Mutation	Effect	Status	rs	Pathogenicity	Final diagnosis	Additional findings
47-1	n.d	*PIEZO1*	NM_001142864.3	c.7367G>A	p.R2456H	Het	rs587776988	5-P	DHSt	
47-2	n.d	*PIEZO1*	NM_001142864.3	c.7367G>A	p.R2456H	Het	rs587776988	5-P	DHSt	
48-1	n.d	**SPTB**	**NM_001355436.1**	**c.2278C>T**	**p.Q760X**	Het		5-P	HS	*SPTA1* a - lely
49-1	n.d	**ALAS2**	**NM_000032.5**	**c.1571A>G**	**p.H524R**	Het		4-LP	XLSA	*SPTA1* a - lely
50-1	n.d	*KCNN4*	NM_002250.2	c.1055G>A	p.R352H	Het	rs774455945	5-P	Gardos channelopathy	*SPTA1* a - lely
51-1	n.d	*KCNN4*	NM_002250.2	c.940T>C	p.S314P	Het		4-LP	Gardos channelopathy	*AK1* p.G50N
52-1	n.d	*KCNN4*	NM_002250.2	c.1055G>A	p.R352H	Het	rs774455945	5-P	Gardos channelopathy	*SPTA1* a - lely
53-1	n.d	*PKLR*	NM_000298.5	c.581G>C	p.R194P	Hom		4-LP	PK deficiency	
54-1	n.d	**PKLR**	**NM_000298.5**	**c.1591C>A**	**p.R531S**	Hom		4-LP	PK deficiency	
55-1	n.d	*TPI1*	NM_000365.5	c.315G>T	p.E105D	Hom	rs121964845	5-P	TPI deficiency	
56-1	n.d	*SEC23B*	NM_006363.4	c.40C>T	p.R14W	Comp het	rs121918222	5-P	CDAII	
		SEC23B	NM_006363.4	c.490delG	p.V164Wfs*3	Comp het	rs776983439	5-P		
57-1	n.d	**SPTB**	**NM_001355436.1**	**c.2647delC**	**p.L883Wfs*15**	Het		5-P	HS	*SPTA1* a - lely
57-2	n.d	**SPTB**	**NM_001355436.1**	**c.2647delC**	**p.L883Wfs*15**	Het		5-P	HS	

Newly detected variants are reported in bold. Het, heterozygous; Comp het, compound heterozygous; Hom, homozygous; Hem, hemizygous; 5-P, pathogenic; 4-LP, likely pathogenic; 3-VUS, variant of uncertain significance; n.d, not determined. The ClinVar accessions for the new variants here reported are SCV001469066-SCV001469086.

ANK1: p.Y1514SfsX33, p.A478CfsX17, p.Q1353X; *SPTA1*: c.3477+1g>c; *EPB42* p.V308L, p.T138I). The remaining new variants were identified in enzyme genes (*PKLR*: p.R531S; *HK1*: p.R12X, p.G451R; *NT5C3A*: p.R22X), and in genes associated with sitosterolemia (*ABCG5*: p.Y458X) and sideroblastic anemia (*ALAS2*: p.H524R). All new variants were classified as pathogenic or likely pathogenic by *in silico* analysis, according with transmission, clinical phenotype and functional laboratory evidences. Only one VUS variant in *SLC4A1* gene was included and considered as pathogenic because clearly fitting with patients phenotype in a family with a suspected combined transmission of HS and HE (case 38-2) (**Figure 2**).

DISCUSSION

The observed sensitivity of NGS in the two groups of patients analyzed is very different, varying from 74% (patients with a specific diagnostic orientation, Group 1) to 35% (hemolytic patients with no diagnosis after laboratory investigations, or not investigated due long shipping conditions, Group 2).

These data are in line with the available literature (Russo et al., 2018; Bianchi et al., 2020; Steinberg-Shemer and Tamary, 2020) and may explain the heterogeneous sensitivity that, independently from the number of genes analyzed, is often reported for these techniques. For example, the overall sensitivity was around 38.6% in a large series of 57 hemolytic patients with no definitive diagnosis (Roy et al., 2016) and 45.8% in a group of undiagnosed hemolytic patients studied with a combination of two-step t-NGS platforms covering about 100 genes (Russo et al., 2018). Different results are found when NGS is performed on well characterized patient subsets, for example in series of HS patients, where the reported sensitivity increased to 70–100% (Chonat et al., 2019; van Vuren et al., 2019; Xue et al., 2019; Vives-Corrons et al., 2021).

In this view, a sensitivity of 75%, as obtained in our Group 1, may be expected for t-NGS, considering some technical limitations of this technique (loss of variant calling, presence of pathogenic variants in not sequenced regions, copy number variations not always revealed), or to the presence of diseases-causing variants in new genes or genes not included in the panels. In such cases, results could be further improved by considering other technical approaches (WES or WGS) (Mansour-Hendili et al., 2020).

In a portion of cases reported in literature up to 45%, the genetic testing changed the initial diagnosis, in particular when transfusion dependence prevented laboratory investigations from being carried out, and the suspect was based only on clinical features (Russo et al., 2018). In our experience, when detailed clinical history is available together with fresh blood samples collected far from transfusions, first and second level laboratory diagnosis for CHAs has a good level of specificity, as demonstrated by the finding of only two patients who changed diagnosis after NGS analysis inside group 1: one patient previously diagnosed as HS had indeed a pathogenic variant in *PIEZO1* gene, and one case with a long-lasting diagnosis of overhydrated stomatocytosis (done on morphological bases, with

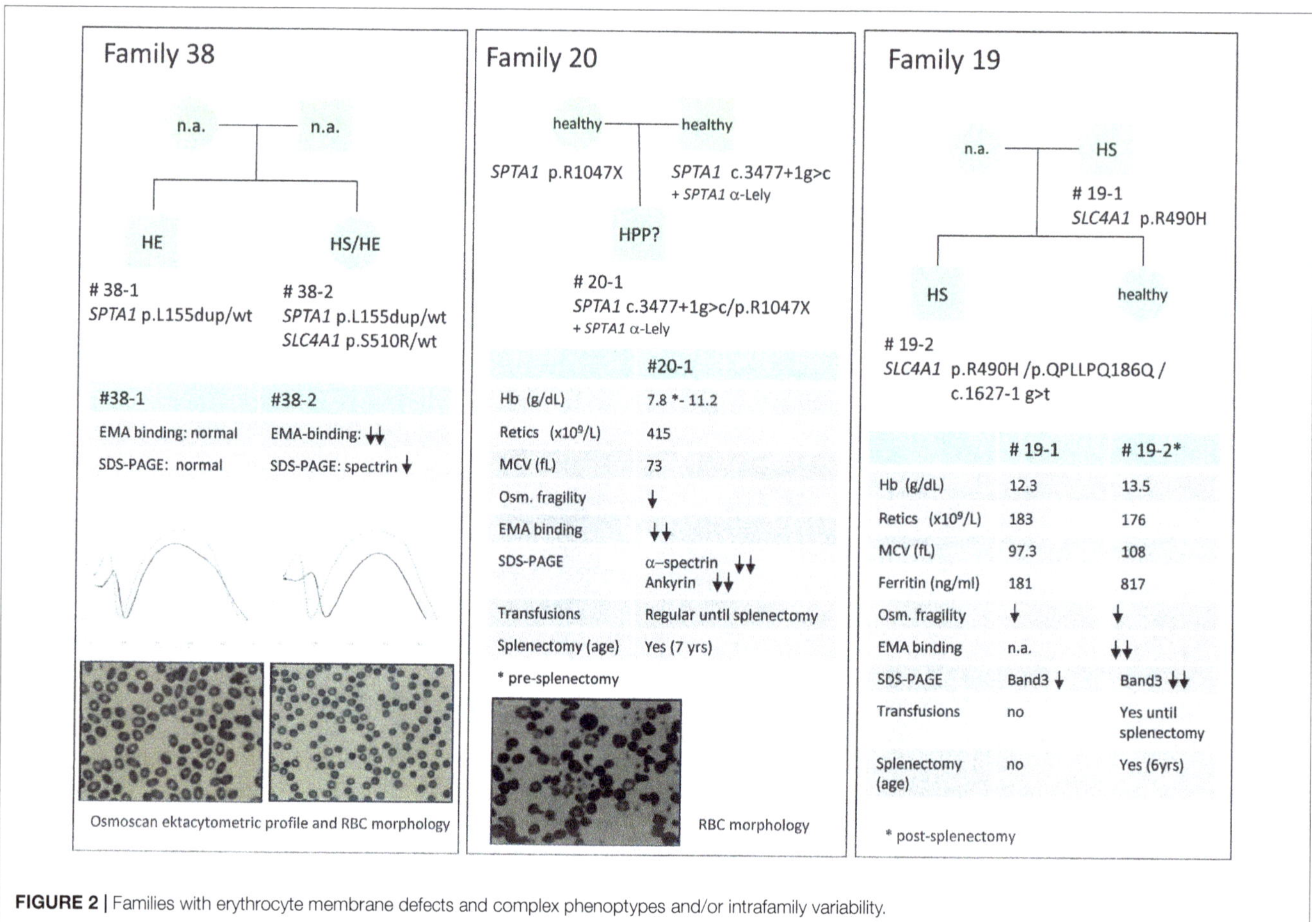

FIGURE 2 | Families with erythrocyte membrane defects and complex phenoptypes and/or intrafamily variability.

18% of stomatocytes and right shift of ektacytometry curve) was indeed a case of familial sitosterolemia with homozygous mutation p.R263Q in *ABCG8* gene (Berge et al., 2000). In both cases the molecular diagnosis changed the clinical management: in the first case the patient was warned about the thrombotic risk associated with splenectomy in hereditary stomatocytosis, and in the second case suggestions were given about dietary reduction of plant sterols consumption, to reduce cardiovascular risk that is associated with familial sitosterolemia (Tzavella et al., 2017). Patients affected by hereditary sitosterolemia may display an extremely variable clinical phenotype, in some cases restricted to abnormal hematological features, leading to misdiagnosis with hereditary stomatocytosis, immune thrombocytopenia (ITP), or Evans syndrome. The mechanisms of hematologic changes in these patients are so far unexplained, and a toxic effect of plant sterols on red blood cells has been hypothezied from *in vitro* experiments (Wang et al., 2012). The finding of two cases in our series, in line with increasing reports of the last years (Wang et al., 2014; Zheng et al., 2019), suggest that sitosterolemia may have an underestimated prevalence, and it should be taken in account in differential diagnosis of CHAs, in particular when stomatocytosis and concomitant macrothrombocytopenia are observed.

Combined defects of red cell membrane and/or metabolism are very rare, and the fact that carriership for a metabolic defect might modify the clinical expression of a membrane defect is still debated (van Zwieten et al., 2015; Fermo et al., 2017). However, the detection of cases with concomitant red cell defects has increased in the last years after the introduction of NGS analysis (Russo et al., 2018; Mansour-Hendili et al., 2020) suggesting that these conditions might not be so rare, also possibly due to protection toward malaria infection. In this series we found the association of G6PD deficiency and *PIEZO1* stomatocytosis in two female patients (mother and daughter) with suspected hereditary stomatocytosis. Among families with membrane defects and complex intra-family phenotypes, we were able to identify a patient with mixed HS/HE who carried two variants in *SPTA1* and *SLC4A1* gene.

Notably, in addition to the disease-causing mutations, four patients also displayed heterozygous variants in other genes associated with hemolytic anemias (see **Table 1**). Although a synergistic effect of heterozygous variants has been hypothesized by some authors (van Zwieten et al., 2015; Pereira et al., 2016), in the presented cases the contribution of these variants to the phenotype remains to be determined.

A careful analysis of Group 2 put in light the limitations of laboratory work-up of CHAs. Only two among the undiagnosed families were found to have HS, suggesting that most of the HS patients can be correctly diagnosed during first and second level screening tools (Fermo et al., 2021). On the other hand, 10 out of the 19 families belonging to this group had an enzyme defect; in five of them enzymopathies were previously excluded because of normal enzyme activity, in the remaining cases the enzyme assays were not performed due to recent transfusions or unavailability of the appropriate sample. RBC enzyme defects are often detected by NGS in cases of unexplained anemia (Roy et al., 2016) or in patients misdiagnosed as CDAII (Shefer Averbuch et al., 2018; Russo et al., 2018), suggesting that this approach may greatly improve the diagnosis of these disorders.

In this series two PK deficient cases carrying known pathogenic variants (p.R486W, p.T384M, and p.V320M) repeatedly showed normal PK activity (14.7 and 18.8 IU/gHb, respectively, reference range 11.9–16.5), in absence of clear explanation (reticulocyte/WBC contamination excluded). Information on sensitivity of spectrophotometric enzyme assays is scanty, in particular for the rarer defects (Bianchi et al., 2019) but, as shown in the present cohort of patients, it will be possibly available in the next future through the systematic analysis of undiagnosed cases by NGS technologies.

Interestingly, we identified a triosephosphate isomerase deficiency in a 7 month child presenting at the time of the study only hemolytic anemia without neuromuscular manifestations. An early diagnosis allowed to inform and prepare the family and to offer bone marrow transplantation (BMT) as a therapeutic option before that neurological damage became evident (Conway et al., 2018); BMT was refused by the parents and the patient developed neurological signs at the age of 18 months. Early diagnosis in red cell disorders by NGS technologies opens interesting landscapes and perspectives in view of future therapeutic approaches for these diseases

	Laboratory testing	Molecular analysis (t-NGS)
HS	EMA-binding test Ectacytometry Others	High molecular heterogenity Consistency with clinical and laboratory features required
HE	Osmotic fragility tests Ectacytometry RBC morphology	High molecular heterogenity Consistency with clinical and laboratory features required
HSt- *PIEZO1*	Rbc morphology; Ektacytometry Always requiring molecular testing to confirm diagnosis	Highly polymorphic gene Functional tests mandatory in presence of new variants
HSt-*KCNN4*	Absence of specific laboratory markers	
RBC enzyme defects	RBC enzyme assay . Always requiring molecular testing to confirm diagnosis	
Familial sitosterolemia	Complete blood count	
Atypical conditions		

FIGURE 3 | Advandages and limitations of laboratory approach vs. t-NGS technologies in congenital hemolytic anemias. The laboratory tests/parameters specifically useful for diagnosis of different diseases are reported. Brilliant green, exhaustive diagnostic approach; Green, requiring other approaches to confirm the diagnosis; gray, insufficient laboratory markers to reach the diagnosis.

(Garcia-Gomez et al., 2016; Grace et al., 2019; Dessy-Rodriguez et al., 2020; Hrizo et al., 2021).

Among enzymopathies, we also identified 3 cases with undiagnosed G6PD deficiency, an heterozygous woman and two men with G6PD variants associated with chronic hemolysis, conditions in some cases underdiagnosed. A correct diagnosis enabled to give specific recommendation regarding contraindication for drugs associated with hemolytic anemia in G6PD -deficient patients (Luzzatto et al., 2020).

Differently from the *PIEZO1* mutated cases, all the *KCNN4* mutated patients belonged to group 2, confirming that molecular testing, and in particular NGS approaches, are the only diagnostic tools able to identify Gardos channelopathy, given the absence of a specific clinical or biological phenotype in this disorder; in fact, although ektacytometry, and some RBC and reticulocytes indices (Picard et al., 2019, 2021) may help to differentiate *PIEZO1* from *KCNN4* cases, no hematological parameter or specific laboratory test enable to differentiate Gardos channelopathy from other CHAs.

In conclusion, by considering the entire cohort here studied, the implementation of diagnostic work-up with t-NGS analysis increased the number of diagnoses of patients with unexplained anemia up to 35%. As reported in **Figure 3**, cytoskeleton defects

(HS/HE) are well diagnosed by laboratory approaches, deserving NGS only to atypical cases; on the other hand, the diagnosis of Gardos channelopathy, some RBC enzyme defects, familial siterosterolemia, X-linked defects in females and other rare and ultra-rare conditions definitely benefits of NGS approaches. However, it should be always considered that a set of functional tests is required to validate the molecular results and assess pathogenicity of the new identified variants, in particular in highly polymorphic genes such as *PIEZO1*.

AUTHOR CONTRIBUTIONS

EF and PB designed the study and performed NGS analysis and molecular studies. CV, AnZ, and AM performed laboratory investigations. WB, AZ, SP, EK, and JG followed-up patients. EF and PB drafted the manuscript, all other authors revised and approved the manuscript. All authors contributed to the article and approved the submitted version.

SUPPLEMENTARY MATERIAL

Supplementary Table 1 | Genes associated with CHAs investigated through a 43 genes-targeted Next Generation Sequencing panel.

REFERENCES

Agarwal, A. M., Nussenzveig, R. H., Reading, N. S., Patel, J. L., Sangle, N., Salama, M. E., et al. (2016). Clinical utility of next-generation sequencing in the diagnosis of hereditary haemolytic anaemias. *Br. J. Haematol.* 174, 806–814. doi: 10.1111/bjh.14131

Berge, K. E., Tian, H., Graf, G. A., Yu, L., Grishin, N. V., Schultz, J., et al. (2000). Accumulation of dietary cholesterol in sitosterolemia caused by mutations in adjacent ABC transporters. *Science* 290, 1771–1775.

Beutler, E., Blume, K. G., Kaplan, J. C., Lohr, G. W., Ramot, B., and Valentine, W. N. (1977). International Committee for Standardization in Haematology: recommended methods for red-cell enzyme analysis. *Br. J. Haematol.* 35, 331–340.

Bianchi, P., Fermo, E., Glader, B., Kanno, H., Agarwal, A., Barcellini, W., et al. (2019). Addressing the diagnostic gaps in pyruvate kinase deficiency: Consensus recommendations on the diagnosis of pyruvate kinase deficiency. *Am. J. Hematol.* 94, 149–161. doi: 10.1002/ajh.25325

Bianchi, P., Fermo, E., Vercellati, C., Marcello, A. P., Porretti, L., Cortelezzi, A., et al. (2012). Diagnostic power of laboratory tests for hereditary spherocytosis: a comparison study in 150 patients grouped according to molecular and clinical characteristics. *Haematologica* 97, 516–523. doi: 10.3324/haematol.2011.052845

Bianchi, P., Vercellati, C., and Fermo, E. (2020). How will next generation sequencing (NGS) improve the diagnosis of congenital hemolytic anemia?. *Ann. Transl. Med.* 8:268. doi: 10.21037/atm.2020.02.151

Choi, H. O., Choi, Q., Kim, J. A., Im, K. O., Park, S. N., Park, Y., et al. (2019). Molecular diagnosis of hereditary spherocytosis by multi-gene target sequencing in Korea: matching with osmotic fragility test and presence of spherocyte. *Orphanet. J. Rare Dis.* 14:114. doi: 10.1186/s13023-019-1070-0

Chonat, S., Risinger, M., Sakthivel, H., Niss, O., Rothman, J. A., Hsieh, L., et al. (2019). The Spectrum of SPTA1-Associated Hereditary Spherocytosis. *Front. Physiol.* 10:815. doi: 10.3389/fphys.2019.00815

Conway, A. J., Brown, F. C., Hortle, E. J., Burgio, G., Foote, S. J., Morton, C. J., et al. (2018). Bone marrow transplantation corrects haemolytic anaemia in a novel ENU mutagenesis mouse model of TPI deficiency. *Dis. Model Mech.* 11:dmm034678. doi: 10.1242/dmm.034678

Del Orbe Barreto, R., Arrizabalaga, B., De la Hoz, A. B., García-Orad, Á, Tejada, M. I, Garcia-Ruiz, J. C., et al. (2016). Detection of new pathogenic mutations in patients with congenital haemolytic anaemia using next-generation sequencing. *Int. J. Lab. Hematol.* 38, 629–338. doi: 10.1111/ijlh.12551

Dessy-Rodriguez, M., Fañanas-Baquero, S. A., Venturi, V., Payán-Pernía, S., Tornador, C., Hernandez, G., et al. (2020). Quintana Bustamante, O. Modelling Congenital Dyserythropoietic Anemia Type II through Gene Editing in Hematopoietic Stem and Progenitor Cells. *Blood* 136(Suppl. 1):27.

Fairbanks, G., Steck, T. L., and Wallach, D. F. H. (1971). Electrophoretic analysis of the major polypeptides of the human erythrocyte membrane. *Biochemistry* 10, 2606–2617.

Fermo, E., Monedero-Alonso, D., Petkova-Kirova, P., Makhro, A., Pérès, L., Bouyer, G., et al. (2020). Gardos channelopathy: functional analysis of a novel KCNN4 variant. *Blood Adv.* 4, 6336–6341. doi: 10.1182/bloodadvances.2020003285

Fermo, E., Vercellati, C., and Bianchi, P. (2021). Screening tools for hereditary hemolytic anemia: new concepts and strategies. *Expert Rev. Hematol.* 1:11. doi: 10.1080/17474086.2021.1886919

Fermo, E., Vercellati, C., Marcello, A. P., Zaninoni, A., Aytac, S., Cetin, M., et al. (2019). Clinical and Molecular Spectrum of Glucose-6-Phosphate Isomerase Deficiency. Report of 12 New Cases. *Front. Physiol.* 10:467. doi: 10.3389/fphys.2019.00467

Fermo, E., Vercellati, C., Marcello, A. P., Zaninoni, A., van Wijk, R., Mirra, N., et al. (2017). Hereditary Xerocytosis due to Mutations in PIEZO1 Gene Associated with Heterozygous Pyruvate Kinase Deficiency and Beta-Thalassemia Trait in Two Unrelated Families. *Case Rep. Hematol.* 2017:2769570. doi: 10.1155/2017/2769570

Garcia-Gomez, M., Calabria, A., Garcia-Bravo, M., Benedicenti, F., Kosinski, P., López-Manzaneda, S., et al. (2016). Safe and Efficient Gene Therapy for Pyruvate Kinase Deficiency. *Mol. Ther.* 24, 1187–1198. doi: 10.1038/mt.2016.87

Grace, R. F., Rose, C., Layton, D. M., Galactéros, F., Barcellini, W., Morton, D. H., et al. (2019). Safety and Efficacy of Mitapivat in Pyruvate Kinase Deficiency. *N. Engl. J. Med.* 381, 933–944. doi: 10.1056/NEJMoa1902678

Hrizo, S. L., Eicher, S. L., Myers, T. D., McGrath, I., Wodrich, A. P. K., Venkatesh, H., et al. (2021). Identification of protein quality control regulators using a Drosophila model of TPI deficiency. *Neurobiol. Dis.* 15:105299. doi: 10.1016/j.nbd.2021.105299

Kedar, P. S., Gupta, V., Dongerdiye, R., Chiddarwar, A., Warang, P. P., and Madkaikar, M. R. (2019a). Molecular diagnosis of unexplained haemolytic anaemia using targeted next-generation sequencing panel revealed (p.Ala337Thr) novel mutation in GPI gene in two Indian patients. *J. Clin. Pathol.* 72, 81–85. doi: 10.1136/jclinpath-2018-205420

Kedar, P. S., Harigae, H., Ito, E., Muramatsu, H., Kojima, S., Okuno, Y., et al. (2019b). Study of pathophysiology and molecular characterization of congenital anemia in India using targeted next-generation sequencing approach. *Int. J. Hematol.* 110, 618–626. doi: 10.1007/s12185-019-02716-9

King, M. J., Garçon, L., Hoyer, J. D., Iolascon, A., Picard, V., Stewart, G., et al. (2015). International Council for Standardization in Haematology.ICSH guidelines for the laboratory diagnosis of nonimmune hereditary red cell membrane disorders. *Int. J. Lab. Hematol.* 37, 304–325. doi: 10.1111/ijlh. 12335

Kopanos, C., Tsiolkas, V., Kouris, A., Chapple, C. E., Albarca Aguilera, M., Meyer, R., et al. (2019). VarSome: the human genomic variant search engine. *Bioinformatics* 35, 1978–1980. doi: 10.1093/bioinformatics/bty897

Laemmli, U. K. (1970). Cleavage of structural proteins during the assembly of the head of bacteriophage T4. *Nature* 227, 680–685.

Luzzatto, L., Ally, M., and Notaro, R. (2020). Glucose-6-phosphate dehydrogenase deficiency. *Blood* 136, 1225–1240. doi: 10.1182/blood.2019000944

Mansour-Hendili, L., Aissat, A., Badaoui, B., Sakka, M., Gameiro, C., Ortonne, V., et al. (2020). Exome sequencing for diagnosis of congenital hemolytic anemia. *Orphanet. J. Rare Dis.* 15:180. doi: 10.1186/s13023-020-01425-5

Mariani, M., Barcellini, W., Vercellati, C., Marcello, A. P., Fermo, E., Pedotti, P., et al. (2008). Clinical and hematologic features of 300 patients affected by hereditary spherocytosis grouped according to the type of the membrane protein defect. *Haematologica* 93, 1310–1317.

Niss, O., Chonat, S., Dagaonkar, N., Almansoori, M. O., Kerr, K., Rogers, Z. R., et al. (2016). Genotype-phenotype correlations in hereditary elliptocytosis and hereditary pyropoikilocytosis. *Blood Cells Mol. Dis.* 61, 4–9. doi: 10.1016/j. bcmd.2016.07.003

Pereira, J., Bento, C., Maco, L., Gonzalez, A., Vagace, J., and Ribeiro, M. L. (2016). Congenital dyserythropoietic anemia associated to a GATA1 mutation aggravated by pyruvate kinase deficiency. *Ann. Hematol.* 95, 1551–1553. doi: 10.1007/s00277-016-2720-0

Picard, V., Guitton, C., Mansour-Hendili, L., Jondeau, B., Bendélac, L., Denguir, M., et al. (2021). Rapid Gardos Hereditary Xerocytosis Diagnosis in 8 Families Using Reticulocyte Indices. *Front. Physiol.* 11:602109. doi: 10.3389/fphys.2020. 602109

Picard, V., Guitton, C., Thuret, I., Rose, C., Bendelac, L., Ghazal, K., et al. (2019). Clinical and biological features in PIEZO1-hereditary xerocytosis and Gardos channelopathy: a retrospective series of 126 patients. *Haematologica* 104, 1554–1564. doi: 10.3324/haematol.2018.20 5328

Richards, S., Aziz, N., Bale, S., Bick, D., Das, S., Gastier-Foster, J., et al. (2015). ACMG Laboratory Quality Assurance Committee. Standards and guidelines for the interpretation of sequence variants: a joint consensus recommendation of the American College of Medical Genetics and Genomics and the Association for Molecular Pathology. *Genet. Med.* 17, 405–424. doi: 10.1038/gim. 2015.30

Roy, N. B. A., Wilson, E. A., Henderson, S., Wray, K., Babbs, C., Okoli, S., et al. (2016). A novel 33-Gene targeted resequencing panel provides accurate, clinical-grade diagnosis and improves patient management for rare inherited anaemias. *Br. J. Haematol.* 175, 318–330. doi: 10.1111/bjh. 14221

Russo, R., Andolfo, I., Manna, F., Gambale, A., Marra, R., Rosato, B. E., et al. (2018). Multi-gene panel testing improves diagnosis and management of patients with hereditary anemias. *Am. J. Hematol.* 93, 672–682. doi: 10.1002/ajh. 25058

Russo, R., Marra, R., Rosato, B. E., Iolascon, A., and Andolfo, I. (2020). Genetics and Genomics Approaches for Diagnosis and Research Into Hereditary Anemias. *Front. Physiol.* 11:613559. doi: 10.3389/fphys.2020.61 3559

Shefer Averbuch, N., Steinberg-Shemer, O., Dgany, O., Krasnov, T., Noy-Lotan, S., Yacobovich, J., et al. (2018). Targeted next generation sequencing for the diagnosis of patients with rare congenital anemias. *Eur. J. Haematol.* 101, 297–304. doi: 10.1111/ejh.13097

Steinberg-Shemer, O., and Tamary, H. (2020). Impact of next-generation sequencing on the diagnosis and treatment of congenital anemias. *Mol. Diagn Ther.* 4, 397–407. doi: 10.1007/s40291-020-00478-3

Svidnicki, M. C. C. M., Zanetta, G. K., Congrains-Castillo, A., Costa, F. F., and Saad, S. T. O. (2020). Targeted next-generation sequencing identified novel mutations associated with hereditary anemias in Brazil. *Ann. Hematol.* 99, 955–962. doi: 10.1007/s00277-020-03986-8

Tzavella, E., Hatzimichael, E., Kostara, C., Bairaktari, E., Elisaf, M., and Tsimihodimos, V. (2017). Sitosterolemia: A multifaceted metabolic disorder with important clinical consequences. *J. Clin. Lipidol.* 11, 1095–1100. doi: 10. 1016/j.jacl.2017.04.116

van Vuren, A., van der Zwaag, B., Huisjes, R., Lak, N., Bierings, M., Gerritsen, E., et al. (2019). The complexity of genotype-phenotype correlations in hereditary spherocytosis: a cohort of 95 patients. *HemaSphere* 3:4.

van Zwieten, R., van Oirschot, B. A., Veldthuis, M., Dobbe, J. G., Streekstra, G. J., van Solinge, W. W., et al. (2015). Partial pyruvate kinase deficiency aggravates the phenotypic expression of band 3 deficiency in a family with hereditary spherocytosis. *Am. J. Hematol.* 90, E35–E39. doi: 10.1002/ajh. 23899

Vives-Corrons, J. L., Krishnevskaya, E., Rodriguez, I. H., and Ancochea, A. (2021). Characterization of hereditary red blood cell membranopathies using combined targeted next-generation sequencing and osmotic gradient ektacytometry. *Int. J. Hematol.* 113, 163–174. doi: 10.1007/s12185-020-03010-9

Wang, G., Cao, L., Wang, Z., Jiang, M., Sun, X., Bai, X., et al. (2012). Macrothrombocytopenia/stomatocytosis specially associated with phytosterolemia. *Clin. Appl. Thromb. Hemost.* 18, 582–587.

Wang, Z., Cao, L., Su, Y., Wang, G., Wang, R., Yu, Z., et al. (2014). Specific macrothrombocytopenia/hemolytic anemia associated with sitosterolemia. *Am. J. Hematol.* 89, 320–324.

Xue, J., He, Q., Xie, X., Su, A., and Cao, S. (2019). Clinical utility of targeted gene enrichment and sequencing technique in the diagnosis of adult hereditary spherocytosis. *Ann. Transl. Med.* 7:527.

Zaninoni, A., Fermo, E., Vercellati, C., Consonni, D., Marcello, A. P., et al. (2018). Use of Laser Assisted Optical Rotational Cell Analyzer (LoRRca MaxSis) in the Diagnosis of RBC Membrane Disorders, Enzyme Defects, and Congenital Dyserythropoietic Anemias: A Monocentric Study on 202 Patients. *Front. Physiol.* 9:451. doi: 10.3389/fphys.2018.00451

Zheng, J., Ma, J., Wu, R. H., Zhang, X., Su, Y., Zhang, R., et al. (2019). Unusual presentations of sitosterolemia limited to hematological abnormalities: A report of four cases presenting with stomatocytic anemia and thrombocytopenia with macrothrombocytes. *Am. J. Hematol.* 94, E124–E127.

6

Hepcidin and Anemia: A Tight Relationship

*Alessia Pagani[1], Antonella Nai[1,2], Laura Silvestri[1,2] and Clara Camaschella[1]**

[1]*Division of Genetics and Cell Biology, San Raffaele Scientific Institute, Milan, Italy,* [2]*Vita-Salute San Raffaele University, Milan, Italy*

**Correspondence:*
Clara Camaschella
camaschella.clara@hsr.it

Hepcidin, the master regulator of systemic iron homeostasis, tightly influences erythrocyte production. High hepcidin levels block intestinal iron absorption and macrophage iron recycling, causing iron restricted erythropoiesis and anemia. Low hepcidin levels favor bone marrow iron supply for hemoglobin synthesis and red blood cells production. Expanded erythropoiesis, as after hemorrhage or erythropoietin treatment, blocks hepcidin through an acute reduction of transferrin saturation and the release of the erythroblast hormone and hepcidin inhibitor erythroferrone. Quantitatively reduced erythropoiesis, limiting iron consumption, increases transferrin saturation and stimulates hepcidin transcription. Deregulation of hepcidin synthesis is associated with anemia in three conditions: iron refractory iron deficiency anemia (IRIDA), the common anemia of acute and chronic inflammatory disorders, and the extremely rare hepcidin-producing adenomas that may develop in the liver of children with an inborn error of glucose metabolism. Inappropriately high levels of hepcidin cause iron-restricted or even iron-deficient erythropoiesis in all these conditions. Patients with IRIDA or anemia of inflammation do not respond to oral iron supplementation and show a delayed or partial response to intravenous iron. In hepcidin-producing adenomas, anemia is reverted by surgery. Other hepcidin-related anemias are the "iron loading anemias" characterized by ineffective erythropoiesis and hepcidin suppression. This group of anemias includes thalassemia syndromes, congenital dyserythropoietic anemias, congenital sideroblastic anemias, and some forms of hemolytic anemias as pyruvate kinase deficiency. The paradigm is non-transfusion-dependent thalassemia where the release of erythroferrone from the expanded pool of immature erythroid cells results in hepcidin suppression and secondary iron overload that in turn worsens ineffective erythropoiesis and anemia. In thalassemia murine models, approaches that induce iron restriction ameliorate both anemia and the iron phenotype. Manipulations of hepcidin might benefit all the above-described anemias. Compounds that antagonize hepcidin or its effect may be useful in inflammation and IRIDA, while hepcidin agonists may improve ineffective erythropoiesis. Correcting ineffective erythropoiesis in animal models ameliorates not only anemia but also iron homeostasis by reducing hepcidin inhibition. Some targeted approaches are now in clinical trials: hopefully they will result in novel treatments for a variety of anemias.

Keywords: anemia, iron, hepcidin, erythropoiesis, inflammation

INTRODUCTION

Anemia is one of the most common disorders worldwide and anemia due to iron deficiency is the prevalent form according to multiple analyses (review in Camaschella, 2019). This type of anemia results from the total body iron deficiency and the inability to supply the large amount of iron that the bone marrow consumes to produce an adequate number of red blood cells in order to maintain tissue oxygenation.

The iron availability is controlled by the liver peptide hormone hepcidin. The body iron increase causes the production of hepcidin, which is released in the circulation and acts on its receptor ferroportin, a transmembrane iron exporter protein highly expressed on enterocyte, macrophages, and hepatocytes. Hepcidin reduces the iron entry to plasma from absorptive duodenal cells and iron recycling macrophages by blocking iron export (Aschemeyer et al., 2018) and by degrading the iron exporter ferroportin (Nemeth et al., 2004). By regulating plasma iron and systemic iron homeostasis, the hepcidin/ferroportin axis strongly affects erythropoiesis, hence the possible development of anemia.

THE IRON-ERYTHROPOIESIS CONNECTION

The process of red blood cells production consumes approximately 80% of circulating iron for hemoglobin synthesis of maturing erythroblasts. Most iron (20–25 mg/daily) is recycled by macrophages, while a limited amount (1–2 mg daily) derives from intestinal absorption. The kidney hormone erythropoietin (EPO) controls the proliferation of erythroid progenitors, especially of CFU-e and at a lower degree of BFU-e, and the early phase of terminal erythropoiesis, while iron needs are increased in the late differentiation stages from proerythroblasts to reticulocyte, for the synthesis of heme incorporated into hemoglobin (Muckenthaler et al., 2017).

Hepcidin regulation requires a crosstalk between liver endothelial sinusoidal cells (LSEC) that produce the bone morphogenetic proteins (BMPs) to activate the BMP-SMAD pathway and hepatocytes that produce and release hepcidin (Babitt et al., 2006; Rausa et al., 2015). BMP6 and BMP2 are the most important BMPs that upregulate hepcidin, while BMP6 expression is iron dependent (Andriopoulos et al., 2009; Meynard et al., 2009) BMP2 appears less iron-responsive (Canali et al., 2017; Koch et al., 2017).

Hepcidin levels are low in absolute iron deficiency and iron deficiency anemia. In these conditions, the iron stores are exhausted and the BMP-SMAD signaling is switched off at multiple levels. First, BMP6 expression is suppressed; next, the activity of TMPRSS6, a protease that cleaves the BMP co-receptor hemojuvelin (Silvestri et al., 2008), is strongly increased (Lakhal et al., 2011); and third, histone deacetylase3 (HDAC3) suppresses the hepcidin locus (Pasricha et al., 2017). In conditions of iron deficiency, the reduction of hepcidin production is an adaptation mechanism that facilitates dietary and pharmacological iron absorption (Camaschella and Pagani, 2018).

When anemia is severe, the coexisting hypoxia stimulates erythropoiesis through increased kidney synthesis and release of EPO. This leads to suppression of hepcidin transcription by erythroferrone (ERFE), an EPO target gene produced by erythroblasts (Kautz et al., 2014), by molecules (e.g., PDGF-BB) released by other tissues (Sonnweber et al., 2014), and likely by soluble components of transferrin receptors (TFR), sTFR1 (Beguin, 2003), and sTFR2 (Pagani et al., 2015). The final aim is to supply enough iron for the needs of an expanded erythropoiesis.

ANEMIAS WITH ABNORMAL HEPCIDIN LEVELS

Anemias may be classified on the basis of hepcidin levels as anemias with high and low hepcidin. It is intuitive that persistently high hepcidin levels, by blocking iron absorption, cause iron deficiency anemia because of decreased iron supply to erythropoiesis. Conversely, ineffective erythropoiesis characterizes the so-called *iron-loading anemias* that have low hepcidin levels and iron overload. These two groups of anemias are the outcome of opposite pathophysiology mechanisms (**Figure 1**). In the first group, anemia is due to the inhibitory effect exerted by hepcidin on iron absorption and recycling that leads to systemic iron deficiency; in the second group, anemia is due to hepcidin suppression by an expanded abnormal erythropoiesis (Camaschella and Nai, 2016).

Anemia Associated With High Hepcidin Levels

This group includes two inherited rare disorders (iron refractory iron deficiency anemia and hepcidin-producing adenomas in an inborn error of glucose metabolism) and an acquired common condition: anemia of inflammation (**Table 1**).

Iron Refractory Iron Deficiency Anemia

Iron refractory iron deficiency anemia (IRIDA) is a rare recessive disorder characterized by hypochromic microcytic anemia, low transferrin saturation, and inappropriately normal/high hepcidin levels. It is caused by mutations of *TMPRSS6* (Finberg et al., 2008), a gene that encodes the type II serine protease, matriptase-2 (Du et al., 2008). Mutations of *TMPRSS6* are spread along the gene and may affect different domains especially the catalytic domain (De Falco et al., 2014). This transmembrane protease, highly expressed in the liver, inhibits hepcidin transcription by cleaving the cell surface BMP co-receptor hemojuvelin, thus attenuating the BMP signaling and hepcidin synthesis (Silvestri et al., 2008). TMPRSS6 function is essential in iron deficiency to allow the compensatory mechanism of increased iron absorption.

IRIDA is present since birth and usually diagnosed in childhood. Compared with classic iron deficiency, iron parameters are atypical and raise the suspicion of the disease. The percent saturation of transferrin is strongly reduced (less than 10%) as in other forms of iron deficiency; however, at variance with iron deficiency, levels of serum ferritin are normal/increased (Camaschella, 2013; De Falco et al., 2013). This reflects an increased ferritin accumulation in macrophages, due to the high hepcidin levels that induce store iron sequestration.

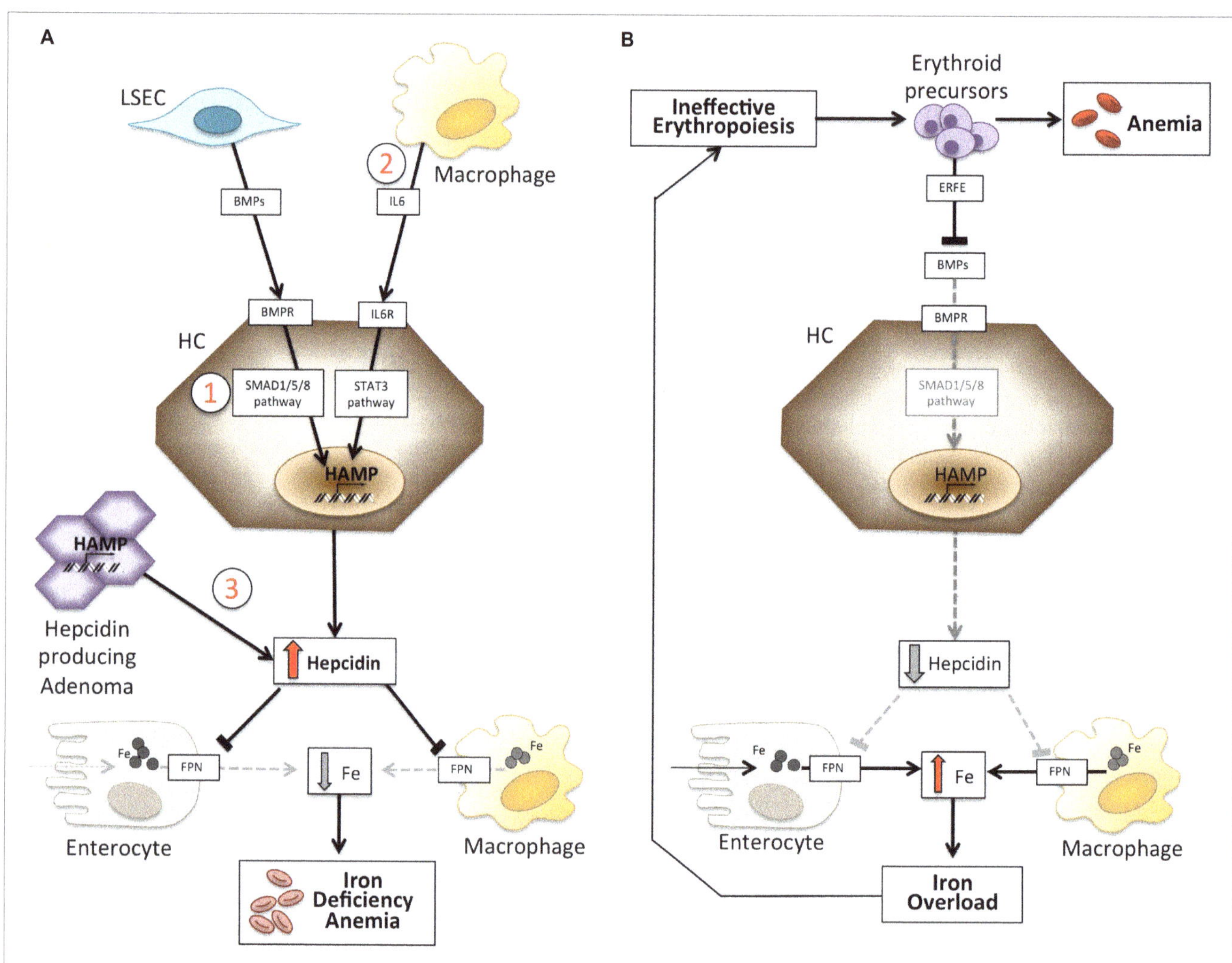

FIGURE 1 | Schematic representation of mechanisms of anemias with high (left panel) and low hepcidin (right panel). Panel **(A)**. Molecular pathogenesis of anemia associated with high hepcidin levels. LSEC, liver sinusoidal endothelial cells producing bone morphogenetic proteins (BMPs); BMPRs, BMP receptors; IL6, interleukin 6; HC, hepatocytes; HAMP, hepcidin gene. Fe, iron; FPN, ferroportin; 1, IRIDA; 2, Anemia of inflammation; 3, hepcidin producing adenoma. Panel **(B)**. Molecular pathogenesis of hepcidin variation in anemias due to ineffective erythropoiesis. ERFE, erythroferrone sequestering BMPs. Other mechanisms inhibiting hepcidin in this type of anemia, as decrease of transferrin saturation and hypoxia, are not shown. See text for details.

None of the tests proposed for IRIDA diagnosis covers 100% of the cases. The genetic test identifies that *TMPRSS6* mutations, that in some cases (non-sense, frame-shift, and splicing mutations), are clearly causal. In other cases, as for previously unreported missense mutations, functional studies are needed to demonstrate causality (Silvestri et al., 2013). However, these tests are scarcely available. Serum hepcidin levels are usually increased/normal, independently of iron deficiency, and consistent with high/normal ferritin. It is important to exclude inflammation by concomitantly dosing C-reactive protein.

Some patients with a phenotype of refractory iron deficiency have been reported to have a single *TMPRSS6* mutated allele; here, the debate is whether they should be considered IRIDA or not. A spectrum of conditions can be envisaged ranging from classic severe IRIDA due to homozygous or compound heterozygous *TMPRSS6* mutations to increased susceptibility to iron deficiency conferred by single mutations/polymorphic changes.

One approach proposed to predict classic IRIDA is hepcidin normalization on other iron parameters, as ratios transferrin saturation (Tsat)/log hepcidin or Tsat/log Ferritin (Donker et al., 2016). According to other authors, most patients with a severe IRIDA phenotype have biallelic *TMPRSS6* mutations and, when unidentified, the second allele may be genetically occult (Heeney et al., 2018). In general terms, subjects with a single allele have a milder phenotype than those with two mutations and respond better to iron treatment (Donker et al., 2016). Interestingly, several *TMPRSS6* SNPs have been shown to provide susceptibility to iron deficiency in some populations (An et al., 2012) and in blood donors (Sorensen et al., 2019).

A digenic inheritance has been reported in a 5-year-old female originally found to have an atypical IRIDA genotype with one *TMPRSS6* (I212T) causal and one (R271Q) silent mutation (De Falco et al., 2014). She was later diagnosed *Fibrodysplasia ossificans progressiva* (FOP), a rare dominant

TABLE 1 | Anemias classified according to hepcidin levels.

High-hepcidin anemias		
Hereditary	OMIM n.	Prevalence
Iron refractory iron deficiency anemia (IRIDA)	#206200	Rare
Hepcidin-producing adenomas*	#232200	Rare
Acquired		
Anemia of acute inflammation		Common**
Anemia of chronic inflammation		Common
(anemia of chronic disease)		
Low-hepcidin anemias		
Hereditary – iron loading anemias	OMIM n.	
β-thalassemia	#613985	Common***
Congenital dyserythropoietic anemia	#224100	Rare
Sideroblastic anemias	#300751	Rare
Acquired		
Low risk MDS with ringed sideroblasts		Rare

*OMIM, online Mendelian Inheritance In Man; MDS, myelodysplastic syndromes.*Described in glycogen-storage-disease 1a.*
***In hospitalized patients and in intensive care units.*
****In people of Mediterranean or southern-east Asian origin.*

disorder with ectopic bone formation in soft tissues due to mutated BMP type I receptor gene *ACVR1*, encoding ALK2 (Shore et al., 2006). The pathological allele *ALK2*[R258S] is constitutively active since the mutation affects the glycine-serine-rich domain of the gene and renders the BMP/SMAD pathway overactive being unable to bind its specific inhibitor FKBP12 (Pagani et al., 2017).

This rare case is especially illustrative. First, since the ALK2 glycine-serine-rich domain interacts with FKBP12 and the mutation destabilizes the binding, it has revealed a previously unsuspected role for FKBP12 as a modulator of liver ALK2 and hepcidin (Colucci et al., 2017). Second, it has led to identify a link between activation of bone and liver BMP type I receptors. Third, the case strengthens the relevance of intact TMPRSS6 in controlling the hepatic BMP/SMAD signaling, since no IRIDA was identified among other FOP patients with the same *ACVR1* mutation and presumably normal *TMPRSS6* (Pagani et al., 2017). Finally, this case is consistent with the concept that *TMPRSS6* haploinsufficiency cannot cause classic IRIDA.

The optimal treatment of IRIDA is undefined. Oral iron is ineffective, since it is not absorbed. The addition of vitamin C allows sporadic response. Intravenous iron induces a partial response usually at a slower rate in comparison with patients with acquired iron deficiency. EPO is ineffective in classic cases (De Falco et al., 2013; Heeney and Finberg, 2014).

Anemia of Hepcidin-Producing Adenomas

This is an extremely rare condition in adult patients affected by *glycogen storage disease 1a*, a recessive disorder due to deficiency of glucose-6 phosphatase, which catalyzes a reaction involved in both glycogenolysis and gluconeogenesis. A common dangerous disease symptom is hypoglycemia. The current treatment leads to prolonged survival of affected children up to adult age with the occurrence of several complications, such as anemia and liver adenomas. Anemia is microcytic and hypochromic, iron deficient, and refractory to oral iron treatment.

Anemia reverted after surgical adenoma resection. Adenoma tissue was found positive for hepcidin mRNA, while normal surrounding tissue showed hepcidin suppression, as expected because of the ectopic uncontrolled hepcidin production (Weinstein et al., 2002). The hematological features of patients resemble those of IRIDA as they share high hepcidin levels as a common mechanism of anemia.

Anemia of Inflammation

Anemia of inflammation (AI), previously known as anemia of chronic diseases, is a moderate normochromic-normocytic anemia that develops in conditions of systemic inflammation and immune activation. It occurs in several common disorders, including chronic infections, autoimmune diseases, advanced cancer, chronic kidney disease, congestive heart failure, chronic obstructive pulmonary disease, anemia of the elderly (at least partly), and graft versus host disease. AI is one of the most common anemias worldwide and the most frequent anemia in hospitalized patients. Acute inflammation contributes to the severity of anemia in intensive care units. Molecular mechanisms underlying AI are multiple and complex. Overproduction of cytokines such as IL1-β, TNF-α, and IL-6 by macrophages and INF-γ by lymphocytes blunts EPO production, impairs the erythropoiesis response, increases hepcidin levels, and may activate erythrophagocytosis, especially in the acute forms (Weiss and Goodnough, 2005; Ganz, 2019).

Hepcidin is activated by IL-6 through IL-6 receptor (IL-6R) and JAK2-STAT3 signaling. Full hepcidin activation requires an active BMP-SMAD pathway because inactivation of BMP signaling decreases hepcidin in animal models of inflammation (Theurl et al., 2011). The deregulation of systemic iron homeostasis causes macrophage iron sequestration and reduced absorption and recycling that leads to low saturation of transferrin and iron restriction of erythropoiesis and other tissues.

Traditional treatment of AI is based on reversibility/control of the underlying disease, whenever possible. If the disease is untreatable and anemia is mild, a careful evaluation of risks-benefits is needed to avoid side effects of any treatment. Pathophysiology-based treatments are limited to erythropoietin-like compounds and iron. The use of erythropoiesis stimulating agents (ESA) suppresses hepcidin by inducing erythropoiesis expansion. This approach is widely used in patients with chronic kidney disease, low-risk myelodysplastic syndromes, and cancer undergoing chemotherapy. However, a careful clinical control is necessary because high doses have cardiovascular side effects. The administration of intravenous iron may relieve iron restriction, caused by ESA-dependent expansion of erythropoiesis. Oral iron is usually ineffective since the high hepcidin levels counteract its intestinal absorption. Inhibitors of prolyl hydroxylase (hypoxia inducible factor, HIF stabilizers) are experimental in chronic kidney disease, to the aim of increasing endogenous EPO. Chronic treatment with red blood cells transfusions is not recommended because of transient effect and adverse reactions; it is limited to severe refractory anemia (Camaschella, 2019; Weiss et al., 2019).

Anemias Associated With Low Hepcidin Levels

Ineffective erythropoiesis and low or inappropriately normal hepcidin levels, with consequent iron overload, are typical features of the *"iron-loading anemias."* The prototype is β-thalassemia, a genetic recessive disease due to β-globin gene mutations that cause anemia and excess α-globin chain production. The latter precipitates as hemichromes in the bone marrow, damaging maturing erythroid precursors and leading to ineffective erythropoiesis. This occurs in non-transfusion-dependent thalassemia or thalassemia intermedia, whose erythropoiesis is characterized by the prevalence of immature cells that release erythroferrone to inhibit liver hepcidin expression. Hepcidin levels are usually greater in transfusion-dependent thalassemia, where endogenous ineffective erythropoiesis is at least partially suppressed by transfusions (Camaschella and Nai, 2016).

Hepcidin suppression is mediated by the increased cytokine erythroferrone (ERFE), a member of the TNF-α family encoded by *ERFE* gene, synthesized by erythroblasts upon EPO stimulation (Kautz et al., 2014). ERFE is released into the circulation and sequesters BMPs, especially BMP6 (Arezes et al., 2018), attenuating the hepcidin signaling in response to iron. In addition, an epigenetic suppression occurs at the hepcidin locus by histone deacetylase HDAC3 (Pasricha et al., 2017). When anemia causes hypoxia, other mediators such as PDGF-BB (Sonnweber et al., 2014), which is released by different cell types, suppress hepcidin.

Hepcidin levels are decreased by a special mechanism in low-risk myelodysplasia with ringed sideroblasts, a clonal disorder due to mutations of the spliceosome gene *SF3B1*. Iron accumulates in mitochondria, leading to ineffective erythropoiesis and systemic iron overload. An abnormally spliced, elongated ERFE protein is more powerful than wild type ERFE in suppressing hepcidin (Bondu et al., 2019) and causing transfusion-independent iron loading.

TARGETED THERAPIES FOR HEPCIDIN-RELATED ANEMIAS

The identification of molecular mechanisms responsible of the previously discussed anemias has stimulated research in developing targeted therapies to replace current symptomatic treatment (Sebastiani et al., 2016; Crielaard et al., 2017). Approaches differ according to the type of anemia and the aim of decreasing or increasing hepcidin levels or their effects (**Table 2**).

Experimental Therapies to Decrease Hepcidin Levels/Increase Ferroportin Function

Except for hepcidin producing tumors, which have to be surgically removed, compounds that antagonize hepcidin or its effects may be useful in all anemias characterized by high hepcidin levels. Their main application would be in chronic inflammatory diseases in order to reverse hypoferremia and anemia. Several experimental therapies aimed at manipulating the hepcidin

TABLE 2 | Experimental therapies targeting the hepcidin-ferroportin axis.

	Mechanism	Compounds
Compounds that decrease hepcidin or increase ferroportin function		
Class I	Reduction of the signaling pathway stimulating hepcidin	Anti IL6-R, anti IL-6
		Anti-BMP6 MoAb*
		BMPR inhibitors
		Anti-HJV MoAb
Class II	Hepcidin binders	Non anticoagulant heparins Anti-HAMP MoAb
Class III	Interfering with hepcidin-FPN interaction	Oligonucelotides aptamers Anti-FPN MoAb, GDP
Compounds that increase hepcidin or decrease ferroportin function		
Class I	Hepcidin mimics	Hepcidin analogues*
Class II	Activating hepcidin Blocking the hepcidin inhibitor Blocking the hepcidin receptor	Minihepcidin BMPs (preclinical studies) Anti-*TMPRSS6* (siRNA, ASO*) FPN Inhibitors*
Class III	Others	Human transferrin infusions Protoporphyrin IX (inhibition of HO) Bone marrow TFR2 inactivation

*BMPR, BMP receptor; HAMP, hepcidin gene; HJV, hemojuvelin; MoAb, monoclonal antibody; FPN, ferroportin; siRNA, small interfering RNA; ASO, antisense oligonucleotides; GDP, guanosine 5' diphosphate; HO, heme oxygenase; TFR2, transferrin receptor 2. Compounds indicated by * are in clinical trials.*

pathway and its function have been investigated in preclinical studies. Hepcidin antagonists are inhibitors of hepcidin synthesis/regulators (Ganz, 2019), hepcidin binders that block its function, and compounds that interfere with hepcidin-ferroportin interaction (**Table 2**). Some compounds are in clinical trials especially in chronic kidney disease (Sheetz et al., 2019). In IRIDA, manipulation of the hepcidin pathway has been proposed in preclinical studies with the use of anti-HJV MoAb (Kovac et al., 2016).

Experimental Therapies to Increase Hepcidin Levels/Decrease Ferroportin Function

Increasing hepcidin levels may not only reduce iron overload but also partially control ineffective erythropoiesis in *iron loading anemias*. β-thalassemia is the most studied among these conditions (Casu et al., 2018; Gupta et al., 2018). Proposed drugs are hepcidin analogs (some in clinical trials), hepcidin modulators, especially TMPRSS6 inhibitors, or compounds that interfere with hepcidin-ferroportin interaction decreasing iron export (**Table 2**).

While compounds that increase hepcidin reduce ineffective erythropoiesis due to the vicious cycle between ineffective erythropoiesis and iron loading (Camaschella and Nai, 2016), drugs that favor erythroid precursor maturation, as the activin receptor IIB ligand trap, luspatercept, not only improve anemia but also ameliorate iron homeostasis by reducing hepcidin inhibition (Piga et al., 2019).

Some targeted approaches now in clinical trials will hopefully result in novel treatments for a variety of anemias.

CONCLUSION

The spectacular advances in understanding the regulation of iron metabolism and hepcidin allowed a better understanding of erythropoiesis control, since together with erythropoietin iron is a fundamental factor for erythroid cells maturation. Conditions that lead to anemia can be associated with high and low hepcidin levels. In both instances, contrasting hepcidin deregulation may ameliorate/ correct anemia in preclinical models, offering new tools that are already or will be soon clinically explored for the treatment of specific anemias.

AUTHOR CONTRIBUTIONS

AP drafted the paper. CC developed the final version. AN and LS contributed to writing and to critical review the manuscript. All the authors approved the final version.

REFERENCES

An, P., Wu, Q., Wang, H., Guan, Y., Mu, M., Liao, Y., et al. (2012). TMPRSS6, but not TF, TFR2 or BMP2 variants are associated with increased risk of iron-deficiency anemia. *Hum. Mol. Genet.* 21, 2124–2131. doi: 10.1093/hmg/dds028

Andriopoulos, B. Jr., Corradini, E., Xia, Y., Faasse, S. A., Chen, S., Grgurevic, L., et al. (2009). BMP6 is a key endogenous regulator of hepcidin expression and iron metabolism. *Nat. Genet.* 41, 482–487. doi: 10.1038/ng.335

Arezes, J., Foy, N., McHugh, K., Sawant, A., Quinkert, D., Terraube, V., et al. (2018). Erythroferrone inhibits the induction of hepcidin by BMP6. *Blood* 132, 1473–1477. doi: 10.1182/blood-2018-06-857995

Aschemeyer, S., Qiao, B., Stefanova, D., Valore, E. V., Sek, A. C., Ruwe, T. A., et al. (2018). Structure-function analysis of ferroportin defines the binding site and an alternative mechanism of action of hepcidin. *Blood.* 131, 899–910. doi: 10.1182/blood-2017-05-786590

Babitt, J. L., Huang, F. W., Wrighting, D. M., Xia, Y., Sidis, Y., Samad, T. A., et al. (2006). Bone morphogenetic protein signaling by hemojuvelin regulates hepcidin expression. *Nat. Genet.* 38, 531–539. doi: 10.1038/ng1777

Beguin, Y. (2003). Soluble transferrin receptor for the evaluation of erythropoiesis and iron status. *Clin. Chim. Acta* 329, 9–22. doi: 10.1016/S0009-8981(03)00005-6

Bondu, S., Alary, A. S., Lefevre, C., Houy, A., Jung, G., Lefebvre, T., et al. (2019). A variant erythroferrone disrupts iron homeostasis in SF3B1-mutated myelodysplastic syndrome. *Sci. Transl. Med.* 11:pii:eaav5467. doi: 10.1126/scitranslmed.aav5467

Camaschella, C. (2013). How I manage patients with atypical microcytic anaemia. *Br. J. Haematol.* 160, 12–24. doi: 10.1111/bjh.12081

Camaschella, C. (2019). Iron deficiency. *Blood* 133, 30–39. doi: 10.1182/blood-2018-05-815944

Camaschella, C., and Nai, A. (2016). Ineffective erythropoiesis and regulation of iron status in iron loading anaemias. *Br. J. Haematol.* 172, 512–523. doi: 10.1111/bjh.13820

Camaschella, C., and Pagani, A. (2018). Advances in understanding iron metabolism and its crosstalk with erythropoiesis. *Br. J. Haematol.* 182, 481–494. doi: 10.1111/bjh.15403

Canali, S., Wang, C. Y., Zumbrennen-Bullough, K. B., Bayer, A., and Babitt, J. L. (2017). Bone morphogenetic protein 2 controls iron homeostasis in mice independent of Bmp6. *Am. J. Hematol.* 92, 1204–1213. doi: 10.1002/ajh.24888

Casu, C., Nemeth, E., and Rivella, S. (2018). Hepcidin agonists as therapeutic tools. *Blood* 131, 1790–1794. doi: 10.1182/blood-2017-11-737411

Colucci, S., Pagani, A., Pettinato, M., Artuso, I., Nai, A., Camaschella, C., et al. (2017). The immunophilin FKBP12 inhibits hepcidin expression by binding the BMP type I receptor ALK2 in hepatocytes. *Blood* 130, 2111–2120. doi: 10.1182/blood-2017-04-780692

Crielaard, B. J., Lammers, T., and Rivella, S. (2017). Targeting iron metabolism in drug discovery and delivery. *Nat. Rev. Drug Discov.* 16, 400–423. doi: 10.1038/nrd.2016.248

De Falco, L., Sanchez, M., Silvestri, L., Kannengiesser, C., Muckenthaler, M. U., Iolascon, A., et al. (2013). Iron refractory iron deficiency anemia. *Haematologica* 98, 845–853. doi: 10.3324/haematol.2012.075515

De Falco, L., Silvestri, L., Kannengiesser, C., Moran, E., Oudin, C., Rausa, M., et al. (2014). Functional and clinical impact of novel TMPRSS6 variants in iron-refractory iron-deficiency anemia patients and genotype-phenotype studies. *Hum. Mutat.* 35, 1321–1329. doi: 10.1002/humu.22632

Donker, A. E., Schaap, C. C., Novotny, V. M., Smeets, R., Peters, T. M., van den Heuvel, B. L., et al. (2016). Iron refractory iron deficiency anemia: a heterogeneous disease that is not always iron refractory. *Am. J. Hematol.* 91, E482–E490. doi: 10.1002/ajh.24561

Du, X., She, E., Gelbart, T., Truksa, J., Lee, P., Xia, Y., et al. (2008). The serine protease TMPRSS6 is required to sense iron deficiency. *Science* 320, 1088–1092. doi: 10.1126/science.1157121

Finberg, K. E., Heeney, M. M., Campagna, D. R., Aydinok, Y., Pearson, H. A., Hartman, K. R., et al. (2008). Mutations in TMPRSS6 cause iron-refractory iron deficiency anemia (IRIDA). *Nat. Genet.* 40, 569–571. doi: 10.1038/ng.130

Ganz, T. (2019). Anemia of inflammation. *N. Engl. J. Med.* 381, 1148–1157. doi: 10.1056/NEJMra1804281

Gupta, R., Musallam, K. M., Taher, A. T., and Rivella, S. (2018). Ineffective erythropoiesis: anemia and iron overload. *Hematol. Oncol. Clin. North Am.* 32, 213–221. doi: 10.1016/j.hoc.2017.11.009

Heeney, M. M., and Finberg, K. E. (2014). Iron-refractory iron deficiency anemia (IRIDA). *Hematol. Oncol. Clin. North Am.* 28, 637–652. doi: 10.1016/j.hoc.2014.04.009

Heeney, M. M., Guo, D., De Falco, L., Campagna, D. R., Olbina, G., Kao, P. P., et al. (2018). Normalizing hepcidin predicts TMPRSS6 mutation status in patients with chronic iron deficiency. *Blood* 132, 448–452. doi: 10.1182/blood-2017-03-773028

Kautz, L., Jung, G., Valore, E. V., Rivella, S., Nemeth, E., and Ganz, T. (2014). Identification of erythroferrone as an erythroid regulator of iron metabolism. *Nat. Genet.* 46, 678–684. doi: 10.1038/ng.2996

Koch, P. S., Olsavszky, V., Ulbrich, F., Sticht, C., Demory, A., Leibing, T., et al. (2017). Angiocrine Bmp2 signaling in murine liver controls normal iron homeostasis. *Blood* 129, 415–419. doi: 10.1182/blood-2016-07-729822

Kovac, S., Boser, P., Cui, Y., Ferring-Appel, D., Casarrubea, D., Huang, L., et al. (2016). Anti-hemojuvelin antibody corrects anemia caused by inappropriately high hepcidin levels. *Haematologica* 101, e173–e176. doi: 10.3324/haematol.2015.140772

Lakhal, S., Schodel, J., Townsend, A. R., Pugh, C. W., Ratcliffe, P. J., and Mole, D. R. (2011). Regulation of type II transmembrane serine proteinase TMPRSS6 by hypoxia-inducible factors: new link between hypoxia signaling and iron homeostasis. *J. Biol. Chem.* 286, 4090–4097. doi: 10.1074/jbc.M110.173096

Meynard, D., Kautz, L., Darnaud, V., Canonne-Hergaux, F., Coppin, H., and Roth, M. P. (2009). Lack of the bone morphogenetic protein BMP6 induces massive iron overload. *Nat. Genet.* 41, 478–481. doi: 10.1038/ng.320

Muckenthaler, M. U., Rivella, S., Hentze, M. W., and Galy, B. (2017). A red carpet for iron metabolism. *Cell* 168, 344–361. doi: 10.1016/j.cell.2016.12.034

Nemeth, E., Tuttle, M. S., Powelson, J., Vaughn, M. B., Donovan, A., Ward, D. M., et al. (2004). Hepcidin regulates cellular iron efflux by binding to ferroportin and inducing its internalization. *Science* 306, 2090–2093. doi: 10.1126/science.1104742

Pagani, A., Colucci, S., Bocciardi, R., Bertamino, M., Dufour, C., Ravazzolo, R., et al. (2017). A new form of IRIDA due to combined heterozygous mutations of TMPRSS6 and ACVR1A encoding the BMP receptor ALK2. *Blood* 129, 3392–3395. doi: 10.1182/blood-2017-03-773481

Pagani, A., Vieillevoye, M., Nai, A., Rausa, M., Ladli, M., Lacombe, C., et al. (2015). Regulation of cell surface transferrin receptor-2 by iron-dependent cleavage and release of a soluble form. *Haematologica* 100, 458–465. doi: 10.3324/haematol.2014.118521

Pasricha, S. R., Lim, P. J., Duarte, T. L., Casu, C., Oosterhuis, D., Mleczko-Sanecka, K., et al. (2017). Hepcidin is regulated by promoter-associated histone acetylation and HDAC3. *Nat. Commun.* 8:403. doi: 10.1038/s41467-017-00500-z

Piga, A., Perrotta, S., Gamberini, M. R., Voskaridou, E., Melpignano, A., Filosa, A., et al. (2019). Luspatercept improves hemoglobin levels and blood transfusion requirements in a study of patients with beta-thalassemia. *Blood* 133, 1279–1289. doi: 10.1182/blood-2018-10-879247

Rausa, M., Pagani, A., Nai, A., Campanella, A., Gilberti, M. E., Apostoli, P., et al. (2015). Bmp6 expression in murine liver non parenchymal cells: a mechanism to control their high iron exporter activity and protect hepatocytes from iron overload? *PLoS One* 10:e0122696. doi: 10.1371/journal.pone.0122696

Sebastiani, G., Wilkinson, N., and Pantopoulos, K. (2016). Pharmacological targeting of the hepcidin/ferroportin axis. *Front. Pharmacol.* 7:160. doi: 10.3389/fphar.2016.00160

Sheetz, M., Barrington, P., Callies, S., Berg, P. H., McColm, J., Marbury, T., et al. (2019). Targeting the hepcidin-ferroportin pathway in anaemia of chronic kidney disease. *Br. J. Clin. Pharmacol.* 85, 935–948. doi: 10.1111/bcp.13877

Shore, E. M., Xu, M., Feldman, G. J., Fenstermacher, D. A., Cho, T. J., Choi, I. H., et al. (2006). A recurrent mutation in the BMP type I receptor ACVR1 causes inherited and sporadic fibrodysplasia ossificans progressiva. *Nat. Genet.* 38, 525–527. doi: 10.1038/ng1783

Silvestri, L., Pagani, A., Nai, A., De Domenico, I., Kaplan, J., and Camaschella, C. (2008). The serine protease matriptase-2 (TMPRSS6) inhibits hepcidin activation by cleaving membrane hemojuvelin. *Cell Metab.* 8, 502–511. doi: 10.1016/j.cmet.2008.09.012

Silvestri, L., Rausa, M., Pagani, A., Nai, A., and Camaschella, C. (2013). How to assess causality of TMPRSS6 mutations? *Hum. Mutat.* 34, 1043–1045. doi: 10.1002/humu.22321

Sonnweber, T., Nachbaur, D., Schroll, A., Nairz, M., Seifert, M., Demetz, E., et al. (2014). Hypoxia induced downregulation of hepcidin is mediated by platelet derived growth factor BB. *Gut* 63, 1951–1959. doi: 10.1136/gutjnl-2013-305317

Sorensen, E., Rigas, A. S., Didriksen, M., Burgdorf, K. S., Thorner, L. W., Pedersen, O. B., et al. (2019). Genetic factors influencing hemoglobin levels in 15,567 blood donors: results from the Danish blood donor study. *Transfusion* 59, 226–231. doi: 10.1111/trf.15075

Theurl, I., Schroll, A., Sonnweber, T., Nairz, M., Theurl, M., Willenbacher, W., et al. (2011). Pharmacologic inhibition of hepcidin expression reverses anemia of chronic inflammation in rats. *Blood* 118, 4977–4984. doi: 10.1182/blood-2011-03-345066

Weinstein, D. A., Roy, C. N., Fleming, M. D., Loda, M. F., Wolfsdorf, J. I., and Andrews, N. C. (2002). Inappropriate expression of hepcidin is associated with iron refractory anemia: implications for the anemia of chronic disease. *Blood* 100, 3776–3781. doi: 10.1182/blood-2002-04-1260

Weiss, G., Ganz, T., and Goodnough, L. T. (2019). Anemia of inflammation. *Blood* 133, 40–50. doi: 10.1182/blood-2018-06-856500

Weiss, G., and Goodnough, L. T. (2005). Anemia of chronic disease. *N. Engl. J. Med.* 352, 1011–1023. doi: 10.1056/NEJMra041809

7

Genetics and Genomics Approaches for Diagnosis and Research into Hereditary Anemias

Roberta Russo[1,2]*, Roberta Marra[1,2], Barbara Eleni Rosato[1,2], Achille Iolascon[1,2] and Immacolata Andolfo[1,2]*

[1] Department of Molecular Medicine and Medical Biotechnologies, University of Naples Federico II, Naples, Italy, [2] CEINGE Biotecnologie Avanzate, Naples, Italy

*Correspondence:
Roberta Russo
roberta.russo@unina.it
Immacolata Andolfo
andolfo@ceinge.unina.it

The hereditary anemias are a relatively heterogeneous set of disorders that can show wide clinical and genetic heterogeneity, which often hampers correct clinical diagnosis. The classical diagnostic workflow for these conditions generally used to start with analysis of the family and personal histories, followed by biochemical and morphological evaluations, and ending with genetic testing. However, the diagnostic framework has changed more recently, and genetic testing is now a suitable approach for differential diagnosis of these patients. There are several approaches to this genetic testing, the choice of which depends on phenotyping, genetic heterogeneity, and gene size. For patients who show complete phenotyping, single-gene testing remains recommended. However, genetic analysis now includes next-generation sequencing, which is generally based on custom-designed targeting panels and whole-exome sequencing. The use of next-generation sequencing also allows the identification of new causative genes, and of polygenic conditions and genetic factors that modify disease severity of hereditary anemias. In the research field, whole-genome sequencing is useful for the identification of non-coding causative mutations, which might account for the disruption of transcriptional factor occupancy sites and *cis*-regulatory elements. Moreover, advances in high-throughput sequencing techniques have now resulted in the identification of genome-wide profiling of the chromatin structures known as the topologically associating domains. These represent a recurrent disease mechanism that exposes genes to inappropriate regulatory elements, causing errors in gene expression. This review focuses on the challenges of diagnosis and research into hereditary anemias, with indications of both the advantages and disadvantages. Finally, we consider the future perspectives for the use of next-generation sequencing technologies in this era of precision medicine.

Keywords: hereditary anemias, next generation sequencing, differential diagnosis, chromatin 3D architecture, genetic test

INTRODUCTION

The hereditary anemias (HAs) represent a particularly heterogeneous group of disorders with rare to low frequency that are characterized by complex genotype–phenotype correlations that remain to be explained. It has only been in recent years that major advances have been made in our understanding of the genetic basis and the pathophysiology of HAs. Indeed, more than 70 genes involved in red blood cell (RBC) physiology have been identified as causative of HAs to date.

Based on clinical manifestations and morphological RBC alterations, HAs can be broadly classified into four different subtypes: (i) disorders of hemoglobin (Hb) synthesis, such as thalassemia and hemoglobinopathies; (ii) hyporegenerative anemias, such as Diamond-Blackfan anemia and congenital dyserythropoietic anemias; (iii) RBC membrane defects that are due to either altered structural organization of their membranes, such as hereditary spherocytosis and hereditary elliptocytosis, or to alterations to membrane transport functions, such as hereditary stomatocytosis; and (iv) non-spherocytic hemolytic anemias due to RBC enzyme defects.

In the last few years, much time and effort has been spent on the identification of the genes and mutations that underlie HAs. The rapid evolution in the diagnostic and research technologies involved now needs to be taken into account. The use of such new techniques has changed the ways in which both diagnosis and research are carried out.

This review focuses on the past, present, and future genetics and genomics approaches for establishment of correct differential diagnosis among these conditions, and for research into new causative/modifier genes and new pathogenetic mechanisms.

CLASSIFICATION OF THE HEREDITARY ANEMIAS

Initially, we will briefly describe the main hallmarks, both clinical and molecular, of the various subtype of HAs. Specific guidance on the pathway to establish correct diagnosis for these conditions has been extensively reviewed elsewhere (Da Costa et al., 2013; Andolfo et al., 2016, 2018b; Gambale et al., 2016; Taher et al., 2018; Bianchi et al., 2019; Grace et al., 2019a; Iolascon et al., 2019, 2020), and so this will not be discussed in any detail in the present review.

Disorders of Hemoglobin Synthesis

The globin disorders can generally be classified as quantitative (e.g., thalassemias) and qualitative (e.g., hemoglobinopathies) defects that lead to hemolysis, and that are defined according to the globin chains. Thalassemia has been shown to be one of the most common genetic disorders worldwide, and it arises as a result of mutations in the α-globin or β-globin genes (Taher et al., 2018). The mutations associated with thalassemia now number over 1,530, and these range from single-nucleotide variations to large genome rearrangements (Higgs et al., 2012; Zhao et al., 2020). α-Thalassemia arises from deletions in the *HBA* genes in ~95% of patients, while the remaining cases arise from point mutations. In contrast, some 95% of β-thalassemias arise either from *HBB* gene point mutations that result in disruption of RNA transcription, processing or stability, or from nonsense mutations that result in the production of abnormal proteins or in nonsense-mediated decay of RNA. Many of the aspects of β-thalassemia pathophysiology are now explained by excess production of α-globin chains, with the result that they can precipitate in RBC precursors (resulting in ineffective erythropoiesis) and mature RBCs (resulting in hemolysis). Structural variants of the globin genes are associated with sickle-cell anemia, hemolysis caused by the unstable Hb, the altered oxygen affinity of Hb, and Hb where

the ferrous (Fe^{2+}) state of the iron cannot be maintained (Rees et al., 1999; Sabath, 2017). The use of various electrophoretic techniques to separate the various Hb states has become the mainstay for the diagnosis of these Hb disorders.

Hypo-Regenerative and Hypo-Productive Anemias

These represent a heterogeneous group of disorders that have effects on the normal differentiation–proliferation pathways at the different steps through the erythroid lineage, and that mainly result in monolinear cytopenia.

Diamond-Blackfan anemia (DBA) is defined by macrocytic moderate or severe anemia that can occur in association with hyporegenerative bone marrow and with reticulocytopenia. Almost half of these patients have physical abnormalities, with 50% showing craniofacial anomalies, and 38% showing defects of the upper limb and hand, which mainly include the thumb. This disease shows mutations in 20 genes for ribosomal proteins, of a total of 80 genes that encode the complete ribosome. For six of these 20 causative genes (i.e., *RPS19, RPS24, RPS26, RPL5, RPL11, RPL35a*), the mutations and deletions comprise 70% of all DBA patients (Da Costa et al., 2018). There is strong evidence that suggests that erythroid blockage occurs between the BFU-e and CFU-e erythroid development stages, although the exact stage remains to be fully defined (Ohene-Abuakwa et al., 2005).

Congenital dyserythropoietic anemias (CDAs) are a large group of hypo-productive anemias that have been classified into five subtypes: types I, II and III CDAs; CDAs related to transcription factors; and variant CDAs. The most common forms are CDA types I and II (Heimpel et al., 2010), which are caused by bi-allelic mutations in the *CDAN1/CDIN1* genes for CDAI (Dgany et al., 2002; Babbs et al., 2013), and in the *SEC23B* gene for CDAII (Bianchi et al., 2009; Schwarz et al., 2009). For many years, the main diagnostic features of these CDAs were morphological abnormalities of bone marrow (e.g., erythroid hyperplasia with binuclearity) or late erythroblast multinuclearity. However, these particular features of CDAs are not always specific; in particular, they are also seen for other acquired conditions that involve erythropoietic stress, including iron deficiency and preterm birth (Iolascon et al., 2020).

Red Blood Cell Membrane Disorders

RBC membrane disorders consist of hemolytic anemias, which can show wide differences in their clinical, morphological, laboratory, and molecular aspects. The main effects that these genetic alterations have relate to lowered RBC deformability and shortened RBC survival (Iolascon et al., 2019). Among these, there are hereditary spherocytosis, hereditary elliptocytosis, hereditary pyropoikilocytosis, and Southeast Asian ovalocytosis, which are caused by altered membrane structural organization. Hereditary spherocytosis is the most frequent form, and it is characterized by phenotypic, locus, and allelic heterogeneity. Indeed, mutations in five genes that encode proteins that are involved in interactions between the cytoskeleton and the RBC phospholipid bilayer are associated with these conditions: *ANK1, SPTA1, SPTB, SLC4A1,* and *EPB42* (Iolascon et al., 2019). The clinical manifestations can range from symptom-free carriers to patients with severe hemolysis, jaundice, reticulocytosis, splenomegaly, and cholelithiasis (Da Costa et al., 2013; Andolfo et al., 2016). Conversely, hereditary RBC membrane disorders can arise from genetic defects of RBC transport proteins, which leads to abnormal cation permeability, and the consequent changes in RBC hydration. Hereditary stomatocytosis represents this wide spectrum of diseases, among which the most frequent form is dehydrated hereditary stomatocytosis (DHS). Cation leaks cause dysregulation of cellular volume, which leads in turn to morphological abnormalities of RBCs, with stomatocytes seen for peripheral blood smears, and the consequent leftward shift in the osmolarity curve seen by ektacytometry for patients with mutation of the *PIEZO1* gene (Andolfo et al., 2018a). As generally seen for all hemolytic conditions, the key symptoms include pallor, fatigue, jaundice, gallstones, and splenomegaly. DHS is an autosomal dominant disease that is caused by gain-of-function mutations in *PIEZO1* (DHS type I) and in *KCNN4*, also known as 'Gardos channelopathy' due to its peculiar characteristics, which include lack of clear signs of RBC dehydration and normal ektacytometric curve (Zarychanski et al., 2012; Andolfo et al., 2013a, 2018b; Rapetti-Mauss et al., 2015; Picard et al., 2019).

Non-sperocytic Hemolytic Anemias Due to RBC Enzyme Defects

Mature RBCs rely exclusively on glycolysis for their energy production. Indeed, mutations to almost any of the glycolytic enzymes can result in hemolytic anemias (van Wijk and van Solinge, 2005). The most common human defect is deficiency of glucose-6-phosphate dehydrogenase (G6PD). About 140 mutations in the *G6PD* gene that have X-linked inheritance can cause G6PD functional variants, with many biochemical and clinical phenotypes seen. Of the clinical phenotypes, the main ones are neonatal jaundice and acute hemolytic anemia; this latter is generally triggered by exogenous agents (Luzzatto et al., 2016). Similarly, pyruvate kinase deficiency is generally the cause of chronic non-spherocytic hemolytic anemia, which is an autosomal recessive disease. More than 250 mutations in the *PKLR* gene that encodes the liver and RBC pyruvate kinase isoforms are known to be causative of this condition. Pyruvate kinase deficiency is characterized by a highly variable clinical spectrum (Grace et al., 2018), with patients with two non-missense mutations more severely affected (Bianchi et al., 2020). Furthermore, there are other more rare enzyme defects that are associated with HAs that involve the following: adenylate kinase, aldolase, phosphofructokinase, phosphoglycerate kinase, glucose phosphate isomerase, glutathione reductase, glutathione synthetase, hexokinase, pyrimidine-50-nucleotidase, and triosephosphate isomerase (Koralkova et al., 2014).

PAST, CURRENT, AND FUTURE OF MOLECULAR TESTING FOR DIAGNOSIS OF HEREDITARY ANEMIAS

Here, we will briefly describe the past, current, and future methodologies for genetic testing of HAs, and highlight the strengths and limitations of these different approaches.

First-Generation Sequencing: Sanger Method

The conventional workflow for diagnosis of these conditions started as first line of investigation with positive familial history, complete blood count, and peripheral blood smear. Then specialized biochemical tests, and eventually bone-marrow aspirate, were required. Finally, genetic testing by Sanger sequencing served as the confirmatory test. Very often, no mutations in the candidate gene were identified by this approach for the genetic heterogeneity of the conditions, which led to confusing or lacking molecular diagnoses.

Currently, genetic testing is used early in the diagnostic workflow of HAs, which removes the need for some of the specialized tests, such as bone marrow biopsies (Roy and Babbs, 2019), especially when the clinical data for the patients are not informative, or when the patient is transfusion dependent. Sanger sequencing was also our starting point, while we now use second-generation sequencing (i.e., NGS), which allows us to move from a monogenic approach to an oligo/multigenic approach (**Figure 1**). For many years, monogenic approaches were used for diagnosis and identification of new causative genes of HAs. For mapping to define the causative gene, there was the need to find families with confirmed Mendelian inheritance that preferably involves multiple generations. The use of linkage approaches associated with the functional mapping of the candidate genes was used to successfully identify the mutations responsible for several HAs, such as in the *SEC23B* gene for CDAII, or in *CDAN1* for CDAI, and in *ABCB6* for familial pseudohyperkalemia (Dgany et al., 2002; Schwarz et al., 2009; Andolfo et al., 2013b).

A critical aspect here is that this traditional approach still has great value, mainly in the diagnostic field. Indeed, the testing of single genes is preferred when: (i) the patient's clinical features are typical of a specific HA; (ii) an association has been established between the disorder and the specific gene; and (iii) there is complete phenotyping (i.e., at the clinical, biochemical, and morphological levels). For example, for G6PD deficiency, starting with single-gene analysis is usually still preferred, because in most of the cases an assured clinical diagnosis can be obtained, and *G6PD* is the only causative gene of this condition. Such single-gene tests show high clinical sensitivity because the phenotype and further findings indicate a single disorder associated with a single gene. Furthermore, one particular interpretive advantage of this single-gene approach is that there is minimal likelihood of uncovering multiple confounding variants that have unknown clinical significance. On the other hand, for many other HAs, the clinical variability and the heterogeneity of the genetic locus are significant enough such that a custom targeting panel and/or whole-exome sequencing (WES) are the most appropriate for efficient and timely molecular diagnosis.

Second-Generation Sequencing: NGS Short Reads

Next-generation sequencing has revolutionized the framework of HA diagnosis (**Figure 1**). Although WES or WGS can provide more information (e.g., identification of non-coding causative mutations, or new causative genes), a drawback remains in that

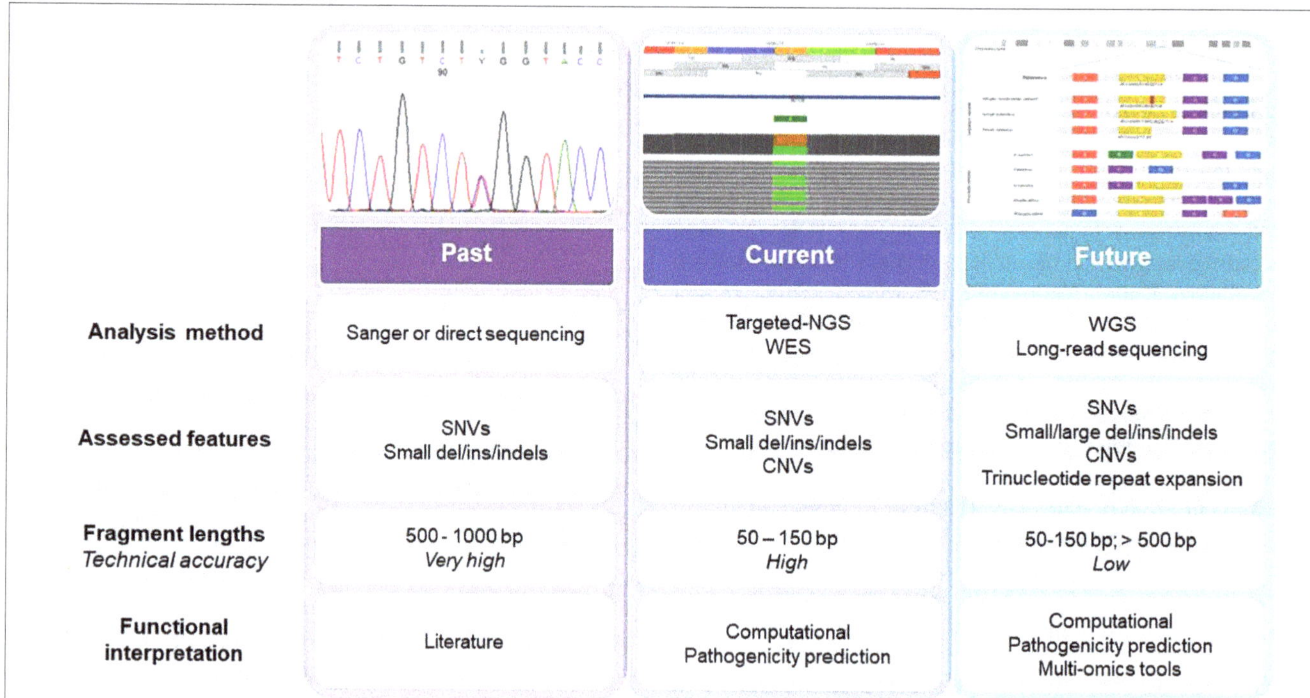

FIGURE 1 | The past, present, and future of genetics and genomics technologies for the diagnosis of and research into hereditary anemias. The diagram shows the current analysis methods for DNA sequencing, while also highlighting the main differences among them. WES, whole-exome sequencing; WGS, whole-genome sequencing; SNV, single nucleotide variant; CNV, copy number variation.

the overall complexity of the data analysis includes a number of variants of unknown significance. The functional tests are therefore crucial to assess the pathogenicity of new variants detected by NGS. Moreover, further problems can arise from the need for adequate coverage of the genes when there are copy number variants and GC−rich regions with low−complexity (Rets et al., 2019). Currently, a targeted-NGS approach is preferred in the diagnostic work-up for these disorders (Agarwal et al., 2016; Del Orbe Barreto et al., 2016; Niss et al., 2016; Roy et al., 2016; Russo et al., 2018; Choi et al., 2019; Kedar et al., 2019;

Svidnicki et al., 2020). Custom panels for HAs include variable numbers of genes (e.g., 50–200) that can provide diagnostic yields of 38–87%, which depends on how many and what types of genes are included, and on the depth of the phenotypic assessment (**Table 1**). The main limitation of the targeted NGS approach is the need for continuous updating of the gene list for each panel, to include all recently identified causative genes. Indeed, the targeted NGS approach does not allow for the identification of new genes, by design. Moreover, the cost of WES is now similar to the cost of the large panels that are required to cover

TABLE 1 | Case series of hereditary anemias diagnosed by targeted next-generation sequencing.

Hereditary anemia subtypes	Families (n)	Patients (n)	Genes (n)	Diagnostic yield (%)[‡]	References
CDA; DBA; sideroblastic anemia; RBC enzymatic defects	57	57	33	38.6	Roy et al., 2016
RBC membrane defects; CDA; RBC enzymatic defects	10	10	40	100[$]	Del Orbe Barreto et al., 2016
RBC membrane defects; RBC enzymatic defects	15	15	28	86.7	Agarwal et al., 2016
RBC membrane defects	13	15	12	100[$]	Niss et al., 2016
RBC membrane defects; CDA; DBA; RBC enzymatic defects	62	74	71	64.9	Russo et al., 2018
RBC membrane defects; CDA; DBA; RBC enzymatic defects	21	21	76	61.9	Shefer Averbuch et al., 2018
BMFS; CDA; RBC enzymatic defects; hematological malignancies	21	21	76	81.0	Kedar et al., 2019
RBC membrane defects; RBC enzymatic defects; hereditary anemia modifiers	59	59	43	84.7	Choi et al., 2019
CDA; RBC membrane defects; RBC enzymatic defects	26	36	35	72.2	Svidnicki et al., 2020

[‡]Ratio of number of diagnosed patients/numbers of tested patients. [$]These studies reported only the number of diagnosed patients. CDA, congenital dyserythropoietic anemia; DBA, Diamond-Blackfan anemia; RBC, red blood cell; BMFS, bone marrow failure syndrome.

the full gamut of potential gene mutations that underpin HAs. Therefore, in the near future, WES with the use of managed variant lists that contain known pathogenic variants will become the most appropriate diagnostic approach for investigation of HAs (excluding the hemoglobinopathies, which require more specialized tests).

The exome is believed to represent 1–2% of the genome; however, it also contains 85% of the mutations that are known to cause disease. WES has been reported to have a diagnostic yield of ∼30–50% when used in clinical diagnostics, which depends on detailed phenotyping (Alkan et al., 2011; Sankaran et al., 2012; Rehm et al., 2013; Hamada et al., 2018). Thus, ∼30–50% of cases will remain undiagnosed following NGS testing. This arises mainly because of incomplete phenotyping of the patients, which reduces the specificity of the data, as supported by a report of the numerous variants of unknown significance (Trujillano et al., 2017). Contrary to the targeted-NGS and WES approaches, no estimate of the diagnostic rate of WGS for HAs has been established yet. In a prospective study with 100 patients referred to a pediatric genetics service, genetic variants that met the clinical diagnostic criteria were identified by WGS in 34% of cases (Stavropoulos et al., 2016).

Next-generation sequencing has many advantageous aspects, which include the low cost and the high speed and yield. Conversely, there are some intrinsic limitations of NGS that can have significant impact on accuracy of these analyses. Considering these as bottlenecks, the most noticeable drawback is the short read length. Short read length has been shown to limit precision in a number of biological studies, in particular for genome assembly and transcriptome analyses. Here, NGS short reads are liable for unresolved complexities that can arise from heterozygosity, GC-rich regions, transposable elements, tandem repeats, and repetitive regions (10 kb–10 Mb or more) interspersed in the genome (Alkan et al., 2011). The use of NGS to sequence polymorphic tandem repeats in the genome can be severely impaired by read length (Mousavi et al., 2019).

Considering the analysis of complex chromosomal rearrangements and structural variants, the analytical approaches of comparative genomic hybridization and multiplex ligation-dependent probe amplification are still widely used for detection of copy-number changes (Minervini et al., 2020). For instance, these techniques are crucial for β-thalassemia diagnosis. Indeed, most β-thalassemia alleles (∼90%) are point mutations, and are thus easily identified through Sanger sequencing or other dedicated methods. As the remaining 10% of the alleles are deletions, these are detectable by comparative genomic hybridization arrays or multiplex ligation-dependent probe amplification (Joly et al., 2014). It is important, however, that the application of all of these techniques should not be considered as mutually exclusive or as consecutive, but rather as synergistic. Indeed, several cases of apparent recessive inheritance due to uniparental isodisomy in probands with homozygous recessive mutations where only one parent is heterozygous for the same variant have been described among patients with HAs (Bogardus et al., 2014; Andolfo et al., 2020).

Third-Generation Sequencing: NGS Long Reads

The use of NGS with improved read lengths represents a milestone in the study of the genetics of HAs. This so-called third-generation sequencing now has two advantages that are crucial: single molecules can be sequenced without the need for PCR amplification, which thus avoids any PCR bias; and secondly, this also generates read lengths that are longer (Furst et al., 2020). The two main systems that are based on this third-generation NGS are from Pacific Biosystems[1] and from Oxford Nanopore Technologies[2]. These both have the advantage of long read lengths, although they also share the disadvantage of high randomly distributed error rates (∼5–20%). Indeed, despite their advantages, such methods that provide long reads can also show high error rates, thus reducing the accuracy of this genome sequencing (Minervini et al., 2020). Recently, nanopore sequencing has been used in a number of fields, such as genomics, epigenomics, and transcriptomics. The third-generation NGS approach overcomes the problems of (i) sequencing of tandem repeats; (ii) detection of complex chromosomal rearrangements and structural variants; (iii) RNA sequencing for detection of alternative splicing/transcripts or RNA isoforms; and (iv) direct sequencing of epigenetic/methylation markers. The direct detection of epigenetic modifications or RNA molecules can remove the need for reverse transcription for RNA sequencing and for bisulfite treatment to decipher methylation. Of note, a strong limitation arising from long reads and their analysis is the computational requirements. During the process of genome assembly, as the number of reads increases, an exponential increase is seen for the number of overlaps computed between the reads. For these reasons, many studies have demonstrated that even with the long reads technologies that have been developed, the relevance of short reads has not been lost yet. Indeed, the long reads approach is not yet used in the diagnosis of HAs.

APPLICATION OF GENETICS AND GENOMICS TO DIFFERENTIAL DIAGNOSIS OF HEREDITARY ANEMIAS: CLINICAL AND THERAPEUTIC IMPLICATIONS

Although the diagnostic workflow for HAs is part of normal clinical practice, it is often very difficult to obtain differential diagnosis and classification. Indeed, there is a wide range of unspecific phenotypes and overlapping phenotypes in patients with different genetic backgrounds. For instance, CDAI/CDAII and DHS share several clinical aspects (such as low reticulocyte counts, and dyserythropoietic features of the bone marrow) and are thus often misdiagnosed, as also for CDAII and hereditary spherocytosis (Andolfo et al., 2018a). Such incorrect diagnosis can critically impact on the follow-up and therapy of these patients. Splenectomy is the most effective surgical

---

[1] www.pacb.com

[2] www.nanoporetech.com

treatment for hereditary spherocytosis, although it might not be necessary for a patient with CDAII, or even worst, it can be contraindicated in DHS because of the risk of severe thrombotic events (Stewart et al., 1996; Iolascon et al., 2017; Andolfo et al., 2018a; Picard et al., 2019).

Next-generation sequencing-based diagnosis has resulted in the modification of initial clinical diagnoses for 10–40% of patients investigated (Iolascon et al., 2020). Moreover, a recent study showed that 45.5% of patients with CDAs had conclusive diagnosis of chronic anemia arising from enzymatic defects, mainly in terms of pyruvate kinase deficiency (Russo et al., 2018). Of note, most of these cases were transfusion-dependent patients, where a classical laboratory approach (e.g., pyruvate kinase enzyme assay) was not performed or was not available. A few years ago, it was demonstrated that an allosteric activator of wild-type and mutant pyruvate kinases, AG-348 (also known as mitapivat) can increase enzymatic activities in the RBCs of these patients, and thus it has been proposed as a novel therapeutic approach for this disease (Kung et al., 2017). Indeed, a phase 2 study of patients with pyruvate kinase deficiency treated with mitapivat was recently completed (ClinicalTrials.gov: NCT02476916), which indicated that the correct identification of these patients might be valuable for guiding their treatment (Grace et al., 2019b). Co-inheritance of multiple conditions or multiple disease-associated variants are further issues that increase the complex scenario of the diagnosis of HAs. For instance, even though ektacytometry is the gold standard for DHS diagnosis, the co-inheritance of β-thalassemia or the splenectomy might modify the shape of the ektacytometry curve, thereby resulting in misdiagnosis (Lazarova et al., 2017; Zaninoni et al., 2018).

APPLICATION OF GENETICS AND GENOMICS TO THE IDENTIFICATION OF HIDDEN DISEASE MECHANISMS

In this section, we will briefly present the main advances and the future perspectives for the use of NGS technologies for identification of genetic modifier variants, non-coding regulatory variants, and genomic structural variants in the erythroid genes. Currently, this is an active and interesting field of research, but it is still far from application to routine clinical diagnostics.

Polygenic Conditions and Genetic Modifiers

Identification of new causative genes has provided improved knowledge of disease etiology; however, there remains the need for better understanding of the genetic factors that can modify HA disease severity. Even diseases that are simple to diagnose can show clinical variability, and it might be that this variability itself involves genetic factors, as so-called "modifier genes" (Genin et al., 2008). One of the major advantages of the NGS approach is the identification of both the polygenic conditions and the modifier variants associated with causative mutations. Although there is no unique definition, we would define such a modifier as

a gene that changes the expression of another gene at a different locus that affects the phenotypic expression of another gene, or that can modify the phenotypic manifestation of a mutant gene while not showing any effects on the normal condition (Genin et al., 2008).

In a recent study, an analysis of patients with CDAII that used a target panel of 81 genes for modifier genes identified modifier variants that could explain the clinical variability of these patients, and even among patients who shared the same pathogenic variants. Among these variants, an *ERFE* gene recurrent low-frequency variant, A260S, was shown in 12.5% of patients with CDAII who had severe phenotypes. *ERFE* encodes erythroferrone, which is a soluble protein that is secreted by erythroid precursors and that suppresses expression of hepcidin. Increased levels of ERFE are associated with the ERFE-A260S variant, which results in large impairment of the regulation pathways for iron at the level of the liver. ERFE-A260S functional characterization in a hepatic cell system showed that it has a modifier role in iron overload through the BMP/SMAD pathway. This was the first description of ERFE polymorphism as a genetic modifier variant (Andolfo et al., 2019).

Next-generation sequencing technologies allowed the identification of co-inheritance of multiple disease genotypes in a patient with DHS1 who had severe iron overload that was caused by bi-allelic mutations in *SEC23B* that had been co-inherited with a *PIEZO1* mutation (Russo et al., 2018). *PIEZO1* is a highly polymorphic gene that shows high tolerance to variations. Recently, we provided evidence that the *de novo* R1864H mutation in the *PIEZO1* gene has a phenotype modifier role that co-segregates with the inherited E2492_L2493dup, where the severe phenotype was seen for a proband with DHS1. Of note, the effects that are mediated by R1864H are mainly evident in modulation of the RBC hydration status. Indeed, we were able to show that this rare missense variant resulted in augmented K$^+$ efflux when it was co-inherited with the duplication, which led in turn to the disease taking on a more severe clinical presentation (Andolfo et al., 2018c).

A recent study on 73 Asian families in an investigated that used NGS-based diagnostic approaches demonstrated that co-inherited G6PD deficiency was seen for 15% of patients with hereditary spherocytosis. *G6PD* variants worsen the phenotype by increasing the transfusion rate. Additionally, *UGT1A1* promoter variant homozygosity (Gilbert syndrome) was shown to lead to a significantly greater mean bilirubin levels, with high frequency of cholelithiasis (30% of the patients with hereditary spherocytosis analyzed) (Aggarwal et al., 2020). Hence, these new genetic technologies have provided useful tools to fill some of the gaps in our understanding of the genetic factors that modify HA disease severity.

Non-coding Genetic Variants: Transcription Factor Binding Sites and *Cis*-Regulatory Elements

Increased knowledge of how transcription factors and *cis*-regulatory elements influence and guide the fine-tuned processes of erythroid differentiation is critical for translation of these

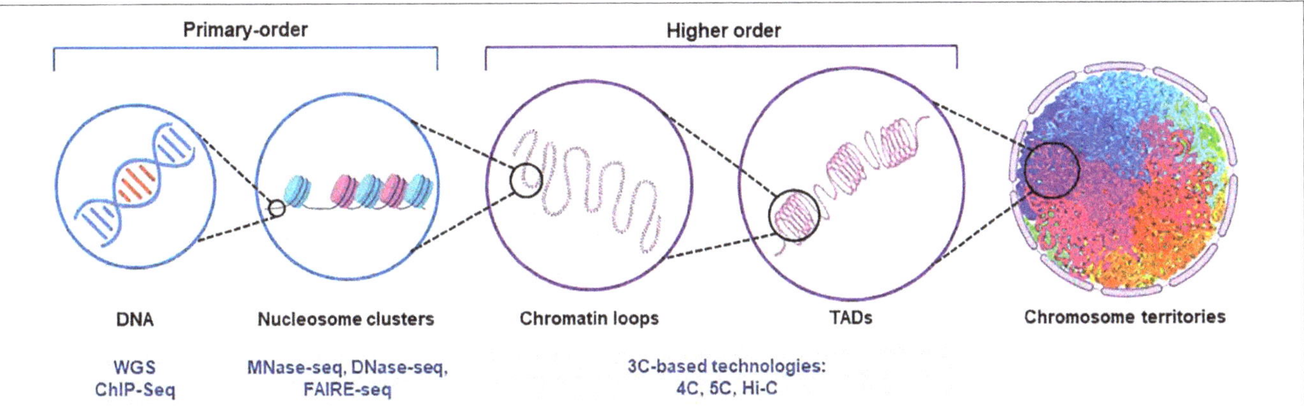

FIGURE 2 | Representation of the structure of chromosomal DNA and the technical procedures for the assessment of chromatin hierarchy. The diagram shows the main features of the organization of chromatin, from those of higher order to those of primary order. The different next-generation-sequencing-based techniques used for each order degree are shown. Whole-genome sequencing (WGS) applications allow identification of both coding and non-coding regulatory variants. Chromatin immunoprecipitation sequencing (ChIP-seq) provides identification of binding sites of DNA-associated proteins, which can be used to map the global binding sites for any given protein (e.g., transcription factors). Endo-exonuclease sequencing (MNase-seq) allows identification of closed chromatin by using an endo-exonuclease to cleave the linker DNA between nucleosomes. DNase I endonuclease sequencing (DNase-seq) identifies open chromatin regions, as accessible regions of the genome are hypersensitive to this activity. Sequencing through formaldehyde-assisted isolation of regulatory elements (FAIRE-seq) is a similar method that identifies open regions of the genome. The chromosome conformation capture (3C)-based technologies (e.g., 4C, 5C, and their variants [Hi-C]) incorporate next-generation sequencing and can provide quantitative measures for intra (*cis*)-chromosomal and inter (*trans*)-chromosomal interactions. TADs, topologically associated domains.

findings from research to possible future diagnosis and treatment of HAs. Although both targeted NGS and WES are currently used in the diagnostic workflow for these disorders, WGS might be useful in the research field to identify the presence of putative non-coding causative mutations (**Figure 2**). Indeed, the mutational disruption of transcription factor occupancy sites has been shown to be a pathogenic mechanism in several hematological disorders. For instance, mutations of the GATA1-binding *cis*-element in an intron of the *ALAS2* gene account for X-linked sideroblastic anemia (Kaneko et al., 2014). Similarly, disruption of the consensus binding motif for GATA1 in the promoter of *PKLR* results in severe pyruvate kinase deficiency (Manco et al., 2000; van Wijk and van Solinge, 2005).

As well as non-coding causative mutations, single nucleotide polymorphisms within *cis*-elements might also result in significant inter-individual differences in hematological parameters, and might influence therapeutic strategies. Genome-wide association studies have identified motif-disrupting single nucleotide polymorphisms in the enhancer of *BCL11A*, a critical repressor of fetal Hb levels that is associated with decreased *BCL11A* expression and elevated fetal Hb (Bauer et al., 2013). Indeed, they proposed the CRISPR–Cas9-mediated *BCL11A* enhancer editing approach in hematopoietic stem cells, as a practicable therapeutic strategy to produce durable fetal Hb in patients with β-hemoglobinopathies (Wu et al., 2019). Similarly, a polymorphism in the 5' upstream region of *GATA1* was described as a genetic modifier in patients with CDAII (Russo et al., 2017).

Along with time-dependent regulation of erythroid gene expression, several studies have described the fine regulation that underlies tissue specificity. Thirty-five percent of genes expressed in erythroid cells have many distinct alternative promoters and first exons (Tan et al., 2006) that regulate tissue-specific isoform selection. For erythrocytes, a high number of cytoskeletal proteins and transcription factors have alternative first exons that can be upstream, positioned close together (e.g., *NRF1*) or tens of kilobases apart (e.g., *EPB4.1*), or downstream of one or more internal coding exons (e.g., *ANK1*). Of note, a 2-kb upstream region of the *ANK1* erythroid promoter leads to selective activation of its transcription in erythroid cells, while this is silent in cell types where alternative tissue-specific *ANK1* promoters are active (Gallagher et al., 2010). This sheds light on the possible pathogenic roles of mutations that do not alter coding sequences, but reduce and alter the transcription of specific isoforms (Yocum et al., 2012).

Genomic Structural Variants: Topologically Associating Domains and Chromatin Occupancy

The assembly of DNA and proteins in chromatin is compact and organized, with the three-dimensional structure showing intricate folding. The chromosomal DNA structure is defined as 'higher order' and 'primary order' based on this folding complexity (Chang et al., 2018). The human genome is characterized by architectural features that are conserved, which include chromatin loops and domains, and nuclear bodies. It also shows non-random patterns, as seen for the location of genes and chromosomes in three-dimensional space (Misteli, 2020). With the advances seen for high-throughput sequencing, profiling of chromatin structures on a genome-wide basis has been made possible. **Figure 2** briefly summarizes the main NGS-based application for the assessment of chromatin hierarchy. Generally, intra-chromosomal interactions are within regions that are known as megabase-sized topologically associating domains (TADs). These TADs have been identified as large-scale (>100 kb), highly self-interacting regions that are in spatial proximity, and that constrain the spread of heterochromatin within the same region. CCCTC-binding factor (CTCF) is enriched at the TAD boundaries, and this transcriptional repressor or insulator prevents communication between different TADs (Dixon et al., 2012). The so-called 'loop extrusion mechanism' that is mediated by binding of cohesin to sites on CTCF was proposed as the formation process for TADs (Fudenberg et al., 2016; Oudelaar et al., 2018; Vian et al., 2018). It is now established that disruption of the TAD boundaries is a recurrent disease mechanism that promotes aberrant gene expression due to exposure of genes to inappropriate regulatory elements (Lupianez et al., 2015; Gonzalez-Sandoval and Gasser, 2016; Sikorska and Sexton, 2020).

Genomic structural variants have been extensively studied for the globin *loci*. Hay and colleagues characterized the mouse α-globin super-enhancer, which contains five enhancer-like elements flanked by two pairs of CTCF binding sites, which define a single TAD (Hay et al., 2016).

Although TADs are strongly conserved between species, a systematic comparative analysis of chromatin occupancy of master regulators between mouse and human was performed, in the transcriptional landscape of erythroid differentiation. It was shown that transcription factor occupancy sites are well conserved across proerythroblasts and cell lines of the same species (e.g., human proerythroblasts *versus* human K562 cells) but less across different species, according to previous observations regarding the differences in timing and expression levels of some constituent erythroid genes (Pishesha et al., 2014). Conversely, substantial divergence across the mouse and human epigenomes have been reported, as demonstrated for *SEC23A* (paralog of *SEC23B*) transcription factor occupancy and histone state. Human *SEC23A* and the surrounding region are in a general state of heterochromatin, whereas the mouse *sec23a* region is open for transcription. This finding provides evidence for why hematopoietic deficiency for *SEC23B* in mice does not result in anemia or other CDAII characteristics, while it does in humans (Khoriaty et al., 2014). Moreover, the SEC23B deficient phenotype in mice can be completely rescued by expression of SEC23A from the endogenous *Sec23b* locus, which indicates that *SEC23A* and *SEC23B* have functional overlap (Khoriaty et al., 2018). In agreement with this overlap in erythroid cells, a report has shown a milder phenotype in two CDAII patients with higher SEC23A levels compared to one CDAII patient with lower SEC23A levels (Russo et al., 2013). These studies also emphasize

the importance of looking at differences in chromatin occupancy between species during the development of a disease model (Ulirsch et al., 2014).

WHERE NEXT FOR GENETICS AND GENOMICS IN HEREDITARY ANEMIAS?

The advent of high-throughput sequencing techniques has provided increased knowledge of the genetic and genomic differences that can be found among individuals. This has gradually led to changes in clinical management and therapeutic plans, which have moved from population-based approaches to providing personalized therapies for individual patients. Genetic approaches now form a routine part of studies of HAs, and they are becoming widespread in clinical practice. On the other hand, the study of non-coding genetic variants and genomic structural variants is very interesting and promising in the search for new pathogenetic mechanisms, but these remain far from being applied in clinical diagnosis.

Testing based on NGS can now provide time-effective diagnosis, while also identifying polygenic conditions and modifier variants. With our increasing knowledge of the genetic features combined with detailed phenotyping of patients with HA, this will facilitate the diagnosis of these patients, to improve their personalized clinical management, while also generating advanced telemedicine tools (Tornador et al., 2019). Third-generation sequencing still represents a revolution for NGS technologies. In terms of HAs, these approaches are still far from diagnostic applications, but they provide a valid approach for the discovery of new pathogenic mechanisms.

In the near future, this transition to personalized medicine provided by the use of genomics for HAs, as in other fields of medicine, should result in the use of individualized treatments through the targeting of the right medication to the right person at the right time, based on their unique profile.

AUTHOR CONTRIBUTIONS

RR, IA, RM, and BR reviewed the literature data and wrote the manuscript. AI revised the manuscript. RR designed the figures. All authors contributed to the article and approved the submitted version.

REFERENCES

Agarwal, A. M., Nussenzveig, R. H., Reading, N. S., Patel, J. L., Sangle, N., Salama, M. E., et al. (2016). Clinical utility of next-generation sequencing in the diagnosis of hereditary haemolytic anaemias. *Br. J. Haematol.* 174, 806–814. doi: 10.1111/bjh.14131

Aggarwal, A., Jamwal, M., Sharma, P., Sachdeva, M. U. S., Bansal, D., Malhotra, P., et al. (2020). Deciphering molecular heterogeneity of Indian families with hereditary spherocytosis using targeted next-generation sequencing: first south asian study. *Br. J. Haematol.* 188, 784–795. doi: 10.1111/bjh.16244

Alkan, C., Sajjadian, S., and Eichler, E. E. (2011). Limitations of next-generation genome sequence assembly. *Nat. Methods* 8, 61–65. doi: 10.1038/nmeth.1527

Andolfo, I., Alper, S. L., De Franceschi, L., Auriemma, C., Russo, R., De Falco, L., et al. (2013a). Multiple clinical forms of dehydrated hereditary stomatocytosis arise from mutations in PIEZO1. *Blood* 121, S3912–S3921.

Andolfo, I., Alper, S. L., Delaunay, J., Auriemma, C., Russo, R., Asci, R., et al. (2013b). Missense mutations in the ABCB6 transporter cause dominant familial pseudohyperkalemia. *Am. J. Hematol.* 88, 66–72. doi: 10.1002/ajh.23357

Andolfo, I., Manna, F., De Rosa, G., Rosato, B. E., Gambale, A., Tomaiuolo, G., et al. (2018c). PIEZO1-R1864H rare variant accounts for a genetic phenotype-modifier role in dehydrated hereditary stomatocytosis. *Haematologica* 103, e94–e97.

Andolfo, I., Martone, S., Ribersani, M., Bianchi, S., Manna, F., Genesio, R., et al. (2020). Apparent recessive inheritance of sideroblastic anemia type 2 due to uniparental isodisomy at the SLC25A38 locus. *Haematologica* 2020:258533.

Andolfo, I., Rosato, B. E., Marra, R., De Rosa, G., Manna, F., Gambale, A., et al. (2019). The BMP-SMAD pathway mediates the impaired hepatic iron metabolism associated with the ERFE-A260S variant. *Am. J. Hematol.* 94, 1227–1235. doi: 10.1002/ajh.25613

Andolfo, I., Russo, R., Gambale, A., and Iolascon, A. (2016). New insights on hereditary erythrocyte membrane defects. *Haematologica* 101, 1284–1294. doi: 10.3324/haematol.2016.142463

Andolfo, I., Russo, R., Gambale, A., and Iolascon, A. (2018a). Hereditary stomatocytosis: an underdiagnosed condition. *Am. J. Hematol.* 93, 107–121. doi: 10.1002/ajh.24929

Andolfo, I., Russo, R., Rosato, B. E., Manna, F., Gambale, A., Brugnara, C., et al. (2018b). Genotype-phenotype correlation and risk stratification in a cohort of 123 hereditary stomatocytosis patients. *Am. J. Hematol.* 93, 1509–1517. doi: 10.1002/ajh.25276

Babbs, C., Roberts, N. A., Sanchez-Pulido, L., McGowan, S. J., Ahmed, M. R., Brown, J. M., et al. (2013). Homozygous mutations in a predicted endonuclease are a novel cause of congenital dyserythropoietic anemia type I. *Haematologica* 98, 1383–1387. doi: 10.3324/haematol.2013.089490

Bauer, D. E., Kamran, S. C., Lessard, S., Xu, J., Fujiwara, Y., Lin, C., et al. (2013). An erythroid enhancer of *BCL11A* subject to genetic variation determines fetal hemoglobin level. *Science* 342, 253–257. doi: 10.1126/science.1242088

Bianchi, P., Fermo, E., Glader, B., Kanno, H., Agarwal, A., Barcellini, W., et al. (2019). Addressing the diagnostic gaps in pyruvate kinase deficiency: consensus recommendations on the diagnosis of pyruvate kinase deficiency. *Am. J. Hematol.* 94, 149–161. doi: 10.1002/ajh.25325

Bianchi, P., Fermo, E., Lezon-Geyda, K., van Beers, E. J., Morton, H. D., Barcellini, W., et al. (2020). Genotype-phenotype correlation and molecular heterogeneity in pyruvate kinase deficiency. *Am. J. Hematol.* 95, 472–482.

Bianchi, P., Fermo, E., Vercellati, C., Boschetti, C., Barcellini, W., Iurlo, A., et al. (2009). Congenital dyserythropoietic anemia type II (CDAII) is caused by mutations in the SEC23B gene. *Hum. Mutat.* 30, 1292–1298. doi: 10.1002/humu.21077

Bogardus, H., Schulz, V. P., Maksimova, Y., Miller, B. A., Li, P., Forget, B. G., et al. (2014). Severe nondominant hereditary spherocytosis due to uniparental isodisomy at the SPTA1 locus. *Haematologica* 99, e168–e170.

Chang, P., Gohain, M., Yen, M. R., and Chen, P. Y. (2018). Computational methods for assessing chromatin hierarchy. *Comput. Struct. Biotechnol. J.* 16, 43–53. doi: 10.1016/j.csbj.2018.02.003

Choi, H. S., Choi, Q., Kim, J. A., Im, K. O., Park, S. N., Park, Y., et al. (2019). Molecular diagnosis of hereditary spherocytosis by multi-gene target sequencing in Korea: matching with osmotic fragility test and presence of spherocyte. *Orphanet. J. Rare Dis.* 14:114.

Da Costa, L., Galimand, J., Fenneteau, O., and Mohandas, N. (2013). Hereditary spherocytosis, elliptocytosis, and other red cell membrane disorders. *Blood Rev.* 27, 167–178. doi: 10.1016/j.blre.2013.04.003

Da Costa, L., Narla, A., and Mohandas, N. (2018). An update on the pathogenesis and diagnosis of diamond-Blackfan anemia. *F1000Res.* 7:F1000.

Del Orbe Barreto, R., Arrizabalaga, B., De la Hoz, A. B., García-Orad, Á., Tejada, M. I., Garcia-Ruiz, J. C., et al. (2016). Detection of new pathogenic mutations in patients with congenital haemolytic anaemia using next-generation sequencing. *Int. J. Lab. Hematol.* 38, 629–638. doi: 10.1111/ijlh.12551

Dgany, O., Avidan, N., Delaunay, J., Krasnov, T., Shalmon, L., Shalev, H., et al. (2002). Congenital dyserythropoietic anemia type I is caused by mutations in codanin-1. *Am. J. Hum. Genet.* 71, 1467–1474.

Dixon, J. R., Selvaraj, S., Yue, F., Kim, A., Li, Y., Shen, Y., et al. (2012). Topological domains in mammalian genomes identified by analysis of chromatin interactions. *Nature* 485, 376–380. doi: 10.1038/nature11082

Fudenberg, G., Imakaev, M., Lu, C., Goloborodko, A., Abdennur, N., and Mirny, L. A. (2016). Formation of chromosomal domains by loop extrusion. *Cell Rep.* 15, 2038–2049. doi: 10.1016/j.celrep.2016.04.085

Furst, D., Tsamadou, C., Neuchel, C., Schrezenmeier, H., Mytilineos, J., and Weinstock, C. (2020). Next-generation sequencing technologies in blood group typing. *Transfus. Med. Hemother.* 47, 4–13. doi: 10.1159/000504765

Gallagher, P. G., Steiner, L. A., Liem, R. I., Owen, A. N., Cline, A. P., Seidel, N. E., et al. (2010). Mutation of a barrier insulator in the human ankyrin-1 gene is associated with hereditary spherocytosis. *J. Clin. Invest.* 120, 4453–4465. doi: 10.1172/jci42240

Gambale, A., Iolascon, A., Andolfo, I., and Russo, R. (2016). Diagnosis and management of congenital dyserythropoietic anemias. *Expert. Rev. Hematol.* 9, 283–296. doi: 10.1586/17474086.2016.1131608

Genin, E., Feingold, J., and Clerget-Darpoux, F. (2008). Identifying modifier genes of monogenic disease: strategies and difficulties. *Hum. Genet.* 124, 357–368. doi: 10.1007/s00439-008-0560-2

Gonzalez-Sandoval, A., and Gasser, S. M. (2016). On TADs and LADs: spatial control over gene expression. *Trends Genet.* 32, 485–495. doi: 10.1016/j.tig.2016.05.004

Grace, R. F., Bianchi, P., van Beers, E. J., Eber, S. W., Glader, B., Yaish, H. M., et al. (2018). Clinical spectrum of pyruvate kinase deficiency: data from the pyruvate kinase deficiency natural history study. *Blood* 131, 2183–2192.

Grace, R. F., Mark Layton, D., and Barcellini, W. (2019a). How we manage patients with pyruvate kinase deficiency. *Br. J. Haematol.* 184, 721–734. doi: 10.1111/bjh.15758

Grace, R. F., Rose, C., Layton, D. M., Galacteros, F., Barcellini, W., Morton, D. H., et al. (2019b). Safety and efficacy of mitapivat in pyruvate kinase deficiency. *N. Engl. J. Med.* 381, 933–944. doi: 10.1056/nejmoa1902678

Hamada, M., Doisaki, S., Okuno, Y., Muramatsu, H., Hama, A., Kawashima, N., et al. (2018). Whole-exome analysis to detect congenital hemolytic anemia mimicking congenital dyserythropoietic anemia. *Int. J. Hematol.* 108, 306–311. doi: 10.1007/s12185-018-2482-7

Hay, D., Hughes, J. R., Babbs, C., Davies, J. O. J., Graham, B. J., Hanssen, L., et al. (2016). Genetic dissection of the alpha-globin super-enhancer in vivo. *Nat. Genet.* 48, 895–903. doi: 10.1038/ng.3605

Heimpel, H., Matuschek, A., Ahmed, M., Bader-Meunier, B., Colita, A., Delaunay, J., et al. (2010). Frequency of congenital dyserythropoietic anemias in europe. *Eur. J. Haematol.* 85, 20–25.

Higgs, D. R., Engel, J. D., and Stamatoyannopoulos, G. (2012). Thalassaemia. *Lancet* 379, 373–383.

Iolascon, A., Andolfo, I., Barcellini, W., Corcione, F., Garcon, L., De Franceschi, L., et al. (2017). Recommendations regarding splenectomy in hereditary hemolytic anemias. *Haematologica* 102, 1304–1313. doi: 10.3324/haematol.2016.161166

Iolascon, A., Andolfo, I., and Russo, R. (2019). Advances in understanding the pathogenesis of red cell membrane disorders. *Br. J. Haematol.* 187, 13–24. doi: 10.1111/bjh.16126

Iolascon, A., Andolfo, I., and Russo, R. (2020). Congenital dyserythropoietic anemias. *Blood* 136, 1274–1283. doi: 10.1182/blood.2019000948

Joly, P., Pondarre, C., and Badens, C. (2014). [Beta-thalassemias: molecular, epidemiological, diagnostical and clinical aspects]. *Ann. Biol. Clin.* 72, 639–668. doi: 10.1684/abc.2014.1015

Kaneko, K., Furuyama, K., Fujiwara, T., Kobayashi, R., Ishida, H., Harigae, H., et al. (2014). Identification of a novel erythroid-specific enhancer for the ALAS2 gene and its loss-of-function mutation which is associated with congenital sideroblastic anemia. *Haematologica* 99, 252–261. doi: 10.3324/haematol.2013.085449

Kedar, P. S., Harigae, H., Ito, E., Muramatsu, H., Kojima, S., Okuno, Y., et al. (2019). Study of pathophysiology and molecular characterization of congenital anemia in India using targeted next-generation sequencing approach. *Int. J. Hematol.* 110, 618–626. doi: 10.1007/s12185-019-02716-9

Khoriaty, R., Hesketh, G. G., Bernard, A., Weyand, A. C., Mellacheruvu, D., Zhu, G., et al. (2018). Functions of the COPII gene paralogs SEC23A and SEC23B are interchangeable in vivo. *Proc. Natl. Acad. Sci. U S A.* 115, E7748–E7757.

Khoriaty, R., Vasievich, M. P., Jones, M., Everett, L., Chase, J., Tao, J., et al. (2014). Absence of a red blood cell phenotype in mice with hematopoietic deficiency of SEC23B. *Mol. Cell. Biol.* 34, 3721–3734. doi: 10.1128/mcb.00287-14

Koralkova, P., van Solinge, W. W., and van Wijk, R. (2014). Rare hereditary red blood cell enzymopathies associated with hemolytic anemia - pathophysiology, clinical aspects, and laboratory diagnosis. *Int. J. Lab. Hematol.* 36, 388–397. doi: 10.1111/ijlh.12223

Kung, C., Hixon, J., Kosinski, P. A., Cianchetta, G., Histen, G., Chen, Y., et al. (2017). AG-348 enhances pyruvate kinase activity in red blood cells from patients with pyruvate kinase deficiency. *Blood* 130, 1347–1356. doi: 10.1182/blood-2016-11-753525

Lazarova, E., Gulbis, B., Oirschot, B. V., and van Wijk, R. (2017). Next-generation osmotic gradient ektacytometry for the diagnosis of hereditary spherocytosis: interlaboratory method validation and experience. *Clin. Chem. Lab. Med.* 55, 394–402.

Lupianez, D. G., Kraft, K., Heinrich, V., Krawitz, P., Brancati, F., Klopocki, E., et al. (2015). Disruptions of topological chromatin domains cause pathogenic rewiring of gene-enhancer interactions. *Cell* 161, 1012–1025. doi: 10.1016/j.cell.2015.04.004

Luzzatto, L., Nannelli, C., and Notaro, R. (2016). Glucose-6-phosphate dehydrogenase deficiency. *Hematol. Oncol. Clin. North Am.* 30, 373–393.

Manco, L., Ribeiro, M. L., Maximo, V., Almeida, H., Costa, A., Freitas, O., et al. (2000). A new PKLR gene mutation in the R-type promoter region affects the gene transcription causing pyruvate kinase deficiency. *Br. J. Haematol.* 110, 993–997. doi: 10.1046/j.1365-2141.2000.02283.x

Minervini, C. F., Cumbo, C., Orsini, P., Anelli, L., Zagaria, A., Specchia, G., et al. (2020). Nanopore sequencing in blood diseases: a wide range of opportunities. *Front. Genet.* 11:76. doi: 10.3389/fgene.2020.00076

Misteli, T. (2020). The self-organizing genome: principles of genome architecture and function. *Cell* 183, 28–45. doi: 10.1016/j.cell.2020.09.014

Mousavi, N., Shleizer-Burko, S., Yanicky, R., and Gymrek, M. (2019). Profiling the genome-wide landscape of tandem repeat expansions. *Nucleic Acids Res.* 47:e90. doi: 10.1093/nar/gkz501

Niss, O., Chonat, S., Dagaonkar, N., Almansoori, M. O., Kerr, K., Rogers, Z. R., et al. (2016). Genotype-phenotype correlations in hereditary elliptocytosis and hereditary pyropoikilocytosis. *Blood Cells Mol. Dis.* 61, 4–9. doi: 10.1016/j.bcmd.2016.07.003

Ohene-Abuakwa, Y., Orfali, K. A., Marius, C., and Ball, S. E. (2005). Two-phase culture in diamond blackfan anemia: localization of erythroid defect. *Blood* 105, 838–846. doi: 10.1182/blood-2004-03-1016

Oudelaar, A. M., Davies, J. O. J., Hanssen, L. L. P., Telenius, J. M., Schwessinger, R., Liu, Y., et al. (2018). Single-allele chromatin interactions identify regulatory hubs in dynamic compartmentalized domains. *Nat. Genet.* 50, 1744–1751. doi: 10.1038/s41588-018-0253-2

Picard, V., Guitton, C. I, Thuret, I., Rose, C., Bendelac, L., Ghazal, K., et al. (2019). Clinical and biological features in PIEZO1-hereditary xerocytosis and Gardos channelopathy: a retrospective series of 126 patients. *Haematologica* 104, 1554–1564. doi: 10.3324/haematol.2018.205328

Pishesha, N., Thiru, P., Shi, J., Eng, J. C., Sankaran, V. G., and Lodish, H. F. (2014). Transcriptional divergence and conservation of human and mouse erythropoiesis. *Proc. Natl. Acad. Sci. U S A.* 111, 4103–4108. doi: 10.1073/pnas.1401598111

Rapetti-Mauss, R., Lacoste, C., Picard, V., Guitton, C., Lombard, E., Loosveld, M., et al. (2015). A mutation in the Gardos channel is associated with hereditary xerocytosis. *Blood* 126, 1273–1280. doi: 10.1182/blood-2015-04-642496

Rees, D. C., Porter, J. B., Clegg, J. B., and Weatherall, D. J. (1999). Why are hemoglobin F levels increased in HbE/beta thalassemia? *Blood* 94, 3199–3204. doi: 10.1182/blood.v94.9.3199.421k19_3199_3204

Rehm, H. L., Bale, S. J., Bayrak-Toydemir, P., Berg, J. S., Brown, K. K., Deignan, J. L., et al. (2013). ACMG clinical laboratory standards for next-generation sequencing. *Genet. Med.* 15, 733–747.

Rets, A., Clayton, A. L., Christensen, R. D., and Agarwal, A. M. (2019). Molecular diagnostic update in hereditary hemolytic anemia and neonatal hyperbilirubinemia. *Int. J. Lab. Hematol.* 41, 95–101. doi: 10.1111/ijlh.13014

Roy, N. B., Wilson, E. A., Henderson, S., Wray, K., Babbs, C., Okoli, S., et al. (2016). A novel 33-Gene targeted resequencing panel provides accurate, clinical-grade diagnosis and improves patient management for rare inherited anaemias. *Br. J. Haematol.* 175, 318–330. doi: 10.1111/bjh.14221

Roy, N. B. A., and Babbs, C. (2019). The pathogenesis, diagnosis and management of congenital dyserythropoietic anaemia type I. *Br. J. Haematol.* 185, 436–449. doi: 10.1111/bjh.15817

Russo, R., Andolfo, I., Gambale, A., De Rosa, G., Manna, F., Arillo, A., et al. (2017). GATA1 erythroid-specific regulation of SEC23B expression and its implication in the pathogenesis of congenital dyserythropoietic anemia type II. *Haematologica* 102, e371–e374. doi: 10.3324/haematol.2016.162966

Russo, R., Andolfo, I., Manna, F., Gambale, A., Marra, R., Rosato, B. E., et al. (2018). Multi-gene panel testing improves diagnosis and management of patients with hereditary anemias. *Am. J. Hematol.* 93, 672–682. doi: 10.1002/ajh.25058

Russo, R., Langella, C., Esposito, M. R., Gambale, A., Vitiello, F., Vallefuoco, F., et al. (2013). Hypomorphic mutations of SEC23B gene account for mild phenotypes of congenital dyserythropoietic anemia type II. *Blood Cells Mol. Dis.* 51, 17–21. doi: 10.1016/j.bcmd.2013.02.003

Sabath, D. E. (2017). Molecular diagnosis of thalassemias and hemoglobinopathies: an ACLPS critical review. *Am. J. Clin. Pathol.* 148, 6–15. doi: 10.1093/ajcp/aqx047

Sankaran, V. G., Ghazvinian, R., Do, R., Thiru, P., Vergilio, J. A., Beggs, A. H., et al. (2012). Exome sequencing identifies GATA1 mutations resulting in diamond-blackfan anemia. *J. Clin. Invest.* 122, 2439–2443. doi: 10.1172/jci63597

Schwarz, K., Iolascon, A., Verissimo, F., Trede, N. S., Horsley, W., Chen, W., et al. (2009). Mutations affecting the secretory COPII coat component SEC23B cause congenital dyserythropoietic anemia type II. *Nat. Genet.* 41, 936–940. doi: 10.1038/ng.405

Shefer Averbuch, N., Steinberg-Shemer, O., Dgany, O., Krasnov, T., Noy-Lotan, S., Yacobovich, J., et al. (2018). Targeted next generation sequencing for the diagnosis of patients with rare congenital anemias. *Eur. J. Haematol.* 101, 297–304. doi: 10.1111/ejh.13097

Sikorska, N., and Sexton, T. (2020). Defining functionally relevant spatial chromatin domains: it is a TAD complicated. *J. Mol. Biol.* 432, 653–664. doi: 10.1016/j.jmb.2019.12.006

Stavropoulos, D. J., Merico, D., Jobling, R., Bowdin, S., Monfared, N., Thiruvahindrapuram, B., et al. (2016). Whole genome sequencing expands diagnostic utility and improves clinical management in pediatric medicine. *NP.J. Genom. Med.* 1:15012.

Stewart, G. W., Amess, J. A., Eber, S. W., Kingswood, C., Lane, P. A., Smith, B. D., et al. (1996). Thrombo-embolic disease after splenectomy for hereditary stomatocytosis. *Br. J. Haematol.* 93, 303–310. doi: 10.1046/j.1365-2141.1996.4881033.x

Svidnicki, M., Zanetta, G. K., Congrains-Castillo, A., Costa, F. F., and Saad, S. T. O. (2020). Targeted next-generation sequencing identified novel mutations associated with hereditary anemias in Brazil. *Ann. Hematol.* 99, 955–962. doi: 10.1007/s00277-020-03986-8

Taher, A. T., Weatherall, D. J., and Cappellini, M. D. (2018). Thalassaemia. *Lancet* 391, 155–167.

Tan, J. S., Mohandas, N., and Conboy, J. G. (2006). High frequency of alternative first exons in erythroid genes suggests a critical role in regulating gene function. *Blood* 107, 2557–2561. doi: 10.1182/blood-2005-07-2957

Tornador, C., Sánchez-Prados, E., Cadenas, B., Russo, R., Venturi, V., Andolfo, I., et al. (2019). CoDysAn: a telemedicine tool to improve awareness and diagnosis for patients with congenital dyserythropoietic anemia. *Front. Physiol.* 10:1063. doi: 10.3389/fphys.2019.01063

Trujillano, D., Bertoli-Avella, A. M., Kumar Kandaswamy, K., Weiss, M. E., Köster, J., Marais, A., et al. (2017). Clinical exome sequencing: results from 2819 samples reflecting 1000 families. *Eur. J. Hum. Genet.* 25, 176–182. doi: 10.1038/ejhg.2016.146

Ulirsch, J. C., Lacy, J. N., An, X., Mohandas, N., Mikkelsen, T. S., and Sankaran, V. G. (2014). Altered chromatin occupancy of master regulators underlies evolutionary divergence in the transcriptional landscape of erythroid differentiation. *PLoS Genet* 10:e1004890. doi: 10.1371/journal.pgen.1004890

van Wijk, R., and van Solinge, W. W. (2005). The energy-less red blood cell is lost: erythrocyte enzyme abnormalities of glycolysis. *Blood* 106, 4034–4042. doi: 10.1182/blood-2005-04-1622

Vian, L., Pekowska, A., Rao, S. S. P., Kieffer-Kwon, K. R., Jung, S., Baranello, L., et al. (2018). The energetics and physiological impact of cohesin extrusion. *Cell* 173, 1165–1178.

Wu, Y., Zeng, J., Roscoe, B. P., Liu, P., Yao, Q., Lazzarotto, C. R., et al. (2019). Highly efficient therapeutic gene editing of human hematopoietic stem cells. *Nat. Med.* 25, 776–783.

Yocum, A. O., Steiner, L. A., Seidel, N. E., Cline, A. P., Rout, E. D., Lin, J. Y., et al. (2012). A tissue-specific chromatin loop activates the erythroid ankyrin-1 promoter. *Blood* 120, 3586–3593. doi: 10.1182/blood-2012-08-450262

Zaninoni, A., Fermo, E., Vercellati, C., Consonni, D., Marcello, A. P., Zanella, A., et al. (2018). Use of laser assisted optical rotational cell analyzer (LoRRca MaxSis) in the diagnosis of rbc membrane disorders, enzyme defects, and congenital dyserythropoietic anemias: a monocentric study on 202 patients. *Front. Physiol.* 9:451. doi: 10.3389/fphys.2018.00451

Zarychanski, R., Schulz, V. P., Houston, B. L., Maksimova, Y., Houston, D. S., Smith, B., et al. (2012). Mutations in the mechanotransduction protein PIEZO1 are associated with hereditary xerocytosis. *Blood* 120, 1908–1915. doi: 10.1182/blood-2012-04-422253

Zhao, J., Li, J., Lai, Q., and Yu, Y. (2020). Combined use of gap-PCR and next-generation sequencing improves thalassaemia carrier screening among premarital adults in china. *J. Clin. Pathol.* 73, 488–492. doi: 10.1136/jclinpath-2019-206339

8

N-Methyl-D-Aspartate Receptors in Hematopoietic Cells: What have we Learned?

Maggie L. Kalev-Zylinska[1,2], James I. Hearn[1], Asya Makhro[3,4] and Anna Bogdanova[3,4]*

[1]*Blood and Cancer Biology Laboratory, Department of Molecular Medicine and Pathology, University of Auckland, Auckland, New Zealand, [2]Department of Pathology and Laboratory Medicine, LabPlus Haematology, Auckland City Hospital, Auckland, New Zealand, [3]Red Blood Cell Research Group, Institute of Veterinary Physiology, Vetsuisse Faculty, University of Zurich, Zürich, Switzerland, [4]Zurich Center for Integrative Human Physiology, University of Zurich, Zürich, Switzerland*

***Correspondence:**
Maggie L. Kalev-Zylinska
m.kalev@auckland.ac.nz

Dr. Heimo Mairbäurl, Heidelberg University Hospital, also contributed to the review process for this manuscript.

The *N*-methyl-D-aspartate receptor (NMDAR) provides a pathway for glutamate-mediated inter-cellular communication, best known for its role in the brain but with multiple examples of functionality in non-neuronal cells. Data previously published by others and us provided *ex vivo* evidence that NMDARs regulate platelet and red blood cell (RBC) production. Here, we summarize what is known about these hematopoietic roles of the NMDAR. Types of NMDAR subunits expressed in megakaryocytes (platelet precursors) and erythroid cells are more commonly found in the developing rather than adult brain, suggesting trophic functions. Nevertheless, similar to their neuronal counterparts, hematopoietic NMDARs function as ion channels, and are permeable to calcium ions (Ca^{2+}). Inhibitors that block open NMDAR (memantine and MK-801) interfere with megakaryocytic maturation and proplatelet formation in primary culture. The effect on proplatelet formation appears to involve Ca^{2+} influx-dependent regulation of the cytoskeletal remodeling. In contrast to normal megakaryocytes, NMDAR effects in leukemic Meg-01 cells are diverted away from differentiation to increase proliferation. NMDAR hypofunction triggers differentiation of Meg-01 cells with the bias toward erythropoiesis. The underlying mechanism involves changes in the intracellular Ca^{2+} homeostasis, cell stress pathways, and hematopoietic transcription factors that upon NMDAR inhibition shift from the predominance of megakaryocytic toward erythroid regulators. This ability of NMDAR to balance both megakaryocytic and erythroid cell fates suggests receptor involvement at the level of a bipotential megakaryocyte-erythroid progenitor. In human erythroid precursors and circulating RBCs, NMDAR regulates intracellular Ca^{2+} homeostasis. NMDAR activity supports survival of early proerythroblasts, and in mature RBCs NMDARs impact cellular hydration state, hemoglobin oxygen affinity, and nitric oxide synthase activity. Overexcitation of NMDAR in mature RBCs leads to Ca^{2+} overload, K^+ loss, RBC dehydration, and oxidative stress, which may contribute to the pathogenesis of sickle cell disease. In summary, there is growing evidence that glutamate-NMDAR signaling regulates megakaryocytic and erythroid cells at different stages of maturation, with some intriguing differences emerging in NMDAR expression and function between normal and diseased cells. NMDAR signaling may provide new therapeutic opportunities in hematological disease, but *in vivo* applicability needs to be confirmed.

Keywords: glutamate, intracellular calcium signaling, megakaryocyte, erythropoiesis, red cells, platelets

INTRODUCTION

This review summarizes what has been learned about the roles of *N*-methyl-D-aspartate receptor (NMDAR) in megakaryocytic and erythroid cells. NMDARs are best known for their functions as glutamate-gated cation channels in the central nervous system (Traynelis et al., 2010). It appears that the NMDAR ion channel functionality is maintained in blood progenitors but NMDAR channel properties and its downstream pathways await further characterization in these cells. This paper starts with a brief overview of glutamate signaling in the brain. On this background, we highlight distinctive features of NMDAR in hematopoietic cells. Other glutamate receptors and mature blood cells are not discussed in detail but the appropriate background is provided to place this emerging field of research in a meaningful context. We describe NMDAR effects on hematopoietic differentiation, including some of our recent observations that suggest a novel role for the receptor in balancing megakaryocytic and erythroid cell fates (Hearn et al., 2020).

CLASSICAL GLUTAMATE-NMDAR AXIS IN THE BRAIN

Glutamate is synthesized from glutamine as a part of normal cellular metabolism in all cells (Yelamanchi et al., 2016). In neurons, vesicular glutamate transporters (VGLUT) pump glutamate into pre-synaptic vesicles (Daikhin and Yudkoff, 2000; Zhou and Danbolt, 2014). Upon membrane depolarization, vesicles fuse with the pre-synaptic plasma membrane and glutamate is released into the synaptic cleft. This process engages soluble *N*-ethyl-maleimide-sensitive factor attachment protein receptor (SNARE) proteins that are activated by Ca^{2+} entry through voltage-gated Ca^{2+} channels. Following release, glutamate concentrations in the synaptic cleft increase markedly, from 2–5 µM to approximately 1.1 mM. While in the synaptic cleft, glutamate activates ionotropic and metabotropic receptors located on the post-synaptic plasma membrane (Reiner and Levitz, 2018). Ionotropic receptors function as ion channels (for Na^+, K^+, and Ca^{2+}), and metabotropic receptors activate G-proteins that modulate ion channels directly and indirectly. The main purpose of the ionic flux is to generate and propagate action potentials characteristic of excitable tissues. The synaptic glutamate signal is terminated by the excitatory amino acid transporters (EAAT) present on astrocytes that remove glutamate from the synaptic cleft (Featherstone, 2010).

Abbreviations: ADP, adenosine diphosphate; AMPA, α-amino-3-hydroxy-5-methyl-4-isoxazolepropionic acid; AP5, D-2-amino-5-phosphonopentanoate; CaMK, Ca^{2+}/calmodulin-dependent kinase; CREB, cAMP response element binding protein; EAAT, excitatory amino acid transporters; EC_{50}, the concentration of an agonist that gives half-maximal response; EPO, erythropoietin; ER, endoplasmic reticulum; ErbB4, epidermal growth factor receptor Erb-B2 receptor tyrosine kinase 4; ERK, extracellular signal-regulated kinase; IC_{50}, the concentration of an inhibitor where the response (or binding) is reduced by half; MAPK, mitogen-activated protein kinase; MEP, megakaryocyte-erythroid progenitor; MK, megakaryocyte; NMDAR, *N*-methyl-D-aspartate receptor; PI3-K, phosphoinositide 3-kinase; PMA, phorbol 12-myristate 13-acetate; PSD, post-synaptic density; RBC, red blood cell; SNARE, soluble *N*-ethyl maleimide-sensitive factor attachment protein receptor; TPO, thrombopoietin; VGLUT, vesicular glutamate transporter.

The family of ionotropic glutamate receptors includes NMDA, α-amino-3-hydroxy-5-methyl-4-isoxazolepropionic acid (AMPA), and kainate receptors, each named after a distinct, synthetic agonist that activates them (Traynelis et al., 2010). AMPA and kainate receptors respond to glutamate first. They mediate intracellular influx of mostly Na^+, which leads to membrane depolarization and if large enough, triggers action potential. Membrane depolarization releases a Mg^{2+} ion blocking the pore of NMDAR, enabling receptor function. This order of events highlights that neuronal NMDAR can activate only when glutamate binding and membrane depolarization coincide (which is named "coincidence detection"). NMDAR-mediated Ca^{2+} influx contributes little to membrane depolarization but modifies synaptic strength through molecular events related to the Ca^{2+} role as "second messenger" (Traynelis et al., 2010; Hansen et al., 2018).

Typical NMDARs are built as tetramers that combine two obligate GluN1 subunits with another two GluN2 (A–D) or GluN3 (A or B) subunits, in various combinations. It is believed that GluN1 subunit is an essential component of all NMDARs, and variable GluN2 and GluN3 subunits are modulatory. NMDAR activation requires binding of L-glutamate on each of the GluN2 subunits, as well as glycine (co-agonist) on the GluN1 and GluN3 subunits. The alternative NMDAR ligands include D- and L- aspartate, homocysteine, homocysteic acid, and D-serine. NMDAR subunit composition varies substantially in different areas of the brain, and changes during development (Monyer et al., 1994; Wenzel et al., 1997). NMDAR subunits define the current amplitude and inactivation time, as well as cation selectivity and the regulation patterns, such as agonist affinity, mechano-sensitivity, Mg^{2+}-sensitivity, and responsiveness to polyamines. GluN2A and GluN2B subunits contribute high channel conductance and relatively fast de-activation kinetics compared to GluN2C- and GluN2D- containing NMDAR (Traynelis et al., 2010). In addition, NMDARs containing GluN2C, GluN2D, and GluN3 subunits display low affinity for Mg^{2+} blocking the pore, making activation of such receptors independent of membrane depolarization (Monyer et al., 1994; Chatterton et al., 2002; Wrighton et al., 2008).

NMDAR sensitivity (EC_{50}) to agonists is high, ranging from 0.4 to 1.7 µM for glutamate (in GluN1–GluN2D and GluN1–GluN2A receptors, respectively), and 0.1 to 2.1 µM for glycine (in GluN1–GluN2D and GluN1–GluN2A receptors, respectively) (Yamakura and Shimoji, 1999). These concentrations lie within the range that is normal in an inactive synaptic cleft. However, all types of NMDAR are extremely sensitive to the inhibition by protons (IC_{50} around 7.4 µM for most of the subunits) (Yamakura and Shimoji, 1999; Low et al., 2003; Cavara et al., 2009), and Zn^{2+} [IC_{50} of 10 nM, 1 µM, and 10 µM for the NMDAR containing GluN2A, GluN2B, and GluN2D, respectively (Gielen et al., 2009)].

NMDAR-mediated Ca^{2+} entry activates a number of intracellular signaling pathways, including Ca^{2+}/calmodulin-dependent kinase (CaMK), mitogen-activated protein kinase (MAPK) [including extracellular signal-regulated kinase (ERK), Jun kinase, and p38 MAPK], and phosphoinositide 3-kinase (PI3K) (Hardingham, 2006). NMDARs regulate activity-dependent gene expression through cAMP response element binding protein (CREB) transcription factor (Hardingham et al., 2001). Other mediators downstream of NMDAR include Ras, Fyn, striatal-enriched protein tyrosine phosphatase, and nitric oxide synthase. Highly coordinated (albeit incompletely elucidated) NMDAR signaling plays critical roles in embryonic brain development and later, in neuronal plasticity, which allows the brain to respond to new experiences and changing environment (Traynelis et al., 2010).

Unexpected Discoveries Outside of the Brain

During the past 10–20 years, NMDARs have been reported in multiple non-neuronal cell types, including hematopoietic (Bozic and Valdivielso, 2015; Hogan-Cann and Anderson, 2016), which raised a principal question of why non-excitable cells need these receptors. We admit this area of research is not very clear, sometimes even controversial, mainly due to the very low abundance of NMDAR in non-neuronal cells. Nevertheless, some progress has been achieved in the characterization of the subunit composition and currents mediated by non-neuronal NMDAR, in particular in red blood cells (RBC) (Makhro et al., 2010), platelets (Kalev-Zylinska et al., 2014), lymphocytes (Fenninger and Jefferies, 2019), and hematopoietic precursors, erythroblasts (Makhro et al., 2013; Hanggi et al., 2014, 2015) and megakaryocytes (Genever et al., 1999; Kamal et al., 2015). Information on the potential physiological role of these receptors is accumulating as well, including in erythroid cells (Makhro et al., 2013, 2016), and megakaryocytes (Hitchcock et al., 2003; Green et al., 2017; Kamal et al., 2018; Hearn et al., 2020). The subsequent sections will focus on the subunit composition, properties and the roles of NMDAR in megakaryocytic and erythroid precursors, and their mature progeny, platelets and RBCs.

PLATELET RESPONSIVENESS TO GLUTAMATE

Peripheral blood platelets store and respond to a number of regulatory molecules best known for their roles in neurotransmission, including serotonin, epinephrine, dopamine, histamine, γ-aminobutyric acid, and glutamate (Todrick et al., 1960; Ponomarev, 2018; Canobbio, 2019). In psychiatric patients, there is evidence of a crosstalk between abnormal NMDAR function in the brain and platelet responsiveness to glutamate (Berk et al., 1999). Platelets bind glutamate with similar kinetics to neurons (Almazov et al., 1988), store it in dense granules, and express AMPA, kainate, and NMDA receptors (Franconi et al., 1996, 1998; Morrell et al., 2008; Sun et al., 2009; Kalev-Zylinska et al., 2014; Green et al., 2017). Although there are variations between studies, all main types of ionotropic glutamate receptors have now been shown to be functional in platelets. Morrell et al. demonstrated that AMPA and kainate (but not NMDA) receptors amplify platelet activation by contributing Na^+ influx to membrane depolarization, but not Ca^{2+} influx (Morrell et al., 2008; Sun et al., 2009). Franconi et al. provided

the first evidence of NMDAR functionality in platelets, demonstrating that NMDARs induce Ca^{2+} influx into platelets but inhibit platelet function in the presence of adenosine diphosphate (ADP) and arachidonic acid (Franconi et al., 1996, 1998). Our own work demonstrated that NMDAR inhibitors (memantine, MK-801, and anti-GluN1 antibodies) interfere with platelet activation, aggregation and thrombus formation *ex vivo* (Kalev-Zylinska et al., 2014; Green et al., 2017). It is likely that methodological differences contributed to variable NMDAR effects between studies.

Intriguingly, in schizophrenia and bipolar disorders that are driven by deregulated NMDAR signaling, platelet Ca^{2+} levels are elevated, including in response to glutamate (Berk et al., 2000; Ruljancic et al., 2013; Harrison et al., 2019). Schizophrenia is characterized by NMDAR hypofunction in the limbic system (Coyle, 2012; Nakazawa et al., 2017), compensated by high glutamate levels and NMDAR hypersensitivity in other areas of the brain (Merritt et al., 2016). The fact that platelets from patients with schizophrenia also show glutamate hypersensitivity further argues that NMDAR functioning in platelets is similar to that in neurons (Berk et al., 2000).

Because platelets have limited protein synthesis, one would expect a similar range of glutamate receptors to be present in megakaryocytes. However, most data thus far indicate regulation of megakaryocytic differentiation by NMDAR, with little or no data on AMPA and kainate receptors (Genever et al., 1999; Hitchcock et al., 2003; Kamal et al., 2018). Nevertheless, electrophysiological recordings from freshly isolated mouse megakaryocytes support expression of functional AMPA receptors in megakaryocytes, most likely GluR2-containing and Ca^{2+}-impermeable (Morrell et al., 2008).

GLUTAMATE AND NMDAR IN MEGAKARYOCYTIC CELLS

Evidence for NMDAR Functionality in Megakaryocytic Cells

The first evidence that NMDARs operate as ion channels in megakaryocytes was obtained by demonstrating that [³H]MK-801 binds to native mouse megakaryocytes *in vivo*. Mice were injected with [³H]MK-801 intracardially, followed by bone marrow examination 15 min later (Genever et al., 1999). Because MK-801 is a non-competitive, use-dependent NMDAR inhibitor that can only bind within an open NMDAR pore (Traynelis et al., 2010), its labeling of megakaryocytes was consistent with the NMDAR function as an ion channel in megakaryocytic cells. Later, we showed that glutamate, NMDA, and glycine induce Ca^{2+} fluxes in Meg-01 cells, and NMDAR antagonists (MK-801, memantine, and AP5 [D-2-amino-5-phosphonopentanoate]) counteract this effect, indicating that NMDARs operate as Ca^{2+} channels in these cells (Kamal et al., 2015, 2018).

Table 1 provides a summary of the NMDAR subunit expression in megakaryocytic and erythroid cells, reported at either protein or transcript level. Unfortunately, testing for GluN proteins in hematopoietic cells has been difficult due to (a) very low abundance, (b) various protein isoforms and post-translational modifications, and (c) the lack of antibodies optimized for use in non-neuronal cells. Human megakaryocytes were first shown to express GluN1 using immunocytochemistry and Western blotting, the latter indicated that GluN1 was non-glycosylated, which may affect NMDAR distribution in the plasma membrane (Genever et al., 1999). Our group demonstrated expression of GluN1, GluN2A, and GluN2D in Meg-01, K-562, and Set-2 cells using flow cytometry and a modified Western blotting procedure that employed membrane enrichment and high-sensitivity peroxidase substrates (Kamal et al., 2015).

The composition of NMDAR in megakaryocytes differs from that in neurons. In the brain, NMDARs are built mostly from GluN1, GluN2A, and/or GluN2B subunits, but human, native and culture-derived megakaryocytes express predominantly GluN2D, with some GluN2A and GluN1 (**Table 1**; Genever et al., 1999; Hitchcock et al., 2003; Kamal et al., 2015, 2018). The dominant expression of GluN2D in normal megakaryocytes will affect NMDAR functioning, however no electrophysiological recordings are available from these cells to document the effect. In other systems, GluN2D-containing NMDAR displays the following differences compared with GluN2A and GluN2B containing receptors: approximately 5-fold higher sensitivity to glutamate, 10-fold higher sensitivity to glycine, 100-fold longer deactivation time, lower conductance, lower Ca^{2+} permeability and weaker Mg^{2+} block (Paoletti et al., 2013; Wyllie et al., 2013; Hansen et al., 2018). The weak Mg^{2+} block suggests that NMDAR in megakaryocytes may not require membrane depolarization to become active; meaning the principle of "coincidence detection" may not apply. The unique functionality of the GluN2D subunit is underscored by its dominant expression in the embryonic and postnatal brain; however, the mechanism through which GluN2D subunits provide trophic effects remains incompletely understood (Watanabe et al., 1992; Akazawa et al., 1994).

For those of us working with mouse models, it is relevant to note that there are differences in the NMDAR expression patterns between human and mouse cells. In contrast to human megakaryocytes (that express only GluN1, GluN2A, and GluN2D), mouse megakaryocytes also express GluN2C and GluN3B (**Table 1**; Kamal et al., 2018). Small numbers of other, yet un-identified mononuclear cells in the mouse bone marrow also express NMDAR, but there is no documented expression in mouse erythroid precursors or mature RBCs (Genever et al., 1999), which differs from human cells (**Table 1**; Makhro et al., 2013; Hanggi et al., 2014, 2015).

In contrast to normal megakaryocytes, patient-derived leukemic megakaryoblasts and megakaryocyte leukemia cell lines (Meg-01, K-562, and Set-2) carry all possible NMDAR subunits, including GluN2B, GluN3A, and GluN3B (**Table 1**; Kamal et al., 2015). Meg-01 and K-562 cell lines are derived from patients with chronic myeloid leukemia in megakaryocytic and myeloid blast crisis respectively, and carry oncogenic *BCR-ABL1* gene fusion (Lozzio and Lozzio, 1975; Ogura et al., 1985). Both Meg-01 and K-562 cell lines express thrombopoietin (TPO) and erythropoietin (EPO) receptors and can be induced to differentiate into megakaryocytic (Ogura et al., 1988, Herrera et al., 1998) and erythroid cells (Andersson et al., 1979; Morle et al., 1992), thus providing experimental models of bipotential megakaryocyte-erythroid progenitors. Set-2 cell line is derived from a leukemic transformation of essential thrombocythemia and carries *JAK2* V617F mutation, an established driver in myeloproliferative neoplasms. Set-2 differentiates spontaneously into megakaryocyte-like cells (Uozumi et al., 2000). Biological characteristics of leukemic cell lines are obviously very different from normal progenitors, which we should keep in mind while interpreting cell line data.

We found that Meg-01 cells are better suited for studies of NMDAR function than K-562 and Set-2 cells, mostly because of their higher levels of NMDAR expression. Upon differentiation with phorbol-12-myristate-13-acetate (PMA), Meg-01 cells up-regulate NMDAR expression further,

TABLE 1 | Expression of NMDAR subunits documented in megakaryocytic and erythroid cells.

	Megakaryocytic cells[1]						Erythroid cells[2]			Mature brain cortex[3]
	Human				Mouse		Human			Mouse and human
	Normal		Leukemic		Normal		Normal – cultured			Normal
GluN subunit	Whole bone marrow	Isolated mature MKs	Cell lines	Patient-derived	Isolated mature MKs	Early cultured MKs	Proerythroblasts	Orthochromatic	Retics /RBC	Neurons
1	+; P	−	+/++; P	+	+	++; P	+	+	+; P	+++; P
2A	+	+	++; P	+	+	++	++; P	+	+; P	++; P
2B	−	−	−/+	+	−	−	−	−	−	++; P
2C	−	−	−/+	++	+	−	+; P	+++; P	++; P	+; P
2D	+	+	+++; P	+++	++	−	++; P	++; P	++; P	+; P
3A	−	−	+	++	−	−	+; P	+; P	+; P	−
3B	−	−	+++	++	++	−	+; P	++; P	++; P	−

[1]Data generated mostly by RT-PCR, conventional and real-time (Genever et al., 1999; Hitchcock et al., 2003; Kamal et al., 2015, 2018).
[2]Data generated by TaqMan quantitative RT-PCR, flow cytometry and immunoblotting (Makhro et al., 2013; Hanggi et al., 2014, 2015).
[3]Shown by multiple techniques. The "+" symbol means expression was demonstrated; the number of "+" signs reflects the level of expression; "−" means expression was not detected. The letter "P" indicates protein expression was documented using flow cytometry or immunostaining in addition to transcript data, on which the semi-quantitative assessment was based. MK, megakaryocyte; Retics, reticulocytes; RBC, red blood cells.

providing a model in which to examine NMDAR involvement in megakaryocytic differentiation (Genever et al., 1999; Kamal et al., 2018).

The role of GluN3 subunits (highly expressed in leukemic cells; **Table 1**) is poorly understood, including in the brain, but its functions have already been described as exquisite, peculiar, unconventional, and transformative (Kehoe et al., 2013; Perez-Otano et al., 2016; Grand et al., 2018). This is because GluN3 subunits do not require glutamate for activation (Nilsson et al., 2007). In GluN1-GluN3 receptors, glycine acts both as the sole agonist binding on GluN3, and provides feedback inhibition through GluN1. In GluN1-GluN2-GluN3 receptors, the presence of GluN3 reduces Mg^{2+} block and Ca^{2+} entry (Matsuda et al., 2002; Cavara and Hollmann, 2008).

Overall, the presence of nonconventional GluN subunits (in particular GluN2D and GluN3) in megakaryocytic cells, normal and leukemic, suggests that NMDAR generates weaker but more sustained Ca^{2+} influx, and may allow stronger modulation by glycine than glutamate, in particular in leukemic cells. There is also a possibility of regulation by metabolic factors through GluN1. This is because megakaryocytic cells express h1-1 to h1-4 GluN1 isoforms, all of type "a" (Kamal et al., 2015). GluN1a isoforms lack the N1-cassette from the N-terminal domain (encoded by exon 5) that if present, reduces inhibition by protons and zinc, and potentiation by polyamines (Traynelis et al., 1995; Yi et al., 2018). Because the bone marrow environment is intrinsically hypoxic (Spencer et al., 2014), this form of metabolic regulation warrants testing in blood progenitors.

Evidence for Regulation of Glutamate Levels

Mouse and human megakaryocytes express a range of molecules for glutamate re-uptake, storage, and release, including VGLUT1, VGLUT2, SNARE, and the high-affinity glutamate re-uptake system, EAAT1, shown at both transcript and protein levels (Genever et al., 1999; Thompson et al., 2010). Fluorimetric measurements of glutamate concentrations in culture media suggest that human and mouse megakaryocytes, and Meg-01 cells release glutamate in a constitutive manner (Thompson et al., 2010); congruently, similar has been shown for ADP packaged together with glutamate in dense granules (Balduini et al., 2012). We do not know why megakaryocytes need glutamate sensitivity but it is a critical question to seek answers to in the future. A possible auto-regulatory loop (autocrine or paracrine) is suggested by the observation that Meg-01 cells release more glutamate upon differentiation with PMA (Thompson et al., 2010). Glutamate regulation may be independent of TPO, as NMDAR expression is maintained in megakaryocytes from c-Mpl-(TPO receptor) knockout mice (Hitchcock et al., 2003).

Unfortunately, there is no information on glutamate concentrations in the interstitial fluid of the bone marrow. In peripheral blood plasma, physiological glutamate levels are usually maintained between 20 and 100 µM (Kiessling et al., 2000), but vary widely depending on the diet (Stegink et al., 1983) and exercise (Makhro et al., 2016). In the interstitial fluids, concentrations of glutamate have been observed to be as

low as 0.5 µM in the masseter muscle of myofascial temporomandibular joint (Castrillon et al., 2010), to as high as blood plasma levels in the vastus lateralis muscle of the lower limb (Gerdle et al., 2016). There is no experimental data on how plasma/interstitial glutamate levels impact endogenous NMDAR in blood/progenitor cells. Paracrine NMDAR activation in neuron-like fashion appears more likely in tissues with low, steady state glutamate concentrations. On the other hand, conditional NMDAR activation *via* local pH changes would be more likely in tissues with higher, plasma-like glutamate concentrations. Different types of NMDAR subunits will also affect cellular sensitivity to glutamate, and modulate the receptor response (as described in section Evidence for NMDAR Functionality in Megakaryocytic Cells).

NMDAR Effects on Megakaryocytic Differentiation

NMDAR channel blockers (memantine and MK-801) induce two types of apparently opposing effects in cultured megakaryocytic cells: inhibition of differentiation in normal megakaryocytes, but induction of differentiation in megakaryocytic leukemia cell lines (**Figure 1Ai–ii**).

When human megakaryocytes are grown from CD34-positive umbilical cord stem cells, the addition of MK-801 inhibits acquisition of megakaryocytic markers (CD61, CD41a, and CD42a), nuclear ploidy and proplatelet formation; however, progenitor proliferation is unaffected (Hitchcock et al., 2003). Similar effects are seen in cultures of mouse hematopoietic progenitors, and in the native bone marrow milieu of mouse bone marrow explants. MK-801 inhibits actin reorganization in mature mouse megakaryocytes, suggesting that NMDAR-mediated Ca^{2+} influx is required for the cytoskeletal remodeling that underlies proplatelet formation (Kamal et al., 2018). This process may be similar to dendritic spine formation arising in response to neuronal NMDAR firing (Furuyashiki et al., 2002). NMDAR links with cytoskeletal elements through post-synaptic density (PSD) proteins such as PSD-95 and Yotiao; both of which are expressed in megakaryocytes, suggesting similar interactions may be possible in hematopoietic cells (Hitchcock et al., 2003).

In contrast to normal megakaryocytes that utilize NMDAR function to assist differentiation, leukemic cell lines (Meg-01, K-562, and Set-2) appear to divert NMDAR activity to increase proliferation (**Figure 1Ai–ii**). In the presence of NMDAR blockers (memantine and MK-801) Meg-01 cells undergo atypical differentiation and accumulate prominent cytoplasmic vacuoles (**Figure 1Aiii**; Kamal et al., 2015, 2018). The opposing NMDAR effects on cellular phenotype between normal and leukemic cells suggest divergence of NMDAR pathways during leukemogenesis to increase cell proliferation.

To get more insights into the mechanism of this divergence, we recently created a model of NMDAR hypofunction in Meg-01 cells using CRISPR-Cas9 mediated knockout of the *GRIN1* gene that encodes the obligate, GluN1 subunit of the NMDAR (Hearn et al., 2020). We found that *GRIN1* deletion caused marked changes in the intracellular Ca^{2+} homeostasis, including higher cytosolic Ca^{2+} levels at baseline but lower

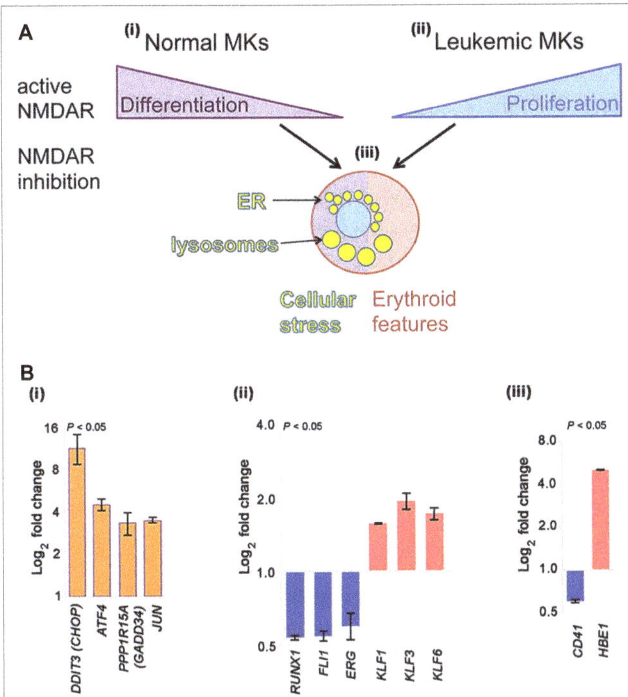

FIGURE 1 | NMDAR effects in normal megakaryocytes and leukemic Meg-01 cells. **(A)** Schematic indicating that in normal megakaryocytes NMDAR activity supports differentiation, in particular proplatelet formation **(i)**. In contrast, in leukemic cell lines NMDARs increase cell proliferation **(ii)**. In both normal and leukemic cells, NMDAR inhibition induces cellular stress response associated with endoplasmic reticulum (ER) dilatation and accumulation of lysosomes **(iii)**. Red shade in a cell reflects features of erythroid differentiation. **(B)** Experimental data showing that CRISPR-Cas9-mediated deletion of *GRIN1* in Meg-01 cells increased expression of ER stress markers (*DDIT3/CHOP, ATF4, PPP1R15A/GADD34,* and *JUN*; **Bi**; orange bars), associated with decreased expression of megakaryocytic transcription factors (*RUNX1, FLI1, ERG*; **Bii**) and megakaryocytic maturation marker, CD41 **(Biii)** (blue bars). Instead, expression of erythroid transcription factors (*KLF1, KLF3, KLF6*; **Bii**) and embryonic hemoglobin (*HBE1*; **Biii**) was increased (red bars). Transcript levels were determined by real-time RT-PCR **(Bi)** and Clariom S microarrays **(Bii–iii)**, as described (Hearn et al., 2020). Statistical significance is shown ($p < 0.05$ for all markers versus unmodified Meg-01 cells set at 1.0, tested by one-way ANOVA with Dunnett post hoc. MK, megakaryocyte.

ER Ca^{2+} release after activation. Deregulated Ca^{2+} handling led to endoplasmic reticulum (ER) stress and induced autophagy. Prominent cytoplasmic vacuoles accumulated in Meg-01-*GRIN1*⁻/⁻ cells and were found to represent dilated ER and lysosomal organelles (**Figure 1Aiii**). Microarray analysis revealed that Meg-01-*GRIN1*⁻/⁻ cells had deregulated expression of transcripts involved in Ca^{2+} metabolism, together with a shift in the pattern of hematopoietic transcription factors toward erythropoiesis (**Figure 1Bi–ii**). In keeping with the pro-erythroid pattern of transcription factors, Meg-01-*GRIN1*⁻/⁻ cells displayed features of erythroid differentiation (**Figure 1Biii**). Our data provide the first evidence that NMDARs comprise an integral component of the Ca^{2+} toolkit in megakaryocytic cells, and argue that intracellular Ca^{2+} homeostasis may be more important than currently recognized for balancing megakaryocytic with erythroid differentiation at the level of a common progenitor (Hearn et al., 2020).

In support of our findings, Kinney et al. provided computational evidence of NMDAR involvement in erythropoiesis (Kinney et al., 2019). The authors analyzed 164 publicly available erythroid microarray datasets using an enhanced CellNet bioinformatics algorithm to delineate key transitional states of erythroid differentiation at high resolution. This approach identified a role for signaling through epidermal growth factor receptor Erb-B2 receptor tyrosine kinase 4 (ErbB4) in erythroid differentiation, which was further validated experimentally in zebrafish, mouse and human models. The authors linked ErbB4 with NMDAR signaling by finding increased levels of *GRIN3B* transcripts, coding for GluN3B, in the reticulocyte gene cluster. A similar link between ErbB4 and NMDAR is well-documented in neurons, where ErbB4 and its neuregulin ligands stabilize synaptic NMDAR (Li et al., 2007). In fact, altered neuregulin 1–ErbB4 signaling is a well-established mechanism of NMDAR hypofunction in schizophrenia (Hahn et al., 2006). Encouragingly, we also found that in Meg-01-*GRIN1*⁻/⁻ cells, transcripts for neuregulin 1 and ErbB receptor feedback inhibitor 1 were up-regulated (1.98- and 2.05-fold, respectively), implying the ErbB4-NMDAR link is maintained during megakaryocytic-erythroid differentiation (Hearn et al., 2020).

The interrogation of publicly available transcriptomic data obtained from human megakaryocyte-erythroid progenitors at a single cell level demonstrated the presence of *GRINA* transcripts (encoding NMDAR-associated protein 1, known to be expressed at relatively high levels) and a scatter of low signals for *GRIN1, GRIN2A, GRIN2C,* and *GRIN2D* (Lu et al., 2018). Deep sequencing in that study was performed with approximately 3 million reads per cell, which captured approximately 6,000 of the most highly expressed transcripts in a cell, which may explain why *GRIN* transcripts were detected at very low levels.

NMDAR FUNCTIONALITY IN ERYTHROID CELLS AND IN THE CIRCULATING RBCs

RBCs sense plasma glutamate levels through the NMDAR. Using radiolabeled antagonist ([³H]MK-801) binding assay, basal activity of NMDARs in RBCs suspended in plasma was shown (Makhro et al., 2013). Supplementation of glutamate to plasma caused further activation of the receptors (Hanggi et al., 2014). Other findings suggest that the shear of flowing blood may also activate NMDAR in RBCs of patients with sickle cell disease (Hanggi et al., 2014, 2015).

NMDAR in Erythroid Precursor Cells

The abundance of NMDARs is particularly high in erythroid precursors and the UT-7/EPO cell line (Makhro et al., 2013; Hanggi et al., 2015). The receptor density decreases from hundreds of thousands per cell in proerythroblasts and erythroblasts cultured from peripheral blood-derived CD34-positive progenitors, to 35 in young human RBCs, and five in mature and senescent RBCs from healthy people (Makhro et al., 2013; Hanggi et al., 2014).

In UT-7/EPO cell line 350,000 NMDARs were detected per cell (Makhro et al., 2010). UT-7/EPO is a subclone of a UT-7 megakaryoblastic leukemia cell line that was maintained in the presence of EPO for more than 6 months to increase erythroid differentiation (Komatsu et al., 1993). In keeping with the NMDAR expression data, the density of currents produced by the NMDAR decreased during differentiation from proerythroblastic to the orthochromatic stage of erythroid progenitors (Hanggi et al., 2015).

Similar to megakaryocytes, the pattern of GluN subunits evolves during erythroid differentiation. Except for GluN2B, all other types of NMDAR subunits have been detected in erythroid cells at either mRNA or protein levels, or both (**Table 1**). During early stages, (proerythroblasts and basophilic erythroblasts) higher levels of GluN2A and GluN2D were shown along with lower levels of GluN3A, GluN3B, and GluN1 (Makhro et al., 2013; Hanggi et al., 2014, 2015). Orthochromatic erythroblasts switched from GluN2A-containing receptors to those predominantly containing GluN2C (Hanggi et al., 2015). As a result, high amplitude, fast, inactivating currents, mediated by the receptor at early stages of erythroid differentiation were replaced by currents of lesser amplitude but longer duration (Hanggi et al., 2015). This change in receptor subunit composition and its function gave rise to a switch in signal transmitted by the NMDAR, and probably, to the alteration in sensitivity to the physiological stimuli. Whereas GluN2A-containing receptors contributed to the modulation of the transmembrane potential, GluN2C/GluN2D-NMDAR mediated Ca^{2+} entry through the channels that remained open for a longer time (Hanggi et al., 2015). Mg^{2+} block is not supposed to control the NMDAR activity in RBCs due to the low transmembrane potential (about -10 mV) and the presence of GluN2C, GluN2D, and GluN3 subunits (Monyer et al., 1994; Wrighton et al., 2008).

Hyperactivation of the NMDAR by repeated stimulation resulted in the channel inactivation and did not affect viability of erythroid precursor cells. However, exposure of erythroid progenitors to the NMDAR channel blockers (MK-801 and memantine) triggered vacuolization and apoptosis, with maximal cell death observed at the early differentiation stages (Hanggi et al., 2014, 2015). This observation is in line with the earlier findings by Miller and Cheung on the importance of Ca^{2+} signaling for EPO-driven effects in precursor cells (Miller and Cheung, 1994; Tong et al., 2008).

NMDAR Function and Physiological Significance in the Circulating RBCs

Relatively low numbers of active NMDAR copies are retained by the circulating RBCs. Young RBCs of healthy humans carry 35 NMDARs per cell on average, whereas mature and senescent cells contain about five receptor copies per cell (Makhro et al., 2013; Hanggi et al., 2014). The NMDAR abundance is 3–4-fold higher in RBCs from patients with sickle cell disease (Hanggi et al., 2014). Activation of NMDAR by exposing the cells to the saturating concentrations of agonists (NMDA and glycine, 300 µM each) results in an acute, transient increase in the intracellular free Ca^{2+} (Makhro et al., 2013). There is striking inter-cellular heterogeneity in responses of the cells

to the NMDAR agonists, including changes in transmembrane currents and Ca^{2+} uptake shown with a fluorescent dye, as well as the level of dehydration and echinocyte formation. These differences cannot be explained solely by the differences in RBC age. Whereas some cells are insensitive to the stimulation, others show a clear response to the NMDAR agonists suggesting inter-cellular heterogeneity in NMDAR numbers/distribution. Along with the changes in RBC volume and density of Ca^{2+} uptake following the NMDAR stimulation, we observed regulation of nitric oxide production in RBCs by the nitric oxide synthase and modulation of the redox state (Makhro et al., 2010).

Physiological responses to the changes in NMDAR activity in the circulating RBCs include regulation of hemoglobin oxygen affinity, cell rheology, and most likely, longevity. Pathophysiological downstream effects associated with NMDAR hyperactivation were revealed *ex vivo* for RBCs of patients with sickle cell disease. These included Ca^{2+} overload, dehydration, and increase in cell density, and oxidative stress (Hanggi et al., 2014). There are no reports of abnormal RBC counts in patients with Alzheimer's disease taking memantine to protect the brain from glutamatergic excitotoxicity (Kavirajan, 2009). One pilot clinical trial was performed at the University Hospital Zurich, in which patients with sickle cell disease received memantine over a year (Makhro et al., 2020; trial identifier NCT02615847), and the other is currently ongoing (trial identifier NCT03247218). These trials provide an opportunity to explore long-term effects of NMDAR inhibition on RBC and platelet production, and cell properties in humans.

CONCLUSIONS AND FUTURE DIRECTIONS

In summary, megakaryocytic and erythroid precursors carry nonconventional NMDAR subunits, therefore NMDAR activity during hematopoiesis may be unique and should be tested. NMDARs regulate megakaryocytic and erythroid differentiation *ex vivo*, and balance both fates during differentiation of Meg-01 cells, suggesting NMDAR role at the level of a bipotential megakaryocyte-erythroid progenitor. NMDAR effects in hematopoietic cells are mediated by Ca^{2+} influx, which in early megakaryoblasts affects transcriptional program of differentiation, and in mature megakaryocytes induces cytoskeletal rearrangements required for proplatelet formation. In contrast to normal progenitors, leukemic cell lines re-direct NMDAR signaling to increase proliferation. The shift in the dominant NMDAR effect in leukemic cells may be at least partially related to different GluN subunits these cells express, which may offer therapeutic opportunities.

In keeping with the proliferative NMDAR effects in leukemic cells, other prominent groups found that GluN2B-containing NMDAR promotes growth of pancreatic tumors (Li and Hanahan, 2013; Li et al., 2018), and enable brain metastases by breast cancer (Zeng et al., 2019). The following molecules acting downstream of NMDAR were shown to assist cancer spread in these studies: CaMKII, MAPK, guanylate-kinase-associated protein, heat shock factor 1, and fragile X mental retardation protein (Li and Hanahan, 2013; Li et al., 2018);

we should examine similar pathways in leukemic cells as they may provide novel therapeutic targets.

In cultured human proerythroblasts, GluN2A-containing NMDAR provides depolarization, and inward Ca^{2+} current of high amplitude and fast inactivation kinetics. The presence of active NMDAR supports survival of early erythroid progenitors, which likely contributes to the EPO-driven signaling; however, this link is still to be demonstrated. Expression of GluN2C and GluN2D in the late erythroid precursors coincides with the onset of hemoglobinization. Thus, the NMDAR role in iron uptake warrants investigation. In mature RBCs, NMDAR regulates basal intracellular Ca^{2+} levels, contributing toward the regulation of cell volume, density, redox balance, and nitric oxide production by RBCs, which most likely contributes to the regulation of RBC longevity and oxygen carrying capacity.

NMDAR signaling can be modulated using small molecules. Memantine is an approved drug for neurological patients and could be repurposed against certain hematological disorders, such as sickle cell disease, megakaryocytic cancers and thrombosis. Preclinical studies have already advanced to stage I clinical trials in sickle cell disease, but a lot more needs to be done to determine if NMDAR modulation could be useful in patients with certain myeloid blood cancers or thrombotic disease. Neurological side effects may limit the use of memantine in hematological patients; therefore, alternative strategies may

need to be considered. These include subunit-specific NMDAR inhibitors, compounds that do not cross the blood-brain-barrier, and drugs that target pathways downstream, or glutamate release upstream of NMDAR.

Considering that the NMDAR role in megakaryocytic cells was first reported in Genever et al. (1999), the progress in this field may be viewed as relatively modest. However, we have reached a state of acceptance that NMDARs provide meaningful biological effects in hematopoietic cells. The field is attracting renewed attention. We await results from the first, stage I clinical trial in patients with sickle cell disease, primarily to establish safety of memantine outside of neurological indications. Further progress into NMDAR role in human leukemia and thrombosis will require studies in more advanced *ex vivo* and *in vivo* models. In addition, the overall principle and purpose of peripheral glutamate signaling needs to be determined. We, thus, invite collaborative approaches engaging experts from multiple disciplines to join us forming an interest group focusing on peripheral glutamate signaling.

AUTHOR CONTRIBUTIONS

MK-Z, AM and AB wrote the paper. JH contributed research data. All authors approved the final version for submission.

REFERENCES

Akazawa, C., Shigemoto, R., Bessho, Y., Nakanishi, S., and Mizuno, N. (1994). Differential expression of five *N*-methyl-D-aspartate receptor subunit mRNAs in the cerebellum of developing and adult rats. *J. Comp. Neurol.* 347, 150–160. doi: 10.1002/cne.903470112

Almazov, V. A., Popov Iu, G., Gorodinskii, A. I., Mikhailova, I. A., and Dambinova, S. A. (1988). The sites of high affinity binding of *L*-[3H]glutamic acid in human platelets. A new type of platelet receptor? *Biokhimiia* 53, 848–852.

Andersson, L. C., Jokinen, M., and Gahmberg, C. G. (1979). Induction of erythroid differentiation in the human leukaemia cell line K562. *Nature* 278, 364–365. doi: 10.1038/278364a0

Balduini, A., Di Buduo, C. A., Malara, A., Lecchi, A., Rebuzzini, P., Currao, M., et al. (2012). Constitutively released adenosine diphosphate regulates proplatelet formation by human megakaryocytes. *Haematologica* 97, 1657–1665. doi: 10.3324/haematol.2011.059212

Berk, M., Plein, H., and Belsham, B. (2000). The specificity of platelet glutamate receptor supersensitivity in psychotic disorders. *Life Sci.* 66, 2427–2432. doi: 10.1016/S0024-3205(00)00573-7

Berk, M., Plein, H., and Csizmadia, T. (1999). Supersensitive platelet glutamate receptors as a possible peripheral marker in schizophrenia. *Int. Clin. Psychopharmacol.* 14, 119–122. doi: 10.1097/00004850-199903000-00009

Bozic, M., and Valdivielso, J. M. (2015). The potential of targeting NMDA receptors outside the CNS. *Expert Opin. Ther. Targets* 19, 399–413. doi: 10.1517/14728222.2014.983900

Canobbio, I. (2019). Blood platelets: circulating mirrors of neurons? *Res. Pract. Thromb. Haemost.* 3, 564–565. doi: 10.1002/rth2.12254

Castrillon, E. E., Ernberg, M., Cairns, B. E., Wang, K., Sessle, B. J., Arendt-Nielsen, L., et al. (2010). Interstitial glutamate concentration is elevated in the masseter muscle of myofascial temporomandibular disorder patients. *J. Orofac. Pain* 24, 350–360.

Cavara, N. A., and Hollmann, M. (2008). Shuffling the deck anew: how NR3 tweaks NMDA receptor function. *Mol. Neurobiol.* 38, 16–26. doi: 10.1007/s12035-008-8029-9

Cavara, N. A., Orth, A., and Hollmann, M. (2009). Effects of NR1 splicing on NR1/NR3B-type excitatory glycine receptors. *BMC Neurosci.* 10:32. doi: 10.1186/1471-2202-10-32

Chatterton, J. E., Awobuluyi, M., Premkumar, L. S., Takahashi, H., Talantova, M., Shin, Y., et al. (2002). Excitatory glycine receptors containing the NR3 family of NMDA receptor subunits. *Nature* 415, 793–798. doi: 10.1038/nature715

Coyle, J. T. (2012). NMDA receptor and schizophrenia: a brief history. *Schizophr. Bull.* 38, 920–926. doi: 10.1093/schbul/sbs076

Daikhin, Y., and Yudkoff, M. (2000). Compartmentation of brain glutamate metabolism in neurons and glia. *J. Nutr.* 130, 1026S–1031S. doi: 10.1093/jn/130.4.1026S

Featherstone, D. E. (2010). Intercellular glutamate signaling in the nervous system and beyond. *ACS Chem. Neurosci.* 1, 4–12. doi: 10.1021/cn900006n

Fenninger, F., and Jefferies, W. A. (2019). What's bred in the bone: calcium channels in lymphocytes. *J. Immunol.* 202, 1021–1030. doi: 10.4049/jimmunol.1800837

Franconi, F., Miceli, M., Alberti, L., Seghieri, G., De Montis, M. G., and Tagliamonte, A. (1998). Further insights into the anti-aggregating activity of NMDA in human platelets. *Br. J. Pharmacol.* 124, 35–40. doi: 10.1038/sj.bjp.0701790

Franconi, F., Miceli, M., De Montis, M. G., Crisafi, E. L., Bennardini, F., and Tagliamonte, A. (1996). NMDA receptors play an anti-aggregating role in human platelets. *Thromb. Haemost.* 76, 84–87.

Furuyashiki, T., Arakawa, Y., Takemoto-Kimura, S., Bito, H., and Narumiya, S. (2002). Multiple spatiotemporal modes of actin reorganization by NMDA receptors and voltage-gated Ca^{2+} channels. *Proc. Natl. Acad. Sci. U. S. A.* 99, 14458–14463. doi: 10.1073/pnas.212148999

Genever, P. G., Wilkinson, D. J., Patton, A. J., Peet, N. M., Hong, Y., Mathur, A., et al. (1999). Expression of a functional *N*-methyl-D-aspartate-type glutamate receptor by bone marrow megakaryocytes. *Blood* 93, 2876–2883. doi: 10.1182/blood.V93.9.2876

Gerdle, B., Ernberg, M., Mannerkorpi, K., Larsson, B., Kosek, E., Christidis, N., et al. (2016). Increased interstitial concentrations of glutamate and pyruvate in vastus lateralis of women with fibromyalgia syndrome are normalized after an exercise intervention—a case-control study. *PLoS One* 11:e0162010. doi: 10.1371/journal.pone.0162010

Gielen, M., Siegler Retchless, B., Mony, L., Johnson, J. W., and Paoletti, P. (2009). Mechanism of differential control of NMDA receptor activity by NR2 subunits. *Nature* 459, 703–707. doi: 10.1038/nature07993

Grand, T., Abi Gerges, S., David, M., Diana, M. A., and Paoletti, P. (2018). Unmasking GluN1/GluN3 excitatory glycine NMDA receptors. *Nat. Commun.* 9:4769. doi: 10.1038/s41467-018-07236-4

Green, T. N., Hamilton, J. R., Morel-Kopp, M. C., Zheng, Z., Chen, T. T., Hearn, J. I., et al. (2017). Inhibition of NMDA receptor function with an anti-GluN1-S2 antibody impairs human platelet function and thrombosis. *Platelets* 28, 799–811. doi: 10.1080/09537104.2017.1280149

Hahn, C. G., Wang, H. Y., Cho, D. S., Talbot, K., Gur, R. E., Berrettini, W. H., et al. (2006). Altered neuregulin 1-erbB4 signaling contributes to NMDA receptor hypofunction in schizophrenia. *Nat. Med.* 12, 824–828. doi: 10.1038/nm1418

Hanggi, P., Makhro, A., Gassmann, M., Schmugge, M., Goede, J. S., Speer, O., et al. (2014). Red blood cells of sickle cell disease patients exhibit abnormally high abundance of N-methyl D-aspartate receptors mediating excessive calcium uptake. *Br. J. Haematol.* 167, 252–264. doi: 10.1111/bjh.13028

Hanggi, P., Telezhkin, V., Kemp, P. J., Schmugge, M., Gassmann, M., Goede, J. S., et al. (2015). Functional plasticity of the N-methyl-D-aspartate receptor in differentiating human erythroid precursor cells. *Am. J. Physiol. Cell Physiol.* 308, C993–C1007. doi: 10.1152/ajpcell.00395.2014

Hansen, K. B., Yi, F., Perszyk, R. E., Furukawa, H., Wollmuth, L. P., Gibb, A. J., et al. (2018). Structure, function, and allosteric modulation of NMDA receptors. *J. Gen. Physiol.* 150, 1081–1105. doi: 10.1085/jgp.201812032

Hardingham, G. E. (2006). Pro-survival signalling from the NMDA receptor. *Biochem. Soc. Trans.* 34, 936–938. doi: 10.1042/BST0340936

Hardingham, G. E., Arnold, F. J., and Bading, H. (2001). Nuclear calcium signaling controls CREB-mediated gene expression triggered by synaptic activity. *Nat. Neurosci.* 4, 261–267. doi: 10.1038/85109

Harrison, P. J., Hall, N., Mould, A., Al-Juffali, N., and Tunbridge, E. M. (2019). Cellular calcium in bipolar disorder: systematic review and meta-analysis. *Mol. Psychiatry.* doi: 10.1038/s41380-019-0622-y

Hearn, J. I., Green, T. N., Chopra, M., Nursalim, Y. N. S., Ladvanszky, L., Knowlton, N., et al. (2020). NMDA receptor hypofunction in Meg-01 cells reveals a role for intracellular calcium homeostasis in balancing megakaryocytic-erythroid differentiation. *Thromb. Haemost.* 120, 671–686. doi: 10.1055/s-0040-1708483

Herrera, R., Hubbell, S., Decker, S., and Petruzzelli, L. (1998). A role for the MEK/MAPK pathway in PMA-induced cell cycle arrest: modulation of megakaryocytic differentiation of K562 cells. *Exp. Cell Res.* 238, 407–414. doi: 10.1006/excr.1997.3847

Hitchcock, I. S., Skerry, T. M., Howard, M. R., and Genever, P. G. (2003). NMDA receptor-mediated regulation of human megakaryocytopoiesis. *Blood* 102, 1254–1259. doi: 10.1182/blood-2002-11-3553

Hogan-Cann, A. D., and Anderson, C. M. (2016). Physiological roles of non-neuronal NMDA receptors. *Trends Pharmacol. Sci.* 37, 750–767. doi: 10.1016/j.tips.2016.05.012

Kalev-Zylinska, M. L., Green, T. N., Morel-Kopp, M. C., Sun, P. P., Park, Y. E., Lasham, A., et al. (2014). N-methyl-D-aspartate receptors amplify activation and aggregation of human platelets. *Thromb. Res.* 133, 837–847. doi: 10.1016/j.thromres.2014.02.011

Kamal, T., Green, T. N., Hearn, J. I., Josefsson, E. C., Morel-Kopp, M. C., Ward, C. M., et al. (2018). N-methyl-D-aspartate receptor mediated calcium influx supports in vitro differentiation of normal mouse megakaryocytes but proliferation of leukemic cell lines. *Res. Pract. Thromb. Haemost.* 2, 125–138. doi: 10.1002/rth2.12068

Kamal, T., Green, T. N., Morel-Kopp, M. C., Ward, C. M., McGregor, A. L., McGlashan, S. R., et al. (2015). Inhibition of glutamate regulated calcium entry into leukemic megakaryoblasts reduces cell proliferation and supports differentiation. *Cell. Signal.* 27, 1860–1872. doi: 10.1016/j.cellsig.2015.05.004

Kavirajan, H. (2009). Memantine: a comprehensive review of safety and efficacy. *Expert Opin. Drug Saf.* 8, 89–109. doi: 10.1517/14740330802528420

Kehoe, L. A., Bernardinelli, Y., and Muller, D. (2013). GluN3A: an NMDA receptor subunit with exquisite properties and functions. *Neural Plast.* 2013:145387. doi: 10.1155/2013/145387

Kiessling, K., Roberts, N., Gibson, J. S., and Ellory, J. C. (2000). A comparison in normal individuals and sickle cell patients of reduced glutathione precursors and their transport between plasma and red cells. *Hematol. J.* 1, 243–249. doi: 10.1038/sj.thj.6200033

Kinney, M. A., Vo, L. T., Frame, J. M., Barragan, J., Conway, A. J., Li, S., et al. (2019). A systems biology pipeline identifies regulatory networks for stem cell engineering. *Nat. Biotechnol.* 37, 810–818. doi: 10.1038/s41587-019-0159-2

Komatsu, N., Yamamoto, M., Fujita, H., Miwa, A., Hatake, K., Endo, T., et al. (1993). Establishment and characterization of an erythropoietin-dependent subline, UT-7/Epo, derived from human leukemia cell line, UT-7. *Blood* 82, 456–464. doi: 10.1182/blood.V82.2.456.456

Li, B., Woo, R. S., Mei, L., and Malinow, R. (2007). The neuregulin-1 receptor erbB4 controls glutamatergic synapse maturation and plasticity. *Neuron* 54, 583–597. doi: 10.1016/j.neuron.2007.03.028

Li, L., and Hanahan, D. (2013). Hijacking the neuronal NMDAR signaling circuit to promote tumor growth and invasion. *Cell* 153, 86–100. doi: 10.1016/j.cell.2013.02.051

Li, L., Zeng, Q., Bhutkar, A., Galvan, J. A., Karamitopoulou, E., Noordermeer, D., et al. (2018). GKAP acts as a genetic modulator of NMDAR signaling to govern invasive tumor growth. *Cancer Cell* 33, 736.e5–751.e5. doi: 10.1016/j.ccell.2018.02.011

Low, C. M., Lyuboslavsky, P., French, A., Le, P., Wyatte, K., Thiel, W. H., et al. (2003). Molecular determinants of proton-sensitive N-methyl-D-aspartate receptor gating. *Mol. Pharmacol.* 63, 1212–1222. doi: 10.1124/mol.63.6.1212

Lozzio, C. B., and Lozzio, B. B. (1975). Human chronic myelogenous leukemia cell-line with positive Philadelphia chromosome. *Blood* 45, 321–334. doi: 10.1182/blood.V45.3.321.321

Lu, Y. C., Sanada, C., Xavier-Ferrucio, J., Wang, L., Zhang, P. X., Grimes, H. L., et al. (2018). The molecular signature of megakaryocyte-erythroid progenitors reveals a role for the cell cycle in fate specification. *Cell Rep.* 25, 2083.e4–2093.e4. doi: 10.1016/j.celrep.2018.10.084

Makhro, A., Hanggi, P., Goede, J. S., Wang, J., Bruggemann, A., Gassmann, M., et al. (2013). N-methyl-D-aspartate receptors in human erythroid precursor cells and in circulating red blood cells contribute to the intracellular calcium regulation. *Am. J. Physiol. Cell Physiol.* 305, C1123–C1138. doi: 10.1152/ajpcell.00031.2013

Makhro, A., Haider, T., Wang, J., Bogdanov, N., Steffen, P., Wagner, C., et al. (2016). Comparing the impact of an acute exercise bout on plasma amino acid composition, intraerythrocytic Ca$^{(2+)}$ handling, and red cell function in athletes and untrained subjects. *Cell Calcium* 60, 235–244. doi: 10.1016/j.ceca.2016.05.005

Makhro, A., Hegemann, I., Seiler, E., Simionato, G., Claveria, V., Bogdanov, N., et al. (2020). A pilot clinical phase II trialMemSID: Acute and durable changes of red blood cells of sickle cell disease patients on memantine treatment. eJHaem. 2020, 1–12. doi: 10.1002/jha2.11

Makhro, A., Wang, J., Vogel, J., Boldyrev, A. A., Gassmann, M., Kaestner, L., et al. (2010). Functional NMDA receptors in rat erythrocytes. *Am. J. Physiol. Cell Physiol.* 298, C1315–C1325. doi: 10.1152/ajpcell.00407.2009

Matsuda, K., Kamiya, Y., Matsuda, S., and Yuzaki, M. (2002). Cloning and characterization of a novel NMDA receptor subunit NR3B: a dominant subunit that reduces calcium permeability. *Brain Res. Mol. Brain Res.* 100, 43–52. doi: 10.1016/S0169-328X(02)00173-0

Merritt, K., Egerton, A., Kempton, M. J., Taylor, M. J., and McGuire, P. K. (2016). Nature of glutamate alterations in schizophrenia: a meta-analysis of proton magnetic resonance spectroscopy studies. *JAMA Psychiat.* 73, 665–674. doi: 10.1001/jamapsychiatry.2016.0442

Miller, B. A., and Cheung, J. Y. (1994). Mechanisms of erythropoietin signal transduction: involvement of calcium channels. *Proc. Soc. Exp. Biol. Med.* 206, 263–267. doi: 10.3181/00379727-206-43756

Monyer, H., Burnashev, N., Laurie, D. J., Sakmann, B., and Seeburg, P. H. (1994). Developmental and regional expression in the rat brain and functional properties of four NMDA receptors. *Neuron* 12, 529–540. doi: 10.1016/0896-6273(94)90210-0

Morle, F., Laverriere, A. C., and Godet, J. (1992). Globin genes are actively transcribed in the human megakaryoblastic leukemia cell line MEG-01. *Blood* 79, 3094–3096. doi: 10.1182/blood.V79.11.3094.3094

Morrell, C. N., Sun, H., Ikeda, M., Beique, J. C., Swaim, A. M., Mason, E., et al. (2008). Glutamate mediates platelet activation through the AMPA receptor. *J. Exp. Med.* 205, 575–584. doi: 10.1084/jem.20071474

Nakazawa, K., Jeevakumar, V., and Nakao, K. (2017). Spatial and temporal boundaries of NMDA receptor hypofunction leading to schizophrenia. *NPJ Schizophr.* 3:7. doi: 10.1038/s41537-016-0003-3

Nilsson, A., Duan, J., Mo-Boquist, L. L., Benedikz, E., and Sundstrom, E. (2007). Characterisation of the human NMDA receptor subunit NR3A glycine binding site. *Neuropharmacology* 52, 1151–1159. doi: 10.1016/j. neuropharm.2006.12.002

Ogura, M., Morishima, Y., Ohno, R., Kato, Y., Hirabayashi, N., Nagura, H., et al. (1985). Establishment of a novel human megakaryoblastic leukemia cell line, MEG-01, with positive Philadelphia chromosome. *Blood* 66, 1384–1392. doi: 10.1182/blood.V66.6.1384.1384

Ogura, M., Morishima, Y., Okumura, M., Hotta, T., Takamoto, S., Ohno, R., et al. (1988). Functional and morphological differentiation induction of a human megakaryoblastic leukemia cell line (MEG-01s) by phorbol diesters. *Blood* 72, 49–60. doi: 10.1182/blood.V72.1.49.49

Paoletti, P., Bellone, C., and Zhou, Q. (2013). NMDA receptor subunit diversity: impact on receptor properties, synaptic plasticity and disease. *Nat. Rev. Neurosci.* 14, 383–400. doi: 10.1038/nrn3504

Perez-Otano, I., Larsen, R. S., and Wesseling, J. F. (2016). Emerging roles of GluN3-containing NMDA receptors in the CNS. *Nat. Rev. Neurosci.* 17, 623–635. doi: 10.1038/nrn.2016.92

Ponomarev, E. D. (2018). Fresh evidence for platelets as neuronal and innate immune cells: their role in the activation, differentiation, and deactivation of Th1, Th17, and Tregs during tissue inflammation. *Front. Immunol.* 9:406. doi: 10.3389/fimmu.2018.00406

Reiner, A., and Levitz, J. (2018). Glutamatergic signaling in the central nervous system: ionotropic and metabotropic receptors in concert. *Neuron* 98, 1080–1098. doi: 10.1016/j.neuron.2018.05.018

Ruljancic, N., Mihanovic, M., Cepelak, I., and Bakliza, A. (2013). Platelet and serum calcium and magnesium concentration in suicidal and non-suicidal schizophrenic patients. *Psychiatry Clin. Neurosci.* 67, 154–159. doi: 10.1111/pcn.12038

Spencer, J. A., Ferraro, F., Roussakis, E., Klein, A., Wu, J., Runnels, J. M., et al. (2014). Direct measurement of local oxygen concentration in the bone marrow of live animals. *Nature* 508, 269–273. doi: 10.1038/nature13034

Stegink, L. D., Baker, G. L., and Filer, L. J. Jr. (1983). Modulating effect of Sustagen on plasma glutamate concentration in humans ingesting monosodium L-glutamate. *Am. J. Clin. Nutr.* 37, 194–200. doi: 10.1093/ajcn/37.2.194

Sun, H., Swaim, A., Herrera, J. E., Becker, D., Becker, L., Srivastava, K., et al. (2009). Platelet kainate receptor signaling promotes thrombosis by stimulating cyclooxygenase activation. *Circ. Res.* 105, 595–603. doi: 10.1161/CIRCRESAHA.109.198861

Thompson, C. J., Schilling, T., Howard, M. R., and Genever, P. G. (2010). SNARE-dependent glutamate release in megakaryocytes. *Exp. Hematol.* 38, 504–515. doi: 10.1016/j.exphem.2010.03.011

Todrick, A., Tait, A. C., and Marshall, E. F. (1960). Blood platelet 5-hydroxytryptamine levels in psychiatric patients. *J. Ment. Sci.* 106, 884–890. doi: 10.1192/bjp.106.444.884

Tong, Q., Hirschler-Laszkiewicz, I., Zhang, W., Conrad, K., Neagley, D. W., Barber, D. L., et al. (2008). TRPC3 is the erythropoietin-regulated calcium channel in human erythroid cells. *J. Biol. Chem.* 283, 10385–10395. doi: 10.1074/jbc.M710231200

Traynelis, S. F., Hartley, M., and Heinemann, S. F. (1995). Control of proton sensitivity of the NMDA receptor by RNA splicing and polyamines. *Science* 268, 873–876. doi: 10.1126/science.7754371

Traynelis, S. F., Wollmuth, L. P., McBain, C. J., Menniti, F. S., Vance, K. M., Ogden, K. K., et al. (2010). Glutamate receptor ion channels: structure, regulation, and function. *Pharmacol. Rev.* 62, 405–496. doi: 10.1124/pr.109.002451

Uozumi, K., Otsuka, M., Ohno, N., Moriyama, T., Suzuki, S., Shimotakahara, S., et al. (2000). Establishment and characterization of a new human megakaryoblastic cell line (SET-2) that spontaneously matures to megakaryocytes and produces platelet-like particles. *Leukemia* 14, 142–152. doi: 10.1038/sj.leu.2401608

Watanabe, M., Inoue, Y., Sakimura, K., and Mishina, M. (1992). Developmental changes in distribution of NMDA receptor channel subunit mRNAs. *Neuroreport* 3, 1138–1140. doi: 10.1097/00001756-199212000-00027

Wenzel, A., Fritschy, J. M., Mohler, H., and Benke, D. (1997). NMDA receptor heterogeneity during postnatal development of the rat brain: differential expression of the NR2A, NR2B, and NR2C subunit proteins. *J. Neurochem.* 68, 469–478. doi: 10.1046/j.1471-4159.1997.68020469.x

Wrighton, D. C., Baker, E. J., Chen, P. E., and Wyllie, D. J. (2008). Mg^{2+} and memantine block of rat recombinant NMDA receptors containing chimeric NR2A/2D subunits expressed in *Xenopus laevis* oocytes. *J. Physiol.* 586, 211–225. doi: 10.1113/jphysiol.2007.143164

Wyllie, D. J., Livesey, M. R., and Hardingham, G. E. (2013). Influence of GluN2 subunit identity on NMDA receptor function. *Neuropharmacology* 74, 4–17. doi: 10.1016/j.neuropharm.2013.01.016

Yamakura, T., and Shimoji, K. (1999). Subunit- and site-specific pharmacology of the NMDA receptor channel. *Prog. Neurobiol.* 59, 279–298. doi: 10.1016/S0301-0082(99)00007-6

Yelamanchi, S. D., Jayaram, S., Thomas, J. K., Gundimeda, S., Khan, A. A., Singhal, A., et al. (2016). A pathway map of glutamate metabolism. *J. Cell Commun. Signal.* 10, 69–75. doi: 10.1007/s12079-015-0315-5

Yi, F., Zachariassen, L. G., Dorsett, K. N., and Hansen, K. B. (2018). Properties of triheteromeric N-methyl-D-aspartate receptors containing two distinct GluN1 isoforms. *Mol. Pharmacol.* 93, 453–467. doi: 10.1124/mol.117.111427

Zeng, Q., Michael, I. P., Zhang, P., Saghafinia, S., Knott, G., Jiao, W., et al. (2019). Synaptic proximity enables NMDAR signalling to promote brain metastasis. *Nature* 573, 526–531. doi: 10.1038/s41586-019-1576-6

Zhou, Y., and Danbolt, N. C. (2014). Glutamate as a neurotransmitter in the healthy brain. *J. Neural Transm.* 121, 799–817. doi: 10.1007/s00702-014-1180-8

9

Inside Out Integrin Activation Mediated by PIEZO1 Signaling in Erythroblasts

*Francesca Aglialoro, Naomi Hofsink, Menno Hofman, Nicole Brandhorst and Emile van den Akker**

Sanquin Research and Landsteiner Laboratory, Department of Haematopoiesis, Amsterdam UMC, University of Amsterdam, Amsterdam, Netherlands

**Correspondence:*
Emile van den Akker
e.vandenakker@sanquin.nl

The non-selective mechanosensitive ion channel PIEZO1 controls erythrocyte volume homeostasis. Different missense gain-of-function mutations in *PIEZO1* gene have been identified that cause Hereditary Xerocytosis (HX), a rare autosomal dominant haemolytic anemia. PIEZO1 expression is not limited to erythrocytes and expression levels are significantly higher in erythroid precursors, hinting to a role in erythropoiesis. During erythropoiesis, interactions between erythroblasts, central macrophages, and extracellular matrix within erythroblastic islands are important. Integrin $\alpha4\beta1$ and $\alpha5\beta1$ present on erythroblasts facilitate such interactions in erythroblastic islands. Here we found that chemical activation of PIEZO1 using Yoda1 leads to increased adhesion to VCAM1 and fibronectin in flowing conditions. Integrin $\alpha4$, $\alpha5$, and $\beta1$ blocking antibodies prevented this PIEZO1-induced adhesion suggesting inside-out activation of integrin on erythroblasts. Blocking the Ca^{2+} dependent Calpain and PKC pathways by using specific inhibitors also blocked increased erythroid adhesion to VCAM1 and fibronectins. Cleavage of Talin was observed as a result of Calpain and PKC activity. In conclusion, PIEZO1 activation results in inside-out integrin activation, facilitated by calcium-dependent activation of PKC and Calpain. The data introduces novel concepts in Ca^{2+} signaling during erythropoiesis with ramification on erythroblastic island homeostasis in health and disease like Hereditary Xerocytosis.

Keywords: PIEZO1, integrins, calcium signaling, integrin $\alpha4\beta1$, erythroblastic island

INTRODUCTION

Human erythropoiesis occurs in the bone marrow in a specific niche called the erythroblastic island, composed of a central macrophage surrounded by differentiating erythrocyte precursors (Bessis, 1958; for review see Chasis and Mohandas, 2008; Manwani and Bieker, 2008; Heideveld et al., 2018). The interaction between the erythroblast and the macrophage is regulating erythroid maturation, proliferation, and survival (Manwani and Bieker, 2008). In addition, central macrophage phagocytes the extruded pyrenocyte (the extruded nucleus surrounded by a plasma membrane) left behind by the newly formed reticulocyte. Interactions between erythroid cells and macrophages are mediated by several adhesion molecules. During erythropoiesis, multiple adhesion molecules are expressed that mediate cell-cell and cell-extracellular matrix interactions, among which specific integrins (Chow et al., 2011; Palis, 2016; Seu et al., 2017). Integrin $\alpha4\beta1$

[very late antigen 4 (VLA-4)] and α5β1 [very late antigen 5 (VLA-5)] expressed on erythroblasts can bind to counter-receptor vascular cell adhesion protein 1 (VCAM1, CD106) present on central macrophages or extracellular matrix (ECM) protein fibronectin, respectively (Sadahira et al., 1995; Chasis and Mohandas, 2008; Tanaka et al., 2009; Chow et al., 2013; Spring et al., 2013; Belay et al., 2017). Treatment with anti VLA-4 resulted in disruption of the island integrity *in vitro* and induces anemia in a mouse model and defects in stress erythropoiesis (Sadahira et al., 1995; Hamamura et al., 1996; Lee et al., 2006) clearly indicating the important role of these niche interactions. The activation state of integrins can be controlled by various processes including regulation via intracellular signaling, so called inside-out activation (Kim et al., 2003; Abram and Lowell, 2009; Hu and Luo, 2013) potentially providing a regulatory point to control erythroid-macrophage interactions. Inside-out activation of integrins is, among other mechanism, also controlled by intracellular Ca^{2+} concentration, with increased Ca^{2+} concentration leading to integrin activation (Jaconi et al., 1991; Schwartz et al., 1993; Shankar et al., 1993; Coppolino et al., 1997; Rowin et al., 1998; Bye et al., 2020; Shu et al., 2020). Several Ca^{2+} responsive proteins have been identified that control this inside-out integrin activation. For instance, the Ca^{2+} dependent Calpain unmasks the β-integrin binding motif of Talins leading to integrin activation (Goksoy et al., 2008; Mchugh et al., 2010). Inside out integrin activation is also mediated by Ca^{2+}-dependent Protein kinase C (PKC; Kirchhofer et al., 1991; Harburger and Calderwood, 2009). One mechanism through which PKC operates is by activating the small GTPase Rap1 facilitating the formation of a complex with Rap1 effector Rap-1 interacting molecule (RIAM) and Talin, which causes integrin activation (Bos, 2005; Watanabe et al., 2008).

We and others have recently identified, a role for the non selective mechanosensitive cation channel, PIEZO1 during erythropoiesis (Caulier et al., 2020) as well as in *in vitro* cultured reticulocytes (Moura et al., 2019). Activation of PIEZO1 leads to Ca^{2+} influx and activation of Ca^{2+} dependent signal transduction among which NFATs, Calcineurin, MAPK, and calcium dependent PKCs (Von Lindern et al., 2000; Lanuti et al., 2006). Besides a role during erythropoiesis, PIEZO1 expression is maintained on erythrocytes where it regulates volume homeostasis. Indeed, activating mutations within PIEZO1 lead to dehydrated erythrocytes termed Hereditary Xerocytosis (HX; Andolfo et al., 2013; Bagriantsev et al., 2014; Cahalan et al., 2015). In endothelial cells activation of PIEZO1 has been associated with integrin activation and increased cell adhesion resulting in integrin-dependent focal adhesion kinase (FAK) signaling in flowing conditions (Albarran-Juarez et al., 2018). Concomitantly, reduced integrin activation was observed in an endothelium-specific PIEZO1 deficient mouse model (Albarran-Juarez et al., 2018). In agreement with this, depletion of PIEZO1 in small lung cancer cell lines also caused decreased integrin activation (Mchugh et al., 2012). PIEZO1 localization at focal adhesion induced integrin activation and FAK signaling regulated by the ECM (Chen et al., 2018). PIEZO1, located at the endoplasmic reticulum (ER), has been associated with integrin activation in epithelial cells (Mchugh et al., 2010). Inactivation of integrin β1 was observed in PIEZO1 siRNA knockdown in epithelial cells, resulting in reduced cell adhesion. Activation of Ca^{2+}-protease Calpain and cleavage of its target Talin was demonstrated after PIEZO1 activation in epithelial cells (Mchugh et al., 2010). Here, we show that activation of PIEZO1 on erythroblasts using the selective agonist Yoda1 causes activation of downstream Ca^{2+}-mediators Calpain and PKC, resulting in inside-out activation of integrins leading to increased adhesive properties of erythroblasts. The results give new insights into the regulation of integrin activation in erythroblasts, which will have consequences for erythroblastic island homeostasis.

MATERIALS AND METHODS

Human Blood Sample

Human blood mononuclear cells were purified by density separation, following manufacturer's protocol (GE Healthcare, Chicago, IL, United States). Informed consent was given in accordance with the Declaration of Helsinki, the Dutch National and Sanquin Internal Ethic Boards.

Erythroblast Cell Culture and Differentiation

Erythroblasts were expanded as previously described (Heshusius et al., 2019). In short, cells were cultured and expanded in presence of EPO (2 UI/mL; ProSpec, East Brunswick, NJ, United States), human recombinant Stem Cell Factor (100 ng/mL, supernatant SCF producing cell line), and dexamethasone (1 μM; Sigma, St. Louis, MO, United States).

Flow Adhesion Assay

Channels of the μ-Slide $VI^{0.4}$ (Ibidi) were coated with 10 μg/ml recombinant human VCAM-1 (R&D systems, Minneapolis, MN, United States) or fibronectin (Sigma Aldrich, St. Louis, Missouri, United States) for 30 min at room temperature. The flow chambers were placed at room temperature on an inverted microscope. Tubes connected a syringe pump, containing CellQuin medium, with one side of the channel and the other side of the channel with a waste reservoir. Tubes and channels were washed with CellQuin medium at a flow speed of 7,5 ml/h. Samples consisted of 4×10^6 erythroblasts and were incubated 10 min at 37°C with or without PKC inhibitor Gö6976 (Tocris, Bristol, United Kingdom), Calpain 1 and 2 inhibitor (Sigma), integrin α4 antibody (1 μg/ml, BD Biosciences, San Josè, CA, United States), β1 (0,5 μg/ml, Abcam, Cambridge, United Kingdom), and α5 (1 μg/ml, BD Biosciences, San Josè, CA, United States), GsMTx4 (Tocris, Bristol, United Kingdom), or Yoda1 (Sigma Aldrich, St. Louis, Missouri, United States). After incubation, samples were injected into the flow with a syringe and the pump was left running for 15 min unless otherwise stated. Nine pictures were taken across the channel after 15 min and attached erythroblasts were counted with ImageJ. Tubes were washed with water and fresh CellQuin medium in between samples. Data was analyzed with Prism 8 (GraphPad, San Diego, CA, United States). Unpaired student *t* test was used to calculate statistical significance between each condition. D'Agostino-Pearson test was used to assess the normality of the samples' distribution.

Flow Cytometry

Erythroblasts were treated with or without inhibitor, Yoda1, or Manganese (Mn^{2+}). Samples were taken at different time points, each containing 200.000 erythroblasts in a volume of 200 µl, and were stained in a 96-well plate. The samples were kept on ice after treatment and during staining. Erythroblasts were incubated for 10 min at 37°C with or without PKC inhibitor Gö6976 (500 nM) or Calpain inhibitor 1 and 2 (1 mM) with gentle mixing. Samples were taken and resuspended into ice-cold PBS. Remaining erythroblasts were incubated at 37°C with or without Yoda1 or Manganese (II) chloride dihydrate solution in water (2 mM, Sigma Aldrich, St. Louis, MO, United States). Samples were taken after 10 and 30 min of incubation and were resuspended into ice-cold PBS. Staining was performed with 30 min incubation in FACS buffer (PBS, 0.5% BSA) with the following antibodies antiCD71 (Miltenyi Biotec, Bergisch Gladbach, Germany), antiCD235a OriGene Technologies, Inc., Rockville, MD, United States), antiCD49d (α4 integrin; BD Biosciences, San Jose, CA, United States), anti β1 integrin unconjugated (Abcam, Cambridge, United Kingdom). Staining was followed by two washing steps with FACS buffer, followed or not by staining with secondary antibody. After the last staining and wash, the cells were resuspended into 200 µl of FACS buffer. Cells were transferred to FACS tubes before measurement on FACSCanto II (BD Biosciences, San Josè, CA, United States). Data was analyzed with FlowJo® software (BD Biosciences, San Josè, CA, United States) and Prism 8 (GraphPad, San Diego, CA, United States).

Calpain Activity Assay

Calpain activity was measured with the Calpain Activity Fluorometric Assay Kit (Sigma Aldrich, St. Louis, MO, United States), following manufacturer protocol. Samples were measured in triplicates. In short, the amount of 2×10^6 erythroblasts was taken per sample. Samples were incubated with or without Yoda1 (1 µM or 5 µM, 10 min, 37°C). After centrifugation, samples were washed in PBS and resuspended into 100 µl Extraction Buffer. Samples were incubated (20 min, on ice) and spun down (10.000 *g*, 1 min). Cell lysate was transferred to a new tube and kept on ice. 10x Reaction buffer and Calpain substrate was added to 85 µl cell lysate and samples were transferred to 96-well plate. After incubation in the dark (1 h, 37°C), samples were measured with 400 nm excitation filter and 505 nm emission filter with a plate reader (BioTek, Winooski, VT, United States). Positive control (1 µl Active Calpain) and negative control (1 µl Calpain inhibitor) were taken along. Data was analyzed with Prism 8. Two-way ANOVA was used to calculate statistical significance between samples.

Western Blot

Cells were lysed in CARIN lysis buffer (20 mM Tris–HCl pH 8.0, 138 mM NaCl, 10 mM EDTA, 100 mM NaF, 1% Nonidet P-40, and 10% glycerol). Following Bradford protein quantification

(Bio-Rad Laboratories, Hercules, CA, United States), lysates were boiled in Laemmli sample buffer [2% sodium dodecyl sulfate (SDS) wt/vol, 10% glycerol, 5%2-mercaptoethanol, 60 mM Tris–HCl pH6.8, and trace amount brome-phenol blue; 3 min, 95°C], subjected to SDS-polyacrylamide gel electrophoresis, blotted using iBlot-PVDF blotting system (Thermo Fisher Scientific, Bleiswijk, Netherlands), and stained as indicated in the figure legends.

Fluorescent Microscopy Imaging

Expression of integrin β1 on erythroblasts was determined with live imaging. Erythroblasts, 25×10^5 in 0.5 ml CellQuin, were incubated with integrin β1 antibody [1:100, clone (P5D2), Abcam, Cambridge, United Kingdom, Alexa Fluor 488], DRAQ5™ (1:2500, Abcam, Cambridge, United Kingdom), and CD235a (Glycophorin A, 1:200, PB, Miltenyi, Bergisch Gladbach, Germany). Microscopy images were taken with an Axiovert 200 microscope (Zeiss, Oberkochen, Germany) with bright field, DAPI, Alexa Fluor 488, and Alexa Fluor 647 filters using 40x/1.3 oil objective. Software ZEN2.3 (Zeiss, Oberkochen, Germany) was used to analyze and convert images.

RESULTS

Integrin Subunits Are Differently Expressed in Erythroblasts

We have previously reported membrane expression of ITGA4 and ITGB1 (VLA-4) in our *in vitro* cultured erythroid cells (Heideveld et al., 2018). To further evaluate the expression of integrin subunits present on erythroblasts, we data-mined RNA-sequencing that we reported previously containing RNA-expression profiles of CD71+/CD235+ erythroblasts (Heshusius et al., 2019). This analysis showed mRNA expression of eleven integrin subunits: six α subunits (*ITGA*) and five β subunits (*ITGB*; **Figure 1A**). At the erythroblast stage mRNA expression of integrin subunits α2B (*ITGA2B*), α4 (*ITGA4*), α5 (*ITGA5*), αE (*ITGAE*), αV (*ITGAV*), α6 (*ITGA6*), and subunits β1 (*ITGB1*), β2 (*ITGB2*), β3 (*ITGB3*), and β4 (*ITGB4*) were observed albeit with significantly different expression levels (**Figure 1A**). At the stage where erythroblast are CD71+/CD235+ (**Figure 1B**), α4 and β1 RNAs are most abundantly expressed followed by *ITGAIIB, ITGAE, ITGA5, ITGB4,* while *ITGAV, ITGB3, ITGB2, and ITG6* are lowly expressed. This confirms the presence of α4β1 (VLA-4; Heideveld et al., 2018) and would further allow α5β1 (VLA-5) and low expression of αVβ1, α6β1, αIIbβ3 (GPIIb/IIIa), αVβ3 (CD61), and αVβ5. Integrins αIIbβ3 (GPIIb/IIIa) and αVβ3 (CD61) are associated with megakaryopoiesis and platelet homeostasis and expression of the beta partner ITGB3 (β3) is significantly downregulated during differentiation (Heshusius et al., 2019). Of note, no interaction partners for the subunits ITGAE, ITGB4, and ITGB2 were detected (**Supplementary Figure 1A**). Expression of the most abundant integrin subunits α4, and β1 was confirmed by flow cytometry in three donors (**Figure 1B** and **Supplementary Figure 1B**) and imaging (**Figure 1B** and **Supplementary Figure 1C**).

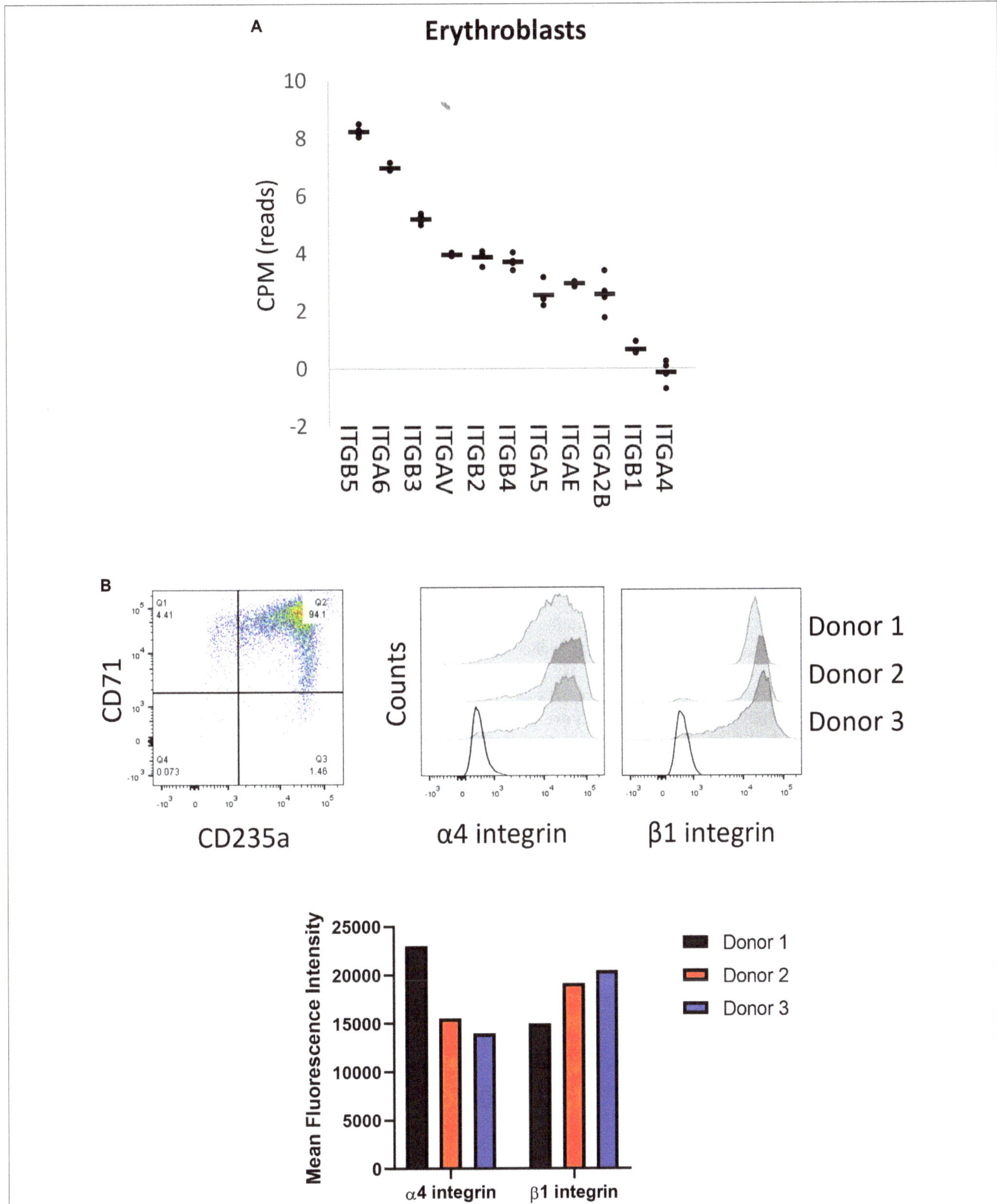

FIGURE 1 | Erythroblasts integrin expression. **(A)** Integrin mRNA expression (count per million) in CD71+/CD235a+ erythroblasts; data mined from Heshusius et al. (2019). Gene expression was found for 11 integrin subunits: six α sub-units (*ITGA*) and five β sub-units (*ITGB*; $n = 4$). **(B)** Dot plot showing a representative CD71/CD235 expression plot depicting the erythroblast stage of analysis. Right histograms and bar graph indicate expression of α4 integrin and β1 integrin in CD71+CD235a+ erythroblasts in 3 donors as measured by flow cytometry.

Activation of PIEZO1 With Yoda1 Leads to Integrin Activation and Increases Erythroblasts Adhesion

Activation of mechanosensitive PIEZO1 has been shown to lead to inside out integrin activation on non-hematopoietic cells and specific Ca^{2+} dependent signal transduction in erythroblasts (Rowin et al., 1998; Mchugh et al., 2010; Caulier et al., 2020). The high abundant integrin subunit $\beta 1$ in combination with $\alpha 4$ (VLA-4) and $\alpha 5$ (VLA-5) on erythroblasts can bind VCAM1 and fibronectin, respectively. The effect of chemical activation of PIEZO1 on integrin activation and adhesion of CD71HighCD235aHigh erythroblasts to VCAM-1 and fibronectin was assessed. Erythroblasts were activated by 2 mM Mn^{2+}, a strong integrin activator, or with different concentration of the PIEZO1 agonist Yoda1 in presence of soluble VCAM1. Compared to untreated, incubation with Mn^{2+} or Yoda1 increased binding to anti-VCAM1 antibody, albeit that activation with Yoda1 was lower compared to Mn^{2+} (**Supplementary Figure 2A**). Note that the expression of total integrin $\beta 1$ remained unchanged throughout the different treatments (**Supplementary Figure 2B**). To assess the adhesive properties of erythroblasts, a flow adhesion assay in which the erythroblasts are flowed within a flow cell that is coated with VCAM1 or fibronectin was used. Increased adhesion to VCAM1 (**Figure 2A**) and fibronectin (**Figure 2B**) was observed when erythroblasts were incubated with the PIEZO1 agonist Yoda1 (**Supplementary Figure 3A**). Of note, a Yoda1 concentration of 1 μM was chosen as short term and long term use of higher concentrations have been shown to be detrimental, both in terms of adhesion to VCAM1 (**Supplementary Figure 3B**), as well as cell viability, respectively (**Supplementary Figure 4**). Of note, different flow speeds did not influence the increased adhesion in Yoda1 treated cells (**Supplementary Figure 3C**). Pre-incubation with an antagonizing $\beta 1$ antibody thus blocking the common beta integrin subunit within VLA-4 and VLA5 results in a significant reduction of both Yoda1-induced and steady state adhesion to VCAM1 or fibronectin (**Figures 2C,D**). Blocking VLA-4 or VLA-5 specifically using anti-$\alpha 4$ or anti-$\alpha 5$ blocking antibodies, respectively, also led to decreased adhesion (**Figures 2C,D**). Note that blocking VLA-4 using $\alpha 4$-integrin antibodies leads to lower inhibition of adhesion to fibronectin compared to blockage of VLA-5, which has been shown before for mouse erythroblasts (Eshghi et al., 2007). Treatment with GsMTx4, a mechanosensitive ion channel inhibitor, resulted in decreased adhesion following Yoda1 treatment but did not revert the basal adhesion of untreated cells (**Supplementary Figure 5**). In conclusion, the results indicate that erythroblast adhesion to VCAM1 and fibronectin is integrin-dependent and can be increased by activating PIEZO1.

Activation of PIEZO1 Leads to Increased Calpain Activity

Ca^{2+} influx caused by PIEZO1 activation has been shown in endothelial cells, erythroid cells and erythrocytes to lead to activation of calcium dependent pathways, among which the positive regulators of integrin activation calpain, PKC and Talin

(Bos, 2005; Watanabe et al., 2008; Mchugh et al., 2010). To demonstrate whether calpain is activated following PIEZO1 activation, cells were treated with 1 μM Yoda1 or 5 μM Yoda1 for 60 min or left untreated. Calpain activity was significantly increased upon Yoda1 treatment compared to untreated cells. No significant difference between 1 and 5 μM Yoda1 was observed suggesting that 1 μM Yoda1 is sufficient to activate the pool of calpain within erythroblasts. Of note, short term treatment with 1 μM Yoda1 or 5 μM Yoda1 did not influence cell viability (**Supplementary Figure 4**). Note that Calpain activity is similar between untreated controls and samples treated with a calpain inhibitor 1 and 2, indicating that in erythroblasts basal calpain activity is low (**Figure 3A**). Calpain cleaves Talin to an activated state mediating inside-out integrin activation (Goksoy et al., 2008; Mchugh et al., 2010). Note that 1 μM Yoda1 led to a modest but clear increase in cleaved activated Talin, depicted in western blot as a lower molecular weight migrating band (**Figure 3B**).

Flow-Induced Adhesion to VCAM1 and Fibronectin Is Dependent on PKC and Calpain

Inside-out integrin activation can be mediated by several signaling cascades among which calpain and PKC-dependent pathways. Ca^{2+} influx through PIEZO1 activation by Yoda1, or in patient-derived erythroblasts with specific activation mutations in PIEZO1, causes activation of PKC-dependent signal transduction that can be partly blocked by inhibiting calcium-dependent PKCs (Caulier et al., 2020). Together with the observation that Calpain is activated after PIEZO1 activation we evaluated whether the increased adhesion is a consequence of Calpain and/or PKC activity leading to integrin activation in Yoda1 treated erythroblasts. Pretreatment with Calpain inhibitor or Gö9676 (an inhibitor of Ca^{2+} dependent PKCs) before Yoda1 incubation decreased adhesion of Yoda1 treated erythroblasts to VCAM1 (**Figure 3C**) and fibronectin (**Figure 3B**) to untreated background adhesion levels. Of note, inhibition of the PKC or Calpain pathway does not lead to complete block of adhesion as observed upon blocking integrin subunits (**Figures 3C,D**). The data indicates that PIEZO1 activation leads to both PKC and Calpain dependent integrin activation. Note that the inhibition of adhesion is primarily affecting the Yoda1 mediated increase in adhesion. This is in contrast to blocking specific integrin subunits, which also blocks the basal level of erythroblast adhesion. This is in agreement with the low level of calpain activation in steady state erythroblasts (**Figure 3A**).

DISCUSSION

Ca^{2+} influx has been linked to integrin activation in several cell types (Jaconi et al., 1991; Schwartz et al., 1993; Shankar et al., 1993; Coppolino et al., 1997; Rowin et al., 1998; Bye et al., 2020; Shu et al., 2020). This report identifies for the first time a role of the mechanosensor PIEZO1 in integrin activation in erythroblasts. We found PIEZO1 activation leads to inside-out VLA4 and VLA5 activation in erythroblasts. This activation is dependent on Ca^{2+} activated

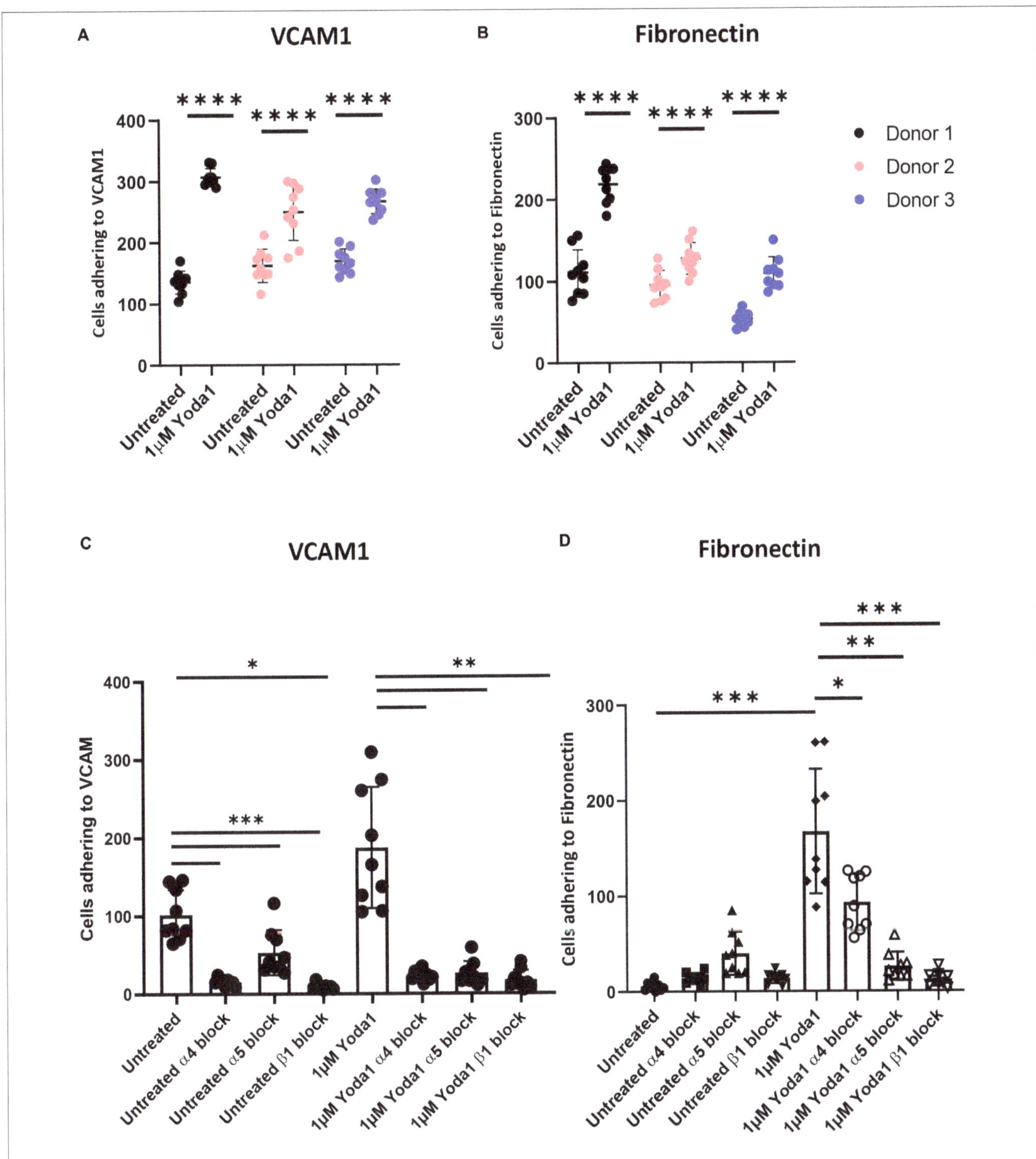

FIGURE 2 | Increased erythroblast binding to VCAM-1 and fibronectin after Yoda1 mediated PIEZO1 stimulation. **(A,B)** Erythroblasts were treated or not with 1 μM of Yoda1 for 10 min and subjected to a flow assay as described in Materials and Methods (7.5 ml/h flow rate). Adhesion of erythroblasts to VCAM **(A)** or fibronectin **(B)** was quantified using imageJ (n = 9 pictures of different donors; data depicted as mean ± SD; ****P < 0.0001; **C,D)** Erythroblasts were pre-treated with anti-α4, anti-α5, or anti-β1 and stimulated or not with 1 μM Yoda1 (10 min) as indicated and flowed over a surface coated with VCAM **(C)** or fibronectin **(D**; data depicted as mean ± SD mean n = 9 images was used to calculate statistic; *P < 0.05, **P < 0.01, ***P < 0.001, and ****P < 0.0001).

PKCs and calpain. Together, the data indicates that erythroid cells can perceive mechanical forces and in response to this activate VLA5 and VLA4. This may have consequences for bone marrow retention and migration, for instance with the central macrophage within the erythroid island, which expresses VCAM1, the ligand of VLA4.

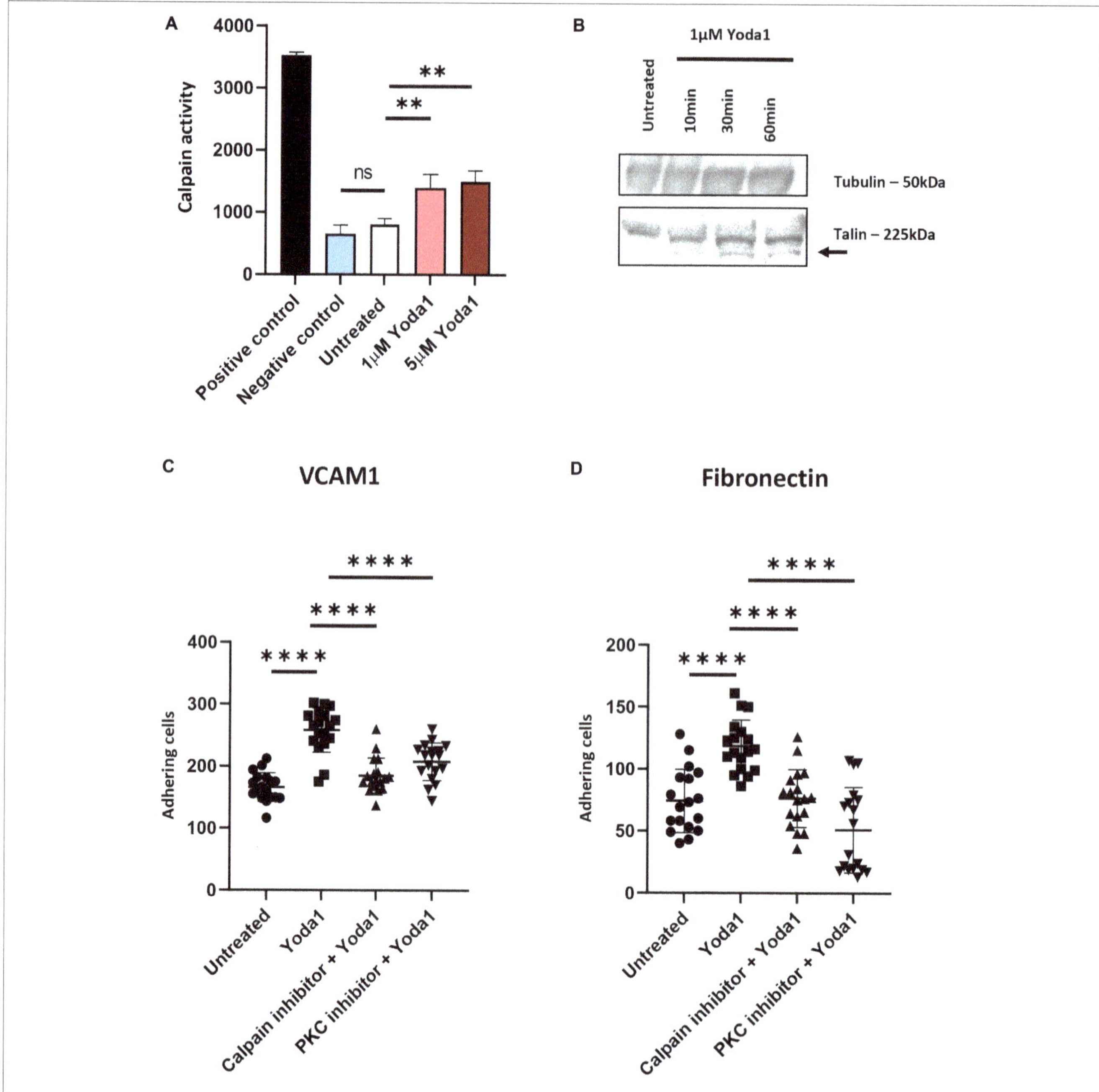

FIGURE 3 | PIEZO1 stimulation with Yoda1 results calpain and PKC dependent integrin activation. **(A)** In untreated and Yoda1 treated erythroblasts, the calpain activity assay measured the absorbance of cleaved calpain substrate at 505 nm, which is quantified using a fluorescence plate reader. Active calpain and inhibited calpain where used as positive and negative controls, respectively (n = 3, **P < 0.01). **(B)** Expression of Talin (±225 kDa) and the cleavage of Talin (±190 kDa; indicated by arrow) in untreated erythroblasts and erythroblasts incubated with 1 µM Yoda1 for 10, 30, and 60 min. The same amount of protein was loaded and Tubulin (±50 kDa) was used as loading control (representative of n = 2). **(C,D)** Erythroblasts were pretreated with PKC inhibitor (500 nM) and calpain inhibitor (1 mM) and stimulated with and without 1 µM Yoda1 (10 min) as indicated and flowed over a surface coated with VCAM **(C)** or fibronectin (**D**; n = 2 donors; data depicted as mean ± SD; mean of n = 18 images (of two different donors) was used to calculate statistic ****P < 0.0001).

The semi-solid bone marrow is occupied with a heterogeneous pool of cells that will perceive and induce particular cellular programs upon shear forces from fluid movements, tensile strain, hydrostatic pressure and other physical forces (Meyer et al., 2016; Kim and Bixel, 2020). The consequences of these forces on haematopoiesis are beginning to emerge. For instance, it has been shown that hydrostatic pressure improves the clonogenic potential and CD34+ cell number in ex-vivo cultures (Kim and Bixel, 2020). PIEZO1 may be one of the sensors that could be involved in perceiving these forces and integrating them into a cellular response, for instance through integrin activation as shown here. Indeed, PIEZO1 has been shown to mediate tissue hydrostatic pressure sensing in various hematopoietic cells including T-cells and monocytes (Solis et al., 2019). More specific to the erythroid system, adhesive interactions facilitated by integrins between (i) the central macrophage and developing erythroblasts, (ii) the erythroblasts themselves, and (iii) the erythroblasts and ECM are important for the support and regulation of erythropoiesis (Vuillet-Gaugler et al., 1990; Hanspal, 1997; Arroyo et al., 1999). Loss of α4 integrin affects erythropoiesis in the mice fetal liver and bone marrow, in terms of decreased cellularity. Moreover, in an *in vitro* system, *ITGA4* null erythroid progenitors display defective proliferation and migration (Arroyo et al., 1999). Ulyanova et al. (2011) observed that *ITGA4*-deficient mice are defective in triggering an optimal stress erythropoiesis response with a decrease in peripheral blood reticulocytes and inability to increase haematocrit after treatment with Phenylhydrazine. The effect was less evident in *ITGA5* deficient mice (Ulyanova et al., 2014). Indeed, integrins have been implicated in regulating the proliferative response of erythroid cells through inhibition of apoptosis by upregulating BCL-XL (Eshghi et al., 2007). Activation of VLA4 and VLA5 via PIEZO1 mechanoactivation in the bone marrow may thus strengthen anti-apoptotic signaling pathways. Interestingly, deletion of α4 integrin caused egress of erythroid cells from the bone marrow during erythroid maturation at homeostasis and during stress erythropoiesis (Ulyanova et al., 2014). This suggests that VLA4 plays a role in maintaining erythroid cells within the bone marrow, which may be strengthened upon PIEZO1-induced VLA4 activation.

The role of PIEZO1 during erythropoiesis has been investigated only recently. Increased activation of PIEZO1 was followed by activation of specific (Ca^{2+} dependent) pathways including NFAT and MAPK (Caulier et al., 2020). We found involvement of the PIEZO1-induced Ca^{2+} dependent Calpain and PKC activation in erythroblasts, which were both essential for PIEZO1-induced VLA4 and VLA5 activation. Although we found increased proteolysis of the calpain target Talin upon Yoda1 treatment, more research needs to be done to identify the intermediate steps between calpain, PKC activation and VLA4/VLA5 activation. We found that erythroblast display a basal level of adhesion toward VCAM1 and fibronectin in flow adhesion assays. Blocking α4-integrin or α5-integrin resulted in complete block of adhesion, including the basal adhesion indicating the complete dependence on VLA4 and

VLA5 for adhesion to VCAM and Fibronectin. Of note, untreated cells adhered in the same manner independently of the flow speed used, supporting a basal adhesion at steady state (**Supplementary Figure 3C**). We find that erythroblasts treated with GsMTx4 inhibited the Yoda1-induced adhesion of erythroblasts. However, it must be noted that GsMTx4 is a general inhibitor of mechanosensing by cells and not specific for PIEZO1, as it incorporates itself into the lipid layer, and allows for partial relaxation upon mechanical stress (Gnanasambandam et al., 2017). Nevertheless, the experiments show that integrin activation is dependent on mechanosensing. Interestingly, inhibiting PKC, or Calpain reduced adhesion to basal levels but did not fully block adhesion as observed upon using anti- α4-integrin or α5-integrin. This may suggest that the basal activation of VLA4 and VLA5 is dependent on other PKC and calpain independent processes. Indeed, inhibition of calpain does not significantly decrease the total calpain activity in erythroblasts suggesting that basal levels of calpain activity in steady state erythroblasts is low. In conclusion, we show PIEZO1-induced inside-out integrin activation, facilitated by Ca^{2+}-dependent activation of PKC and Calpain. This knowledge could be of crucial importance in the investigation of the role of Ca^{2+} during erythropoiesis, and in particular of the dynamics of the erythroblastic island, both in health and disease (such as HX). Such a condition could lead to stress erythropoiesis (which would explain the reticulocytosis observed in these patients), a situation where support of the erythroblastic island microenvironment has been shown to be important.

AUTHOR CONTRIBUTIONS

FA designed the experimental setup, performed and designed the experiments, and wrote the manuscript. NH, MH, and NB designed and performed the specific experiments. EA designed the experimental setup, and supervised and edited the manuscript. All authors contributed to the article and approved the submitted version.

ACKNOWLEDGMENTS

We would like to thank the Sanquin Central Facility for flow cytometry support.

REFERENCES

Abram, C. L., and Lowell, C. A. (2009). The ins and outs of leukocyte integrin signaling. *Annu. Rev. Immunol.* 27, 339–362. doi: 10.1146/annurev.immunol.021908.132554

Albarran-Juarez, J., Iring, A., Wang, S., Joseph, S., Grimm, M., Strilic, B., et al. (2018). Piezo1 and Gq/G11 promote endothelial inflammation depending on flow pattern and integrin activation. *J. Exp. Med.* 215, 2655–2672. doi: 10.1084/jem.20180483

Andolfo, I., Alper, S. L., De Franceschi, L., Auriemma, C., Russo, R., De Falco, L., et al. (2013). Multiple clinical forms of dehydrated hereditary stomatocytosis arise from mutations in PIEZO1. *Blood* 121, 3925–3935. doi: 10.1182/blood-2013-02-482489

Arroyo, A. G., Yang, J. T., Rayburn, H., and Hynes, R. O. (1999). Alpha4 integrins regulate the proliferation/differentiation balance of multilineage hematopoietic progenitors in vivo. *Immunity* 11, 555–566. doi: 10.1016/s1074-7613(00)80131-4

Bagriantsev, S. N., Gracheva, E. O., and Gallagher, P. G. (2014). Piezo proteins: regulators of mechanosensation and other cellular processes. *J. Biol. Chem.* 289, 31673–31681. doi: 10.1074/jbc.r114.612697

Belay, E., Hayes, B. J., Blau, C. A., and Torok-Storb, B. (2017). Human cord blood and bone marrow CD34+ cells generate macrophages that support erythroid islands. *PLoS One* 12:e0171096. doi: 10.1371/journal.pone.0171096

Bessis, M. (1958). Erythroblastic island, functional unity of bone marrow. *Rev. Hematol.* 13, 8–11.

Bos, J. L. (2005). Linking rap to cell adhesion. *Curr. Opin. Cell Biol.* 17, 123–128. doi: 10.1016/j.ceb.2005.02.009

Bye, A. P., Gibbins, J. M., and Mahaut-Smith, M. P. (2020). Ca(2+) waves coordinate purinergic receptor-evoked integrin activation and polarization. *Sci. Signal.* 13:eaav7354. doi: 10.1126/scisignal.aav7354

Cahalan, S. M., Lukacs, V., Ranade, S. S., Chien, S., Bandell, M., and Patapoutian, A. (2015). Piezo1 links mechanical forces to red blood cell volume. *eLife* 4:e07370.

Caulier, A., Jankovsky, N., Demont, Y., Ouled-Haddou, H., Demagny, J., Guitton, C., et al. (2020). PIEZO1 activation delays erythroid differentiation of normal and hereditary xerocytosis-derived human progenitor cells. *Haematologica* 105, 610–622. doi: 10.3324/haematol.2019.218503

Chasis, J. A., and Mohandas, N. (2008). Erythroblastic islands: niches for erythropoiesis. *Blood* 112, 470–478. doi: 10.1182/blood-2008-03-077883

Chen, X., Wanggou, S., Bodalia, A., Zhu, M., Dong, W., Fan, J. J., et al. (2018). A feedforward mechanism mediated by mechanosensitive ion channel PIEZO1 and tissue mechanics promotes glioma aggression. *Neuron* 100, 799.e7–815.e7.

Chow, A., Huggins, M., Ahmed, J., Hashimoto, D., Lucas, D., Kunisaki, Y., et al. (2013). CD169(+) macrophages provide a niche promoting erythropoiesis under homeostasis and stress. *Nat. Med.* 19, 429–436. doi: 10.1038/nm.3057

Chow, A., Lucas, D., Hidalgo, A., Mendez-Ferrer, S., Hashimoto, D., Scheiermann, C., et al. (2011). Bone marrow CD169+ macrophages promote the retention of hematopoietic stem and progenitor cells in the mesenchymal stem cell niche. *J. Exp. Med.* 208, 261–271. doi: 10.1084/jem.20101688

Coppolino, M. G., Woodside, M. J., Demaurex, N., Grinstein, S., St-Arnaud, R., and Dedhar, S. (1997). Calreticulin is essential for integrin-mediated calcium signalling and cell adhesion. *Nature* 386, 843–847. doi: 10.1038/386843a0

Eshghi, S., Vogelezang, M. G., Hynes, R. O., Griffith, L. G., and Lodish, H. F. (2007). Alpha4beta1 integrin and erythropoietin mediate temporally distinct steps in erythropoiesis: integrins in red cell development. *J. Cell Biol.* 177, 871–880. doi: 10.1083/jcb.200702080

Gnanasambandam, R., Ghatak, C., Yasmann, A., Nishizawa, K., Sachs, F., Ladokhin, A. S., et al. (2017). GsMTx4: mechanism of inhibiting mechanosensitive ion channels. *Biophys. J.* 112, 31–45. doi: 10.1016/j.bpj.2016.11.013

Goksoy, E., Ma, Y. Q., Wang, X., Kong, X., Perera, D., Plow, E. F., et al. (2008). Structural basis for the autoinhibition of talin in regulating integrin activation. *Mol. Cell* 31, 124–133. doi: 10.1016/j.molcel.2008.06.011

Hamamura, K., Matsuda, H., Takeuchi, Y., Habu, S., Yagita, H., and Okumura, K. (1996). A critical role of VLA-4 in erythropoiesis in vivo. *Blood* 87, 2513–2517. doi: 10.1182/blood.v87.6.2513.bloodjournal8762513

Hanspal, M. (1997). Importance of cell-cell interactions in regulation of erythropoiesis. *Curr. Opin. Hematol.* 4, 142–147. doi: 10.1097/00062752-199704020-00011

Harburger, D. S., and Calderwood, D. A. (2009). Integrin signalling at a glance. *J. Cell Sci.* 122, 159–163. doi: 10.1242/jcs.018093

Heideveld, E., Hampton-O'neil, L. A., Cross, S. J., Van Alphen, F. P. J., Van Den Biggelaar, M., and Van Den Akker, E. (2018). Glucocorticoids induce differentiation of monocytes towards macrophages that share functional and phenotypical aspects with erythroblastic island macrophages. *Haematologica* 103, 395–405. doi: 10.3324/haematol.2017.179341

Heshusius, S., Heideveld, E., Burger, P., Thiel-Valkhof, M., Sellink, E., Varga, E., et al. (2019). Large-scale in vitro production of red blood cells from human peripheral blood mononuclear cells. *Blood Adv.* 3, 3337–3350. doi: 10.1182/bloodadvances.2019000689

Hu, P., and Luo, B. H. (2013). Integrin bi-directional signaling across the plasma membrane. *J. Cell Physiol.* 228, 306–312. doi: 10.1002/jcp.24154

Jaconi, M. E., Theler, J. M., Schlegel, W., Appel, R. D., Wright, S. D., and Lew, P. D. (1991). Multiple elevations of cytosolic-free Ca2+ in human neutrophils: initiation by adherence receptors of the integrin family. *J. Cell Biol.* 112, 1249–1257. doi: 10.1083/jcb.112.6.1249

Kim, J., and Bixel, M. G. (2020). Intravital multiphoton imaging of the bone and bone marrow environment. *Cytometry A* 97, 496–503. doi: 10.1002/cyto.a.23937

Kim, M., Carman, C. V., and Springer, T. A. (2003). Bidirectional transmembrane signaling by cytoplasmic domain separation in integrins. *Science* 301, 1720–1725. doi: 10.1126/science.1084174

Kirchhofer, D., Grzesiak, J., and Pierschbacher, M. D. (1991). Calcium as a potential physiological regulator of integrin-mediated cell adhesion. *J. Biol. Chem.* 266, 4471–4477.

Lanuti, P., Bertagnolo, V., Gaspari, A. R., Ciccocioppo, F., Pierdomenico, L., Bascelli, A., et al. (2006). Parallel regulation of PKC-alpha and PKC-delta characterizes the occurrence of erythroid differentiation from human primary hematopoietic progenitors. *Exp. Hematol.* 34, 1624–1634. doi: 10.1016/j.exphem.2006.07.018

Lee, G., Lo, A., Short, S. A., Mankelow, T. J., Spring, F., Parsons, S. F., et al. (2006). Targeted gene deletion demonstrates that the cell adhesion molecule ICAM-4 is critical for erythroblastic island formation. *Blood* 108, 2064–2071. doi: 10.1182/blood-2006-03-006759

Manwani, D., and Bieker, J. J. (2008). The erythroblastic island. *Curr. Top. Dev. Biol.* 82, 23–53. doi: 10.1016/s0070-2153(07)00002-6

Mchugh, B. J., Buttery, R., Lad, Y., Banks, S., Haslett, C., and Sethi, T. (2010). Integrin activation by Fam38A uses a novel mechanism of R-Ras targeting to the endoplasmic reticulum. *J. Cell Sci.* 123, 51–61. doi: 10.1242/jcs.056424

Mchugh, B. J., Murdoch, A., Haslett, C., and Sethi, T. (2012). Loss of the integrin-activating transmembrane protein Fam38A (Piezo1) promotes a switch to a reduced integrin-dependent mode of cell migration. *PLoS One* 7:e40346. doi: 10.1371/journal.pone.0040346

Meyer, M. B., Benkusky, N. A., Sen, B., Rubin, J., and Pike, J. W. (2016). Epigenetic plasticity drives adipogenic and osteogenic differentiation of marrow-derived mesenchymal stem cells. *J. Biol. Chem.* 291, 17829–17847. doi: 10.1074/jbc.m116.736538

Moura, P. L., Hawley, B. R., Dobbe, J. G. G., Streekstra, G. J., Rab, M. A. E., Bianchi, P., et al. (2019). PIEZO1 gain-of-function mutations delay reticulocyte maturation in hereditary xerocytosis. *Haematologica* 105, e268–e271. doi: 10.3324/haematol.2019.231159

Palis, J. (2016). Interaction of the macrophage and primitive erythroid lineages in the mammalian embryo. *Front. Immunol.* 7:669. doi: 10.3389/fimmu.2016.00669

Rowin, M. E., Whatley, R. E., Yednock, T., and Bohnsack, J. F. (1998). Intracellular calcium requirements for beta1 integrin activation. *J. Cell. Physiol.* 175, 193–202. doi: 10.1002/(sici)1097-4652(199805)175:2<193::aid-jcp9>3.0.co;2-j

Sadahira, Y., Yoshino, T., and Monobe, Y. (1995). Very late activation antigen 4-vascular cell adhesion molecule 1 interaction is involved in the formation of erythroblastic islands. *J. Exp. Med.* 181, 411–415. doi: 10.1084/jem.181.1.411

Schwartz, M. A., Brown, E. J., and Fazeli, B. (1993). A 50-kDa integrin-associated protein is required for integrin-regulated calcium entry in endothelial cells. *J. Biol. Chem.* 268, 19931–19934.

Seu, K. G., Papoin, J., Fessler, R., Hom, J., Huang, G., Mohandas, N., et al. (2017). Unraveling macrophage heterogeneity in erythroblastic islands. *Front. Immunol.* 8:1140. doi: 10.3389/fimmu.2017.01140

Shankar, G., Davison, I., Helfrich, M. H., Mason, W. T., and Horton, M. A. (1993). Integrin receptor-mediated mobilisation of intranuclear calcium in rat osteoclasts. *J. Cell Sci.* 105(Pt 1), 61–68.

Shu, X., Li, N., Huang, D., Zhang, Y., Lu, S., and Long, M. (2020). Mechanical strength determines Ca(2+) transients triggered by the engagement of beta2 integrins to their ligands. *Exp. Cell Res.* 387:111807. doi: 10.1016/j.yexcr.2019.111807

Solis, A. G., Bielecki, P., Steach, H. R., Sharma, L., Harman, C. C. D., Yun, S., et al. (2019). Mechanosensation of cyclical force by PIEZO1 is essential for innate immunity. *Nature* 573, 69–74. doi: 10.1038/s41586-019-1485-8

Spring, F. A., Griffiths, R. E., Mankelow, T. J., Agnew, C., Parsons, S. F., and Anstee, D. J. (2013). Tetraspanins CD81 and CD82 facilitate alpha4beta1-mediated adhesion of human erythroblasts to vascular cell adhesion molecule-1. *PLoS One* 8:e62654. doi: 10.1371/journal.pone.0062654

Tanaka, R., Owaki, T., Kamiya, S., Matsunaga, T., Shimoda, K., Kodama, H., et al. (2009). VLA-5-mediated adhesion to fibronectin accelerates hemin-stimulated erythroid differentiation of K562 cells through induction of VLA-4 expression. *J. Biol. Chem.* 284, 19817–19825. doi: 10.1074/jbc.m109.009860

Ulyanova, T., Jiang, Y., Padilla, S., Nakamoto, B., and Papayannopoulou, T. (2011). Combinatorial and distinct roles of alpha(5) and alpha(4) integrins in stress erythropoiesis in mice. *Blood* 117, 975–985. doi: 10.1182/blood-2010-05-283218

Ulyanova, T., Padilla, S. M., and Papayannopoulou, T. (2014). Stage-specific functional roles of integrins in murine erythropoiesis. *Exp. Hematol.* 42, 404.e–409.e.

Von Lindern, M., Parren-Van Amelsvoort, M., Van Dijk, T., Deiner, E., Van Den Akker, E., Van Emst-De Vries, S., et al. (2000). Protein kinase C alpha controls erythropoietin receptor signaling. *J. Biol. Chem.* 275, 34719–34727.

Vuillet-Gaugler, M. H., Breton-Gorius, J., Vainchenker, W., Guichard, J., Leroy, C., Tchernia, G., et al. (1990). Loss of attachment to fibronectin with terminal human erythroid differentiation. *Blood* 75, 865–873. doi: 10.1182/blood.v75.4.865.bloodjournal 754865

Watanabe, N., Bodin, L., Pandey, M., Krause, M., Coughlin, S., Boussiotis, V. A., et al. (2008). Mechanisms and consequences of agonist-induced talin recruitment to platelet integrin alphaIIbbeta3. *J Cell Biol.* 181, 1211–1222. doi: 10.1083/jcb.200803094

10

Blood Rheology: Key Parameters, Impact on Blood Flow, Role in Sickle Cell Disease and Effects of Exercise

*Elie Nader[1,2†], Sarah Skinner[1,2†], Marc Romana[2,3,4], Romain Fort[1,2,5], Nathalie Lemonne[6], Nicolas Guillot[7], Alexandra Gauthier[1,2,8], Sophie Antoine-Jonville[9], Céline Renoux[1,2,10], Marie-Dominique Hardy-Dessources[2,3,4], Emeric Stauffer[1,2,11], Philippe Joly[1,2,10], Yves Bertrand[8] and Philippe Connes[1,2]**

[1] Laboratory LIBM EA7424, Team "Vascular Biology and Red Blood Cell", University of Lyon 1, Lyon, France, [2] Laboratory of Excellence GR-Ex, Paris, France, [3] Biologie Intégrée du Globule Rouge, Université de Paris, UMR_S1134, BIGR, INSERM, F-75015, Paris, France, [4] Biologie Intégrée du Globule Rouge, The Université des Antilles, UMR_S1134, BIGR, F- 97157, Pointe-a-Pitre, France, [5] Département de Médecine, Hôpital Edouard Herriot, Hospices Civils de Lyon, Lyon, France, [6] Unité Transversale de la Drépanocytose, Hôpital de Pointe-a-Pitre, Hôpital Ricou, Pointe-a-Pitre, France, [7] Laboratoire Carmen INSERM 1060, INSA Lyon, Université Claude Bernard Lyon 1, Université de Lyon, Villeurbanne, France, [8] d'Hématologie et d'Oncologie Pédiatrique, Hospices Civils de Lyon, Lyon, France, [9] Laboratoire ACTES EA 3596, The Université des Antilles, Pointe-a-Pitre, France, [10] Laboratoire de Biochimie et de Biologie Moleiculaire, UF de Biochimie des Pathologies Eirythrocytaires, Centre de Biologie et de Pathologie Est, Hospices Civils de Lyon, Lyon, France, [11] Centre de Médecine du Sommeil et des Maladies Respiratoires, Hospices Civils de Lyon, Hôpital de la Croix Rousse, Lyon, France

**Correspondence:*
Philippe Connes
pconnes@yahoo.fr;
philippe.connes@univ-lyon1.fr
[†] These authors have contributed equally to this work

Blood viscosity is an important determinant of local flow characteristics, which exhibits shear thinning behavior: it decreases exponentially with increasing shear rates. Both hematocrit and plasma viscosity influence blood viscosity. The shear thinning property of blood is mainly attributed to red blood cell (RBC) rheological properties. RBC aggregation occurs at low shear rates, and increases blood viscosity and depends on both cellular (RBC aggregability) and plasma factors. Blood flow in the microcirculation is highly dependent on the ability of RBC to deform, but RBC deformability also affects blood flow in the macrocirculation since a loss of deformability causes a rise in blood viscosity. Indeed, any changes in one or several of these parameters may affect blood viscosity differently. Poiseuille's Law predicts that any increase in blood viscosity should cause a rise in vascular resistance. However, blood viscosity, through its effects on wall shear stress, is a key modulator of nitric oxide (NO) production by the endothelial NO-synthase. Indeed, any increase in blood viscosity should promote vasodilation. This is the case in healthy individuals when vascular function is intact and able to adapt to blood rheological strains. However, in sickle cell disease (SCD) vascular function is impaired. In this context, any increase in blood viscosity can promote vaso-occlusive like events. We previously showed that sickle cell patients with high blood viscosity usually have more frequent vaso-occlusive crises than those with low blood viscosity. However, while the deformability of RBC decreases during acute vaso-occlusive events in SCD, patients with the highest RBC deformability at steady-state have a higher risk of developing frequent painful vaso-occlusive crises. This paradox seems to be due to the fact that in SCD RBC with the highest deformability are also the most adherent, which would trigger vaso-occlusion. While acute, intense exercise may increase blood viscosity in healthy individuals, recent works conducted in sickle cell patients have shown that light cycling exercise did not cause dramatic changes in blood rheology. Moreover, regular physical exercise has been shown to decrease blood viscosity in sickle cell mice, which could be beneficial for adequate blood flow and tissue perfusion.

Keywords: blood rheology, red blood cell deformability, red blood cell aggregation, sickle cell disease, exercise

BLOOD FLOW RESISTANCE AND THE CARDIOVASCULAR SYSTEM

Flow velocity in a given tube depends on pressure and flow resistance. According to Poiseuille's Law (Poiseuille, 1835), flow resistance depends on the geometry of the tube [length (L) and radius of the tube (r)] and the fluid's viscosity (η), and is calculated using the following formula:

$$R = \frac{8\eta\ L}{\pi\ r^4}$$

When applying Poiseuille's Law to the cardiovascular system, one must consider the radius and the length of the vessels, and the viscosity of the blood. The dimensions of the vascular system (most notably the radius, which is raised to the fourth power) play a more important role in determining vascular resistance than blood viscosity does. However, several works conducted in the past 10–15 years have shown that, in a physiological context, the parameters of this equation cannot be considered to be truly independent of each other. This is because vessels are not rigid tubes; they can change their diameters in response to various physiological stimuli. One of the most important molecules that promotes an augmentation in vascular diameter (i.e., vasodilation) is nitric oxide (NO). Martini et al. (2005), Tsai et al. (2005), Intaglietta (2009), and Sriram et al. (2012) showed that mild to moderate increases in hematocrit and blood viscosity did not result in a rise in vascular resistance or blood pressure, but actually caused the opposite effect. They also showed that increasing blood viscosity promoted the activation of endothelial NO-synthase through shear stress-dependent mechanisms, resulting in higher NO production, compensatory vasodilation, and decreased arterial pressure. However, evidence shows that these vascular adaptations can only occur in a functioning vascular system with a healthy endothelium. When vascular dysfunction is present, vasodilation is impaired. Therefore, a rise in blood viscosity is not accompanied by an increase in vasodilation. As a result, vascular resistance and arterial pressure increase (Vazquez et al., 2010; Salazar Vazquez et al., 2011). Although the role of blood viscosity in vascular adaptations is often ignored, these studies clearly demonstrate that vascular geometry and blood viscosity should not be considered separately when studying the regulation of vascular resistance in healthy populations or in people with cardiovascular diseases.

BLOOD IS NOT A SIMPLE FLUID

Whole blood is a two-phase liquid, composed of cellular elements suspended in plasma, an aqueous solution containing organic molecules, proteins, and salts (Baskurt and Meiselman, 2003). The cellular phase of blood includes, erythrocytes, leukocytes, and platelets. White blood cells and platelets can affect blood rheology, but under normal conditions, red blood cells (RBCs) have the biggest influence (Pop et al., 2002). Blood rheological properties are determined by the physical properties of these two phases and their relative contribution to total blood volume.

Blood is a non-Newtonian, shear thinning fluid with thixotropic and viscoelastic properties. Many cardiovascular handbooks consider blood viscosity values between 3.5 and 5.5 cP to be normal. However, blood viscosity cannot be summarized by a single value. Due to the shear thinning property of blood, which is dependent on RBC rheological properties, the viscosity of this fluid changes depending on the hemodynamic conditions. The same blood can have a viscosity value of 60 cP at a shear rate of $0.1\ s^{-1}$, whereas the viscosity would be 5 or 6 cP at a shear rate of $200\ s^{-1}$. This means that blood viscosity is different in the large arteries, the veins, and the microcirculation, where shear rate can vary from few s^{-1} to more than $1000\ s^{-1}$ (Connes et al., 2016). Blood viscosity depends on several factors: hematocrit, plasma viscosity, the ability of RBCs to deform under flow, and RBC aggregation-disaggregation properties (Baskurt and Meiselman, 2003; Cokelet and Meiselman, 2007).

Effect of Hematocrit

Whole blood viscosity is dependent on the number (and volume) of erythrocytes in the blood, and is thus linearly related to hematocrit (Chien et al., 1975).The impact of hematocrit on blood viscosity is much higher at low shear rates (veins for instance) than at high shear rate (arteries for instance) (Cokelet and Meiselman, 2007). At high shear rate, it is estimated that a rise of hematocrit of one unit would cause an increase of blood viscosity of 4% (if RBC rheological properties remain the same).

Plasma Viscosity

Plasma is a newtonian fluid, which means that its viscosity does not vary with shear rate. The viscosity of plasma is dependent on the concentration of plasma proteins, such as fibrinogen, α1-globulins, α2-globulins, β-globulins, and γ-globulins (Connes et al., 2008). Any elevation in the concentration of these proteins can cause plasma, and thus whole blood, viscosity to increase (Kesmarky et al., 2008). Normal plasma at 37 degrees Celsius has a viscosity of around 1.2–1.3 cP, but these values may be higher in various inflammatory, metabolic, or cardiovascular diseases (Kesmarky et al., 2008). Furthermore, increased plasma viscosity is associated with higher rates of adverse clinical events in unstable angina pectoris and stroke (Kesmarky et al., 2008).

RBC Deformability

Red blood cell deformability is another important determinant of blood viscosity. RBC deformability depends on several factors, including internal (cytosolic) viscosity (mainly determined by the mean cell hemoglobin concentration), membrane viscoelasticity (which is dependent on cytoskeleton proteins and lipid bilayer properties), and the surface-area-to-volume ratio (also called cell sphericity) (Clark et al., 1983; Renoux et al., 2019). At low shear rates, rigid RBCs are less likely to aggregate than deformable RBCs. Therefore, a loss of RBC deformability at very low shear rates (less than 1 s^{-1}) results in a decrease in blood viscosity (Chien et al., 1970). In contrast, at shear rates above 1 s^{-1}, a decrease in RBC deformability causes blood viscosity to increase (Chien et al., 1970).

Initial experiments done to analyze RBC deformability in blood flow were conducted in the 1970s and 1980s using microtubes (Goldsmith et al., 1972) and rheoscopes (Fischer et al., 1978). These studies demonstrated that, as shear stress increases, normal RBCs align with the direction of flow by deforming into an elliptical shape via a "tank tread-like" motion of the cell membrane around the cytoplasm (Schmid-Schonbein et al., 1969; Goldsmith et al., 1972; Fischer et al., 1978). Rigid RBCs, on the other hand, cannot properly deform into an ellipse and remain perpendicular to blood flow, consequently increasing vascular resistance. In these fundamental experiments, RBCs were suspended in solutions, such as dextrose, with higher viscosities than the internal viscosity of a RBC (Goldsmith et al., 1972; Fischer et al., 1978). However, when a RBC is flowing in plasma *in vivo*, the plasma viscosity is lower than the viscosity of the erythrocyte's cytosol. This is important because recent experiments reveal that the "tank-treading" behavior of erythrocytes does not occur when RBCs are suspended in solutions with lower viscosities that are more similar to plasma viscosity *in vivo* (Dupire et al., 2012). Instead, observations by Lanotte et al. (2016) showed that RBCs display a wide variety of cell shapes for any given flow condition. For example, erythrocytes in dilute suspensions behave like rigid oblate ellipsoids at low shear rates (<1 s^{-1}). Then, as shear rates increase, the erythrocytes successively tumble, roll, and deform into stomatocytes, and eventually adopt highly deformed poly-lobed. The findings of Lanotte et al. (2016) suggest that the pathological alterations of multiple parameters, including plasma composition, erythrocyte cytosol viscosity, and/or membrane mechanical properties could contribute to pathological blood rheology and flow. Further research should be conducted to determine how decreased RBC deformability could affect RBC shape transitions.

Red blood cell deformability is also a key determinant of blood flow in the microcirculation. RBCs are bi-concave disks with an average diameter of around 7–8 μm. Capillaries can have a diameter of less than 5 μm. Therefore, RBC must be highly deformable to pass through the narrowest vessels of the microcirculation. A 15% decrease in RBC deformability has been shown to cause a 75% increase in whole flow resistance in isolated perfused rat hind limbs (Baskurt et al., 2004). Moreover, when dog lungs were perfused with rigid RBC, whole pulmonary arterial pressure increased, and the main increase was localized in the microcirculation (Hakim, 1988; Raj et al., 1991). Indeed, any decrease of RBC deformability may affect flow resistance, tissue perfusion, and oxygenation (Parthasarathi and Lipowsky, 1999).

RBC Aggregation

Red blood cell aggregation is the reversible formation of three-dimensional stacks of RBC, called "rouleaux," which takes place at low shear rates. This unique process requires low energy and is reversible under high shear rate conditions. RBC aggregation depends on both plasma and cellular factors. Initially, most research on RBC aggregation formation was focused on the effects of protein levels, polymer type, and concentration. More recent research has shown that RBC cellular properties can also modulate a cell's intrinsic tendency to aggregate (termed RBC aggregability) (Baskurt and Meiselman, 2009). For example, RBC surface properties, including surface charge and glycocalyx depth, also play an important role in this process. Two models have been proposed to explain RBC aggregation mechanisms (Meiselman et al., 2007), the depletion model and the bridging model. The depletion model suggests that RBC aggregates are formed due to osmotic pressure from surrounding plasma proteins or other macromolecules. The bridging model suggests that aggregates form due to "crossbridges," made of plasma proteins or other macromolecules (Rampling et al., 2004). Fibrinogen is the most physiologically relevant macromolecule that promotes RBC aggregation (Rampling et al., 2004), but other molecules, including thrombospondin and the von-Willebrand factor, also play a role (Nader et al., 2017).

The impact of RBC aggregation on blood flow, tissue perfusion and vascular resistance is complex and depends on the vascular areas where RBC aggregates are flowing. RBC aggregates usually form in low shear rate areas, such as in veins or in bifurcations. Therefore, increased RBC aggregation would cause a dramatic increase of blood viscosity in these zones. RBC aggregates disaggregate in high shear rate areas, such as in arteries and arterioles. However, it has been demonstrated that some RBC aggregates can persist in large arteries and affect flow dynamics. Increased RBC aggregation has been shown to promote RBC axial migration in these vessels, which in turn increases the cell free layer width (Baskurt and Meiselman, 2007). This latter phenomenon has three main consequences: (i) a decrease of apparent dynamic blood viscosity and of flow resistance, (ii) a decrease of wall shear stress, which in turn results in a lower activation of endothelial NO-synthase, lower NO production, and less vasodilation, and (iii) an increase of plasma skimming phenomena at bifurcations, which in turn lowers microcirculatory hematocrit and blood viscosity (Fharaeus and Fahraeaus–Lindqvist effects).

At the microcirculatory level, persisting RBC aggregates may increase pre-capillary resistance. Experiments performed on large glass tubes also shown that the consequences of RBC aggregation on flow resistance are dependent on the orientation of the tube (vertical vs. horizontal) (Cokelet and Goldsmith, 1991). In horizontal tubes, elevated RBC aggregation increases RBC sedimentation, thereby increasing flow resistance. On the other hand, in vertical tubes RBC aggregation increases RBC axial migration and facilitates blood flow. For this reason, the consequences of increased RBC aggregation on blood flow are difficult to predict. However, experiments done on whole organs, such as guinea pig hearts, reported that gradually increasing RBC aggregation causes a 3-phase evolution of blood flow resistance (Yalcin et al., 2005). First, when RBC aggregation was increased by 50–100%, the authors observed a rise in blood flow resistance. Next, a 100–150% increase in RBC aggregation caused blood flow resistance to decrease. Finally, an increase in RBC aggregation of over 150% caused blood flow resistance to increase once again. Overall, predicting the consequences of increased RBC aggregation *in vivo* appears to be complex. However, clinical works performed to study aggregation in cardiovascular, metabolic, or inflammatory diseases consistently show that people with these diseases generally have higher RBC aggregation levels than healthy individuals, and the elevated aggregation contributes to the development of adverse disease outcomes (Totsimon et al., 2017; Biro et al., 2018; Brun et al., 2018; Ko et al., 2018; Lapoumeroulie et al., 2019; Piecuch et al., 2019; Sheremet'ev et al., 2019).

SICKLE CELL DISEASE, BLOOD RHEOLOGY AND VASCULAR DYSFUNCTION

Sickle cell disease (SCD) is the most prevalent genetic disease in the world. Sickle cell anemia (SCA) is by far the most common form of SCD, followed by hemoglobin SC disease (SC) (Piel et al., 2010).It is estimated that more than 300,000 children are born each year with a severe inherited hemoglobinopathy, over 80% of these infants are born in low-or middle-income countries, and approximately 220,000 are affected by SCA (Piel et al., 2010).

Sickle Cell Anemia as a Hemorheological Disease

Sickle cell anemia is caused by a single nucleotide mutation (adenine → thymine) in exon I of the β-globin gene. This point mutation (rs334 T) leads to the production of sickle hemoglobin (HbS), due to the substitution of valine for glutamic acid at the sixth position of the β-globin chain. The hydrophobic residue of valine associates with other hydrophobic residues, which causes HbS molecules to aggregate, forming fibrous precipitates when hemoglobin is deoxygenated. This phenomenon is called "HbS polymerization," and is responsible for the characteristic shape change, termed "sickling," of RBCs. Sickle RBC are rigid, and therefore do not easily flow through the microcirculation, causing frequent vaso-occlusive episodes in affected patients. In addition, when RBCs lose their deformability, they become more fragile and prone to hemolysis, which is the root cause of chronic hemolytic anemia in SCA (Connes et al., 2014).

Although the loss of RBC deformability is a fundamental characteristic of SCA, patients exhibit varying degrees of RBC rigidity, which can differentially affect SCA disease severity and complications (Renoux et al., 2016). Newborns usually have almost normal RBC deformability because they still have a high percentage of fetal hemoglobin (HbF) (Renoux et al., 2016). However, when patients become older HbF is replaced by HbS, and, as a result, RBC deformability decreases (Renoux et al., 2016). Hydroxyurea (HU) therapy can affect RBC deformability because HU stimulates HbF synthesis, thereby improving RBC deformability, although the deformability values still remain lower than in the healthy population (Lemonne et al., 2015). In individuals who are not under HU therapy, the presence or absence of alpha-thalassemia can also modulate RBC deformability (Renoux et al., 2017), as the inheritance of alpha-thalassemia results in decreased production of HbS (Rees et al., 2010) and thus lower HbS polymerization.

Rigid, sickle deformed RBC can cause pre-capillary obstruction and contribute to vaso-occlusion (Rees et al., 2010). Furthermore, RBC deformability further decreases during acute vaso-occlusive events. However, patients (adults and children over 5 years of age) who have the highest RBC deformability at steady-state, and who are not under HU therapy, actually have a higher risk of developing frequent painful vaso-occlusive crises, as well as osteonecrosis (Lamarre et al., 2012; Lemonne et al., 2013; Renoux et al., 2017). This paradox seems to be due to the fact that in SCA, RBC with the highest deformability are also the most adherent, which would trigger vaso-occlusion (Ballas et al., 1988). On the other hand, individuals who have improved RBC deformability as a result of HU therapy do not have an elevated risk of vaso-occlusive crisis because HU inhibits RBC, platelets and WBC adhesion to the vascular wall via several mechanisms (Bartolucci et al., 2010; Laurance et al., 2011; Chaar et al., 2014; Pallis et al., 2014; Verger et al., 2014). RBC aggregation is usually lower in SCA patients because of low hematocrit and the inability of irreversible sickle cell to form aggregates. However, the formed RBCs aggregates are reported to be much more robust than in healthy population which could further alter blood flow and tissue perfusion at the pre and post capillary levels (Tripette et al., 2009; Connes et al., 2014).

Sickle cell anemia patients with the highest degree of chronic hemolysis (i.e., those with the most severe anemia) are usually characterized by a severe and chronic reduction of RBC deformability as well as the presence of robust RBC aggregates, and seem to be more prone to develop complications such as leg ulcers, priapism and glomerulopathy (Bartolucci et al., 2012; Connes et al., 2013a; Lamarre et al., 2014). In contrast, patients with the highest RBC deformability tend to have less severe anemia and increased blood viscosity, which could increase the risk of developing frequent vaso-occlusive like complications (Nebor et al., 2011; Lamarre et al., 2012; Charlot et al., 2016). Surprisingly, a recent study did not show a further rise of blood viscosity during acute painful vaso-occlusive crisis (Lapoumeroulie et al., 2019). Instead, the authors found a decrease of RBC deformability, as a result of massive RBC sickling (Ballas and Smith, 1992), and a rise in RBC aggregation and RBC aggregates robustness, which would probably impair blood flow into the microcirculation (Lapoumeroulie et al., 2019).

Vascular Function Cannot Compensate for the Hemorheological Alterations in SCA

As previously discussed, recent works have demonstrated that increased blood viscosity does not systematically cause a rise in vascular resistance in healthy individuals (Vazquez et al., 2010; Salazar Vazquez et al., 2011). Instead, increasing blood viscosity can facilitate vasodilation and decrease vascular resistance through shear stress-dependent mechanisms, which increase NO production. However, when endothelial/vascular dysfunction is present, like in SCA (Kato et al., 2007; Ataga et al., 2016; Charlot et al., 2016), vasodilation is impaired. Therefore, a rise in blood viscosity cannot be fully compensated for, and could increase the risk of frequent vaso-occlusive crises (Lemonne et al., 2012; Charlot et al., 2016).

Chronic hemolysis is highly implicated in the development of vascular dysfunction in SCA. Once hemoglobin is released into the plasma it autoxidizes, forming free heme, iron, and reactive oxygen species, which cause eNOS uncoupling, and decrease NO bioavailability (Kato et al., 2007).These alterations limit the relaxation of vascular smooth muscle, and contribute to the onset of vaso-occlusive crises (Hebbel et al., 2004). In addition, hemolysis releases the arginase contained in erythrocytes into the plasma. The free arginase hydrolyzes arginine, which is the NO precursor, to ornithine and urea, thereby exacerbating the decrease of NO bioavailability (Kato et al., 2007).

Xanthine oxidase (XO) also contributes to vascular dysfunction in SCA. Aslan et al. (2001) demonstrated that episodes of intrahepatic hypoxia-reoxygenation, which can occur in SCA, induce the release of plasma XO. The released XO can impair vascular function by binding to the vessel luminal cells. This creates an oxidative milieu, which results in NO scavenging via an oxygen free radical-dependent mechanism. Furthermore, Mockesch et al. (2017) recently showed that loss of microvascular function in children with SCA was significantly associated with both nitrotyrosine and markers of systemic oxidative stress. These findings confirm the important roles oxidative stress and NO scavenging play in the development of vascular dysfunction in SCA.

Circulating extra-vesicles, such as microparticles (Camus et al., 2012, 2015) and exosomes (Khalyfa et al., 2016), are also thought to play a role in the development of vascular dysfunction in SCA. Indeed, Camus et al. (2012) demonstrated that infusion of *in vitro*-generated RBC microparticles caused kidney vaso-occlusion in sickle cell mice. Additionally, Khalyfa et al. (2016) demonstrated that exosomes isolated from SCA patients with frequent vaso-occlusive crises decreased endothelial permeability and promoted P-selectin expression on cultured endothelial cells. The exosomes isolated from SCA patients with frequent vaso-occlusive crises also significantly increased the adhesion of monocytes to the vascular wall in mice, compared with exosomes isolated from SCA patients with a less severe phenotype. Overall, these works suggest that

therapies focusing both on blood rheology and vascular function could be helpful to decrease the clinical severity of SCA patients.

EXERCISE AND BLOOD RHEOLOGY IN HEALTHY INDIVIDUALS AND INDIVIDUALS WITH SCA

Blood rheology plays an important role in the regulation of tissue perfusion at rest and during exercise. For example, RBCs need to be highly deformable to easily flow through small capillaries and transport oxygen to the tissues (Parthasarathi and Lipowsky, 1999). Any changes in RBC rheological properties during exercise may affect blood viscosity (Baskurt and Meiselman, 2003), which in turn may impact blood flow and exercise performance (Connes et al., 2013b; Waltz et al., 2015). Several investigators have reported significant correlations between blood fluidity and indices of physical fitness in sportsmen, such as time of endurance until exhaustion, aerobic working capacity at 170 W (W170), and maximal oxygen consumption (VO_{2max}) (Ernst et al., 1985; Brun et al., 1986, 1989, 1995).

The Effect of Acute Cycling Exercise on Blood Viscosity and Its Determinants in Healthy Individuals

Most of the studies conducted in the eighties and nineties to investigate the acute effects of exercise on blood rheology focused on cycling efforts performed by moderately trained subjects (Brun et al., 1998). These studies reported that acute cycling exercise caused blood viscosity measured at high shear rate (i.e., $100–200 \text{ s}^{-1}$) to increase by more than 15% above pre-exercise values (Brun et al., 1998; Connes et al., 2004a, 2013b; **Figure 1**). This increase in blood viscosity has been attributed to changes in plasma viscosity, hematocrit, and RBC rheological properties (Brun et al., 1998).

Multiple studies have shown that an acute bout of maximal or submaximal cycling exercise may cause plasma viscosity to increase by 10–12%, compared to resting values (Ernst, 1985; Ernst et al., 1991b; Brun et al., 1994b, 1998; Bouix et al., 1998; Connes et al., 2004b). This increase has been attributed to a rise in plasma protein content, such as fibrinogen, α1-globulins, α2-globulins, β-globulins, and γ-globulins during exercise (Nosadova, 1977; Convertino et al., 1981; Vandewalle et al., 1988; Wood et al., 1991).

Several studies have also shown that an acute cycling exercise can result in a 10–12% (i.e., 3–4 units) rise in hematocrit, when compared to pre-exercise values (Brun et al., 1994b; Bouix et al., 1998; Connes et al., 2004a). The effect of elevated hematocrit on blood viscosity at high shear rates has been well described (Cokelet and Meiselman, 2007). A 1-unit increase in hematocrit can cause a 4% increase in blood viscosity. This rise in hematocrit has been attributed to several mechanisms such as fluid shift between intra- and extra-vascular spaces (Sjogaard et al., 1985; Ploutz-Snyder et al., 1995), dehydration (Nosadova, 1977; Stephenson and Kolka, 1988) the release of sequestered

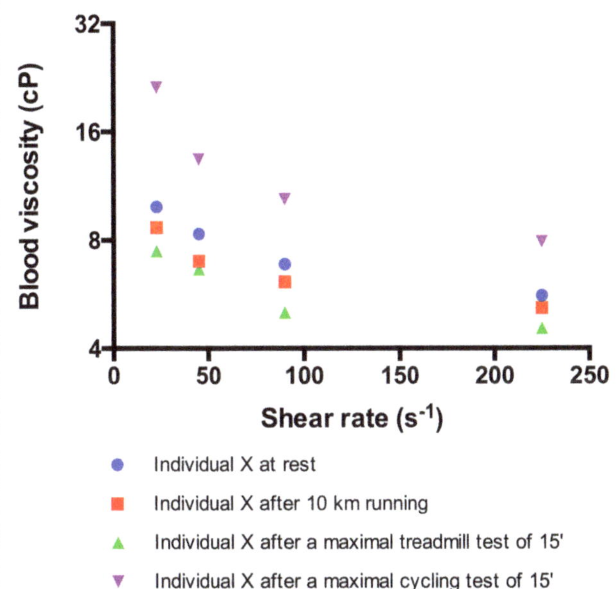

FIGURE 1 | Effects of different kind of exercise on blood viscosity measured at several shear rates in the same trained subject (maximal oxygen consumption, VO₂max = 64 ml/kg/min). The maximal treadmill and cycling tests consisted in a progressive and maximal exercise conducted until VO₂max and performed in laboratory conditions (temperature: 24°C). The 10 km running was performed outdoor (20°C) and at the highest intensity the subject could run. Indeed, the subject ran at 77% of his maximal aerobic velocity determined on the treadmill and reached a heart rate of 92% of his maximal heart rate. The subject did not drink water during each test. For the three tests, blood was sampled immediately at the end of the exercise and blood viscosity was measured with the same cone-plate viscometer within 1 h after sampling.

RBCs from the spleen (Isbister, 1997), and water trapping in muscle (Ploutz-Snyder et al., 1995).

Acute cycling exercise can also modulate RBC deformability. Several authors have reported a decrease in RBC deformability during and immediately after exercise (Reinhart et al., 1983; Galea and Davidson, 1985; Gueguen-Duchesne et al., 1987; Brun et al., 1993, 1994a; Oostenbrug et al., 1997; Yalcin et al., 2000, 2003). This decrease could be a consequence of lactic acid and reactive oxygen species production that occurs during exercise (Connes et al., 2013b). The accumulation of lactate ions and the decrease in pH would promote RBC dehydration through the activation of several RBC cationic channels, leading to RBC shrinkage and decreased RBC deformability (Van Beaumont et al., 1981; Lipovac et al., 1985; Smith et al., 1997; Connes et al., 2004c). Moreover, several groups have suggested that oxidative stress could also play a role in the decreased RBC deformability through the oxidation of membrane lipid and protein components (Senturk et al., 2005a,b; Connes et al., 2013b). However, the magnitude of the change in RBC deformability that occurs during exercise seems to depend on the training status and physical fitness of the subjects. For instance, Senturk et al. (2005a) reported that a short, progressive, maximal cycling exercise test promoted oxidative stress in both sedentary and well-trained subjects, but they only observed a decrease in RBC deformability and an increase of RBC fragility in sedentary individuals.

Very few studies have observed RBC aggregation during cycling exercise, and the results of these studies are very heterogeneous (Connes et al., 2013b). Yalcin et al. (2003) found that RBC aggregation decreased after a Wingate exercise test. In contrast, Ernst et al. (1991b) observed a transient increase in RBC aggregation above pre-exercise values after 1 h of pedaling at a heart rate of 150 beats min⁻¹. RBC aggregation then returned to baseline during the 2 h following the cycling test. Varlet-Marie et al. (2003) and Connes et al. (2007) found no change in RBC aggregation in response to submaximal/maximal cycling exercise. The discrepancies observed are not very well understood and might be related to the (1) kind of exercise performed (short versus prolonged exercise), (2) time of measurement during exercise (i.e., immediately at the end of exercise or few minutes after), (3) time delay for measurement after sampling, and (4) procedure used for RBC aggregation measurement (adequate oxygenation and adjusted hematocrit before measurement or not).

On the whole, these changes lead to a rise in blood viscosity during acute cycling exercise. Large elevations in blood viscosity have generally been considered to be deleterious for the cardiovascular system (Brun et al., 1998; Connes et al., 2013b). However, Connes et al. (2012) reported a positive correlation between the magnitude of change in blood viscosity and the magnitude of change in NO end-stable products, and a negative correlation between the magnitude of change in blood viscosity and the magnitude of change in vascular resistance. These findings suggest that increasing blood viscosity during exercise could be a way to stimulate endothelium-dependent NO production through shear stress-related mechanisms. In healthy individuals, this would result in compensatory vasodilation, thereby preventing any large increases in vascular resistance.

Acute Running Exercise, Blood Viscosity and Its Determinants in Healthy Individuals

Surprisingly, in contrast to cycling exercises, running exercises, such as a marathon or a 10-km race, do not cause a rise in blood viscosity (Neuhaus et al., 1992; Tripette et al., 2011; **Figure 1**). The main reason is that hematocrit and plasma viscosity usually remain very stable during these kinds of efforts, despite dehydration (Galea and Davidson, 1985; Neuhaus et al., 1992; Neuhaus and Gaehtgens, 1994). It has been hypothesized that the lack of change in hematocrit could be explained by repeated RBC foot strike hemolysis during running (Neuhaus et al., 1992; Neuhaus and Gaehtgens, 1994; Tripette et al., 2011; Connes et al., 2013b). However, a recent study in which highly trained subjects performed a progressive and maximal treadmill test did not observe any signs of RBC damage or eryptosis (Nader et al., 2018). However, the picture could be slightly different during ultra-running events. For instance, Robach et al. (2014) reported increased hemolysis and large plasma volume expansion immediately after a 166-km long mountain ultra-endurance marathon with 9500 m of altitude gain/loss. The impact of these changes on blood rheology has not been investigated and further studies are needed to address this question.

Surprisingly, Nader et al. (2018) found a slight but significant increase in RBC deformability in response to a short and maximal exercise test. Indeed, although hematocrit increased, blood viscosity remained unchanged (and tended to decrease), as has been observed in previous studies (Neuhaus and Gaehtgens, 1994; Tripette et al., 2011). The findings suggest that the slight increase in RBC deformability could have compensated for the rise in hematocrit observed during short running events, resulting in a lack of change in blood viscosity. The impact of ultra-run events on RBC deformability is unknown. Several other groups found a slight increase in RBC deformability during running (Suhr et al., 2012) and cycling (Connes et al., 2009) exercises. Although the exact reasons for these findings are unknown, it has been suggested that increased NO production during exercise could increase RBC deformability. One of the first works to suggest that NO could affect RBC deformability was the study of Starzyk et al. (1997), which demonstrated that intravenous infusion of L-NAME (an eNOS inhibitor) in rats caused a reduction of RBC deformability. In addition, Bor-Kucukatay et al. (2003) reported that several eNOS inhibitors also decreased RBC deformability, suggesting that basal release of NO actively maintains RBC deformability. While extracellular sources of NO may impact the deformability of RBC, several works suggest that endogenous RBC NO synthesis may also modulate RBC deformability (Kleinbongard et al., 2006). Suhr et al. (2012) demonstrated that an acute running exercise induced shear stress activation of RBC NOS (increased RBC NOS phosphorylation at Ser1177) via the PI3-kinase/Akt kinase pathway, leading to increased NO production by the RBC, which was critical to maintaining RBC deformability during exercise. Grau et al. (2013) further extended these findings by showing that RBC NOS activation by pharmacological treatment (insulin) increased RBC NO content and improved RBC deformability through direct S-nitrosylation of cytoskeleton proteins, most likely the α- and β-spectrins. In contrast, the use of RBC NOS inhibitors [wortmannin or L-N5-(1-Iminoethyl)-ornithin] resulted in a decrease of RBC NOS Ser1177 phosphorylation, NO content, cytoskeleton protein S-nitrosylation, and RBC deformability.

Impact of Physiological Changes During Acute Exercise on Blood Viscosity

As discussed above, acute cycling exercise would lead to a rise in blood viscosity while running exercise would be characterized by a lack of change in blood viscosity compared to the pre-exercise values. Although methodological aspects could partly explain some of these differences, such as the use of different viscometers (cone-plate viscometer, couette viscometer, capillary viscometer, etc.) or the use of different shear rates (low, moderate, and high shear rate viscosity are affected by different RBC rheological parameters), the picture is probably a little bit more complex. Several physiological changes occur during exercise, which may affect blood viscosity.

To avoid large dehydration during exercise, sportsmen usually drink water or carbohydrate-rich drinks. However, most of the studies performed in the field of blood rheology have been done in laboratory conditions where water is not allowed during the various exercise tests and water loss cannot be compensated by water intake. Tripette et al. (2010) and Diaw et al. (2014) previously tested the impact of add-libitum hydration and water deprivation on blood viscosity during a prolonged submaximal exercise and a soccer game, respectively. Healthy individuals and subjects carrying sickle cell trait were included. While hydration during exercise was able to decrease blood viscosity below the pre-exercise levels in sickle cell trait carriers, blood viscosity increased similarly in healthy individuals in both the "hydration" and "water deprivation" conditions. The rate of dehydration was around 1.5–2% in these studies, which is not very severe. There is a growing interest from the runners community to participate to very prolonged race (>100 kms), sometimes performed at high altitude. The impact of the environment and higher dehydration rate on blood viscosity are unknown and further studies are needed.

Exercise is accompanied by a rise cardiac output and blood flow leading to an increase of shear rate values in the vascular system. For instance, shear rate in the femoral artery has been reported to increase from 60 s^{-1} at rest to 200–250 s^{-1} during exercise (Gonzales et al., 2009). Blood is a shear-thinning fluid, meaning that its viscosity decreases when shear rate increases. Connes et al. (2013b) shown that 15 min of cycling exercise performed at a submaximal intensity increased blood viscosity measured at 90 s^{-1}. However, when the viscosity of the blood sampled at the end of exercise was measured at 225 s^{-1} (which reflects the shear rate reached during exercise), the mean value was almost identical to the viscosity of the blood sampled before exercise and measured at 90 s^{-1}. Indeed, the effects of hemoconcentration, increased plasma viscosity, decreased RBC deformability and increased RBC aggregation on blood viscosity during exercise could be counterbalanced by the effects of increasing shear rate (Connes et al., 2013b). However, the slight remaining changes in blood viscosity observed during exercise were still correlated with the magnitude of changes in NO-end products concentration, suggesting that blood viscosity plays a role in promoting NO production during exercise (Connes et al., 2013b).

Core temperature rises during exercise and it is well known that blood viscosity also depends on temperature (Baskurt et al., 2009). A recent elegant study shown that when the changes in temperature occurring during exercise are took into account in the measurements of blood viscosity, there is no difference between the pre- and post-exercise values (Buono et al., 2016). The slight physiological hyperthermia was shown to increase RBC deformability, which compensated the rise in hematocrit and resulted in a lack of change in blood viscosity. The authors estimated that the combined effects of increasing shear rate and hyperthermia during exercise could decrease blood viscosity by 31% below the pre-exercise levels, in spite of the exercise induced hemoconcentration (Buono et al., 2016). These findings clearly show that studies in the field of exercise hemorheology should consider the effects of various physiological factors for better interpretation about the role of blood viscosity in the cardiovascular adaptations and physical fitness.

Long Term Effects of Exercise on Blood Rheology in Healthy Individuals

Chronic exercise (endurance or resistance exercise) usually decreases blood viscosity (Brun et al., 1998; Romain et al., 2011; Kilic-Toprak et al., 2012). One reason for this change is that plasma volume increases few hours or several days after a single bout of exercise (Fellmann, 1992; Brun et al., 1998), resulting in an "autohemodilution" (Ernst, 1987; Ernst et al., 1991a). Repeated exercise bouts on consecutive days leads to a chronic "autohemodilution" resulting in a low baseline hematocrit – low baseline viscosity pattern (Ernst, 1987; Brun et al., 1998; Kilic-Toprak et al., 2012). The magnitude of plasma volume expansion ranges from 9 to 25%, corresponding to an additional 300 to 700 ml of plasma. It has been shown that the larger the reduction in plasma volume during exercise, the greater the subsequent plasma volume expansion (Fellmann, 1992). The amount of water ingested during and after exercise, as well several fluid-regulating hormones (aldosterone, vasopressin and the atrial natriuretic factor) and the level of plasma proteins, influence the degree of the plasma volume expansion after exercise. Plasma volume expansion is responsible for the decrease in plasma viscosity that contributes to the decrease in blood viscosity (Ernst, 1987).

Exercise training also induces RBC rheological adaptations (Brun et al., 2010; Kilic-Toprak et al., 2012). Ernst (1987) reported increased RBC deformability in athletes compared to sedentary subjects, a finding later confirmed by Smith et al. (1999). A 3-month longitudinal study of initially untrained healthy volunteers who performed regular training also showed a decrease of blood viscosity and a rise of RBC deformability (Ernst, 1987). Few studies have been conducted to determine why RBC deformability improves in healthy individuals after chronic exercise. However, Smith et al. (1999) and Tomschi et al. (2018) found a higher proportion of young deformable RBCs in athletes than in untrained subjects. The hemorheological benefits induced by regular exercise are suspected to contribute to the improvements in cardiovascular health that are induced by training programs in patients with cardiovascular disorders (Sandor et al., 2014).

Acute and Chronic Effects of Exercise in SCA

The metabolic changes occurring during exercise may promote HbS polymerization, RBC sickling, oxidative stress and inflammation. For this reason, physicians have generally been reluctant to promote physical activity for individuals with SCA (Connes, 2010; Waltz and Connes, 2014; Chirico et al., 2016; Martin et al., 2018). However, since regular physical activity has been shown to provide health benefits in various chronic diseases, several groups have begun to study the effects of acute and chronic exercise in individuals and mice with SCA.

Sickle cell anemia patients have decreased aerobic physical fitness compared to the general population (Connes et al., 2011). This is probably due to a combination of several factors, including chronic anemia, reduced muscle mass and strength, abnormal cardiac function, gas exchange abnormalities, mechanical ventilation limitations and peripheral vascular impairment (Callahan et al., 2002; Dougherty et al., 2011; Liem et al., 2015; Badawy et al., 2018; Merlet et al., 2019). A recent study reported negative associations between the oxygen uptake efficiency slope (an index of aerobic physical fitness) and hematocrit, RBC deformability, and RBC aggregate strength in adults with SCA (Charlot et al., 2015). These biological parameters were also associated with the ability of recover from a short submaximal exercise (Charlot et al., 2015). In another study, Waltz et al. (2013) showed that high levels of anemia, low fetal hemoglobin levels, and low RBC deformability were independent predictors of a low 6-min walking test performance.

Blood rheological abnormalities play a key role in the pathophysiology of SCA. For this reason, several works have investigated the effects of acute exercise on various biological parameters to determine what kind of exercise could be considered completely safe for individuals with SCA. In a study performed in Ivory Coast, 17 patients with SCA performed a 20 min moderate exercise (45 Watts) with blood sampling before and at the end of exercise (Balayssac-Siransy et al., 2011; Faes et al., 2014). Despite an increase in the percentages of dense RBC at the end of the effort, blood viscosity and soluble forms of P- and E-selectin remained unchanged compared to the pre-exercise level (Balayssac-Siransy et al., 2011; Faes et al., 2014). Slight increases of the plasma soluble forms of VCAM-1 and ICAM-1 were noted at the end of the exercise in SCA patients, suggesting slight endothelial activation (Faes et al., 2014). On the other hand, another study by Liem et al. (2015) reported higher concentrations of soluble VCAM-1 in the plasma of SCA patients at rest compared to a control group, but progressive and maximal exercise tests did not induce any further rise of VCAM-1. An additional study by Waltz et al. (2012) evaluated blood rheological parameters in subjects with SCA following a short (10–12 min) progressive submaximal cycling exercise, conducted until the first ventilatory threshold was reached. The exercise caused no changes in hematocrit, white blood cell count, blood viscosity, RBC deformability, or RBC aggregation. Furthermore, the strength of RBC aggregates decreased 2 and 3 days after the exercise. This delayed effect of exercise on RBC aggregate strength could be beneficial from a clinical point of view since this parameter is associated with risk of acute chest syndrome (Lamarre et al., 2012), and is increased during vaso-occlusive crises (Lapoumeroulie et al., 2019). Grau et al. (2019) recently confirmed that a short progressive and submaximal cycling exercise had no deleterious effect on RBC deformability and showed that this kind of effort does not exacerbate hemolysis. Finally, Barbeau et al. (2001) previously reported that the repetition of 30 min of moderate exercise for three consecutive days increased plasma NO concentrations in subjects with SCA. This could be viewed as a positive effect since NO bioavailability is reduced in SCA (Kato et al., 2007). It is important to note that none of these studies reported any clinical complications immediately or several days after the exercise (Barbeau et al., 2001; Balayssac-Siransy et al., 2011; Waltz et al., 2012; Faes et al., 2014; Grau et al., 2019). Overall, the results of these studies suggest that mild-to-moderate intensity exercise is probably safe in SCA, but longer or high-intensity

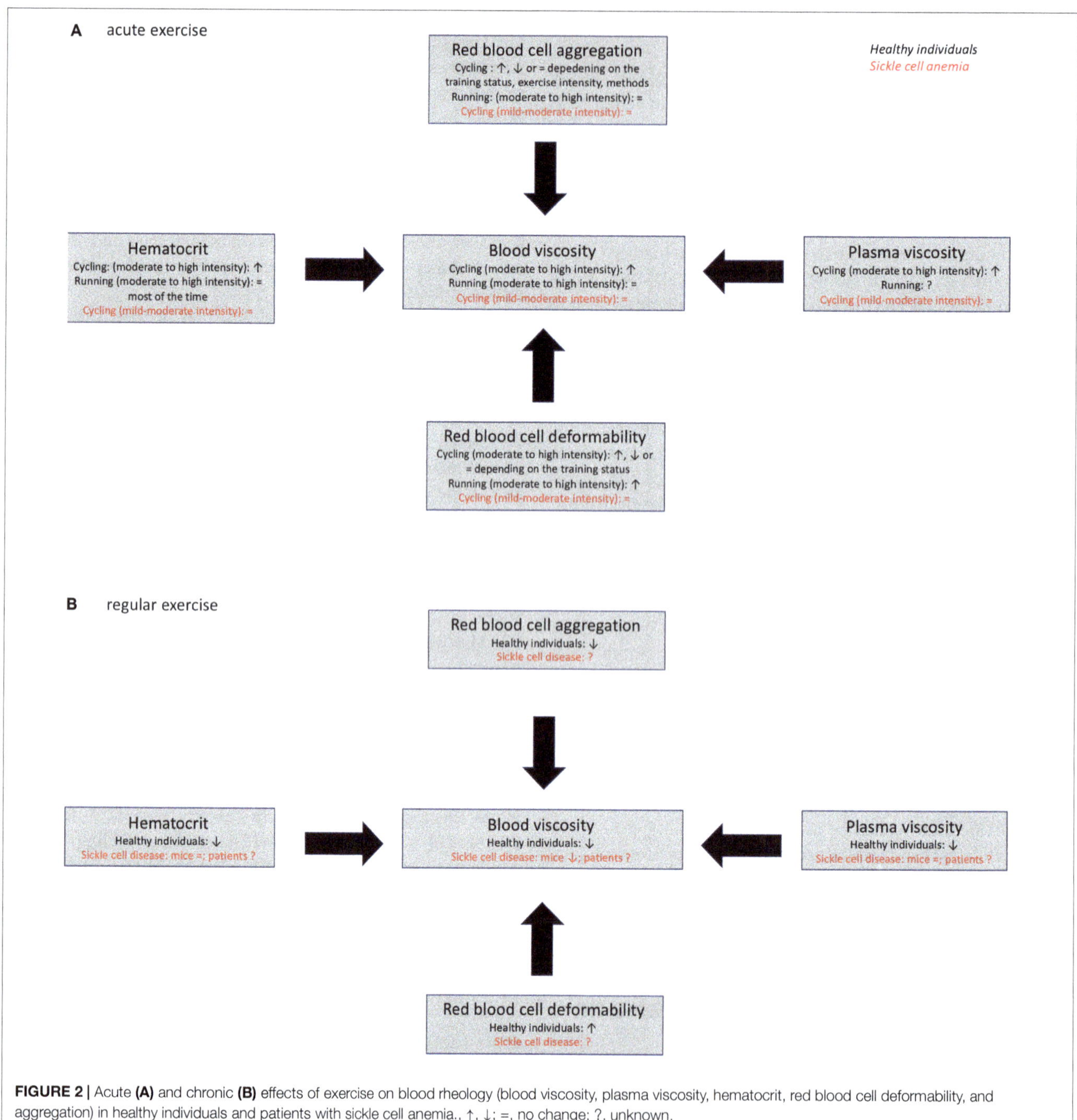

FIGURE 2 | Acute **(A)** and chronic **(B)** effects of exercise on blood rheology (blood viscosity, plasma viscosity, hematocrit, red blood cell deformability, and aggregation) in healthy individuals and patients with sickle cell anemia., ↑, ↓; =, no change; ?, unknown.

exercise should be avoided, or recommended only on a case-by-case basis (Martin et al., 2018). Still, some patients may experience moderate to severe hemoglobin desaturation, even during low-intensity or submaximal exercises, like the 6-min walking test (Waltz et al., 2013). Hemoglobin desaturation can promote RBC sickling in conditions of prolonged hypoxemic stress. Therefore, SCA patients should always be screened for hemoglobin desaturation during exercise tests to identify whether the person is at risk of experiencing exercise-induced hypoxemia during submaximal efforts.

The accumulating evidence showing that acute exercise could be safe for SCA patients prompted several groups to test the effects of regular exercise in mice and humans with SCA. Studies conducted in sickle SAD mice showed that 8 weeks of voluntary wheel running decreased blood viscosity (Faes et al., 2015), limited systemic oxidative stress (Charrin et al., 2015), and decreased pulmonary endothelial activation in response to an hypoxic-reoxygenation stimulus (Aufradet et al., 2014). An additional study conducted by Charrin et al. (2018) evaluated the effects of 8 weeks of aerobic training (1 h/day, 5 days/week) in a more severe mice model of SCA (Townes mice). The chronic exercise decreased several markers of systemic inflammation, including white blood cell count, plasma Th1/Th2 cytokine ratio, and Interleukin-1β level, and reduced the occurrence of splenomegaly. Furthermore, two other studies have also shown that endurance training improved muscle function in Townes SCA mice (Chatel et al., 2018; Gouraud et al., 2019).

Presently, only a few studies have been conducted to evaluate chronic exercise in people with SCA. Omwanghe et al. (2017) recently reported that 90% of children with SCA participate in physical education and 48% participate in sports. These findings illustrate that children with SCA do moderate to vigorous intensity physical activity for short durations. The same research group also prescribed three home exercise sessions per week, for 12 weeks, to 13 children with SCA. Results showed that 77% of subjects completed 89% of the prescribed sessions without any exercise-related adverse events. These results indicate that regular moderate exercise is safe and feasible in children with SCA. In an additional study by Gellen et al. (2018), adults with SCA performed 45 min of exercise three times a week for 8 weeks. No adverse events were reported in the study period, confirming that regular physical activity can be safe for people with SCA. Moreover, the SCA subjects' power outputs measured at 4 mmol/L blood lactate significantly increased after the 8 weeks training period, indicating improved physical fitness. Currently, no human studies have been conducted to determine whether chronic exercise can modulate the biological parameters (i.e., blood rheology, hematology, inflammation, oxidative stress) that cause the various acute complications experienced by people with SCA. However, the current evidence that exists suggests that moderate-intensity endurance-exercise training could potentially be used as a beneficial therapeutic strategy for patients with SCA (Gellen et al., 2018).

CONCLUSION

In conclusion, whole blood viscosity is a physiological parameter that should be considered when studying vascular resistance in healthy populations or in patients affected by various diseases. Plasma viscosity, hematocrit, RBC deformability, and

RBC aggregation are all factors that modulate blood viscosity. Alterations in any of these factors can modify blood flow resistance in the vasculature and alter tissue perfusion. In healthy individuals, the vascular system can adapt to elevations in blood viscosity because increased shear stress results in endothelium-dependent NO production. However, in individuals with vascular dysfunction, the vessels are not able to effectively vasodilate. Therefore, elevated blood viscosity can increase vascular resistance.

Tissue perfusion plays a key role during exercise and exercise can modulate blood viscosity and RBC rheology. The effects of acute exercise in healthy individuals are dependent on exercise modality, intensity, duration and physical fitness of the subjects. Most of the studies have found that acute cycling exercise increases blood viscosity by decreasing RBC deformability and increasing hematocrit and plasma viscosity (**Figure 2**). In contrast, acute running exercise is suggested to not cause any change in blood viscosity because of unaltered hematocrit (**Figure 2**). Furthermore, chronic exercise has been shown to decrease blood viscosity by increasing RBC deformability and decreasing hematocrit because of chronic auto-hemodilution (**Figure 2**).

Sickle cell anemia is a hereditary disease that causes pathological changes in blood rheology that ultimately contribute to the development of vascular dysfunction and disease complications. Several studies have shown that light to moderate intensity acute exercise is well-tolerated in individuals with SCA (**Figure 2**). Additionally, chronic exercise has been shown to cause positive hemorheological adaptations that could play a role in the positive benefits that result from training programs in patients with various cardiovascular disorders (**Figure 2**). Therefore, additional studies should be carried out to determine whether chronic exercise could improve blood rheological profiles in individuals with SCA, and decrease the severity of disease complications.

AUTHOR CONTRIBUTIONS

EN, SS, and PC wrote the first version of the manuscript. All authors read and approved the final version of the manuscript.

REFERENCES

Aslan, M., Ryan, T. M., Adler, B., Townes, T. M., Parks, D. A., Thompson, J. A., et al. (2001). Oxygen radical inhibition of nitric oxide-dependent vascular function in sickle cell disease. *Proc. Natl. Acad. Sci. U.S.A.* 98, 15215–15220. doi: 10.1073/pnas.221292098

Ataga, K. I., Derebail, V. K., Caughey, M., Elsherif, L., Shen, J. H., Jones, S. K., et al. (2016). Albuminuria is associated with endothelial dysfunction and elevated plasma endothelin-1 in sickle cell anemia. *PLoS One* 11:e0162652. doi: 10.1371/journal.pone.0162652

Aufradet, E., Douillard, A., Charrin, E., Romdhani, A., De Souza, G., Bessaad, A., et al. (2014). Physical activity limits pulmonary endothelial activation in sickle cell SAD mice. *Blood* 123, 2745–2747. doi: 10.1182/blood-2013-10-534982

Badawy, S. M., Payne, A. B., Rodeghier, M. J., and Liem, R. I. (2018). Exercise capacity and clinical outcomes in adults followed in the cooperative study of

sickle cell disease (CSSCD). *Eur. J. Haematol.* 101, 532–541. doi: 10.1111/ejh.13140

Balayssac-Siransy, E., Connes, P., Tuo, N., Danho, C., Diaw, M., Sanogo, I., et al. (2011). Mild haemorheological changes induced by a moderate endurance exercise in patients with sickle cell anaemia. *Br. J. Haematol.* 154, 398–407. doi: 10.1111/j.1365-2141.2011.08728.x

Ballas, S. K., Larner, J., Smith, E. D., Surrey, S., Schwartz, E., and Rappaport, E. F. (1988). Rheologic predictors of the severity of the painful sickle cell crisis. *Blood* 72, 1216–1223.

Ballas, S. K., and Smith, E. D. (1992). Red blood cell changes during the evolution of the sickle cell painful crisis. *Blood* 79, 2154–2163.

Barbeau, P., Woods, K. F., Ramsey, L. T., Litaker, M. S., Pollock, D. M., Pollock, J. S., et al. (2001). Exercise in sickle cell anemia: effect on inflammatory and vasoactive mediators. *Endothelium* 8, 147–155.

Bartolucci, P., Brugnara, C., Teixeira-Pinto, A., Pissard, S., Moradkhani, K., Jouault, H., et al. (2012). Erythrocyte density in sickle cell syndromes is associated with specific clinical manifestations and hemolysis. *Blood* 120, 3136–3141. doi: 10.1182/blood-2012-04-424184

Bartolucci, P., Chaar, V., Picot, J., Bachir, D., Habibi, A., Fauroux, C., et al. (2010). Decreased sickle red blood cell adhesion to laminin by hydroxyurea is associated with inhibition of Lu/BCAM protein phosphorylation. *Blood* 116, 2152–2159. doi: 10.1182/blood-2009-12-257444

Baskurt, O., and Meiselman, H. J. (2007). "In vivo hemorheology," in *Handbook of Hemorheology and Hemodynamics*, eds O. K. Baskurt, M. R. Hardeman, M. W. Rampling, and H. J. Meiselman, (Amsterdam: IOS Press), 322–338.

Baskurt, O. K., Boynard, M., Cokelet, G. C., Connes, P., Cooke, B. M., Forconi, S., et al. (2009). New guidelines for hemorheological laboratory techniques. *Clin. Hemorheol. Microcirc* 42, 75–97. doi: 10.3233/CH-2009-1202

Baskurt, O. K., and Meiselman, H. J. (2003). Blood rheology and hemodynamics. *Semin. Thromb. Hemost.* 29, 435–450. doi: 10.1055/s-2003-44551

Baskurt, O. K., and Meiselman, H. J. (2009). Red blood cell "aggregability". *Clin. Hemorheol. Microcirc* 43, 353–354. doi: 10.3233/CH-2009-1255

Baskurt, O. K., Yalcin, O., and Meiselman, H. J. (2004). Hemorheology and vascular control mechanisms. *Clin. Hemorheol. Microcirc* 30, 169–178.

Biro, K., Sandor, B., Kovacs, D., Csiszar, B., Vekasi, J., Totsimon, K., et al. (2018). Lower limb ischemia and microrheological alterations in patients with diabetic retinopathy. *Clin. Hemorheol. Microcirc.* 69, 23–35. doi: 10.3233/CH-189103

Bor-Kucukatay, M., Wenby, R. B., Meiselman, H. J., and Baskurt, O. K. (2003). Effects of nitric oxide on red blood cell deformability. *Am. J. Physiol. Heart Circ. Physiol.* 284, H1577–H1584. doi: 10.1152/ajpheart.00665.2002

Bouix, D., Peyreigne, C., Raynaud, E., Monnier, J. F., Micallef, J. P., and Brun, J. F. (1998). Relationships among body composition, hemorheology and exercise performance in rugbymen. *Clin. Hemorheol. Microcirc* 19, 245–254.

Brun, J. F., Criqui, C., and Orsetti, A. (1986). Paramètres hémorhéologiques et exercice physique. *Sports Med. Acta.* 12, 56–60.

Brun, J. F., Fons, C., Supparo, C., Mallard, C., and Orsetti, A. (1993). Could exercise-induced increase in blood viscosity at high shear rate be entirely explained by hematocrit and plasma viscosity changes? *Clin. Hemorheol.* 13, 187–199.

Brun, J. F., Khaled, S., Raynaud, E., Bouix, D., Micallef, J. P., and Orsetti, A. (1998). The triphasic effects of exercise on blood rheology: which relevance to physiology and pathophysiology? *Clin. Hemorheol. Microcirc.* 19, 89–104.

Brun, J. F., Micallef, J. P., and Orsetti, A. (1994a). Hemorheologic effects of light prolonged exercise. *Clin. Hemorheol.* 14, 807–818.

Brun, J. F., Supparo, C., Mallard, C., and Orsetti, A. (1994b). Low values of resting blood viscosity and erythrocyte aggregation are associated with lower increases in blood lactate during submaximal exercise. *Clin. Hemorheol.* 14, 105–116.

Brun, J. F., Sekkat, M., Lagoueyte, C., Fédou, C., and Orsetti, A. (1989). Relashionships between fitness and blood viscosity in untrained normal short children. *Clin. Hemorheol.* 9, 953–963.

Brun, J. F., Supparo, C., Rama, D., Benezis, C., and Orsetti, A. (1995). Maximal oxygen uptake and lactate thresholds during exercise are related to blood viscosity and erythrocyte aggregation in professional football players. *Clin. Hemorheol.* 15, 201–212.

Brun, J. F., Varlet-Marie, E., Connes, P., and Aloulou, I. (2010). Hemorheological alterations related to training and overtraining. *Biorheology* 47, 95–115. doi: 10.3233/BIR-2010-0563

Brun, J. F., Varlet-Marie, E., Richou, M., Mercier, J., and Raynaud De Mauverger, E. (2018). Blood rheology as a mirror of endocrine and metabolic homeostasis in health and disease1. *Clin. Hemorheol. Microcirc* 69, 239–265. doi: 10.3233/CH-189124

Buono, M. J., Krippes, T., Kolkhorst, F. W., Williams, A. T., and Cabrales, P. (2016). Increases in core temperature counterbalance effects of haemoconcentration on blood viscosity during prolonged exercise in the heat. *Exp. Physiol.* 101, 332–342. doi: 10.1113/EP085504

Callahan, L. A., Woods, K. F., Mensah, G. A., Ramsey, L. T., Barbeau, P., and Gutin, B. (2002). Cardiopulmonary responses to exercise in women with sickle cell anemia. *Am. J. Respir. Crit. Care Med.* 165, 1309–1316. doi: 10.1164/rccm.2002036

Camus, S. M., De Moraes, J. A., Bonnin, P., Abbyad, P., Le Jeune, S., Lionnet, F., et al. (2015). Circulating cell membrane microparticles transfer heme to endothelial cells and trigger vasoocclusions in sickle cell disease. *Blood* 125, 3805–3814. doi: 10.1182/blood-2014-07-589283

Camus, S. M., Gausseres, B., Bonnin, P., Loufrani, L., Grimaud, L., Charue, D., et al. (2012). Erythrocyte microparticles can induce kidney vaso-occlusions in a murine model of sickle cell disease. *Blood* 120, 5050–5058. doi: 10.1182/blood-2012-02-413138

Chaar, V., Laurance, S., Lapoumeroulie, C., Cochet, S., De Grandis, M., Colin, Y., et al. (2014). Hydroxycarbamide decreases sickle reticulocyte adhesion to resting endothelium by inhibiting endothelial lutheran/basal cell adhesion molecule (Lu/BCAM) through phosphodiesterase 4A activation. *J. Biol. Chem.* 289, 11512–11521. doi: 10.1074/jbc.M113.506121

Charlot, K., Romana, M., Moeckesch, B., Jumet, S., Waltz, X., Divialle-Doumdo, L., et al. (2016). Which side of the balance determines the frequency of vaso-occlusive crises in children with sickle cell anemia: Blood viscosity or microvascular dysfunction? *Blood Cells Mol. Dis.* 56, 41–45. doi: 10.1016/j.bcmd.2015.10.005

Charlot, K., Waltz, X., Hedreville, M., Sinnapah, S., Lemonne, N., Etienne-Julan, M., et al. (2015). Impaired oxygen uptake efficiency slope and off-transient kinetics of pulmonary oxygen uptake in sickle cell anemia are associated with hemorheological abnormalities. *Clin. Hemorheol. Microcirc.* 60, 413–421. doi: 10.3233/CH-141891

Charrin, E., Aufradet, E., Douillard, A., Romdhani, A., Souza, G. D., Bessaad, A., et al. (2015). Oxidative stress is decreased in physically active sickle cell SAD mice. *Br. J. Haematol.* 168, 747–756. doi: 10.1111/bjh.13207

Charrin, E., Dube, J. J., Connes, P., Pialoux, V., Ghosh, S., Faes, C., et al. (2018). Moderate exercise training decreases inflammation in transgenic sickle cell mice. *Blood Cells Mol. Dis.* 69, 45–52. doi: 10.1016/j.bcmd.2017.06.002

Chatel, B., Messonnier, L. A., Barge, Q., Vilmen, C., Noirez, P., Bernard, M., et al. (2018). Endurance training reduces exercise-induced acidosis and improves muscle function in a mouse model of sickle cell disease. *Mol. Genet. Metab.* 123, 400–410. doi: 10.1016/j.ymgme.2017.11.010

Chien, S., King, R. G., Skalak, R., Usami, S., and Copley, A. L. (1975). Viscoelastic properties of human blood and red cell suspensions. *Biorheology* 12, 341–346.

Chien, S., Usami, S., Dellenback, R. J., and Gregersen, M. I. (1970). Shear-dependent deformation of erythrocytes in rheology of human blood. *Am. J. Physiol.* 219, 136–142. doi: 10.1152/ajplegacy.1970.219.1.136

Chirico, E. N., Faes, C., Connes, P., Canet-Soulas, E., Martin, C., and Pialoux, V. (2016). Role of exercise-induced oxidative stress in sickle cell trait and disease. *Sports Med.* 46, 629–639. doi: 10.1007/s40279-015-0447-z

Clark, M. R., Mohandas, N., and Shohet, S. B. (1983). Osmotic gradient ektacytometry: comprehensive characterization of red cell volume and surface maintenance. *Blood* 61, 899–910.

Cokelet, G. R., and Goldsmith, H. L. (1991). Decreased hydrodynamic resistance in the two-phase flow of blood through small vertical tubes at low flow rates. *Circ. Res.* 68, 1–17. doi: 10.1161/01.res.68.1.1

Cokelet, G. R., and Meiselman, H. J. (2007). "Macro- and micro-rheological properties of blood," in *Handbook of Hemorheology and Hemodynamics*, eds O. K. Baskurt, M. R. Hardeman, M. W. Rampling, and H. J. Meiselman, (Amsterdam: IOS Press), 45–71.

Connes, P. (2010). Hemorheology and exercise: effects of warm environments and potential consequences for sickle cell trait carriers. *Scand. J. Med. Sci. Sports* 20(Suppl. 3), 48–52. doi: 10.1111/j.1600-0838.2010.01208.x

Connes, P., Alexy, T., Detterich, J., Romana, M., Hardy-Dessources, M. D., and Ballas, S. K. (2016). The role of blood rheology in sickle cell disease. *Blood Rev.* 30, 111–118. doi: 10.1016/j.blre.2015.08.005

Connes, P., Bouix, D., Durand, F., Kippelen, P., Mercier, J., Prefaut, C., et al. (2004a). Is hemoglobin desaturation related to blood viscosity in athletes during exercise? *Int. J. Sports Med.* 25, 569–574.

Connes, P., Bouix, D., Py, G., Caillaud, C., Kippelen, P., Brun, J. F., et al. (2004b). Does exercise-induced hypoxemia modify lactate influx into erythrocytes and hemorheological parameters in athletes? *J. Appl. Physiol.* 97, 1053–1058.

Connes, P., Bouix, D., Py, G., Préfaut, C., Mercier, J., Brun, J. F., et al. (2004c). Opposite effects of in vitro lactate on erythrocyte deformability in athletes and untrained subjects. *Clin. Hemorheol. Microcirc* 31, 311–318.

Connes, P., Caillaud, C., Py, G., Mercier, J., Hue, O., and Brun, J. F. (2007). Maximal exercise and lactate do not change red blood cell aggregation in well trained athletes. *Clin. Hemorheol. Microcirc* 36, 319–326.

Connes, P., Hue, O., Tripette, J., and Hardy-Dessources, M. D. (2008). Blood rheology abnormalities and vascular cell adhesion mechanisms in sickle cell trait carriers during exercise. *Clin. Hemorheol. Microcirc* 39, 179–184.

Connes, P., Lamarre, Y., Hardy-Dessources, M. D., Lemonne, N., Waltz, X., Mougenel, D., et al. (2013a). Decreased hematocrit-to-viscosity ratio and increased lactate dehydrogenase level in patients with sickle cell anemia and recurrent leg ulcers. *PLoS One* 8:e79680. doi: 10.1371/journal.pone.0079680

Connes, P., Simmonds, M. J., Brun, J. F., and Baskurt, O. K. (2013b). Exercise hemorheology: classical data, recent findings and unresolved issues. *Clin. Hemorheol. Microcirc* 53, 187–199. doi: 10.3233/CH-2012-1643

Connes, P., Lamarre, Y., Waltz, X., Ballas, S. K., Lemonne, N., Etienne-Julan, M., et al. (2014). Haemolysis and abnormal haemorheology in sickle cell anaemia. *Br. J. Haematol.* 165, 564–572. doi: 10.1111/bjh.12786

Connes, P., Machado, R., Hue, O., and Reid, H. (2011). Exercise limitation, exercise testing and exercise recommendations in sickle cell anemia. *Clin. Hemorheol. Microcirc.* 49, 151–163. doi: 10.3233/CH-2011-1465

Connes, P., Pichon, A., Hardy-Dessources, M. D., Waltz, X., Lamarre, Y., Simmonds, M. J., et al. (2012). Blood viscosity and hemodynamics during exercise. *Clin. Hemorheol. Microcirc.* 51, 101–109. doi: 10.3233/CH-2011-1515

Connes, P., Tripette, J., Mukisi-Mukaza, M., Baskurt, O. K., Toth, K., Meiselman, H. J., et al. (2009). Relationships between hemodynamic, hemorheological and metabolic responses during exercise. *Biorheology* 46, 133–143. doi: 10.3233/BIR-2009-0529

Convertino, V. A., Keil, L. C., Bernauer, E. M., and Greenleaf, J. E. (1981). Plasma volume, osmolality, vasopressin, and renin activity during graded exercise in man. *J. Appl. Physiol. Respir. Environ. Exerc. Physiol.* 50, 123–128. doi: 10.1152/jappl.1981.50.1.123

Diaw, M., Samb, A., Diop, S., Sall, N. D., Ba, A., Cisse, F., et al. (2014). Effects of hydration and water deprivation on blood viscosity during a soccer game in sickle cell trait carriers. *Br. J. Sports Med.* 48, 326–331. doi: 10.1136/bjsports-2012-091038

Dougherty, K. A., Schall, J. I., Rovner, A. J., Stallings, V. A., and Zemel, B. S. (2011). Attenuated maximal muscle strength and peak power in children with sickle cell disease. *J. Pediatr. Hematol. Oncol.* 33, 93–97. doi: 10.1097/MPH.0b013e318200ef49

Dupire, J., Socol, M., and Viallat, A. (2012). Full dynamics of a red blood cell in shear flow. *Proc. Natl. Acad. Sci. U.S.A.* 109, 20808–20813. doi: 10.1073/pnas.1210236109

Ernst, E. (1985). Changes in blood rheology produced by exercise. *J. Am. Med. Assoc.* 253, 2962–2963.

Ernst, E. (1987). Influence of regular physical activity on blood rheology. *Eur. Heart J.* 8(Suppl. G), 59–62.

Ernst, E., Danburger, L., and Saradeth, T. (1991a). Changes in plasma volume after prolonged endurance exercise. *Med. Sci. Sports Exerc.* 23:884.

Ernst, E., Daburger, L., and Saradeth, T. (1991b). The kinetics of blood rheology during and after prolonged standardized exercise. *Clin Hemorheol.* 11, 429–439.

Ernst, E., Matrai, A., Aschenbrenner, E., Will, V., and Schmidlechner, C. (1985). Relationship between fitness and blood fluidity. *Clin. Hemorheol.* 5, 507–510.

Faes, C., Balayssac-Siransy, E., Connes, P., Hivert, L., Danho, C., Bogui, P., et al. (2014). Moderate endurance exercise in patients with sickle cell anaemia: effects on oxidative stress and endothelial activation. *Br. J. Haematol.* 164, 124–130. doi: 10.1111/bjh.12594

Faes, C., Charrin, E., Connes, P., Pialoux, V., and Martin, C. (2015). Chronic physical activity limits blood rheology alterations in transgenic SAD mice. *Am. J. Hematol.* 90, E32–E33. doi: 10.1002/ajh.23896

Fellmann, N. (1992). Hormonal and plasma volume alterations following endurance exercise. A brief review. *Sports Med.* 13, 37–49. doi: 10.2165/00007256-199213010-00004

Fischer, T. M., Stohr-Lissen, M., and Schmid-Schonbein, H. (1978). The red cell as a fluid droplet: tank tread-like motion of the human erythrocyte membrane in shear flow. *Science* 202, 894–896. doi: 10.1126/science.715448

Galea, G., and Davidson, R. J. (1985). Hemorrheology of marathon running. *Int. J. Sports Med.* 6, 136–138. doi: 10.1055/s-2008-1025826

Gellen, B., Messonnier, L. A., Galacteros, F., Audureau, E., Merlet, A. N., Rupp, T., et al. (2018). Moderate-intensity endurance-exercise training in patients with sickle-cell disease without severe chronic complications (EXDRE): an open-label randomised controlled trial. *Lancet Haematol.* 5, e554–e562. doi: 10.1016/S2352-3026(18)30163-7

Goldsmith, H. L., Frank, J. M., and MacIntosh, C. (1972). Flow behaviour of erythrocytes. Rotation and deformation in dilute suspensions. *Proc. R. Soc. B.* 182, 351–384.

Gonzales, J. U., Parker, B. A., Ridout, S. J., Smithmyer, S. L., and Proctor, D. N. (2009). Femoral shear rate response to knee extensor exercise: an age and sex comparison. *Biorheology* 46, 145–154. doi: 10.3233/BIR-2009-0535

Gouraud, E., Charrin, E., Dube, J. J., Ofori-Acquah, S. F., Martin, C., Skinner, S., et al. (2019). Effects of individualized treadmill endurance training on oxidative stress in skeletal muscles of transgenic sickle mice. *Oxid. Med. Cell Longev.* 2019:3765643. doi: 10.1155/2019/3765643

Grau, M., Jerke, M., Nader, E., Schenk, A., Renoux, C., Collins, B., et al. (2019). Effect of acute exercise on RBC deformability and RBC nitric oxide synthase signalling pathway in young sickle cell anaemia patients. *Sci. Rep.* 9:11813. doi: 10.1038/s41598-019-48364-1

Grau, M., Pauly, S., Ali, J., Walpurgis, K., Thevis, M., Bloch, W., et al. (2013). RBC-NOS-dependent S-nitrosylation of cytoskeletal proteins improves RBC deformability. *PLoS One* 8:e56759. doi: 10.1371/journal.pone.0056759

Gueguen-Duchesne, M., Durand, F., Beillot, J., Dezier, J., Rochcongar, P., Legoff, M., et al. (1987). Could maximal exercise be a hemorheological risk factor? *Clin. Hemorheol.* 7, 418.

Hakim, T. S. (1988). Erythrocyte deformability and segmental pulmonary vascular resistance: osmolarity and heat treatment. *J. Appl. Physiol.* 65, 1634–1641. doi: 10.1152/jappl.1988.65.4.1634

Hebbel, R. P., Osarogiagbon, R., and Kaul, D. (2004). The endothelial biology of sickle cell disease: inflammation and a chronic vasculopathy. *Microcirculation* 11, 129–151.

Intaglietta, M. (2009). Increased blood viscosity: disease, adaptation or treatment? *Clin. Hemorheol. Microcirc.* 42, 305–306. doi: 10.3233/CH-2009-1236

Isbister, J. P. (1997). Physiology and pathophysiology of blood volume regulation. *Transfus. Sci.* 18, 409–423.

Kato, G. J., Gladwin, M. T., and Steinberg, M. H. (2007). Deconstructing sickle cell disease: reappraisal of the role of hemolysis in the development of clinical subphenotypes. *Blood Rev.* 21, 37–47. doi: 10.1016/j.blre.2006.07.001

Kesmarky, G., Kenyeres, P., Rabai, M., and Toth, K. (2008). Plasma viscosity: a forgotten variable. *Clin. Hemorheol. Microcirc.* 39, 243–246.

Khalyfa, A., Khalyfa, A. A., Akbarpour, M., Connes, P., Romana, M., Lapping-Carr, G., et al. (2016). Extracellular microvesicle microRNAs in children with sickle cell anaemia with divergent clinical phenotypes. *Br. J. Haematol.* 174, 786–798. doi: 10.1111/bjh.14104

Kilic-Toprak, E., Ardic, F., Erken, G., Unver-Kocak, F., Kucukatay, V., and Bor-Kucukatay, M. (2012). Hemorheological responses to progressive resistance exercise training in healthy young males. *Med. Sci. Monit.* 18, CR351–CR360. doi: 10.12659/msm.882878

Kleinbongard, P., Schulz, R., Rassaf, T., Lauer, T., Dejam, A., Jax, T., et al. (2006). Red blood cells express a functional endothelial nitric oxide synthase. *Blood* 107, 2943–2951. doi: 10.1182/blood-2005-10-3992

Ko, E., Youn, J. M., Park, H. S., Song, M., Koh, K. H., and Lim, C. H. (2018). Early red blood cell abnormalities as a clinical variable in sepsis diagnosis. *Clin. Hemorheol. Microcirc.* 70, 355–363. doi: 10.3233/CH-180430

Lamarre, Y., Romana, M., Lemonne, N., Hardy-Dessources, M. D., Tarer, V., Mougenel, D., et al. (2014). Alpha thalassemia protects sickle cell anemia patients from macro-albuminuria through its effects on red blood cell rheological properties. *Clin. Hemorheol. Microcirc.* 57, 63–72. doi: 10.3233/CH-131772

Lamarre, Y., Romana, M., Waltz, X., Lalanne-Mistrih, M. L., Tressieres, B., Divialle-Doumdo, L., et al. (2012). Hemorheological risk factors of acute chest syndrome and painful vaso-occlusive crisis in children with sickle cell disease. *Haematologica* 97, 1641–1647. doi: 10.3324/haematol.2012.066670

Lanotte, L., Mauer, J., Mendez, S., Fedosov, D. A., Fromental, J. M., Claveria, V., et al. (2016). Red cells' dynamic morphologies govern blood shear thinning under microcirculatory flow conditions. *Proc. Natl. Acad. Sci. U.S.A.* 113, 13289–13294. doi: 10.1073/pnas.1608074113

Lapoumeroulie, C., Connes, P., El Hoss, S., Hierso, R., Charlot, K., Lemonne, N., et al. (2019). New insights into red cell rheology and adhesion in patients with sickle cell anaemia during vaso-occlusive crises. *Br. J. Haematol.* 185, 991–994. doi: 10.1111/bjh.15686

Laurance, S., Lansiaux, P., Pellay, F. X., Hauchecorne, M., Benecke, A., Elion, J., et al. (2011). Differential modulation of adhesion molecule expression by hydroxycarbamide in human endothelial cells from the micro- and macrocirculation: potential implications in sickle cell disease vasoocclusive events. *Haematologica* 96, 534–542. doi: 10.3324/haematol.2010.026740

Lemonne, N., Charlot, K., Waltz, X., Ballas, S. K., Lamarre, Y., Lee, K., et al. (2015). Hydroxyurea treatment does not increase blood viscosity and improves red blood cell rheology in sickle cell anemia. *Haematologica* 100, e383–e386. doi: 10.3324/haematol.2015.130435

Lemonne, N., Connes, P., Romana, M., Vent-Schmidt, J., Bourhis, V., Lamarre, Y., et al. (2012). Increased blood viscosity and red blood cell aggregation in a patient with sickle cell anemia and smoldering myeloma. *Am. J. Hematol.* 87:E129. doi: 10.1002/ajh.23312

Lemonne, N., Lamarre, Y., Romana, M., Mukisi-Mukaza, M., Hardy-Dessources, M. D., Tarer, V., et al. (2013). Does increased red blood cell deformability raise the risk for osteonecrosis in sickle cell anemia? *Blood* 121, 3054–3056. doi: 10.1182/blood-2013-01-480277

Liem, R. I., Reddy, M., Pelligra, S. A., Savant, A. P., Fernhall, B., Rodeghier, M., et al. (2015). Reduced fitness and abnormal cardiopulmonary responses to maximal exercise testing in children and young adults with sickle cell anemia. *Physiol. Rep.* 3:e12338. doi: 10.14814/phy2.12338

Lipovac, V., Gavella, M., Turck, Z., and Skrabalo, Z. (1985). Influence of lactate on the insulin action on red blood cell filterability. *Clin. Hemorheol.* 5, 421–428.

Martin, C., Pialoux, V., Faes, C., Charrin, E., Skinner, S., and Connes, P. (2018). Does physical activity increase or decrease the risk of sickle cell disease complications? *Br. J. Sports Med.* 52, 214–218. doi: 10.1136/bjsports-2015-095317

Martini, J., Carpentier, B., Negrete, A. C., Frangos, J. A., and Intaglietta, M. (2005). Paradoxical hypotension following increased hematocrit and blood viscosity. *Am. J. Physiol. Heart Circ. Physiol.* 289, H2136–H2143.

Meiselman, H. J., Neu, B., Rampling, M. W., and Baskurt, O. K. (2007). RBC aggregation: laboratory data and models. *Indian J. Exp. Biol.* 45, 9–17.

Merlet, A. N., Chatel, B., Hourde, C., Ravelojaona, M., Bendahan, D., Feasson, L., et al. (2019). How sickle cell disease impairs skeletal muscle function: implications in daily life. *Med. Sci. Sports Exerc.* 51, 4–11. doi: 10.1249/MSS.0000000000001757

Mockesch, B., Connes, P., Charlot, K., Skinner, S., Hardy-Dessources, M. D., Romana, M., et al. (2017). Association between oxidative stress and vascular reactivity in children with sickle cell anaemia and sickle haemoglobin C disease. *Br. J. Haematol.* 178, 468–475. doi: 10.1111/bjh.14693

Nader, E., Connes, P., Lamarre, Y., Renoux, C., Joly, P., Hardy-Dessources, M. D., et al. (2017). Plasmapheresis may improve clinical condition in sickle cell disease through its effects on red blood cell rheology. *Am. J. Hematol.* 92, E629–E630. doi: 10.1002/ajh.24870

Nader, E., Guillot, N., Lavorel, L., Hancco, I., Fort, R., Stauffer, E., et al. (2018). Eryptosis and hemorheological responses to maximal exercise in athletes: comparison between running and cycling. *Scand. J. Med. Sci. Sports* 28, 1532–1540. doi: 10.1111/sms.13059

Nebor, D., Bowers, A., Hardy-Dessources, M. D., Knight-Madden, J., Romana, M., Reid, H., et al. (2011). Frequency of pain crises in sickle cell anemia and its relationship with the sympatho-vagal balance, blood viscosity and inflammation. *Haematologica* 96, 1589–1594. doi: 10.3324/haematol.2011.047365

Neuhaus, D., Behn, C., and Gaehtgens, P. (1992). Haemorheology and exercise: intrinsic flow properties of blood in marathon running. *Int. J. Sports Med.* 13, 506–511. doi: 10.1055/s-2007-1021307

Neuhaus, D., and Gaehtgens, P. (1994). Haemorrheology and long term exercise. *Sports Med.* 18, 10–21. doi: 10.2165/00007256-199418010-00003

Nosadova, J. (1977). The changes in hematocrit, hemoglobin, plasma volume and proteins during and after different types of exercise. *Eur. J. Appl. Physiol.* 36, 223–230.

Omwanghe, O. A., Muntz, D. S., Kwon, S., Montgomery, S., Kemiki, O., Hsu, L. L., et al. (2017). Self-Reported physical activity and exercise patterns in children with sickle cell disease. *Pediatr. Exerc. Sci.* 29, 388–395. doi: 10.1123/pes.2016-0276

Oostenbrug, G. S., Mensink, R. P., Hardeman, M. R., De Vries, T., Brouns, F., and Hornstra, G. (1997). Exercise performance, red blood cell deformability, and lipid peroxidation: effects of fish oil and vitamin E. *J. Appl. Physiol.* 83, 746–752.

Pallis, F. R., Conran, N., Fertrin, K. Y., Olalla Saad, S. T., Costa, F. F., and Franco-Penteado, C. F. (2014). Hydroxycarbamide reduces eosinophil adhesion and degranulation in sickle cell anaemia patients. *Br. J. Haematol.* 164, 286–295. doi: 10.1111/bjh.12628

Parthasarathi, K., and Lipowsky, H. H. (1999). Capillary recruitment in response to tissue hypoxia and its dependence on red blood cell deformability. *Am. J. Physiol.* 277, H2145–H2157. doi: 10.1152/ajpheart.1999.277.6.H2145

Piecuch, J., Mertas, A., Nowowiejska-Wiewiora, A., Zurawel, R., Gregorczyn, S., Czuba, Z., et al. (2019). The relationship between the rheological behavior of RBCs and angiogenesis in the morbidly obese. *Clin. Hemorheol. Microcirc.* 71, 95–102. doi: 10.3233/CH-180420

Piel, F. B., Patil, A. P., Howes, R. E., Nyangiri, O. A., Gething, P. W., Williams, T. N., et al. (2010). Global distribution of the sickle cell gene and geographical confirmation of the malaria hypothesis. *Nat. Commun.* 1:104. doi: 10.1038/ncomms1104

Ploutz-Snyder, L. L., Convertino, V. A., and Dudley, G. A. (1995). Resistance exercise-induced fluid shifts: change in active muscle size and plasma volume. *Am. J. Physiol.* 269, R536–R543.

Poiseuille, J. L. M. (1835). Recherches sur les causes du mouvement du sang dans les vaisseaux capillaires. *C R Acad. Sci. Paris* 1, 554–560.

Pop, G. A., Duncker, D. J., Gardien, M., Vranckx, P., Versluis, S., Hasan, D., et al. (2002). The clinical significance of whole blood viscosity in (cardio)vascular medicine. *Neth. Heart J.* 10, 512–516.

Raj, J. U., Kaapa, P., Hillyard, R., and Anderson, J. (1991). Pulmonary vascular pressure profile in adult ferrets: measurements in vivo and in isolated lungs. *Acta Physiol. Scand.* 142, 41–48. doi: 10.1111/j.1748-1716.1991.tb09126.x

Rampling, M. W., Meiselman, H. J., Neu, B., and Baskurt, O. K. (2004). Influence of cell-specific factors on red blood cell aggregation. *Biorheology* 41, 91–112.

Rees, D. C., Williams, T. N., and Gladwin, M. T. (2010). Sickle-cell disease. *Lancet* 376, 2018–2031. doi: 10.1016/S0140-6736(10)61029-X

Reinhart, W. H., Staubli, M., and Straub, P. W. (1983). Impaired red cell filterability with elimination of old red blood cells during a 100-km race. *J. Appl. Physiol. Respir. Environ. Exerc. Physiol.* 54, 827–830. doi: 10.1152/jappl.1983.54.3.827

Renoux, C., Connes, P., Nader, E., Skinner, S., Faes, C., Petras, M., et al. (2017). Alpha-thalassaemia promotes frequent vaso-occlusive crises in children with sickle cell anaemia through haemorheological changes. *Pediatr. Blood Cancer* 64:e26455. doi: 10.1002/pbc.26455

Renoux, C., Faivre, M., Bessaa, A., Da Costa, L., Joly, P., Gauthier, A., et al. (2019). Impact of surface-area-to-volume ratio, internal viscosity and membrane viscoelasticity on red blood cell deformability measured in isotonic condition. *Sci. Rep.* 9:6771. doi: 10.1038/s41598-019-43200-y

Renoux, C., Romana, M., Joly, P., Ferdinand, S., Faes, C., Lemonne, N., et al. (2016). Effect of age on blood rheology in sickle cell anaemia and sickle cell haemoglobin C disease: a cross-sectional study. *PLoS One* 11:e0158182. doi: 10.1371/journal.pone.0158182

Robach, P., Boisson, R. C., Vincent, L., Lundby, C., Moutereau, S., Gergele, L., et al. (2014). Hemolysis induced by an extreme mountain ultra-marathon is not associated with a decrease in total red blood cell volume. *Scand. J. Med. Sci. Sports.* 24, 18–27. doi: 10.1111/j.1600-0838.2012.01481.x

Romain, A. J., Brun, J. F., Varlet-Marie, E., and Raynaud De Mauverger, E. (2011). Effects of exercise training on blood rheology: a meta-analysis. *Clin. Hemorheol. Microcirc.* 49, 199–205. doi: 10.3233/CH-2011-1469

Salazar Vazquez, B. Y., Cabrales, P., Tsai, A. G., and Intaglietta, M. (2011). Nonlinear cardiovascular regulation consequent to changes in blood viscosity. *Clin. Hemorheol. Microcirc.* 49, 29–36. doi: 10.3233/CH-2011-1454

Sandor, B., Nagy, A., Toth, A., Rabai, M., Mezey, B., Csatho, A., et al. (2014). Effects of moderate aerobic exercise training on hemorheological and laboratory parameters in ischemic heart disease patients. *PLoS One* 9:e110751. doi: 10.1371/journal.pone.0110751

Schmid-Schonbein, H., Wells, R. E., and Goldstone, J. (1969). [Model experiments on erythrocyte rheology]. *Pflugers. Arch.* 312, R39–R40.

Senturk, U. K., Gunduz, F., Kuru, O., Kocer, G., Ozkaya, Y. G., Yesilkaya, A., et al. (2005a). Exercise-induced oxidative stress leads hemolysis in sedentary but not trained humans. *J. Appl. Physiol.* 99, 1434–1441.

Senturk, U. K., Yalcin, O., Gunduz, F., Kuru, O., Meiselman, H. J., and Baskurt, O. K. (2005b). Effect of antioxidant vitamin treatment on the time course of hematological and hemorheological alterations after an exhausting exercise episode in human subjects. *J. Appl. Physiol.* 98, 1272–1279.

Sheremet'ev, Y. A., Popovicheva, A. N., Rogozin, M. M., and Levin, G. Y. (2019). Red blood cell aggregation, disaggregation and aggregate morphology in autologous plasma and serum in diabetic foot disease. *Clin. Hemorheol. Microcirc.* 72, 221–227. doi: 10.3233/CH-180405

Sjogaard, G., Adams, R. P., and Saltin, B. (1985). Water and ion shifts in skeletal muscle of humans with intense dynamic knee extension. *Am. J. Physiol.* 248, R190–R196.

Smith, J. A., Martin, D. T., Telford, R. D., and Ballas, S. K. (1999). Greater erythrocyte deformability in world-class endurance athletes. *Am. J. Physiol.* 276, H2188–H2193. doi: 10.1152/ajpheart.1999.276.6.H2188

Smith, J. A., Telford, R. D., Kolbuch-Braddon, M., and Weidemann, M. J. (1997). Lactate/H+ uptake by red blood cells during exercise alters their physical properties. *Eur. J. Appl. Physiol. Occup. Physiol.* 75, 54–61.

Sriram, K., Salazar Vazquez, B. Y., Tsai, A. G., Cabrales, P., Intaglietta, M., and Tartakovsky, D. M. (2012). Autoregulation and mechanotransduction control the arteriolar response to small changes in hematocrit. *Am. J. Physiol. Heart Circ. Physiol.* 303, H1096–H1106. doi: 10.1152/ajpheart.00438.2012

Starzyk, D., Korbut, R., and Gryglewski, R. J. (1997). The role of nitric oxide in regulation of deformability of red blood cells in acute phase of endotoxaemia in rats. *J. Physiol. Pharmacol.* 48, 731–735.

Stephenson, L. A., and Kolka, M. A. (1988). Plasma volume during heat stress and exercise in women. *Eur. J. Appl. Physiol. Occup. Physiol.* 57, 373–381.

Suhr, F., Brenig, J., Muller, R., Behrens, H., Bloch, W., and Grau, M. (2012). Moderate exercise promotes human RBC-NOS activity, NO production and deformability through Akt kinase pathway. *PLoS One* 7:e45982. doi: 10.1371/journal.pone.0045982

Tomschi, F., Bizjak, D., Bloch, W., Latsch, J., Predel, H. G., and Grau, M. (2018). Deformability of different red blood cell populations and viscosity of differently trained young men in response to intensive and moderate running. *Clin. Hemorheol. Microcirc.* 69, 503–514. doi: 10.3233/CH-189202

Totsimon, K., Biro, K., Szabo, Z. E., Toth, K., Kenyeres, P., and Marton, Z. (2017). The relationship between hemorheological parameters and mortality in critically ill patients with and without sepsis. *Clin. Hemorheol. Microcirc.* 65, 119–129. doi: 10.3233/CH-16136

Tripette, J., Alexy, T., Hardy-Dessources, M. D., Mougenel, D., Beltan, E., Chalabi, T., et al. (2009). Red blood cell aggregation, aggregate strength and oxygen transport potential of blood are abnormal in both homozygous sickle cell anemia and sickle-hemoglobin C disease. *Haematologica* 94, 1060–1065. doi: 10.3324/haematol.2008.005371

Tripette, J., Hardy-Dessources, M. D., Beltan, E., Sanouiller, A., Bangou, J., Chalabi, T., et al. (2011). Endurance running trial in tropical environment: a blood rheological study. *Clin. Hemorheol. Microcirc* 47, 261–268. doi: 10.3233/CH-2011-1388

Tripette, J., Loko, G., Samb, A., Gogh, B. D., Sewade, E., Seck, D., et al. (2010). Effects of hydration and dehydration on blood rheology in sickle cell trait carriers during exercise. *Am. J. Physiol. Heart Circ. Physiol.* 299, H908–H914. doi: 10.1152/ajpheart.00298.2010

Tsai, A. G., Acero, C., Nance, P. R., Cabrales, P., Frangos, J. A., Buerk, D. G., et al. (2005). Elevated plasma viscosity in extreme hemodilution increases perivascular nitric oxide concentration and microvascular perfusion. *Am. J. Physiol. Heart Circ. Physiol.* 288, H1730–H1739. doi: 10.1152/ajpheart.00998.2004

Van Beaumont, W., Underkofler, S., and Van Beaumont, S. (1981). Erythrocyte volume, plasma volume, and acid-base changes in exercise and heat dehydration. *J. Appl. Physiol. Respir. Environ. Exerc. Physiol.* 50, 1255–1262. doi: 10.1152/jappl.1981.50.6.1255

Vandewalle, H., Lacombe, C., Lelievre, J. C., and Poirot, C. (1988). Blood viscosity after a 1-h submaximal exercise with and without drinking. *Int. J. Sports Med.* 9, 104–107.

Varlet-Marie, E., Gaudard, A., Monnier, J. F., Micallef, J. P., Mercier, J., Bressolle, F., et al. (2003). Reduction of red blood cell disaggregability during submaximal exercise: relationship with fibrinogen levels. *Clin. Hemorheol. Microcirc* 28, 139–149.

Vazquez, B. Y., Vazquez, M. A., Jaquez, M. G., Huemoeller, A. H., Intaglietta, M., and Cabrales, P. (2010). Blood pressure directly correlates with blood viscosity in diabetes type 1 children but not in normals. *Clin. Hemorheol. Microcirc* 44, 55–61. doi: 10.3233/CH-2010-1252

Verger, E., Schoevaert, D., Carrivain, P., Victor, J. M., Lapoumeroulie, C., and Elion, J. (2014). Prior exposure of endothelial cells to hydroxycarbamide alters the flow dynamics and adhesion of sickle red blood cells. *Clin. Hemorheol. Microcirc* 57, 9–22. doi: 10.3233/CH-131762

Waltz, X., and Connes, P. (2014). Pathophysiology and physical activity in patients with sickle cell anemia. *Mov. Sport Sci. Sci. Motricité* 83, 41–47. doi: 10.1051/sm/2013105

Waltz, X., Hardy-Dessources, M. D., Lemonne, N., Mougenel, D., Lalanne-Mistrih, M. L., Lamarre, Y., et al. (2015). Is there a relationship between the hematocrit-to-viscosity ratio and microvascular oxygenation in brain and muscle? *Clin. Hemorheol. Microcirc.* 59, 37–43. doi: 10.3233/CH-131742

Waltz, X., Hedreville, M., Sinnapah, S., Lamarre, Y., Soter, V., Lemonne, N., et al. (2012). Delayed beneficial effect of acute exercise on red blood cell aggregate strength in patients with sickle cell anemia. *Clin. Hemorheol. Microcirc.* 52, 15–26. doi: 10.3233/CH-2012-1540

Waltz, X., Romana, M., Lalanne-Mistrih, M. L., Machado, R. F., Lamarre, Y., Tarer, V., et al. (2013). Hematologic and hemorheological determinants of resting and exercise-induced hemoglobin oxygen desaturation in children with sickle cell disease. *Haematologica* 98, 1039–1044. doi: 10.3324/haematol.2013.083576

Wood, S. C., Doyle, M. P., and Appenzeller, O. (1991). Effects of endurance training and long distance running on blood viscosity. *Med. Sci. Sports Exerc.* 23, 1265–1269.

Yalcin, O., Bor-Kucukatay, M., Senturk, U. K., and Baskurt, O. K. (2000). Effects of swimming exercise on red blood cell rheology in trained and untrained rats. *J. Appl. Physiol.* 88, 2074–2080.

Yalcin, O., Erman, A., Muratli, S., Bor-Kucukatay, M., and Baskurt, O. K. (2003). Time course of hemorheological alterations after heavy anaerobic exercise in untrained human subjects. *J. Appl. Physiol.* 94, 997–1002.

Yalcin, O., Meiselman, H. J., Armstrong, J. K., and Baskurt, O. K. (2005). Effect of enhanced red blood cell aggregation on blood flow resistance in an isolated-perfused guinea pig heart preparation. *Biorheology* 42, 511–520.

11

Ektacytometry Analysis of Post-Splenectomy Red Blood Cell Properties Identifies Cell Membrane Stability Test as a Novel Biomarker of Membrane Health in Hereditary Spherocytosis

*M. C. Berrevoets[1], J. Bos[1], R. Huisjes[1], T. H. Merkx[1], B. A. van Oirschot[1], W. W. van Solinge[1], J. W. Verweij[2], M. Y. A. Lindeboom[2], E. J. van Beers[3], M. Bartels[3†], R. van Wijk[1†] and M. A. E. Rab[1,3]**

[1] Central Diagnostic Laboratory-Research, University Medical Center Utrecht, Utrecht University, Utrecht, Netherlands, [2] Department of Pediatric Surgery, University Medical Center Utrecht, Utrecht University, Utrecht, Netherlands, [3] Van Creveldkliniek, University Medical Center Utrecht, Utrecht University, Utrecht, Netherlands

Correspondence:
M. A. E. Rab
m.a.e.rab@umcutrecht.nl

[†] These authors have contributed equally and share penultimate authorship

Hereditary spherocytosis (HS) is the most common form of hereditary chronic hemolytic anemia. It is caused by mutations in red blood cell (RBC) membrane and cytoskeletal proteins, which compromise membrane integrity, leading to vesiculation. Eventually, this leads to entrapment of poorly deformable spherocytes in the spleen. Splenectomy is a procedure often performed in HS. The clinical benefit results from removing the primary site of destruction, thereby improving RBC survival. But whether changes in RBC properties contribute to the clinical benefit of splenectomy is unknown. In this study we used ektacytometry to investigate the longitudinal effects of splenectomy on RBC properties in five well-characterized HS patients at four different time points and in a case-control cohort of 26 HS patients. Osmotic gradient ektacytometry showed that splenectomy resulted in improved intracellular viscosity (hydration state) whereas total surface area and surface-to-volume ratio remained essentially unchanged. The cell membrane stability test (CMST), which assesses the *in vitro* response to shear stress, showed that after splenectomy, HS RBCs had partly regained the ability to shed membrane, a property of healthy RBCs, which was confirmed in the case-control cohort. In particular the CMST holds promise as a novel biomarker in HS that reflects RBC membrane health and may be used to asses treatment response in HS.

Keywords: hereditary spherocytosis, splenectomy, deformability, ektacytometry, biomarker, red blood cell, hemolytic anemia

INTRODUCTION

Hereditary spherocytosis (HS) is a heterogeneous group of inherited anemias that originates from defective anchoring of transmembrane proteins to the cytoskeletal network of the red blood cell (RBC). The defective anchoring is predominantly caused by a mutation in the genes coding for ankyrin (ANK1), α-spectrin (SPTA1), β-spectrin (SPTB), band-3 (SLC4A1), or protein 4.2 (EPB42) (Perrotta et al., 2008). These mutations compromise the vertical linkages between the lipid bilayer and the cytoskeletal network, leading to destabilization of the membrane, increased vesiculation and subsequent membrane loss. The progressive membrane loss leads to formation of dense spherical-shaped RBCs (spherocytes) with reduced deformability (Chasis et al., 1988; Eber and Lux, 2004; Huisjes et al., 2018).

The spleen plays an intricate role in the pathophysiology of HS. Normally, this organ functions as a quality control for RBCs. During the 120-day lifespan of healthy RBCs, membrane surface area, surface area-to-volume ratio, and deformability decrease because of release of essentially hemoglobin-free microvesicles. RBCs with increased density and reduced deformability are eventually trapped in the narrow endothelial slits of the spleen, leading to clearance of aged RBCs (Eber and Lux, 2004; Mebius and Kraal, 2005). The compromised vertical linkages in HS accelerate the loss of membrane and deformability, leading to premature destruction of RBCs in the spleen. Therefore, splenectomy is an effective treatment, and removal of the primary site of RBC destruction generally improves clinical symptoms (Musser et al., 1984; Eber and Lux, 2004; Perrotta et al., 2008). Nevertheless, the risks and benefits should be carefully assessed as splenectomy results in a permanently increased risk of infections caused by encapsulated bacteria and long term risk for cardiovascular events (Perrotta et al., 2008; Schilling et al., 2008).

The effects of splenectomy on RBC rheology and RBC related parameters in HS have been studied to a limited extent. It is known that splenectomy improves the RBC count, hemoglobin (Hb) levels, and hematocrit, and that it reduces mean corpuscular hemoglobin concentration (MCHC) and the percentage of reticulocytes (Reliene et al., 2002; Li et al., 2016; Zaninoni et al., 2018; Huisjes et al., 2020). On a cellular level it has been shown that the size of RBCs increases following splenectomy, and that microspherocytes can no longer be detected (Sugihara et al., 1984). However, splenectomy has little effect on correcting the cytoskeletal membrane defect (Reliene et al., 2002). More recent studies have shown that RBC deformability as measured by osmotic gradient ektacytometry was not improved after splenectomy (Zaninoni et al., 2018; Huisjes et al., 2020). An important limitation of these studies was the fact that they compared cohorts of splenectomized and non-splenectomized patients; longitudinal studies on the response to splenectomy of individual HS patients are scarce (Li et al., 2016).

In this study, we investigated individual responses to splenectomy in a group of five HS patients, with particular focus on RBC functional properties as determined by ektacytometry. Our results indicate that the Cell Membrane Stability Test (CMST), which measures the RBCs response to high shear stress, is able to detect substantial functional improvement of the RBC membrane after splenectomy, by showing a partly restored ability to shed membrane, a feature of healthy RBCs. We suggest that the CMST represents a novel biomarker of RBC membrane health in HS, and may be used to assess the efficacy of treatment.

MATERIALS AND METHODS

Patients

Two groups of patients were enrolled in this study. The first group consisted of five patients (one male and four females, aged between 13 and 43 years) diagnosed with HS, and scheduled to undergo splenectomy. Detailed characteristics are provided in **Supplementary Table 1**. Left-over material of blood collected before splenectomy and at different time points after splenectomy (i.e., 1 week, 1 month and ≥3 months) was used for laboratory measurements. Informed consent was obtained from all patients and/or legal guardians. The second group consisted of a patient cohort of 26 HS patients: 18 non-splenectomized patients, eight patients who underwent splenectomy ≥ 1 year prior to enrollment, and 26 healthy controls (HC). Blood samples of this cohort were obtained after inclusion in the CoMMiTMenT-study which was approved by the Medical Ethical Research Board of the University Medical Center Utrecht, Netherlands (15/426 M) or from anonymized left-over material. Blood from HC individuals was obtained by means of the institutional blood donor service.

Surgical Procedure

Laparoscopic total splenectomy was performed in all five patients. The patients were positioned in right lateral decubitus. Four trocars were used. The lesser sac was entered and the short gastric vessels were divided. After full mobilization of the spleen, the hilar vessels were controlled by using a linear cutting stapler. The spleen was extracted using a retrieval bag.

Laboratory Parameters

Routine hematological laboratory parameters were analyzed on an Abbott Cell-Dyn Sapphire hematology analyzer (Abbott Diagnostics Division, Santa Clara, CA, United States).

Ektacytometry

Deformability of RBCs was measured with the Lorrca (Laser Optical Rotational Red Cell Analyzer, RR Mechatronics, Zwaag, Netherlands). In this ektacytometer, RBCs are exposed to shear stress in a viscous solution (Elon-Iso), forcing the cells to elongate into an elliptical shape. The diffraction pattern that is generated by a laser beam is measured by a camera. The vertical axis (A) and the horizontal axis (B) of the ellipse are used to calculate the elongation index (EI) by the formula $(A\text{-}B)/(A+B)$. The EI reflects the deformability of the total population of RBCs.

Osmotic Gradient Ektacytometry

Osmotic gradient ektacytometry measurements of RBCs of HC and HS patients before and after splenectomy were obtained using the osmoscan module on the Lorrca according to the manufacturer's instructions and as described elsewhere (DaCosta et al., 2016; Lazarova et al., 2017). Briefly, whole blood was standardized to a fixed RBC count of $1,000 \times 10^6$ and mixed with 5 mL of Elon-Iso (RR Mechatronics). RBCs in the viscous solution (Elon-Iso) were exposed to an osmolarity gradient from approximately 60 mOsmol/L to 600 mOsmol/L, while shear stress was kept constant (30 Pa).

Cell Membrane Stability Test

The CMST was performed using the CMST module on the ektacytometer. To perform a CMST, whole blood was standardized to a fixed RBC count of 200×10^6 and mixed with 5 mL of Elon-Iso. In the CMST RBCs are exposed to a shear stress of 100 Pa for 3,600 s (1 h) while the EI is continuously measured. The change in the elongation index (ΔEI) was calculated by determining the median of the first and the last 100 s of the CMST and subsequently calculating the difference between the medians. The ΔEI depicts the capacity of the RBCs to shed membrane and resist shear stress.

Microscopic analysis on a subset of samples was performed with the use of a camera microscope (1/1.8″ Sony CMOS Global IMX265LLR imaging sensor, long working distance VS-Technology 50X Plan LWD, VS-MS-COL tube) which was placed on outside of the rotating cup of the Lorrca. A power-LED flash (415 nm) coupled to a fiber-optic, in bright field illumination, from the inside of the cup into a 45 degrees mirror and diffusor lens directed at the microscope, was used for proper lighting of the RBCs. A flash time of 214 ns was used to get less than 1% motion blur. The rotating cup was modified with 15 thin and small glass windows circumferential in the cup. Images were taken with Image Capture software during a CMST measurement.

Density Separation

To assess the effect of splenectomy on the composition of the RBC population a density separation was carried out before splenectomy and approximately 1 month after in one HS patient. A total of 20 mL whole blood was placed on top of three layers with different percentage percoll (GE Healthcare) 1.130 ± 0.005 g/mL in eight different columns (2 mL whole blood/column). RBCs were fractioned according to density (i.e., cellular age) using this density gradient of percoll with addition of HEPES, NaCl, KCL, and NaOH as described in detail elsewhere (Rennie et al., 1979). Cells were centrifuged at $1,665 \times g$ for 15 min, after which four fractions could be obtained (**Supplementary Figure 1A**). Fraction 1, containing the RBCs with the lowest density, was present on top of the 59% percoll layer, and was only present and subsequently obtained from the pre-splenectomy blood sample (**Supplementary Figure 1A**). Because of the limited amount of RBCs in fraction 1 only a subset of measurements could be performed. Fraction 2 was obtained from the top of the 70% percoll layer. Fraction 3 was obtained from the top of the 78% percoll layer. The 4th fraction, containing the most dense RBCs was obtained from the bottom of the tube (**Supplementary Figure 1A**).

Digital Microscopy

Peripheral blood smears were analyzed using the CellaVision digital microscope DM96 (software 5.0.1 build11). Analysis was performed using a neuronal network which classifies RBCs based on morphological characteristics such as shape, color, and texture. Spherocytes, microcytes and macrocytes (%) were calculated as a percentage of total RBCs as quantified by the software (Huisjes et al., 2017).

Statistical Analysis

All the data were analyzed using GraphPad Prism version 8.3.0 for Windows (GraphPad Software, San Diego, CA, United States). A paired T-test (two-tailed) was used to assess the values before and after splenectomy. An one-way ANOVA, with post-hoc Tukey analysis was used to assess the differences between HC and HS patients, and non-splenectomized and splenectomized HS patients. In addition, a correlation analysis between RBC related parameters and the change in elongation index (ΔEI) was conducted using the Spearman's rank correlation coefficient. A p-value below 0.05 was considered statistically significant.

RESULTS

Routine Hematological Parameters Show a Decrease in RBC Density and Increased RBC Homogeneity After Splenectomy

Following splenectomy, routine hematology parameters showed a significant improvement in RBC count, Hb, and reticulocyte count after 1 and ≥ 3 months (**Figures 1A–C** and **Supplementary Table 1**). Mean corpuscular volume (MCV) increased in the first week after splenectomy, but returned to pre-surgery levels in the following months (**Figure 1D**). This suggests that cell volume is not altered by splenectomy. At the same time, we observed a significant decrease in MCHC and the percentage of hyperchromic cells, indicating a decline in RBC density following splenectomy (**Figures 1E,F**). At the same time, spherocytes and microcytes as assessed by digital microscopy also declined, except for spherocytes in patient 1 (**Supplementary Table 1**). When RBCs of one HS patient were separated according to density (**Supplementary Figure 1A**), we noted that the MCHC before splenectomy seemed determined mainly by density fraction 4, containing the most dense RBCs, whereas after splenectomy density fractions 3 and 4 seem to contribute equally to the MCHC (**Supplementary Figure 1B**), This implies a more homogeneous RBC population after splenectomy, which was also reflected by a more equal distribution of the different fractions after splenectomy (**Supplementary Figure 1A**). Hyperchromic cells were predominantly present in fraction 4 before splenectomy, and their number decreased substantially after splenectomy (**Supplementary Figure 1C**).

Osmotic Gradient Ektacytometry-Derived Parameters Indicate Increased Cellular Hydration After Splenectomy

Osmotic gradient ektacytometry was performed to determine the effect of splenectomy on RBC total surface area (EI_{max}), surface area-to-volume ratio (O_{min}), and RBC hydration state (O_{hyper}). In addition, the area under the curve was calculated (Van Vuren et al., 2019). A representative curve of the effect of splenectomy after 1 month is shown in **Figure 2A**.

Post-splenectomy values for O_{min} and EI_{max} did not change significantly compared to pre-splenectomy values, indicating that after splenectomy HS RBCs still had reduced surface area-to-volume ratio and total surface area (**Figures 2B,C**,

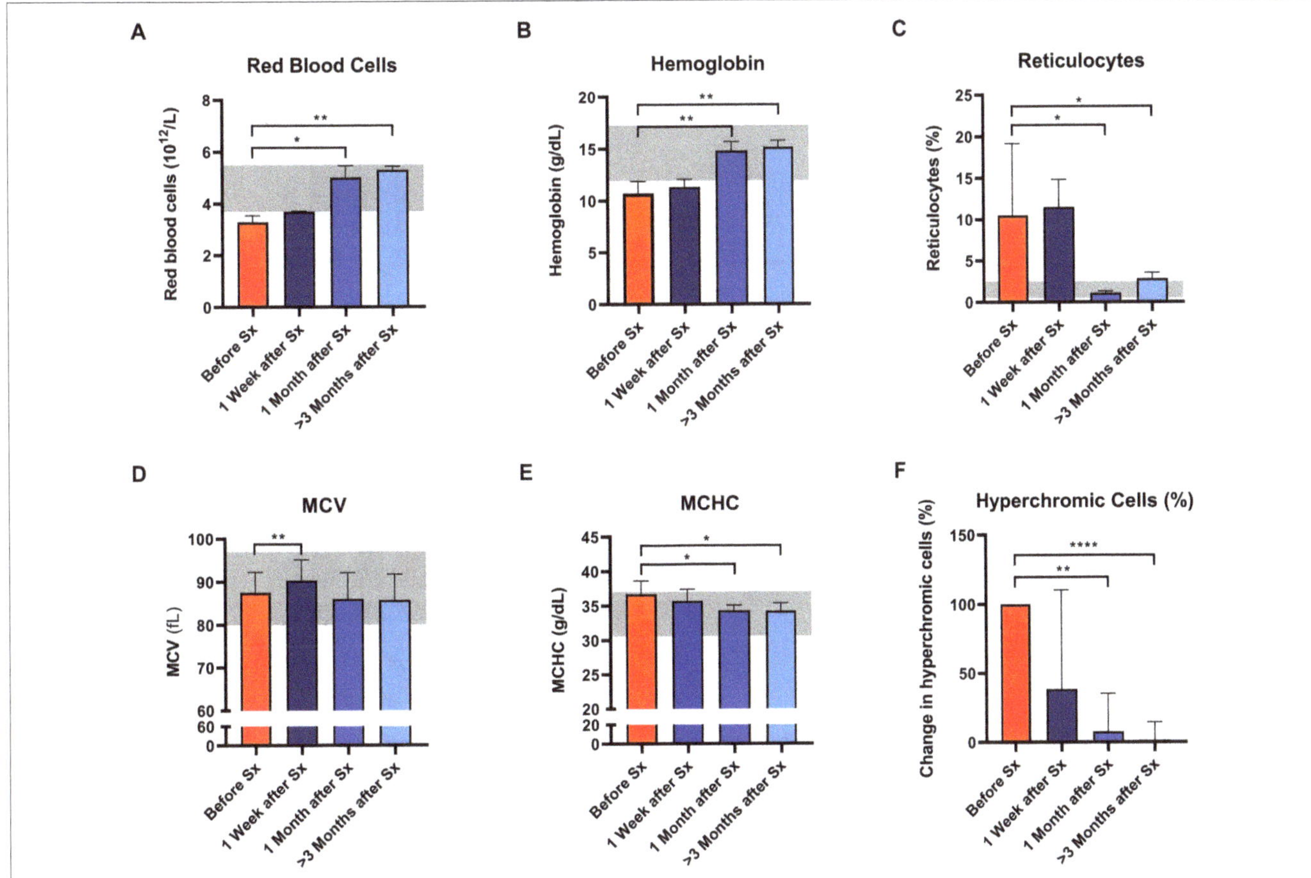

FIGURE 1 | Red blood cell related parameters of patients with hereditary spherocytosis (HS) before and after splenectomy (Sx). Whole blood of HS patients ($n = 5$) was analyzed before and 1 week, 1 and 3 months after splenectomy. **(A)** Red Blood Cells significantly increased after splenectomy. **(B)** Mean hemoglobin values significantly increased after 1 and 3 months after splenectomy. **(C)** Mean values of reticulocytes decreased 1 and 3 months after splenectomy. **(D)** mean values of mean corpuscular volume (MCV) increased significantly 1 week after splenectomy but decreased 1 and 3 months after splenectomy. **(E)** Mean values of mean corpuscular hemoglobin concentration (MCHC) decreased significantly 1 and 3 months after splenectomy. **(F)** Mean change in hyperchromic cells (%) after splenectomy. The laboratory reference ranges (2SD) of the University Medical Center Utrecht (UMCU) are depicted in the light gray area. Error bars represent standard deviation. ****$p < 0.0001$, ***$p < 0.001$, **$p < 0.01$, and *$p \leq 0.05$.

respectively). In contrast, the hydration state or cytoplasmic viscosity (O_{hyper}) showed a significant increase toward normal values after already 1 week and this was maintained after 1 and ≥ 3 months (**Figure 2D**). This is in line with the decrease in number of hyperchromic cells and MCHC (**Figures 1E,F**). The increase in O_{hyper} was accompanied by an increase in the AUC, although values remained lower than normal controls (**Figure 2E**).

Additional osmotic gradient ektacytometry measurements on the different density fractions showed that following splenectomy the variability between the curves from each density fraction is less, again indicating a more homogeneous RBC population (**Supplementary Figures 1D,E**). These analyses also showed that only O_{hyper} of density fraction 4 improved. Therefore, the increase in O_{hyper} after splenectomy seemed mainly determined by this fraction, containing the most dense RBCs. O_{hyper} of fractions 2 and 3 decreased after splenectomy, most presumably due to a decrease in reticulocytes in both fractions (**Supplementary Figures 1D–G**).

The Cell Membrane Stability Test Reveals That HS RBCs Have Regained the Ability to Shed Membrane After Splenectomy, Reflecting Improved Membrane Health

We next investigated RBC rigidity and its ability to respond to mechanical stress by performing CMST measurements. The CMST exposes RBCs to a high, supraphysiological, shear stress (100 Pa) for the duration of 1 h. Healthy RBCs showed a gradual decrease in deformability under these conditions (Representative curve, **Figure 3A**), reflected by a negative ΔEI. The loss of deformability likely results from increased vesiculation *in vitro* and consequent membrane loss under shear. In contrast, HS RBCs showed a significantly lower ΔEI before splenectomy compared to HC ($p < 0.01$, **Figure 3B**; representative curve **Figure 3A**). This suggests that RBCs of HS patients are more rigid and less able to shed membrane *in vitro* compared to healthy RBCs. After splenectomy, ΔEI increased until it was no longer significantly different from HC ≥ 3 months after

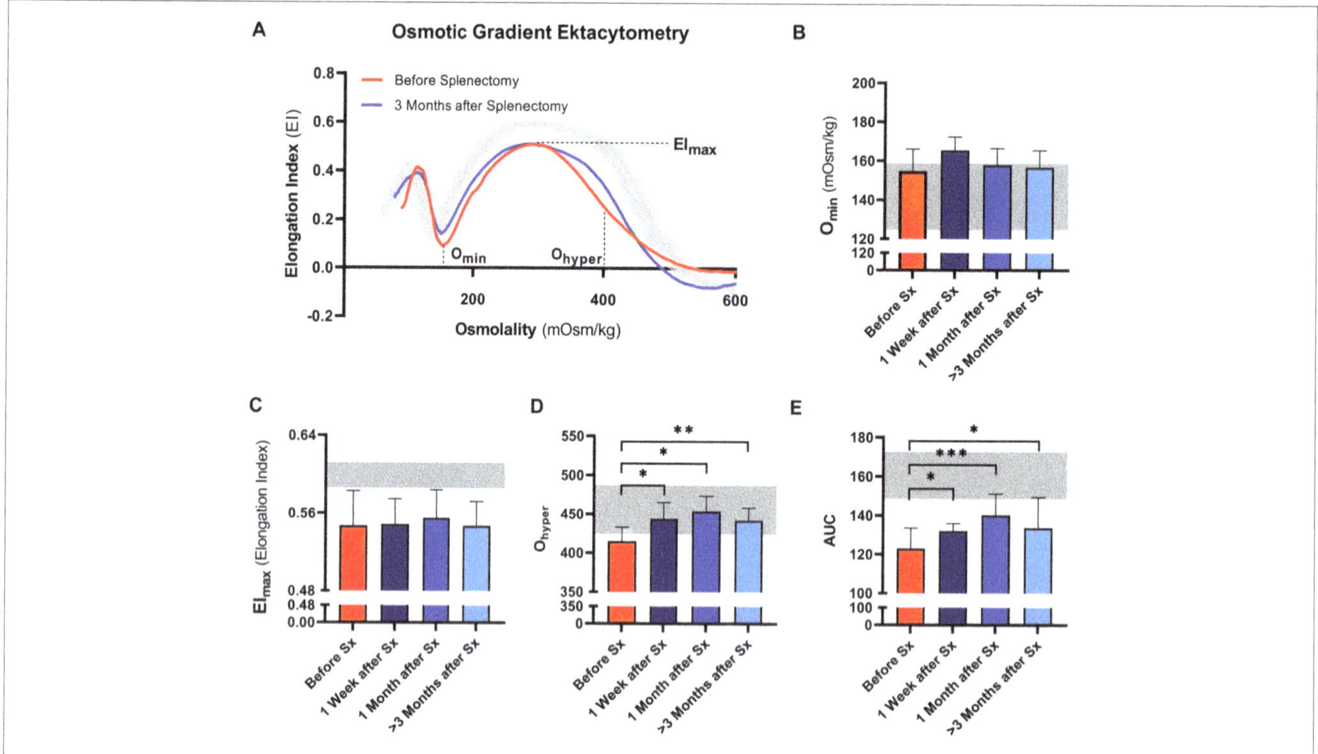

FIGURE 2 | Osmotic gradient ektacytometry curve (Osmoscan) and the corresponding parameters of 5 patients with HS before and after splenectomy (Sx). **(A)** Representative example of the osmotic gradient ektacytometry curve in P03 before and 3 months after splenectomy. The osmotic gradient ektacytometry curves of healthy controls ($n = 40$) are depicted by the gray lines. The changes in **(B)** the surface area to volume ratio of red blood cells (Omin), **(C)** the maximum deformability (Elmax), **(D)** the hydration state or cytoplasmic viscosity (Ohyper) and **(E)** the area under the curve (AUC) are depicted above. The HS patients ($n = 5$) were grouped and the results are displayed as the mean (SD) of the combined values. The mean (2SD) of healthy controls ($n = 74$) are depicted in the light gray area **(B–E)**. Error bars represent standard deviation. ***$p < 0.001$, **$p < 0.01$, *$p \leq 0.05$.

splenectomy (**Figure 3B**). Hence, splenectomy results in a less rigid cell population that has for a large part regained the ability to shed membrane. These findings were strengthened by microscopic analysis of RBCs during the CMST. **Figure 3C** shows how RBCs obtained from a HC were fully elongated and elliptical at the start of the CMST ($t = 10$ s), then turning into dense and less elongated RBCs at the end of the measurement ($t = 3,590$ s). In contrast, RBCs of an HS patient without splenectomy were already dense and unable to elongate fully at the start of the CMST, remaining like this throughout the measurement. Notably, reticulocyte count and hyperchromic cells both correlated with ΔEI ($r = 0.660$, $p < 0.01$ and $r = 0.668$, $p < 0.01$, respectively, **Figures 3D,E**), indicating that these cells strongly influence the outcome of CMST measurements.

Additional CMST measurements on the density fractions suggest that the increase in ΔEI post-splenectomy are determined by density fractions 2 and 3 (**Supplementary Figure 1H**).

Membrane-Shedding as Measured by the CMST Represents a Novel Pathophysiological Property of HS RBCs

To further explore its added value we performed CMST measurements on a large HS cohort consisting of 18 non-splenectomized, 8 splenectomized patients, and 26 HCs. These results confirmed the findings observed in our longitudinal study, showing that the CMST was able to distinguish splenectomized HS patients from non-splenectomized HS patients ($p = 0.028$), in addition to the clear distinction between HS patients in general and HCs (both $p < 0.001$). Similarly, also in this large cohort there was a correlation of ΔEI and reticulocyte count ($R = 0.606$, $p < 0.01$, **Figure 4F**) and hyperchromic cells ($R = 0.521$, $p < 0.01$ **Figure 4G**).

We next evaluated osmotic gradient ektacytometry measurements in this case-control cohort. In agreement with our findings in the longitudinal study (**Figure 2B**), O_{min} was not different in splenectomized patients (**Figure 4B**). Also EI_{max} and AUC were not significantly different in splenectomized HS patients (**Figures 4C,E**). In contrast, O_{hyper} was the only parameter that showed improvement when comparing splenectomized to non-splenectomized HS patients ($p < 0.01$, **Figure 4D**). Furthermore, O_{hyper} correlated with hyperchromic cells although less clear than ΔEI ($R = -0.461$, $p < 0.05$, **Supplementary Figure 2**), but not with reticulocyte count (**Figure 4H**). Importantly, ΔEI and O_{hyper} showed no correlation (**Figure 4I**), suggesting that both biomarkers reflect different features of HS RBCs.

Together, these findings confirm that a decreased ability to shed membrane *in vitro* as measured by the CMST is

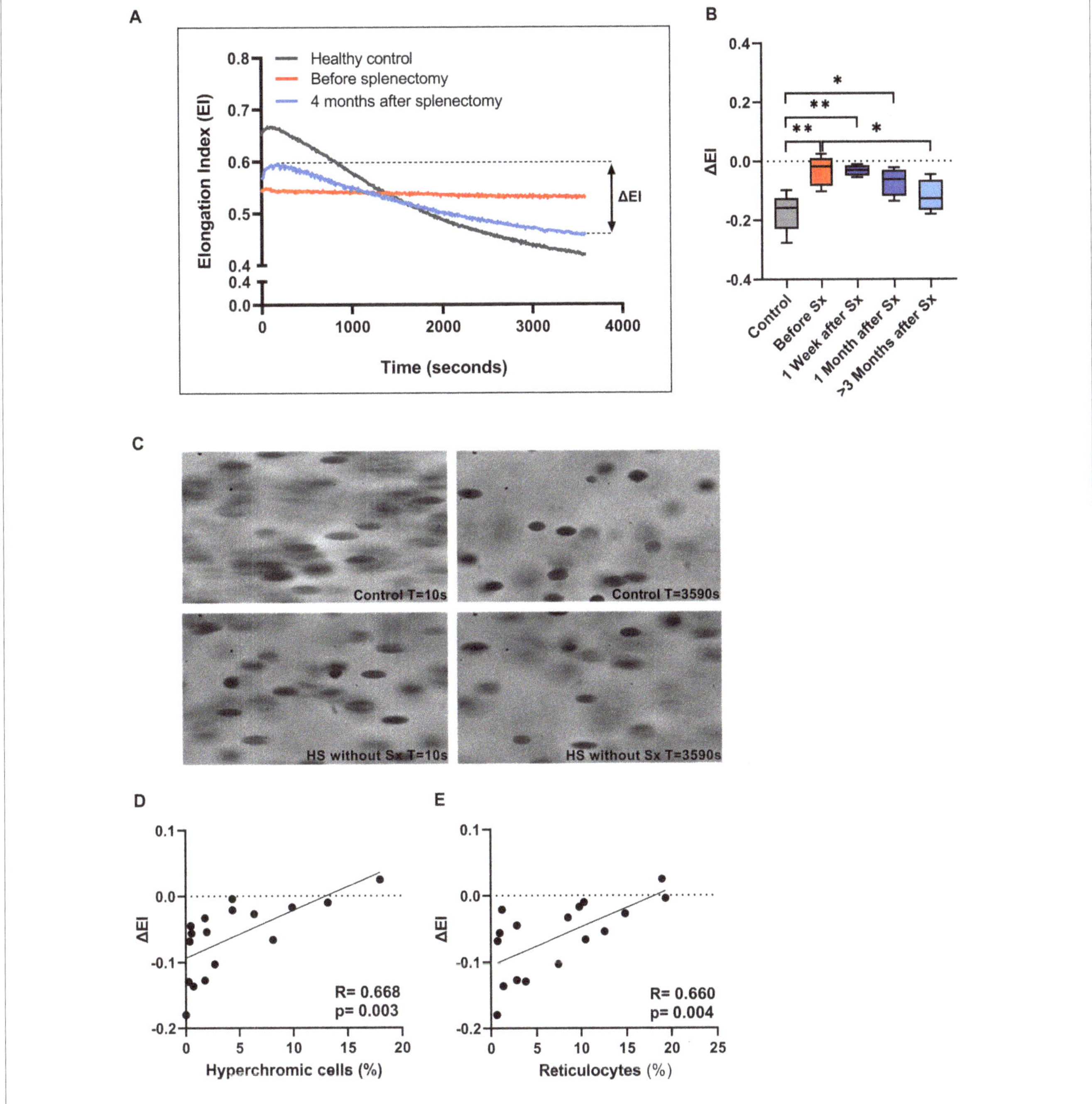

FIGURE 3 | The Cell membrane stability test (CMST) and the calculated parameter (ΔEI) improves after splenectomy in a longitudinal study of 5 HS patients. **(A)** Representative example of the CMST before (red line) and 4 months after splenectomy (light blue line) compared to a healthy control (dark gray line). **(B)** The mean values of ΔEI of HS patients ($n = 5$) before and after splenectomy. **(C)** Microscopic images of the RBCs in the Lorrca during a CMST. Start of the measurement ($T = 10$ s) compared to the end of the measurement ($T = 3{,}590$ s) in a control and HS patient without splenectomy. **(D)** Correlation between ΔEI and hyperchromic red blood cells. **(E)** Correlation between ΔEI and reticulocytes (%). Error bars represent standard deviation. $^{**}p < 0.01$, $^{*}p \leq 0.05$. Sx, splenectomy; HS, hereditary spherocytosis; T, time; s, seconds.

a novel pathophysiological feature of HS RBCs. It likely reflects membrane health and improves after splenectomy, thereby rendering a novel biomarker that is distinct from the improved density/cell hydration as measured by osmotic gradient ektacytometry.

DISCUSSION

In the present study, we report on the longitudinal effects of splenectomy in five HS patients. We specifically focused on cellular properties related to membrane health with the

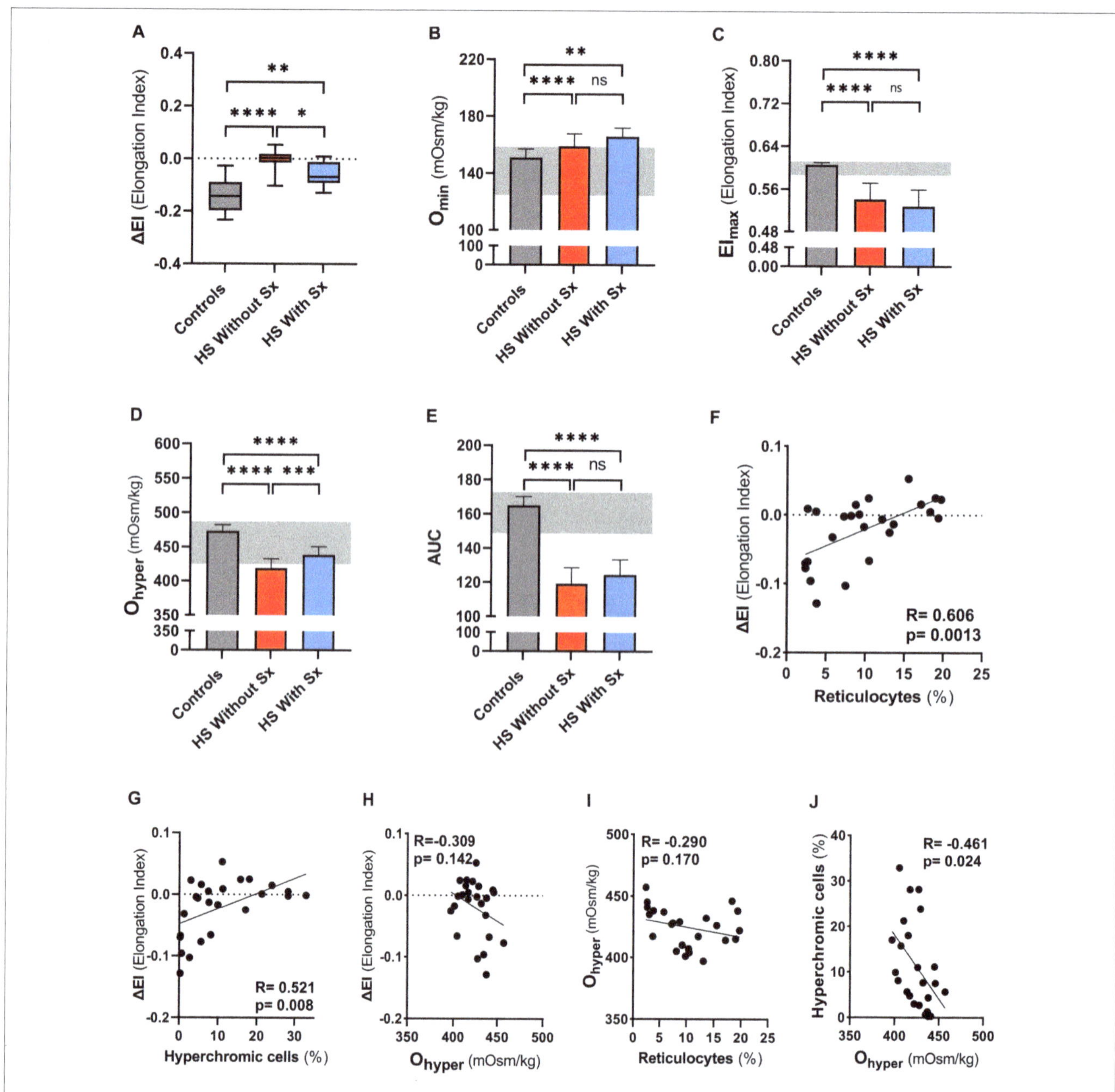

FIGURE 4 | Cell membrane stability test (CMST) and the calculated parameter (ΔEI) shows improvement in splenectomized patients with HS in a case-control study. CMST-derived parameter ΔEI [panels **(A,F,G,H)**], osmotic gradient ektacytometry-derived parameters [panels **(B–E,H,I,J)**], reticulocytes and hyperchromic cells were assessed in 18 non-splenectomized HS patients, 8 non-splenectomized HS patients and 26 healthy controls (HC). **(A)** Mean ΔEI of HS patients without splenectomy compared to patients with splenectomy, both groups were compared to HCs. **(B)** Mean values of Omin are increased in splenectomized patients compared to HCs or non-splenectomized HS. **(C)** Mean values of Elmax are decreased in splenectomized HS patient RBCs, compared to HC and non-splenectomized HS RBCs. **(D)** Mean values of Ohyper are significantly different between the 3 groups mentioned above. **(E)** Mean values of AUC show no significant differences between non-splenectomized and splenectomized HS RBCs. **(F)** Linear correlation between reticulocytes (%) and ΔEI of HS patients. **(G)** Linear correlation between hyperchromic cells (%) and ΔEI of HS patients. **(H)** Linear correlation between Ohyper and ΔEI. **(I)** Linear correlation between reticulocytes (%) and Ohyper. **(J)** Linear correlation between hyperchromic cells and Ohyper. Error bars represent standard deviation. ****$p < 0.0001$, ***$p < 0.001$, **$p < 0.01$, *$p \leq 0.05$; ns, non-significant. Sx, splenectomy; HS, hereditary spherocytosis; AUC, area under the curve.

use of two different forms of ektacytometry: osmotic gradient ektacytometry and the CMST. In particular the CMST results revealed a novel feature of HS RBCs, i.e., the loss of the ability to shed membrane, and improvement of this *in vitro* cellular property following splenectomy. Membrane-shedding capacity in this test is assessed by the loss of deformability that occurs during prolonged exposure of RBCs to high shear stress. We suggest that improved membrane-shedding capacity after splenectomy reflects improved RBC membrane health, and ΔEI as measured by the CMST may thus serve as a novel clinically relevant biomarker.

The longitudinally observed increase in RBC count, Hb and reticulocyte count after splenectomy corresponds well with results from previous studies where splenectomized and non-splenectomized patient groups were compared (Zaninoni et al., 2018; Huisjes et al., 2020). In addition, our patients also showed a decrease in MCHC directly after splenectomy, which continued to decrease in the following months. This implicates that the internal viscosity or cellular density of HS RBCs is reduced after splenectomy, which is supported by the reduction in the percentage of hyperchromic cells. Little is known about the effect of splenectomy on *in vivo* RBC vesiculation in HS, an important pathophysiological feature, but previous studies demonstrated that RBC vesiculation caused an increase in internal viscosity (MCHC) through shedding of RBC-derived microvesicles (Bosch et al., 1994; Alaarg et al., 2013; Bosman, 2013). Hence, both the decrease in MCHC and hyperchromic cells could indicate that *in vivo* vesiculation of RBCs in HS is reduced after splenectomy. In turn this could explain the observed improvement in ΔEI in the CMST after splenectomy, which reflects improved ability to shed membrane *in vitro*.

More detailed analysis of the longitudinal effects of splenectomy was obtained by osmotic gradient ektacytometry. This technique is generally considered as the gold standard in the diagnosis of HS (DaCosta et al., 2016; Lazarova et al., 2017; Llaudet-Planas et al., 2018; Zaninoni et al., 2018), and its parameters EI_{max}, O_{min}, and O_{hyper} are considered biomarkers of, respectively, total membrane surface area, surface area to volume ratio, and RBC hydration status. Upon splenectomy most of these parameters were not affected, only O_{hyper} was significantly increased (**Figures 2D, 4D**). O_{hyper} and MCHC are known to have an inverse correlation with each other (Zaninoni et al., 2018). Both the increase in O_{hyper} and the decrease in MCHC indicate that splenectomy improves the hydration state/intracellular viscosity. In line with this, an increase in the AUC was observed in the longitudinal cohort, whereas AUC remained unchanged after splenectomy in the case-control cohort. Our findings partly contradict with previous studies were both O_{hyper} and AUC remained unaltered after splenectomy (Zaninoni et al., 2018; Huisjes et al., 2020). This could be explained by the non-longitudinal design of these latter studies in which individual differences in response to splenectomy become less apparent.

We next investigated the effect of splenectomy on the ability of HS RBCs to respond to mechanical stress. For this we used the CMST, an ektacytometry based test that was previously used to study membrane stability in HS by studying resealed RBC ghosts (Mohandas et al., 1982; Chasis and Mohandas, 1986). We demonstrate that RBCs from non-splenectomized HS patients show no or only a modest decrease in EI after prolonged exposure to shear stress, in contrast to HC RBCs which display a substantial decrease in EI under these conditions. We hypothesize that HS RBCs from non-splenectomized HS patients are more dense and rigid due to *in vivo* vesiculation that is accelerated by the spleen, and, therefore are less able to shed membrane *in vitro* (Waugh and La Celle, 1980). This is confirmed by the microscopic evaluation of HS RBCs during CMST measurement, which shows that, in contrast to healthy RBCs, HS RBCs do not change morphologically (**Figure 3C**). Further support for this hypothesis is obtained from the significant correlation of ΔEI and hyperchromic cells (**Figure 3D**).

Following splenectomy HS RBCs showed an increase in ΔEI (**Figure 3B**) and after more than 3 months ΔEI was not significantly different compared to HC. This suggests that after splenectomy HS RBCs have regained part of the ability to shed membrane *in vitro*, which may be related to a partly restored ability of HS RBCs for *de novo* synthesis of lipids (Cooper and Jandl, 1969; Sugihara et al., 1984; Takashi and Yoshihito, 1984). The increase in ΔEI could also indicate a change in RBC population due to the absence of the quality control function of the spleen; Instead of shedding micro vesicles *in vivo* in the spleen (i.e., splenic conditioning), this now occurs *in vitro* under the supraphysiological conditions in the CMST. In both cases cellular characteristics as obtained by the CMST measurements indicate an improvement after splenectomy, in a sense that they behave more like normal healthy RBCs. Furthermore, the degree of increase in ΔEI after splenectomy we observed could also be dependent on genetic defect (Ingrosso et al., 1996; Reliene et al., 2002). Our findings were strengthened by CMST results on a large cohort of HS patients that showed that this test was able to discriminate between splenectomized and non-splenectomized HS patients (**Figure 4A**). This implies that the CMST may represent a novel biomarker of HS, which was further supported by the correlation of ΔEI and % reticulocytes ($r = 0.66$, $p = 0.004$). The ability to distinguish splenectomized patients from non-splenectomized patients can be valuable in an era where different types of splenectomy are explored. Partial splenectomy, either through embolization of splenic arteries or through (laparoscopic) removal of a part of the spleen, might ameliorate symptoms and improve anemia while maintaining splenic phagocytic function (Tchernia et al., 1993; Pratl et al., 2008). However, data of several small studies are inconclusive regarding the remaining immunological capacity of the spleen after partial splenectomy even though hemolysis is decreased (Guizzetti, 2016). Larger studies that also include functional analysis of the spleen and functional analysis of RBCs, i.e., the CMST, are warranted to accurately assess the efficacy of (partial) splenectomy or embolization, and to investigate whether changes in RBC properties or the nature of the underlying molecular defect (Ingrosso et al., 1996; Reliene et al., 2002) contribute to the clinical benefit of splenectomy.

In conclusion, we report on the longitudinal effects of splenectomy on HS RBC characteristics and function as studied by ektacytometry. Our data shows that before splenectomy the HS RBC population is more heterogeneous, cells are more rigid, have increased intracellular viscosity and reduced deformability. Functional analysis of HS RBCs using osmotic gradient ektacytometry and CMST further shows that splenectomy improves the hydration state of HS RBCs and allows cells to regain the ability to shed membrane. In particular the CMST reflects an yet-undescribed distinct RBC characteristic and holds promise as a novel biomarker for membrane health in HS that could be helpful, together with a comprehensive clinical evaluation and appropriate follow-up, to assess the effect of different treatments such as embolization and (partial) splenectomy, and that may be related to clinical severity given the correlation of ΔEI and reticulocyte count. Larger studies are warranted to establish if the CMST can be used to improve the assessment of clinical severity and/or is able to contribute to a better understanding of phenotypic differences in HS.

AUTHOR CONTRIBUTIONS

RH, MB, RW, and MR designed the study. RH, JV, ML, MB, EB, and MR collected clinical and laboratory data. JB, RH, TM, BO, and MR performed laboratory experiments. MCB, RW, and MR analyzed the data and wrote the manuscript. All authors edited the manuscript and approved the final version.

ACKNOWLEDGMENTS

The authors would like to thank all patients that donated blood for this study.

REFERENCES

Alaarg, A., Schiffelers, R. M., Van Solinge, W. W., and Van Wijk, R. (2013). Red blood cell vesiculation in hereditary hemolytic anemia. *Front. Physiol.* 4:365. doi: 10.3389/fphys.2013.00365

Bosch, F. H., Werre, J. M., Schipper, L., Roerdinkholder-Stoelwinder, B., Huls, T., Willekens, F. L., et al. (1994). Determinants of red blood cell deformability in relation to cell age. *Eur. J. Haematol.* 52, 35–41.

Bosman, G. J. C. G. M. (2013). Survival of red blood cells after transfusion: processes and consequences. *Front. Physiol.* 4:376. doi: 10.3389/fphys.2013.00376

Chasis, J. A., Agre, P., and Mohandas, N. (1988). Decreased membrane mechanical stability and in vivo loss of surface area reflect spectrin deficiencies in hereditary spherocytosis. *J. Clin. Invest.* 82, 617–623. doi: 10.1172/JCI113640

Chasis, J. A., and Mohandas, N. (1986). Erythrocyte membrane deformability and stability: two distinct membrane properties that are independently regulated by skeletal protein associations. *J. Cell Biol.* 103, 343–350. doi: 10.1083/jcb.103.2.343

Cooper, R. A., and Jandl, J. H. (1969). The role of membrane lipids in the survival of red cells in hereditary spherocytosis. *J. Clin. Invest.* 48, 736–744. doi: 10.1172/JCI106031

DaCosta, L., Suner, L., Galimand, J., Bonnel, A., Pascreau, T., Couque, N., et al. (2016). Diagnostic tool for red blood cell membrane disorders: assessment of a new generation ektacytometer. *Blood Cells Mol. Dis.* 56, 9–22. doi: 10.1016/j.bcmd.2015.09.001

Eber, S. W., and Lux, S. E. (2004). Hereditary spherocytosis–defects in proteins that connect the membrane skeleton to the lipid bilayer. *Semin. Hematol.* 41, 118–141.

Guizzetti, L. (2016). Total versus partial splenectomy in pediatric hereditary spherocytosis: a systematic review and meta-analysis. *Pediatr. Blood Cancer* 63, 1713–1722. doi: 10.1002/pbc.26106

Huisjes, R., Bogdanova, A., van Solinge, W. W., Schiffelers, R. M., Kaestner, L., and van Wijk, R. (2018). Squeezing for life - Properties of red blood cell deformability. *Front. Physiol.* 9:656. doi: 10.3389/fphys.2018.00656

Huisjes, R., Makhro, A., Llaudet-Planas, E., Hertz, L., Petkova-Kirova, P., Verhagen, L. P., et al. (2020). Density, heterogeneity and deformability of red cells as markers of clinical severity in hereditary spherocytosis. *Haematologica* 105, 338–347. doi: 10.3324/haematol.2018.188151

Huisjes, R., Solinge, W. W., Levin, M. D., Wijk, R., and Riedl, J. A. (2017). Digital microscopy as a screening tool for the diagnosis of hereditary hemolytic anemia. *Int. J. Lab. Hematol.* 40, 159–168. doi: 10.1111/ijlh.12758

Ingrosso, D., D'Angelo, S., Perrotta, S., d'Urzo, G., Iolascon, A., Perna, A. F., et al. (1996). Cytoskeletal behaviour in spectrin and in band 3 deficient spherocytic red cells: evidence for a differentiated splenic conditioning role. *Br. J. Haematol.* 93, 38–41. doi: 10.1046/j.1365-2141.1996.451990.x

Lazarova, E., Gulbis, B., van Oirschot, B., and van Wijk, R. (2017). Next-generation osmotic gradient ektacytometry for the diagnosis of hereditary spherocytosis: interlaboratory method validation and experience. *Clin. Chem. Lab. Med.* 55, 394–402. doi: 10.1515/cclm-2016-0290

Li, Y., Lu, L., and Li, J. (2016). Topological structures and membrane nanostructures of erythrocytes after splenectomy in hereditary spherocytosis patients via atomic force microscopy. *Cell Biochem. Biophys.* 74, 365–371. doi: 10.1007/s12013-016-0755-4

Llaudet-Planas, E., Vives-Corrons, J. L., Rizzuto, V., Gómez-Ramírez, P., Navarro, J. S., Sibina, M. T. C., et al. (2018). Osmotic gradient ektacytometry: a valuable screening test for hereditary spherocytosis and other red blood cell membrane disorders. *Int. J. Lab. Hematol.* 40, 94–102. doi: 10.1111/ijlh.12746

Mebius, R. E., and Kraal, G. (2005). Structure and function of the spleen. *Nat. Rev. Immunol.* 5, 606–616. doi: 10.1038/nri1669

Mohandas, N., Clark, M. R., Health, B. P., Rossi, M., Wolfe, L. C., Lux, S. E., et al. (1982). A technique to detect reduced mechanical stability of red cell membranes: relevance to elliptocytic disorders. *Blood* 59, 768–774.

Musser, G., Lazar, G., Hocking, W., and Busuttil, W. (1984). Splenectomy for hematologic disease. The UCLA experience with 306 patients. *Ann. Surg.* 200, 40–45. doi: 10.1097/00000658-198407000-00006

Perrotta, S., Gallagher, P. G., and Mohandas, N. (2008). Hereditary spherocytosis. *Lancet* 372, 1411–1426. doi: 10.1016/S0140-6736(08)61588-3

Pratl, B., Benesch, M., Lackner, H., Portugaller, H. R., Pusswald, B., Sovinz, P., et al. (2008). Partial splenic embolization in children with hereditary spherocytosis. *Eur. J. Haematol.* 80, 76–80. doi: 10.1111/j.1600-0609.2007.00979.x

Reliene, R., Mariani, M., Zanella, A., Reinhart, W. H., Ribeiro, M. L., del Giudice, E. M., et al. (2002). Splenectomy prolongs *in vivo* survival of erythrocytes differently in spectrin/ankyrin- and band 3-deficient hereditary spherocytosis. *Blood* 100, 2208–2215. doi: 10.1182/blood.v100.6.2208.h81802002208_2208_2215

Rennie, C. M., Thompson, S., Parker, A. C., and Maddy, A. (1979). Human erythrocyte fractionation in "percoll" density gradients. *Clin. Chim. Acta* 98, 119–125. doi: 10.1016/0009-8981(79)90172-4

Schilling, R. F., Gangnon, R. E., and Traver, M. I. (2008). Delayed adverse vascular events after splenectomy in hereditary spherocytosis. *J. Thromb. Haemost.* 6, 1289–1295. doi: 10.1111/j.1538-7836.2008.03024.x

Sugihara, T., Miyashima, K., and Yawata, Y. (1984). Disappearance of microspherocytes in peripheral circulation and normalization of decreased lipids in plasma and in red cells of patients with hereditary spherocytosis after splenectomy. *Am. J. Hematol.* 17, 129–139.

Takashi, S., and Yoshihito, Y. (1984). Observations on plasma and red cell lipids in hereditary spherocytosis. *Clin. Chim. Acta* 137, 227–232. doi: 10.1016/0009-8981(84)90182-7

Tchernia, G., Gauthier, F., Mielot, F., Dommergues, J. P., Yvart, J., Chasis, J. A., et al. (1993). Initial assessment of the beneficial effect of partial splenectomy in hereditary spherocytosis. *Blood* 81, 2014–2020. doi: 10.1016/B978-0-323-05226-9.50014-1

Van Vuren, A., Van Der Zwaag, B., Huisjes, R., Lak, N., Bierings, M., Gerritsen, E., et al. (2019). The complexity of genotype-phenotype correlations in hereditary spherocytosis: a cohort of 95 patients: genotype-phenotype correlation in hereditary spherocytosis. *HemaSphere* 3:e276. doi: 10.1097/HS9.0000000000000276

Waugh, R. E., and La Celle, P. L. (1980). Abnormalities in the membrane material properties of hereditary spherocytes. *J. Biomech. Eng.* 102:240. doi: 10.1115/1. 3149580

Zaninoni, A., Fermo, E., Vercellati, C., Consonni, D., Marcello, A. P., Zanella, A., et al. (2018). Use of laser assisted optical rotational cell analyzer (LoRRca MaxSis) in the diagnosis of RBC membrane disorders, enzyme defects, and congenital dyserythropoietic anemias: a monocentric study on 202 patients. *Front. Physiol.* 9:451. doi: 10.3389/fphys.2018.00451

12

Metabolomics of Endurance Capacity in World Tour Professional Cyclists

Iñigo San-Millán[1,2,3], Davide Stefanoni[4], Janel L. Martinez[2], Kirk C. Hansen[4], Angelo D'Alessandro[4] and Travis Nemkov[4]**

[1] Department of Human Physiology and Nutrition, University of Colorado Colorado Springs, Colorado Springs, CO, United States, [2] Division of Endocrinology, Metabolism and Diabetes, Department of Medicine, University of Colorado Anschutz Medical Campus, Aurora, CO, United States, [3] Department of Research and Development, UAE Team Emirates, Abu Dhabi, United Arab Emirates, [4] Department of Biochemistry and Molecular Genetics, University of Colorado Anschutz Medical Campus, Aurora, CO, United States

****Correspondence:***
Iñigo San-Millán
inigo.sanmillan@cuanschutz.edu
Travis Nemkov
travis.nemkov@cuanschutz.edu

The study of elite athletes provides a unique opportunity to define the upper limits of human physiology and performance. Across a variety of sports, these individuals have trained to optimize the physiological parameters of their bodies in order to compete on the world stage. To characterize endurance capacity, techniques such as heart rate monitoring, indirect calorimetry, and whole blood lactate measurement have provided insight into oxygen utilization, and substrate utilization and preference, as well as total metabolic capacity. However, while these techniques enable the measurement of individual, representative variables critical for sports performance, they lack the molecular resolution that is needed to understand which metabolic adaptations are necessary to influence these metrics. Recent advancements in mass spectrometry-based analytical approaches have enabled the measurement of hundreds to thousands of metabolites in a single analysis. Here we employed targeted and untargeted metabolomics approaches to investigate whole blood responses to exercise in elite World Tour (including Tour de France) professional cyclists before and after a graded maximal physiological test. As cyclists within this group demonstrated varying blood lactate accumulation as a function of power output, which is an indicator of performance, we compared metabolic profiles with respect to lactate production to identify adaptations associated with physiological performance. We report that numerous metabolic adaptations occur within this physically elite population ($n = 21$ males, 28.2 ± 4.7 years old) in association with the rate of lactate accumulation during cycling. Correlation of metabolite values with lactate accumulation has revealed metabolic adaptations that occur in conjunction with improved endurance capacity. In this population, cycling induced increases in tricarboxylic acid (TCA) cycle metabolites and Coenzyme A precursors. These responses occurred proportionally to lactate accumulation, suggesting a link between enhanced mitochondrial networks and the ability to sustain higher workloads. In association with lactate accumulation, altered levels of amino acids before and after exercise point to adaptations that confer unique substrate preference for energy production or to promote more rapid recovery. Cyclists with slower lactate accumulation also have higher levels of basal oxidative stress markers, suggesting long term physiological adaptations in these individuals that support their premier competitive status in worldwide competitions.

Keywords: metabolomics, elite athletes, endurance, exercise, lactate, oxidative stress, amino acid metabolism, mitochondrial metabolism

INTRODUCTION

The physiological and metabolic response to exercise has fascinated scientists for centuries. In 1784, Antoine Lavoisier and Pierre-Simon Laplace designed an ice-calorimeter to measure the amount of heat emitted during combustion and respiration, which enabled the measurement of oxygen (O_2) consumption during exercise (Underwood, 1944; Karamanou and Androutsos, 2013). Since then, maximal oxygen consumption (VO_2max) has been considered the gold standard to measure cardiorespiratory fitness (Shephard, 1984). In the last two decades, the monitoring of physiological and metabolic response to exercise using this technique has become increasingly popular in the area of sports medicine and performance, fostered in part by studies involving elite professional athletes. These studies have shaped programs for individualized training, recovery and nutritional regimes and have been based traditionally on VO_2max as the representative parameter due to its relatively easy measurement by indirect calorimetry using commercially available metabolic carts (Jensen et al., 2002).

Despite widespread use, laboratory physiological testing is not ubiquitous as not all athletes have access to advanced exercise laboratory facilities. In addition, respiration-based measurements are not always reproducible in monitoring performance, due to the multiple unmeasured variables that influence this the measured output. As such, additional markers of training status that are more acutely tied to cellular and tissue metabolism have been proposed. Lactate is one such metabolic biomarker that enables measurement of metabolic responses to exercise and serves as a surrogate for muscle stress. Although it had been considered a waste product of anaerobic metabolism, foundational studies by Brooks and colleagues began to shift this decades-old paradigm by demonstrating a higher degree of lactate consumption under fully aerobic conditions than previously thought using isotope tracing studies in rats (Brooks, 1986). Since then, lactate has been shown to be a major fuel source for the body, and even possesses hormone-like properties (Brooks, 2018).

Whole blood lactate measurement has been widely accepted as a valid method to evaluate the metabolic responses to exercise (Jacobs, 1986; Billat, 1996). This measurement also serves as a monitor of training status, since well-trained athletes have a higher lactate clearance capacity and decreased blood lactate levels (Donovan and Brooks, 1983; Bergman et al., 1999) likely due to enhanced mitochondrial function resulting from training (McDermott and Bonen, 1993; Dubouchaud et al., 2000). In addition to ease of measurement using portable equipment, these findings have promoted lactate monitoring as a viable readout for personalized athletic training programs.

To expand upon metabolic programs that may elicit lactate production, studies into substrate utilization (such as fat and carbohydrate oxidation) during exercise using indirect calorimetry have provided improved resolution of exercise physiology (Frayn, 1983). The subsequent combination of blood lactate measurement and substrate utilization during exercise has thus offered novel approaches to measure metabolic function and, indirectly, mitochondrial function and metabolic flexibility (San-Millán and Brooks, 2018). However, despite growing interest into the cellular responses to exercise, these techniques only enable the monitoring of a limited number of parameters thereby hampering our full understanding of metabolic responses in exercise physiology.

The field of metabolomics has emerged strongly in the last decade in many areas of scientific research as a powerful tool to precisely measure metabolic pathways at the cellular and systematic level (D'Alessandro, 2019). Chromatography-based separation of metabolites combined with improved scanning speeds and accuracy of mass detectors has greatly enhanced the breadth and depth of metabolite coverage monitored in a single experiment. These techniques are amenable to measuring metabolites within any biological matrix (Nemkov et al., 2017) and have been applied to multiple exercise-focused studies into metabolic response to exercise that varies by type, intensity, and duration (reviewed in Sakaguchi et al., 2019).

Given the predictive importance of lactate measurement in sports physiology and the robustness of mass spectrometry-based metabolomics, we hypothesized that identification of metabolic pathways altered in conjunction with lactate production will reveal molecular mechanisms of enhanced physical performance. To test this hypothesis, we measured metabolite levels in whole blood samples isolated from a team of elite professional cyclists before and after a graded maximal physiological test and compared metabolic profiles with respect to a lactate-dependent performance cutoff (PC). Expansion of knowledge around lactate-associated metabolic traits could serve to improve multiple aspects of individualized training, nutrition, overtraining, injury prevention and rehabilitation.

MATERIALS AND METHODS

Twenty-one international-level World Tour professional male cyclists (Tour de France Level) performed a graded exercise test to exhaustion on an electrically controlled resistance leg cycle

TABLE 1 | Physiological parameters of male cyclists used in this study.

Age (year)	28.2 ± 4.7
Height (cm)	179.2 ± 7.6
Weight (kg)	70.7 ± 6.7
Body fat (%)	10.4 ± 0.7

ergometer (Elite, Suito, Italy). Physical parameters are included in **Table 1**. After a 15 min warm-up, participants started leg cycling at a low intensity of 2.0 W kg^{-1} of body weight. Exercise intensity was increased 0.5 W kg^{-1} every 10 min as previously described (San-Millán et al., 2009). Power output, heart rate and lactate were measured throughout the entire test and recorded every 10 min including at the end of the test. All study procedures were conducted in accordance with the Declaration of Helsinki and in accordance with a predefined protocol that was approved by all researchers and the Colorado Multiple Institutional Review Board (COMIRB 17-1281). Written informed consent was obtained from all subjects.

Blood Lactate Concentration Measurement

At the end of every intensity stage throughout the graded training period, a sample of capillary blood was collected to analyze both intra- and extra-cellular levels of L-lactate (Lactate Plus, Nova Biomedical, Waltham, MA, United States). Heart rate was monitored during the whole test with a heart monitor (Polar S725x, Polar Electro, Kempele, Finland).

Subjects Group Classification – Performance Cutoff

Cyclists were separated into two groups based on the blood lactate concentration at an exercise intensity of 5.0 W kg^{-1}, designated as the PC. Cyclists with PC lactate levels below the group average of 5 mmol L^{-1} were classified as the Gold group, while cyclists above the average were classified in the Silver group (**Figure 2A**).

Metabolomics Assessment

Sample Collection

Whole blood samples were collected using the Touch-Activated Phlebotomy (TAP) device (Seventh Sense Biosystems, Medford, MA) as previously described (Catala et al., 2018). Due to a limited 100 μL sample volume collected with the TAP device, and to assess the utility of whole blood analyses with the goal of simplifying sample isolation by circumventing the need for centrifugation, whole blood samples were frozen in dry ice within 15 min of isolation and stored at -80°C until analysis. Prior to LC-MS analysis, samples were placed on ice and re-suspended with nine volumes of ice cold methanol:acetonitrile:water (5:3:2, v:v). Suspensions were vortexed continuously for 30 min at 4°C. Insoluble material was removed by centrifugation at 18,000 g for 10 min at 4°C and supernatants were isolated for metabolomics analysis by UHPLC-MS. The extract was then dried down under speed vacuum and re-suspended in an equal volume of 0.1% formic acid for analysis.

UHPLC-MS Analysis

Analyses were performed as previously published (Nemkov et al., 2017; Reisz et al., 2019). Briefly, the analytical platform employs a Vanquish UHPLC system (Thermo Fisher Scientific, San Jose, CA, United States) coupled online to a Q Exactive mass spectrometer (Thermo Fisher Scientific, San Jose, CA, United States). The (semi)polar extracts were resolved over a Kinetex C18 column, 2.1 mm× 150 mm, 1.7 μm particle size (Phenomenex, Torrance, CA, United States) equipped with a guard column (SecurityGuard^{TM} Ultracartridge – UHPLC C18 for 2.1 mm ID Columns – AJO-8782 – Phenomenex, Torrance, CA, United States) using an aqueous phase (A) of water and 0.1% formic acid and a mobile phase (B) of acetonitrile and 0.1% formic acid for positive ion polarity mode, and an aqueous phase (A) of water:acetonitrile (95:5) with 1 mM ammonium acetate and a mobile phase (B) of acetonitrile:water (95:5) with 1 mM ammonium acetate for negative ion polarity mode. The Q Exactive mass spectrometer (Thermo Fisher Scientific, San Jose, CA, United States) was operated independently in positive or negative ion mode, scanning in Full MS mode (2 μscans) from 60 to 900 m/z at 70,000 resolution, with 4 kV spray voltage, 45 sheath gas, 15 auxiliary gas. Calibration was performed prior to analysis using the Pierce^{TM} Positive and Negative Ion Calibration Solutions (Thermo Fisher Scientific). Acquired data was then converted from.raw to.mzXML file format using Mass Matrix (Cleveland, OH, United States). Samples were analyzed in randomized order with a technical mixture injected after every 15 samples to qualify instrument performance. Metabolite assignments, isotopologue distributions, and correction for expected natural abundances of deuterium, ^{13}C, and ^{15}N isotopes were performed using MAVEN (Princeton, NJ, United States) (Melamud et al., 2010). Discovery mode alignment, feature identification, and data filtering was performed using Compound Discoverer 2.0 (Thermo Fisher Scientific).

Graphs, heat maps and statistical analyses (either T-Test or ANOVA), metabolic pathway analysis, partial least squares discriminant analysis (PLS-DA) and hierarchical clustering was performed using the MetaboAnalyst 4.0 package (Chong et al., 2018). XY graphs were plotted through GraphPad Prism 8 (GraphPad Software Inc., La Jolla, CA, United States).

RESULTS

The Whole Blood Metabolome Changes Significantly in Response to Exercise

In order to determine systemic metabolic changes during intense cycling, untargeted metabolomics analyses using LC-MS/MS were performed on whole blood samples isolated from cyclists before (Pre) and after (Post) a graded exercise test with cycling intensity normalized to individual cyclist body mass (**Figure 1A**). Of the 2,790 putative compounds detected in discovery mode analysis (ranging from Level 1 to Level 4 confidence, Schrimpe-Rutledge et al., 2016) 355 metabolites were manually identified using accurate intact mass, isotopic pattern, fragmentation and an in-house standard library. These

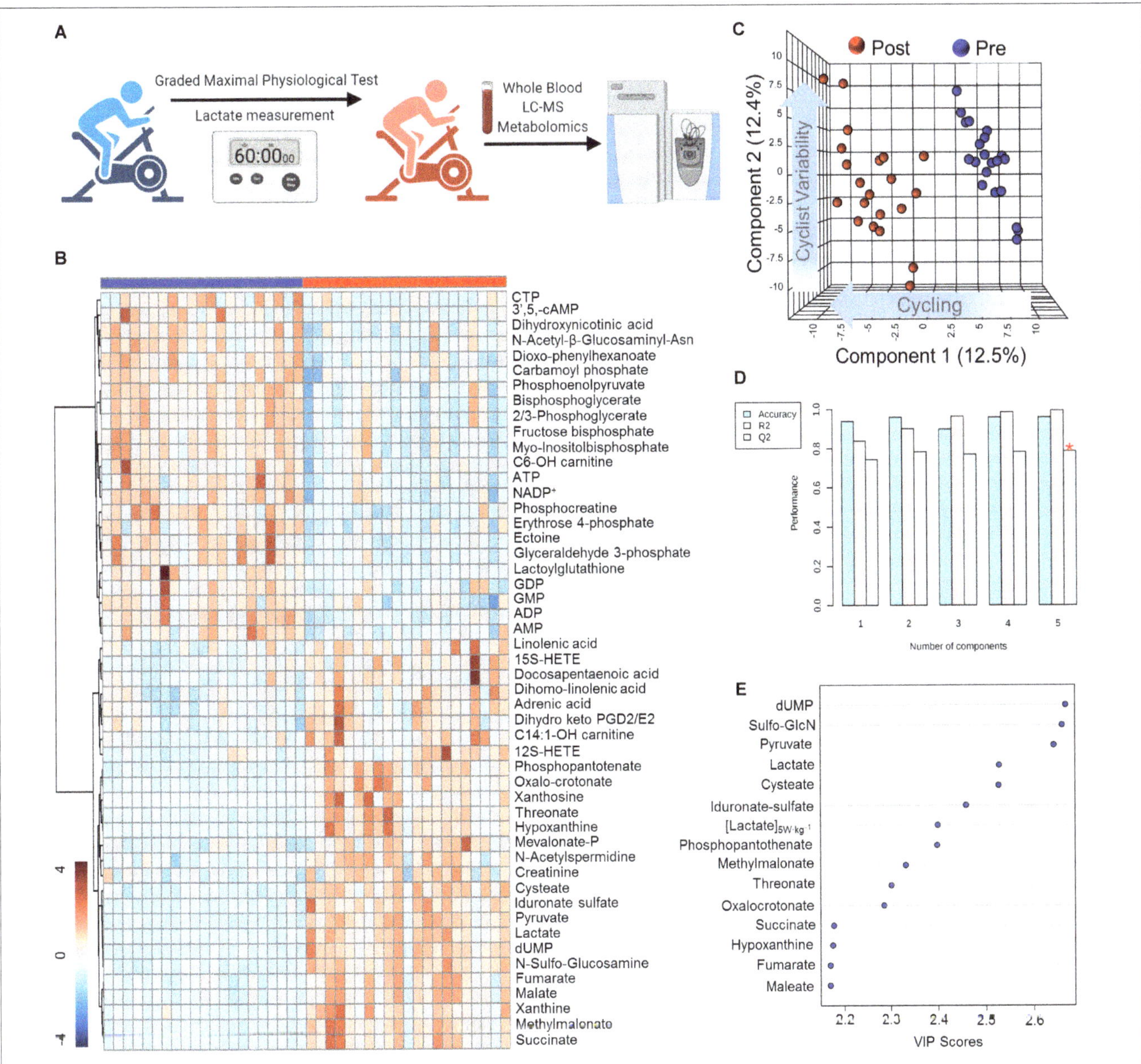

FIGURE 1 | High-Throughput Metabolomics of Endurance Training in Elite Professional Cyclists. **(A)** Blood samples were drawn from subjects using a TAP[TM] push-button blood collection device before and after 1 h of a graded exercise test. Samples were extracted for hydrophilic or non-polar metabolites and analyzed using a high-throughput LC-MS based metabolomics platform. **(B)** Hierarchical Clustering Analysis of the top 50 T-Test significant metabolites is shown as a heat map and are Z-score normalized, with the color-coded gradient depicted in the lower left corner. **(C)** Partial Least Squares Discriminant Analysis (PLS-DA) shows a distinct clustering pattern of baseline (Pre, blue) and post-test (Post, blue) samples along Component 1 axis (12.5% variance explained), while inter-individual variability is described by the Component 2 axis (12.4% variance explained). **(D)** PLS-DA cross validation details are shown for the first 5 components. **(E)** Parameter with the top 15 variable importance in projection (VIP) scores are shown.

metabolites were systematically analyzed by multivariate analyses such as Principal Component Analysis (PCA; **Figure 1B**) and Hierarchical Clustering Analysis (HCA, **Figure 1C**). A table including the molecular weight, retention time, polarity of detection, and raw peak area top values is provided in **Supplementary Table 1**. The top 50 significant metabolites by a two-tailed Student's T-Test were hierarchically clustered to illustrate acute metabolic responses to cycling, showing significant changes for metabolites involved in central energy metabolism (glycolysis, TCA cycle), nucleotide homeostasis, and lipid metabolism (**Figure 1B**). Partially supervised PLS-DA distinguished samples by time point across Component 1 (12.5% of metabolic difference described) and demonstrated variability of metabolic responses amongst cyclists along the Component 2 axis (12.4% of the total metabolic difference described) (**Figure 1C**). The PLS-DA model demonstrated a high degree of accuracy using just the first component ($R2 = 0.84$, $Q2 = 0.74$, Accuracy = 0.92, **Figure 1D**). A larger degree of biological variability was noted in the Post time point cluster relative to the Pre time point, indicating that the cyclists' responses to strenuous exercise varied within this group. The top 15 metabolites contributing to the clustering pattern include mitochondrial and glycolytic metabolites, indicating importance of these pathways (**Figure 1E**). A heatmap of metabolite values for the entire manually curated dataset is provided in **Supplementary Figure 1**.

Metabolite Abundance Before and After Exercise Associates With Lactate Production in Response to Workload

Interval measurements of whole blood lactate levels in response to workload serves as a personalized metric for cycling performance (Jacobs, 1986). The training status of an individual cyclist can be determined in large part based on the power output (measured in watts per kilogram, $W\,kg^{-1}$) at which lactate is produced more rapidly than it is consumed and begins to accumulate, also referred to as the lactate threshold (Brooks, 1985). As training status improves, the lactate threshold increases allowing the cyclist to exert a higher workload for longer periods of time prior to the onset of fatigue. Lactate levels in response to power output varied significantly amongst this group of cyclists. While all but 1 of the 21 cyclists monitored in this study were able to maintain a power output of $5\,W\,kg^{-1}$, 14 cyclists reached $5.5\,W\,kg^{-1}$, and only three cyclists were able to maintain $6\,W\,kg^{-1}$ for 10 min (**Figure 2A**). To understand how the network between metabolic pathways differs in these cyclists (a concept we have previously referred to as metabolic linkage (D'Alessandro et al., 2017) we first classified the cyclists into two groups based on whether blood lactate concentration was higher (Gold) or lower (Silver) than the overall average at the designated PC of $5\,W\,kg^{-1}$ (**Figure 2A**). Using the metabolomics dataset from the Gold group, a Spearman's correlation matrix was prepared and hierarchically clustered to identify regions of the metabolome that are positively and negatively correlated (**Figure 2B**, left panel). A Spearman's correlation matrix of metabolite levels in the Silver group was then prepared and plotted according to the same hierarchical order of the Gold group, and the difference in Spearman's correlation coefficients for each relationship was calculated to highlight changes in metabolic networks between the two groups (**Figure 2B**, central and right panel, respectively). The Spearman's correlation matrix values are provided in **Supplementary Table 2**. This metabolic network analysis revealed numerous significant associations between blood lactate concentration measured at $5\,W\,kg^{-1}$ and metabolite levels before and after the fitness test (**Figure 2C**). Most of the significant correlations observed in this analysis are metabolites central to energy homeostasis and involved in glycolysis, the TCA cycle, purine homeostasis, and oxidative stress. While the top two negative correlates with lactate at $5\,W\,kg^{-1}$ were the tyrosine catabolite 4-hydroxyphenylacetylglycine (R = -0.58, p = 5.92×10^{-5}) and the late glycolysis intermediate phosphoenolpyruvate (R = -0.58, p = 6.55×10^{-5}), the top two positive correlates with peri-test lactate production were the polyamine intermediate N-acetylspermidine (R = 0.62, p = 1.06×10^{-5}) and post-test lactate (R = 0.59, p = 3.26×10^{-5}) (**Figure 2D**).

Furthermore, these two groups appeared to be distinguishable before the initiation of the graded-exercise test, suggesting that the metabolic status prior to exercise could theoretically predict performance (**Supplementary Figure 2**).

Performance Cutoff Is Associated With Basal Oxidative Stress

Lactate production is principally fueled by the metabolic routing of glucose through Embden–Meyerhof–Parnas glycolysis and the pentose phosphate pathway (PPP). In both the Gold and Silver groups, while whole blood glucose increased during exercise due to ongoing glycogenolysis for energy generation, all measured glycolytic intermediates significantly decreased in conjunction with accumulation of end-stage products pyruvate and lactate (**Figure 3A**). Although insignificant by T-test, glucose levels trend higher both before and after exercise, along with higher post-pyruvate and post-lactate levels in the Gold group, indicating an increased glycolytic capacity in these cyclists.

The PPP serves as a primary source of NADPH, which helps manage oxidative stress by functioning as a co-factor for glutathione reductase to convert oxidized glutathione (GSSG) to its reduced state reduced glutathione (GSH). Cycling resulted in overall decreases in PPP intermediates, indicating a preference of glucose metabolism through glycolysis for energy generation (**Figure 3B**). However, PPP intermediates 6-phosphogluconolactone, 6-phosphogluconate, and erythrose 4-phosphate were observed at higher levels overall in the Silver cycling group. This trend suggests increased activation of NADPH generating pathways to cope with the oxidative stress that arises during exercise. Interestingly, oxidative stress markers including oxidized glutathione (GSSG) and methionine sulfoxide (Met-O) were both significantly higher in the Gold group both pre- and post-test, suggesting alterations to oxidative stress and signaling pathways in these individuals (**Figure 3C**). Since handling of these samples was performed *in situ* in the exercise facility, a potential role of iatrogenic intervention on these measurements cannot be ruled out. As such, increased basal levels of these metabolites can be alternatively interpreted as a larger redox reservoir for the athletes in the Gold group at baseline and post-exercise.

Lactate to Pyruvate Ratio Is Associated With Endurance Capacity

Through the activity of GADPH, NAD^+ is reduced to NADH during the conversion of glyceraldehyde 3-phosphate into 1,3-bisphosphoglycerate. Under normal homeostasis in general, and especially in cases of high glycolytic flux that is required during high intensity exercise, lactate dehydrogenase oxidizes NADH back to NAD^+ in the conversion of pyruvate to lactate, thereby maintaining necessary levels of the cofactor for the continuation of glycolysis. Cyclists in the Gold group had a higher post-test lactate-to-pyruvate ratio, which is proportional to $NADH/NAD^+$ and a marker of glycolytic capacity. Interestingly, this ratio was lower at before the test, suggesting increased basal pyruvate oxidation capacity in Gold group cyclists due to larger mitochondrial networks (**Figure 3A**, lower right corner).

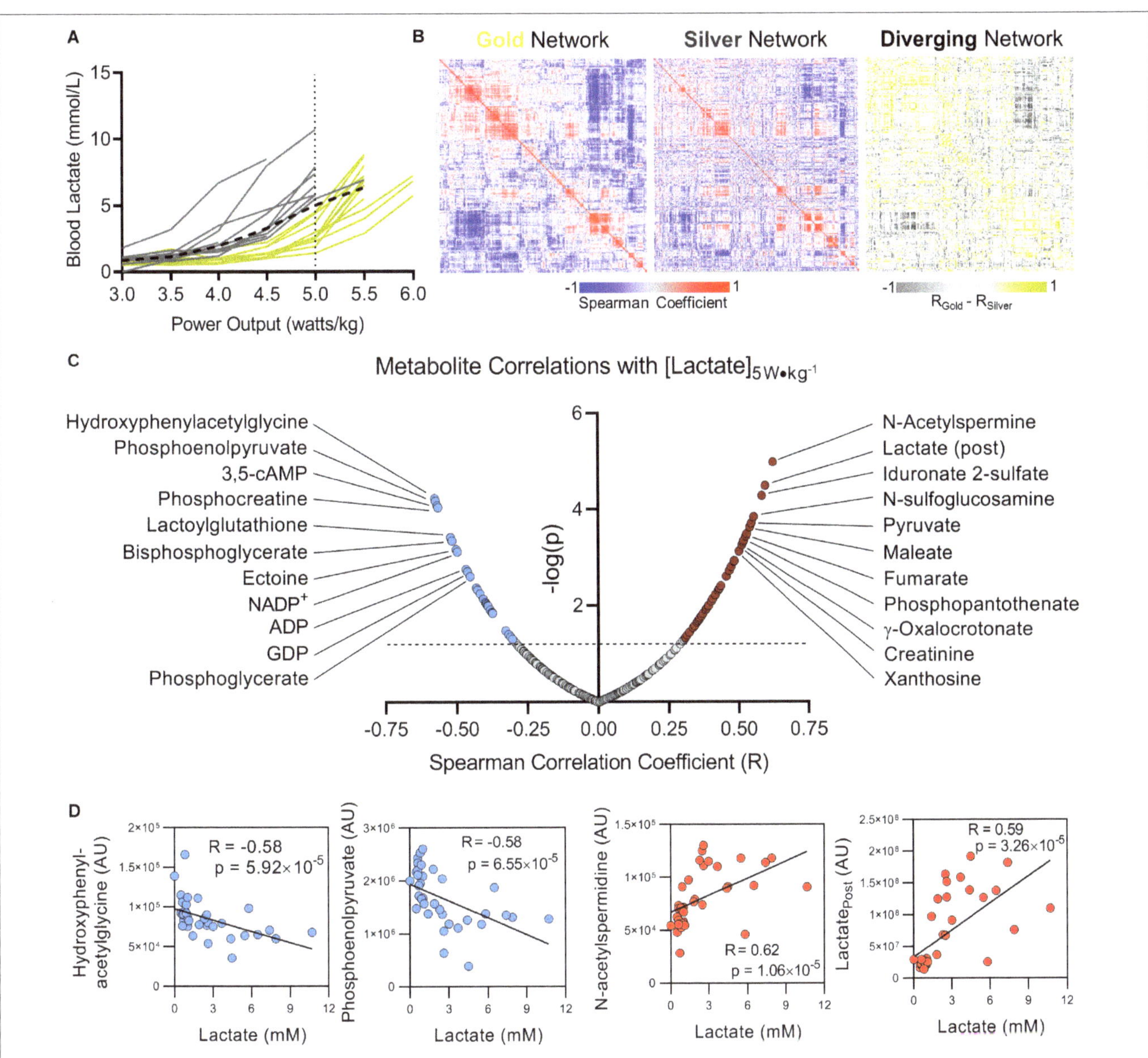

FIGURE 2 | Blood lactate levels as a function of power output can metabolically distinguish cyclists. **(A)** Using blood lactate levels (in mM) at a power output of 5 W kg^{-1} (the maximum output of which most riders achieved, referred to as the performance cutoff, or PC), cyclists were assigned to the Gold or Silver performance groups for having blood lactate levels below or above the group average, respectively (group average indicated with a black dashed line). **(B)** A Spearman Rank Correlation matrix of metabolites measured in the Gold group was Hierarchically clustered and plotted as a heat map **(left)**. A Spearman Rank Correlation matrix of metabolites measured in the Silver group was then plotted according to the same hierarchical order determined for the Gold group **(center)**. The difference of Spearman correlation coefficients between the Gold and Silver group (Gold minus Silver) was then plotted as a heat map according to the same hierarchical order determined for the Gold group. Positive Spearman differences are colored gold, and negative Spearman differences are colored silver. **(C)** The top significant positive (red) and negative (blue) metabolite correlates with blood lactate concentration measured at 5 W kg^{-1} are shown, with the Spearman Correlation Coefficient (R) on the x-axis, and -log(p) on the y-axis. **(D)** Individual plots are shown for the top 2 metabolites with negative (blue, Hydroxyphenylacetylglycine and Phosphoenolpyruvate) and positive (red, N-acetylspermidine and Lactate post) Spearman correlations with blood lactate concentration measured at 5 W kg^{-1}. Associated Spearman Correlation Coefficients and p-values are provided for each. Statistically significant correlations with log(p) > 1.3 ($p < 0.05$) are indicated above the dashed line.

Coenzyme A Synthesis Distinguishes Lactate Production Capacity

In addition to its dehydrogenation into lactate for the regeneration of NAD$^+$, pyruvate can also enter the mitochondria where it undergoes oxidative decarboxylation to form acetyl coenzyme A (acetyl-CoA). Exercise stimulated significant increases in many TCA cycle intermediates including α-ketoglutarate, succinate, fumarate, and malate (**Figure 4A**). The magnitude of increased TCA cycle intermediates in blood was larger in the Gold group, suggesting that these cyclists have a

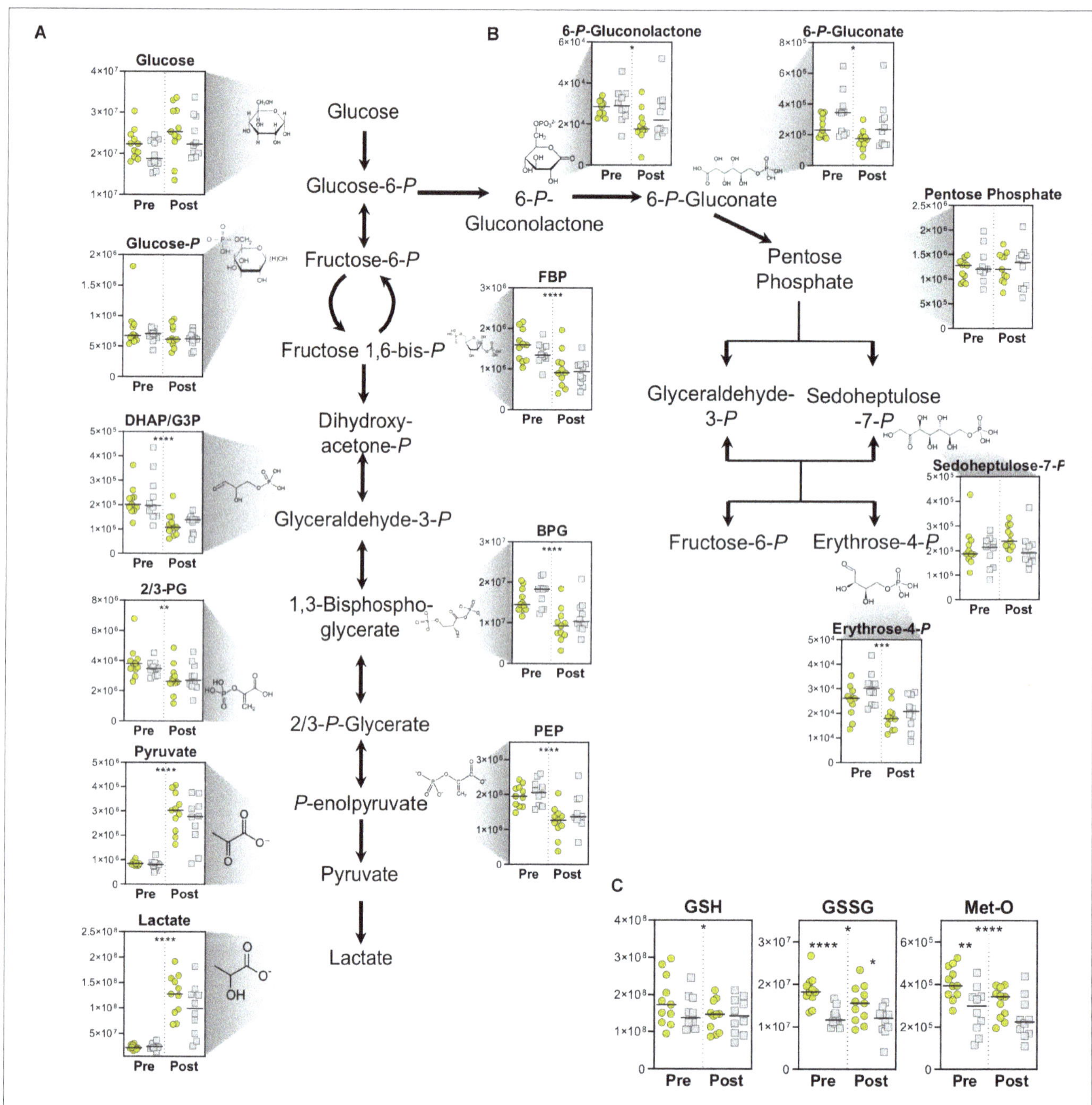

FIGURE 3 | Energy and Redox Metabolism. Metabolite abundances (y-axis, Peak Area (AU)) are plotted for **(A)** Glycolysis, **(B)** pentose phosphate pathway, and **(C)** Reduced (GSH) and oxidized (GSSG) glutathione, as well as oxidative-stress derived methionine sulfoxide (Met-O). Samples from the Gold (○) and Silver (□) groups are shown Pre and Post exercise test (divided by a dotted line). p-values from a two-tailed paired T-test of comparisons between the Pre and Post time points using combined Gold/Silver group values are shown above the dashed line. p-values from a two-tailed unpaired homoscedastic T-test comparing the Gold and Silver groups at each time point are shown on the respective side of the dashed line. *$p < 0.05$; **$p < 0.01$; ***$p < 0.001$; ****$p < 0.0001$.

larger mitochondrial capacity to generate energy during exertion, possibly due to increased mitochondrial content in tissues such as skeletal muscle. In addition, this group demonstrated significantly higher levels of the coenzyme A (CoA) precursor 4'-phospho-pantetheine at baseline, and significantly higher levels of the upstream precursor, 4'-phospho-pantothenate after exercise. These trends point to enhancements in CoA synthesis as a metabolic adaptation that occurs in association with lower PC lactate levels.

In addition to its conjugation to carbohydrate-derived acetate, CoA also plays a critical role in mobilizing fatty acids for oxidation into acetyl-CoA. During aerobic periods of exercise, catecholamine and endocrine signaling results in the lipase-mediated liberation of fatty acids from di- and triacylglycerides

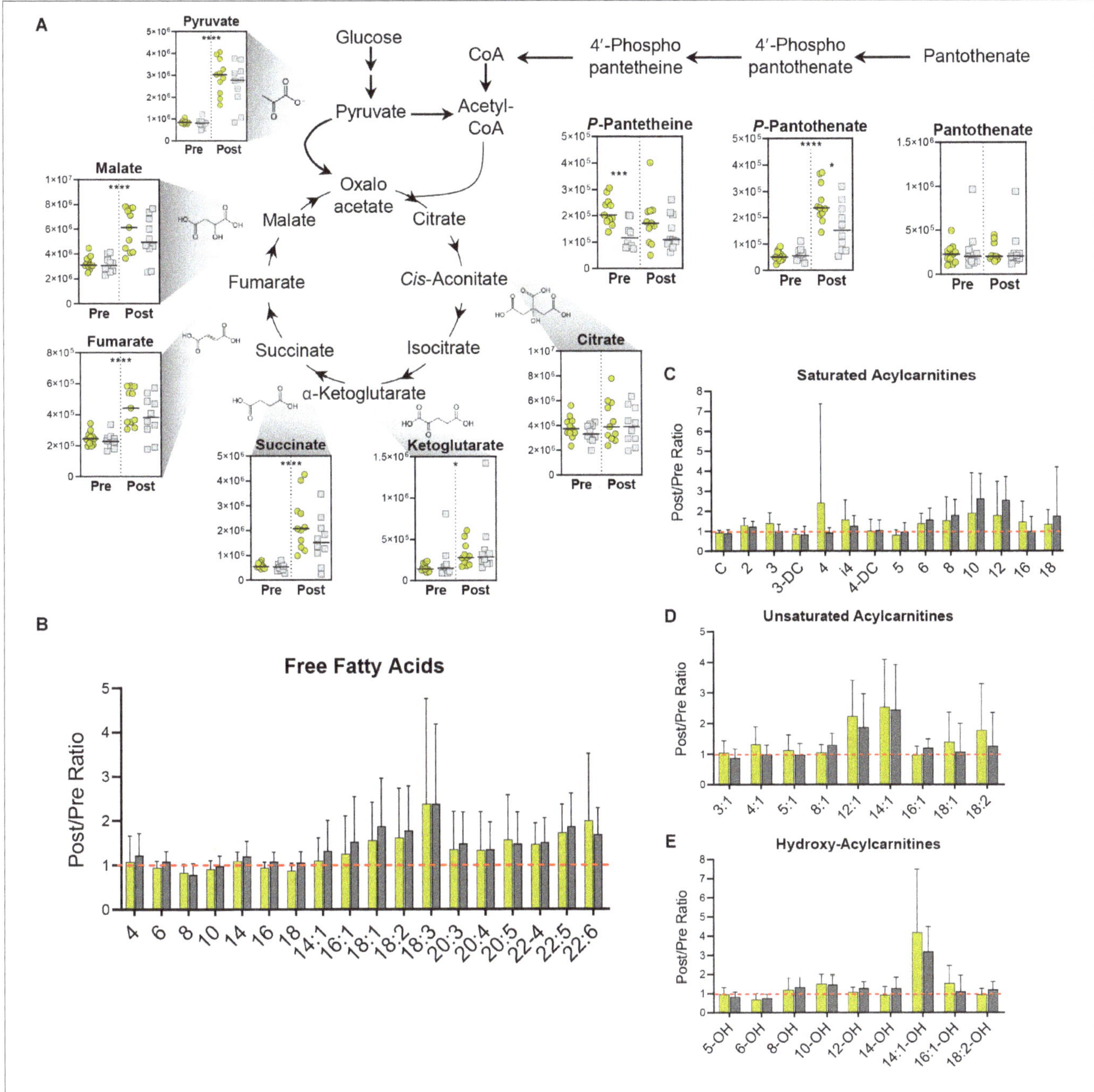

FIGURE 4 | Mitochondrial Metabolism. **(A)** TCA Cycle metabolite abundances [y-axis, Peak Area (AU)] are plotted. Samples from the Gold (○) and Silver (□) groups are shown Pre and Post exercise test (divided by a dotted line). p-values from a two-tailed paired T-test of comparisons between the Pre and Post time points using combined Gold/Silver group values are shown above the dashed line. p-values from a two-tailed unpaired homoscedastic T-test comparing the Gold and Silver groups at each time point are shown on the respective side of the dashed line. Intra-subject ratios between the Post and Pre timepoints for both Gold and Silver groups are shown for **(B)** free fatty acids, **(C)** saturated acylcarnitines, **(D)** unsaturated acylcarnitines, and **(E)** hydroxy acylcarnitines. *$p < 0.05$; **$p < 0.01$; ***$p < 0.001$; ****$p < 0.0001$.

stored in lipid droplets of muscle tissue and adipocytes. These fatty acids are chaperoned through circulation by albumin and transferred to tissues, where they are reversibly conjugated to carnitine and CoA for oxidation in the mitochondria. During the graded exercise test, all cyclists mobilized similar levels of free fatty acids (**Figure 4B**) and produced similar amounts of acylcarnitines (**Figures 4C–E**).

Strenuous Exercise Depletes Energy-Rich Nucleoside Phosphates and Leads to Accumulation of Purine Degradation Products

As expected, high energy phosphate compounds are depleted during exercise. Utilization of phosphocreatine pools maintains a steady state of creatine, and fuels the significantly increased pools of creatinine (**Figure 5B**). The conversion of creatine to creatinine is an irreversible, non-enzymatic process that is favored in low-pH and high-temperature environments (Lempert, 1959) both of which onset during exercise. As creatinine levels serve as a rough measure of muscle mass (Virgili et al., 1994) higher basal levels in the Gold group indicate increased muscle mass that is associated with lower PC lactate (**Figure 5B**). In addition to dietary sources, creatine can also be synthesized from arginine by Arginine:Glycine amidinotransferase (AGAT), producing ornithine in the process. While arginine levels decrease overall during exercise, ornithine

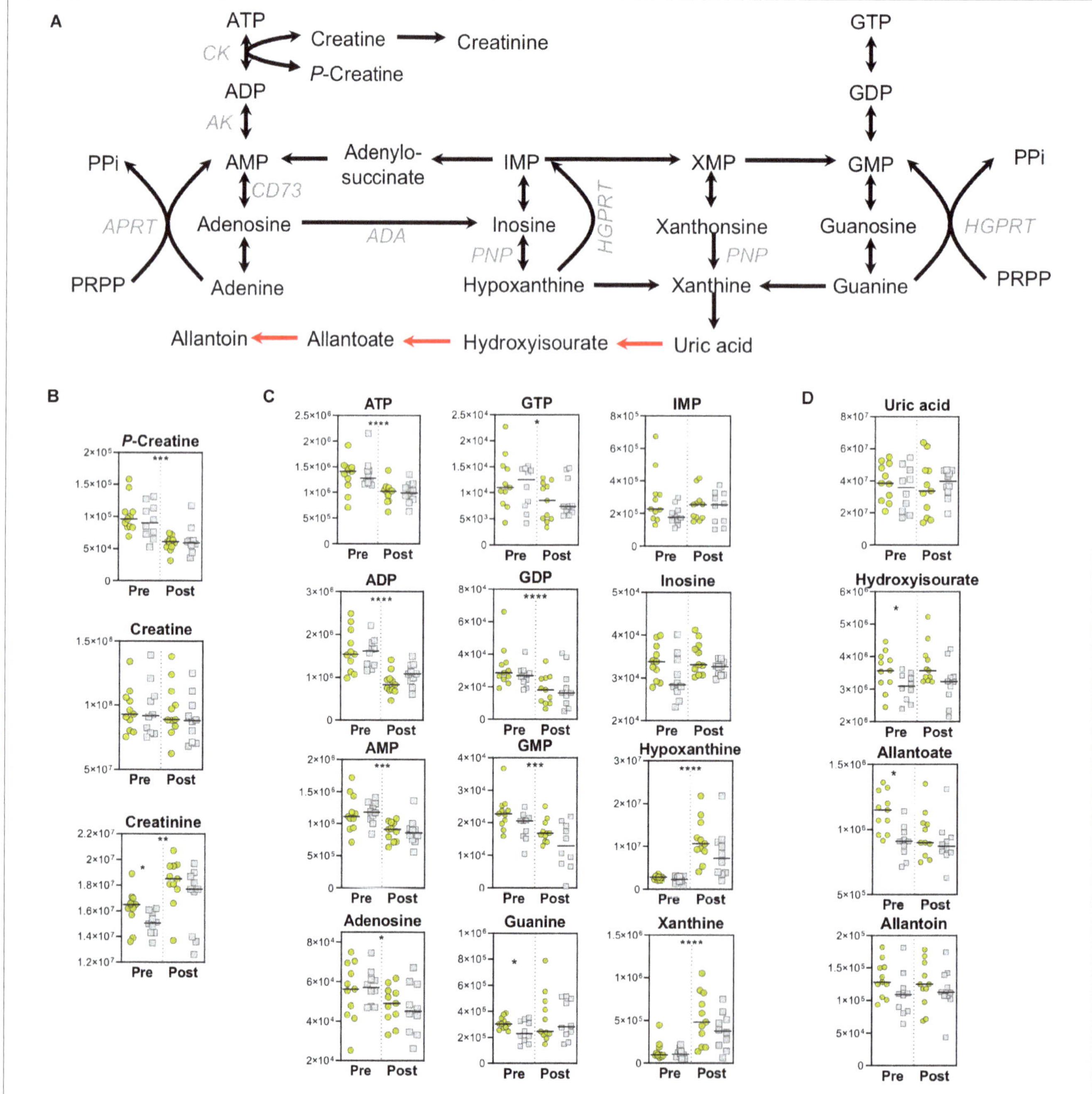

FIGURE 5 | High-Energy Phosphates and Purine Metabolism. **(A)** A pathway map for high-energy phosphate and purine homeostasis is shown. Spontaneous, non-enzymatic reactions are shown in red arrows. Metabolite abundances [y-axis, Peak Area (AU)] are plotted for **(B)** Creatine homeostasis, **(C)** Purine salvage, and **(D)** Purine catabolism. Samples from the Gold (○) and Silver (□) groups are shown Pre and Post exercise test (divided by a dotted line). p-values from a two-tailed paired T-test of comparisons between the Pre and Post time points using combined Gold/Silver group values are shown above the dashed line. p-values from a two-tailed unpaired homoscedastic T-test comparing the Gold and Silver groups at each time point are shown on the respective side of the dashed line. $^{*}p < 0.05$; $^{**}p < 0.01$; $^{***}p < 0.001$; $^{****}p < 0.0001$.

levels are higher at baseline in the Gold group, suggesting alterations to nitrogen metabolism that may serve to differentially fuel creatine pools (**Supplementary Figure 2**).

In like fashion to the depletion of phosphocreatine, high-energy pools of ATP and GTP are also utilized, resulting in increased levels of purine catabolites and intermediates of the salvage and deamination pathways, including hypoxanthine (**Figure 5C**). Despite increased levels of the purine catabolite xanthine, the end stage metabolite of enzymatic purine catabolism, uric acid, did not increase (**Figure 5D**). Although humans do not contain uricase, which converts uric acid into 5-hydroxyisourate and allantoin, this process can proceed spontaneously through the mediation of reactive oxygen species (ROS). As such, ROS-driven production of allantoin is a marker of oxidative stress in red blood cells (Kand'ár and Záková, 2008). Higher baseline levels of 5-hydroxyisourate and allantoate were observed in the Gold cyclists, suggesting higher levels of oxidative stress in these individuals that could be due to adapted metabolic pathways that result in lower PC lactate levels (**Figure 5D**).

Amino Acid Utilization, in Addition to Performance Cutoff Can Distinguish Cyclists

In addition to energy and redox metabolites involved in glycolysis, the PPP, and the TCA cycle, amino acid levels at both baseline and after exercise differ based on PC lactate. At baseline, circulating levels of phenylalanine, lysine, asparagine, serine, threonine, valine, tryptophan, and tyrosine were significantly higher in Gold group cyclists (**Figure 6A**). In addition, the post-exercise levels of most amino acids were lower only in the Gold group, especially for isoleucine, leucine, and asparagine (**Figure 6B**). Given that elevated protein synthesis during this exercise period is unlikely to explain these changes due to both time- and energetic-constraints, it is possible that these amino acids are catabolized preferentially in the Gold group for ATP production (**Figure 6C**). In support, acylcarnitines involved in amino acid catabolism including C3-carnitine and C4-carnitine appear to increase more in this group (**Figure 4C**).

Untargeted Metabolomics Identifies Distinct Tyrosine Metabolic Profiles in Relation to Performance Cutoff

In an attempt to improve and streamline the analysis of complex biological systems, models have been developed to catalog the system of metabolic reactions (Thiele et al., 2013). Using large unannotated metabolomic datasets, these models hold the potential to efficiently identify networks of metabolic pathways that may be up or downregulated in response to a biological stimulus or setting. In order to gain better insight into basal metabolic differences in cyclists with varying PC levels, we applied an unbiased network analysis using the Recon2 model to the 2,790 putative compounds identified in this study. With a specific focus on compounds that showed distinct levels in the Gold versus Silver groups at baseline (**Figure 7A**), pathway analysis revealed the significant enrichment of multiple regions of

metabolism including tyrosine, biopterin, ascorbate, nicotinate, and glycine/serine/alanine/threonine metabolism (**Figure 7B**). The most significantly enriched pathway, tyrosine metabolism, is interesting given that it is responsible for catecholamine synthesis, as well as the production of metabolites that can fuel the TCA cycle directly, such as fumarate, or indirectly, such as the ketone body, acetoacetate, and maleate. Subsequent manual interrogation of this pathway highlighted that many of the contained metabolites are elevated prior to exercise in the Gold group, and many are depleted more so in this group than in the Silver group (**Figure 7C**).

DISCUSSION

Studies on metabolic adaptations to exercise training in humans have been ongoing for decades. While early studies observed elevated activities and levels of glycolytic and electron transport chain proteins in response to endurance training (Gollnick et al., 1973; Holloszy et al., 1977) more recent work has elucidated the molecular mechanisms of such adaptations [reviewed in Hawley et al. (2018), Hackney (2019)]. Endurance training increases the capacity for lactate consumption in the mitochondria, thus increasing the lactate threshold and athletic output (Brooks, 2018). Lactate clearance capacity is enhanced through increased mitochondrial biogenesis (Little et al., 2010) which is upregulated at the transcriptional level in response to endurance training (Mahoney et al., 2005; Perry et al., 2010) and is controlled in large part by the key mediator, PGC-1α (Kim et al., 2015; Stepto et al., 2012). Increased mitochondrial networks driven by biogenesis, promotion of mitochondrial-rich Type I skeletal muscle fibers (Joyner and Coyle, 2008) and overall upregulations of muscle tissue synthesis, subsequently drive demand for substrates in addition to glucose and lactate including fatty acids and, to a lesser extent, amino acids. Lactate itself may regulate transcriptional activity (San-Millán et al., 2020) possibly through post-translation modifications of histones that alter the epigenomic landscape (Zhang et al., 2019). Future studies in sports and exercise science should focus on the extent of which these latter mechanisms apply to endurance training-mediated physiological adaptations.

The application of metabolomics to understand physiological responses to exercise has led to great advancements in the field of sports physiology over the past decade (Sakaguchi et al., 2019). These investigations have explored the effects of exercise intensity and workout duration on metabolism, as well as metabolic adaptations to chronic training in male and female populations over a range of training status (though female populations have been underrepresented in studies up to the present day). While the vast majority of studies involved amateur subjects with various levels of training, few have included elite or professional athletes assessed either at baseline (single sample) (Al-Khelaifi et al., 2018) or over a 10-day period (Knab et al., 2013). In this study, we analyzed the metabolome of whole blood in 21 international-level World Tour professional male cyclists using samples taken before and after a graded exercise test to exhaustion on a leg cycle ergometer. By normalizing

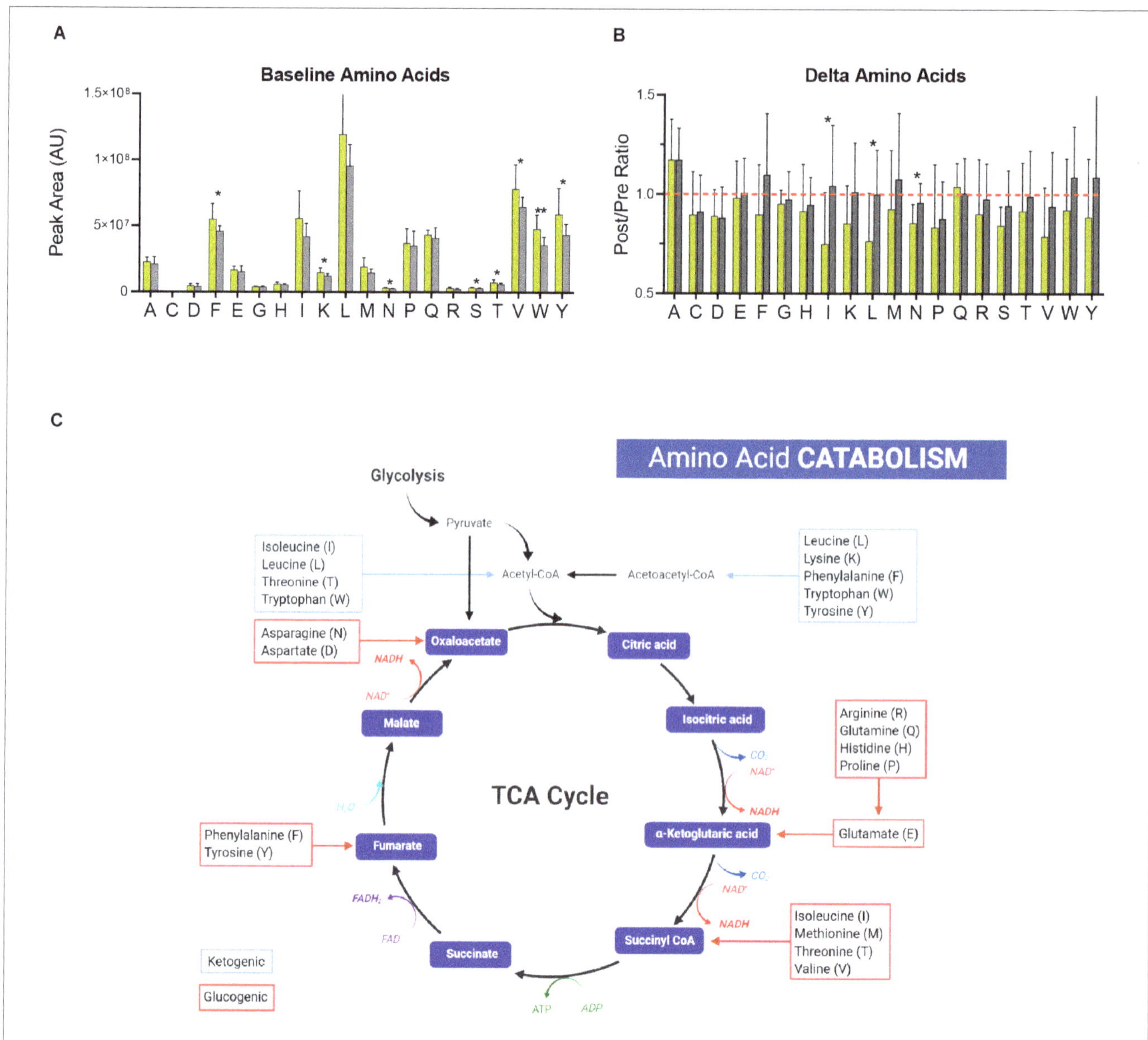

FIGURE 6 | Amino Acid Catabolism. **(A)** The baseline (pre-training) levels of amino acids (labeled with single letter code) are shown for the Gold and Silver performance groups. **(B)** Fold changes of amino acid levels (Post/Pre) for both performance groups are shown. A ratio of 1 is indicated by a dashed red line. **(C)** A generalized schematic of amino acid catabolism through the TCA cycle is shown. Glucogenic amino acids, which have the capacity to fuel gluconeogenesis, are listed in red boxes. Ketogenic amino acids, which contribute to acetoacetate and acetyl-CoA pools, are listed in blue boxes. p-values from a two-tailed Student's T-test are shown as *$p < 0.05$; **$p < 0.01$; ***$p < 0.001$; ****$p < 0.0001$.

power output to body mass, all cyclists exerted similar effort relative to their body type. Furthermore, the use of identical leg cycle ergometers in an indoor, climate-controlled environment, enabled a controlled monitoring of metabolic responses to exertion in this elite population.

Previously published metabolomics studies have analyzed other biological matrices including serum, plasma, saliva, and urine. While these matrices possess unique advantages in their own right, such as the ability to measure freely circulating metabolites between tissues as is the case with plasma, they disregard the metabolism of circulating blood cells by nature of their collection methods. As appreciation is growing with respect to blood cell adaptations to exercise (Bizjak et al., 2019; de Oliveira Ottone et al., 2019; Estruel-Amades et al., 2019; Shi et al., 2019; Uchida et al., 2019), whole blood analysis enables a more complete representation of physiology. Furthermore, this sample collection is less labor intensive than serum or plasma isolation. The introduction of TAP capillary blood collection devices have enabled field blood collection that does not necessitate traditional venipuncture blood draw (Blicharz et al., 2018) yet does provide comparable quantitative metabolomics data (Catala et al., 2018). Thus, TAP use makes possible rapid sample collection from large subject cohorts, such as participants in cycling races or marathons, thereby fostering metabolomics

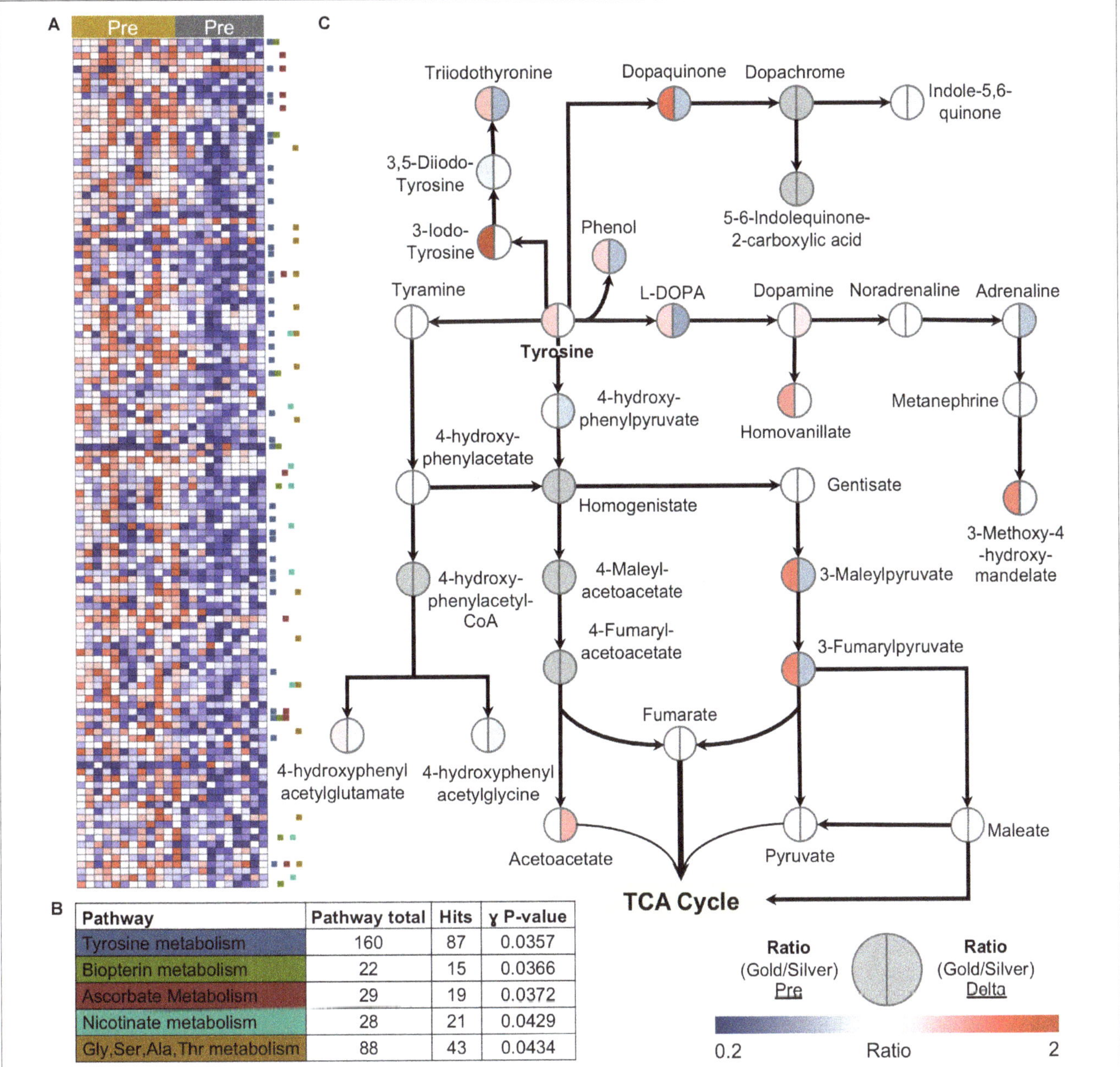

Pathway	Pathway total	Hits	γ P-value
Tyrosine metabolism	160	87	0.0357
Biopterin metabolism	22	15	0.0366
Ascorbate Metabolism	29	19	0.0372
Nicotinate metabolism	28	21	0.0429
Gly, Ser, Ala, Thr metabolism	88	43	0.0434

FIGURE 7 | Unbiased metabolomic pathway analysis reveals a link between tyrosine metabolism and lactate levels at performance cutoff. **(A)** A heatmap of features significantly different between the Gold and Silver groups at baseline is shown. **(B)** Pathway analysis using the Recon2 model identified the top five pathways enriched in this discovery dataset. **(C)** Manually validated features within Tyrosine metabolism are shown. The ratio of median values between Gold and Silver groups at baseline is shown on the left side of the circle, while the ratio of the median Delta values (calculated as the Post/Pre fold change) between Gold and Silver groups is shown on the right side of the circle. Values are plotted according to the color scale shown in the lower right.

analyses of active exercise in substantially larger populations than has been previously performed.

Use of whole blood lactate measurements taken during the graded exercise test allowed us to analyze metabolic states of individuals respective to lactate levels at the PC. One advantage of this approach is that it highlighted unique metabolic characteristics between two distinct groups classified based on the amount of lactate produced at a specific power output. This point should be acknowledged when comparing the best performing

athletes (Gold group) in this study to the rest of the cohort (Silver group) at the post-exercise time point. However, it is interesting to appreciate that this classification also revealed significant differences at baseline, even prior to the graded exercise test. As the measurements of blood lactate accumulation during a short, graded exercise test can discriminate performance in different groups of cyclists (San-Millán et al., 2009) these results indicate that metabolomic measurements at baseline and during graded exercise testing may serve to expand the predictive qualities of lactate measurement alone. In support, it is noteworthy that the World-Tour cycling season started two weeks after our testing was performed. Many cyclists in the Gold group ended up winning or reaching the podium in the first few races of the season, while riders in the Silver group did not show a great level of performance at the beginning of the year. These results highlight the use of our metabolomics platform as a powerful tool to monitor training status and predict athletic performance.

In agreement with many metabolomics studies, we observed increases in TCA cycle intermediates immediately following exercise. These increases may result from mild ischemia that occurs once oxygen demand exceeds the rate in which it can be supplied by erythrocytes (Zhang et al., 2018). Notably, cyclists with lower PC lactate levels tended to have higher levels of circulating TCA cycle metabolites following exercise. As lactate consumption rate is a marker of mitochondrial content in skeletal muscle (owing in most part to slow-twitch, Type I fibers), TCA cycle metabolite concentrations may also serve as a measure of muscle composition. Surprisingly, cyclists with lower PC lactate had significantly higher levels of the CoA precursor 4'-phospho-pantetheine at baseline, and 4'-phospho-pantothenate after exercise, suggesting upregulation of CoA biosynthesis in these athletes. Notably, these metabolites are intracellular and not readily measurable in plasma. This upregulation may represent a natural adaptation of which previously proposed supplementation strategies have attempted to achieve (Williams, 1989). However, pantothenic acid supplementation alone was shown to have no effect on cycling performance (Webster, 1998) indicating that improved bioavailability, in addition to other adaptations, may be required for the body to upregulate this system.

The preference for carbon sources to fuel the TCA cycle may also depend on lactate threshold. All cyclists exhibited similar mobilization of fatty acids and corresponding acylcarnitines for beta oxidation, of which the medium and long-chain forms tend to be released into circulation predominantly from skeletal muscle during exercise (Makrecka-Kuka et al., 2017). However, cyclists with lower PC lactate in the Gold group demonstrated slightly higher levels of short chain acylcarnitines such as propionylcarnitine and butyrlcarnitine. Propionylcarnitine in particular is released into circulation from the hepato-splanchnic bed after exercise (Xu et al., 2016) pointing to potential liver-mediated adaptations to training that support improved endurance capacity. While short- and odd-chain acylcarnitines can be formed be formed as end products of even- and odd-chain fatty acid oxidation, they are also abundantly produced during the catabolism of certain amino acids, such as branched-chain amino acids (BCAA). Indeed, Gold group cyclists had higher levels of multiple amino acids at baseline, and lower levels of BCAAs leucine, isoleucine, as well as asparagine, after exercise. Amino acids provide small stores of ATP during prolonged exercise (Phillips et al., 1993) and may be dependent on glycogen dynamics (Jackman et al., 1997), which are lactate-dependent (Brooks, 1986). Higher amino acid utilization may also reflect orthogonal mechanisms responsible for protein synthesis (Rundqvist et al., 2013), exercise recovery (Saunders et al., 2007), and overall exercise performance (Kephart et al., 2016). Indeed,

skeletal muscle transport of amino acids has been shown to increase in response to training (Roberson et al., 2018).

Amino acid catabolism may serve additional endocrine purposes as well. The upregulation of tyrosine metabolism that is associated with lower PC lactate levels was a particularly interesting finding. Tyrosine is utilized in multiple pathways including protein synthesis, anaplerosis of the TCA cycle, and as a precursor for catecholamine biosynthesis. Catecholamines, including dopamine, norepinephrine, and epinephrine, serve an important role during exercise by mediating sypathoadrenal system function. In addition to cardiovascular and respiratory responses (Zouhal et al., 2008) they modulate hepatic glucose production (Sigal et al., 1996) release of fatty acids from triglycerides (Wahrenberg et al., 1991) and fatigue (Foley and Fleshner, 2008). Indeed, the ratio of dopamine to serotonin decreases with fatigue (Bailey et al., 1993). A study of human metabolic responses to starvation, which offers insight into the physiology of extreme nutrient depletion, observed increased tryptophan consumption (the precursor to serotonin), as well as increased levels of circulating tyrosine, possibly due to decreased utilization (Steinhauser et al., 2018). Meanwhile, pharmacological modulation of serotonin receptors appears to affect time to exhaustion (Meeusen and De Meirleir, 1995). Conversely, aerobic exercise training in animal studies has been show to increase dopamine levels in various brain regions (Foley and Fleshner, 2008). Neuroendocrine control mechanisms of fatigue may have evolved to prevent overexertion (Cordeiro et al., 2017) and as such, these mechanisms should be linked to other mechanisms of endurance capacity in the body. Of note, tyrosine metabolism by monoamine oxidase enzymes is regulated by oxidant stress and glutathione levels (Maker et al., 1981) further suggesting a link between dopamine metabolism, lactate metabolism, and oxidative stress observed in the present study. In this view, it is interesting to note that dopamine (and catecholamines in general) is a direct ROS scavenger (Zhong et al., 2019) and a vasomodulator (Murphy, 2000). As such, dopamine may represent an underappreciated contributor to exercise performance.

In addition to catecholamine biosynthesis, tyrosine could be re-routed for ketogenic roles to provide energy in high endurance trained states. Indeed, levels of tyrosine metabolites were basally higher in the Gold cyclists, and consumed to a larger extent during exertion in conjunction with accumulated levels of acetoacetate. Considering that past reports on tyrosine supplementation have provided mixed results (Chinevere et al., 2002; Tumilty et al., 2011; Watson et al., 2012) future studies using isotopically-labeled tyrosine will help to disentangle its possible catabolic routes with regards to fatigue, lactate clearance capacity, and overall training status.

Glycolysis has long been known as a principal energy generating pathway in tissues due to its high rates of ATP generation under anaerobic conditions. Increased glycolytic markers have also been identified in plasma during exercise (Jacobs et al., 2014). While lactate production as a function of output has been shown to discriminate cyclists of differing training status (San-Millán et al., 2009) all cyclists reached a point of exhaustion just prior to whole blood sampling for metabolomics. As such, only a few glycolytic intermediates trended with rates of lactate accumulation, but none significantly differed between the two groups. Of note, the intra-subject lactate-to-pyruvate ratio is potentially illustrative of performance capacity. Given that this ratio represents an indirect measure of the $NAD^+/NADH$ ratio, inverse trends between the two groups at baseline and post-test time points is suggestive of specific glycolytic capacity that can drive performance during bouts of high-intensity cycling. Indeed, training strategies centered on high-altitude simulation have shown to be beneficial in promoting glycolysis and work capacity in cyclists (Terrados et al., 1988).

Unlike glycolytic comparisons, the observations of lower PPP intermediates in conjunction with significantly higher levels of oxidant-associated metabolites including oxidized glutathione (GSSG), Met-O, 5-hydroxyisourate and allantoate in the Gold group point to an increased level of basal oxidative stress. Considering that glutathione and PPP intermediates are intracellular metabolites, it is likely that the major contributors to these pools in whole blood are red blood cells (RBCs), by far the most abundant cell in both circulation and the human body (Sender et al., 2016). While exercise promotes erythropoiesis due to increased oxygen demand and need to replenish RBC populations in response to elevated hemolysis (reviewed in Mairbäurl, 2013) effects on RBC circulatory lifespan have only recently been appreciated (Bizjak et al., 2019). One of the hallmarks of RBC aging both in circulation (Lutz and Bogdanova, 2013) and in the blood bank (Nemkov et al., 2015; Yoshida et al., 2019) an environment that accelerates RBC aging (D'Alessandro et al., 2015) is the accumulation of oxidative stress markers including oxidized glutathione and allantoate. Thus, while exercise is known to cause localized inflammation and oxidative stress, systemic markers of such may indeed be indicative of altered hemostasis. In addition to gas exchange, RBC play a multitude of roles in human physiology including regulation of vascular tone, circulatory glucose, lactate, and amino acid content, and catecholamine transport (Nemkov et al., 2018). While it is likely that these characteristics are also important factors in exercise, future work is needed to elucidate the mechanisms RBC use to control physical performance and responses to exercise.

CONCLUSION

Here, we report a metabolomics-based investigation of elite, world-competing professional cyclists. Although this group represents a relatively distinct population from the perspective of physical fitness, the robustness and breadth of coverage provided by high-throughput metabolomics highlighted differences with regards to energy and amino acid metabolism, and oxidative stress. These measurements complemented and expanded upon the utility of lactate clearance capacity, which has served as a gold standard to monitor athletic training status. While this study design focused on differences in cyclist oxidative

capacity, additional studies can be designed to emphasize the contribution of alternative purely anaerobic energy systems that are needed for very high intensity cycling and sprinting. Future studies using metabolomics-based methodologies should expand steady state measurements by increasing time points, applied to additional forms of exercise that include varying intensity levels and longer endurance periods, and in additional populations with a wider range of physical fitness that are gender- and age-balanced. Despite temporal sampling, steady state measurements lack the molecular resolution necessary to determine metabolic flux. The use of stable isotope tracing with various substrates in the future will allow for a better understanding of the preference for and rates of utilization. Finally, we observed responses in the levels of intracellular metabolites, which in part arise from erythrocytes as these are substantially the most abundant cell population in whole blood. However, while whole blood analysis offers a comprehensive view of physiology, future studies should focus on understanding contribution of different blood components to overall metabolic response to exercise.

AUTHOR CONTRIBUTIONS

IS-M, TN, and AD'A designed the experiments. IS-M and TN collected samples. TN, AD'A, and DS acquired and processed the data. TN prepared the figures. TN and IS-M wrote the first draft of the manuscript. All authors commented on the final preparation of the manuscript.

SUPPLEMENTARY MATERIAL

FIGURE S1 | Partial Least Squares Discriminant Analysis (PLS-DA) to determine metabolic co-variance between the Gold and Silver cycling groups is shown, along with the metabolites that most strongly contribute to the clustering pattern based on variable importance in projection (VIP) scores.

FIGURE S2 | Urea Cycle metabolism. Samples from the Gold (○) and Silver (□) groups are shown Pre and Post exercise test (divided by a dotted line). p-values from a two-tailed paired T-test of comparisons between the Pre and Post time points using combined Gold/Silver group values are shown above the dashed line. p-values from a two-tailed unpaired homoscedastic T-test comparing the Gold and Silver groups at each time point are shown on the respective side of the dashed line. $^*p < 0.05$; $^{**}p < 0.01$; $^{***}p < 0.001$; $^{****}p < 0.0001$.

TABLE S1 | A table of the physiological and raw metabolomics data is provided. For the metabolomics data, information is included for the compound name, Compound ID in either KEGG or HMDB, the observed parent mass, median retention time, polarity of detection, and peak area top values in arbitrary units.

TABLE S2 | A table of Spearman Correlation Coefficients for the Gold Network, Silver Network, and the Diverging Network (i.e., the difference between the Gold and Silver Networks) is provided.

DATA SHEET S1 | A heat map of metabolite hierarchical clustering analysis is shown. Rows and columns were hierarchically clustered according to 1 minus the Spearman rank correlation. Metabolite values are depicted as Z-scores, with the values color coded from row minimum to maximum on a gradient from blue to red, respectively.

REFERENCES

Al-Khelaifi, F., Diboun, I., Donati, F., Botrè, F., Alsayrafi, M., Georgakopoulos, C., et al. (2018). A pilot study comparing the metabolic profiles of elite-level athletes from different sporting disciplines. *Sports Med. Open* 4:2. doi: 10.1186/s40798-017-0114-z

Bailey, S. P., Davis, J. M., and Ahlborn, E. N. (1993). Neuroendocrine and substrate responses to altered brain 5-HT activity during prolonged exercise to fatigue. *J. Appl. Physiol.* 74, 3006–3012. doi: 10.1152/jappl.1993.74.6.3006

Bergman, B. C., Wolfel, E. E., Butterfield, G. E., Lopaschuk, G. D., Casazza, G. A., Horning, M. A., et al. (1999). Active muscle and whole body lactate kinetics after endurance training in men. *J. Appl. Physiol.* 87, 1684–1696. doi: 10.1152/jappl.1999.87.5.1684

Billat, L. V. (1996). Use of blood lactate measurements for prediction of exercise performance and for control of training. Recommendations for long-distance running. *Sports Med. Auckl.* 22, 157–175. doi: 10.2165/00007256-199622030-00003

Bizjak, D. A., Tomschi, F., Bales, G., Nader, E., Romana, M., Connes, P., et al. (2019). Does endurance training improve red blood cell aging and hemorheology in moderate-trained healthy individuals?. *J. Sport Health Sci.* 10:S2095254619300225. doi: 10.1016/j.jshs.2019.02.002

Blicharz, T. M., Gong, P., Bunner, B. M., Chu, L. L., Leonard, K. M., Wakefield, J. A., et al. (2018). Microneedle-based device for the one-step painless collection of capillary blood samples. *Nat. Biomed. Eng.* 2, 151–157. doi: 10.1038/s41551-018-0194-1

Brooks, G. A. (1985). Anaerobic threshold: review of the concept and directions for future research. *Med. Sci. Sports Exerc.* 17, 22–34.

Brooks, G. A. (1986). The lactate shuttle during exercise and recovery. *Med. Sci. Sports Exerc.* 18, 360–368. doi: 10.1249/00005768-198606000-00019

Brooks, G. A. (2018). The science and translation of lactate shuttle theory. *Cell Metab.* 27, 757–785. doi: 10.1016/j.cmet.2018.03.008

Catala, A., Culp-Hill, R., Nemkov, T., and D'Alessandro, A. (2018). Quantitative metabolomics comparison of traditional blood draws and TAP capillary blood collection. *Metab. Off. J. Metab. Soc.* 14:100. doi: 10.1007/s11306-018-1395-z

Chinevere, T. D., Sawyer, R. D., Creer, A. R., Conlee, R. K., and Parcell, A. C. (2002). Effects of l -tyrosine and carbohydrate ingestion on endurance exercise performance. *J. Appl. Physiol.* 93, 1590–1597. doi: 10.1152/japplphysiol.00625.2001

Chong, J., Soufan, O., Li, C., Caraus, I., Li, S., Bourque, G., et al. (2018). MetaboAnalyst 4.0: towards more transparent and integrative metabolomics analysis. *Nucleic Acids Res.* 46, W486–W494. doi: 10.1093/nar/gky310

Cordeiro, L. M. S., Rabelo, P. C. R., Moraes, M. M., Teixeira-Coelho, F., Coimbra, C. C., Wanner, S. P., et al. (2017). Physical exercise-induced fatigue: the role of serotonergic and dopaminergic systems. *Braz. J. Med. Biol. Res. Rev. Bras. Pesqui. Medicas E Biol.* 50:e6432. doi: 10.1590/1414-431X20176432

D'Alessandro, A. (2019). *High-Throughput Metabolomics: Methods and Protocols.* New York, NY: Springer.

D'Alessandro, A., Nemkov, T., Kelher, M., West, F. B., Schwindt, R. K., Banerjee, A., et al. (2015). Routine storage of red blood cell (RBC) units in additive solution-3: a comprehensive investigation of the RBC metabolome: metabolomics of AS-3 RBCs. *Transfusion* 55, 1155–1168. doi: 10.1111/trf.12975

D'Alessandro, A., Nemkov, T., Reisz, J., Dzieciatkowska, M., Wither, M. J., and Hansen, K. C. (2017). Omics markers of the red cell storage lesion and metabolic linkage. *Blood Transfus. Trasfus.* 15, 137–144. doi: 10.2450/2017.0341-16

de Oliveira Ottone, V., Costa, K., Tossige-Gomes, R., de Matos, M., Brito-Melo, G., Magalhaes, F., et al. (2019). Late neutrophil priming following a single session of high-intensity interval exercise. *Int. J. Sports Med.* 40, 171–179. doi: 10.1055/a-0810-8533

Donovan, C. M., and Brooks, G. A. (1983). Endurance training affects lactate clearance, not lactate production. *Am. J. Physiol.* 244, E83–E92. doi: 10.1152/ajpendo.1983.244.1.E83

Dubouchaud, H., Butterfield, G. E., Wolfel, E. E., Bergman, B. C., and Brooks, G. A. (2000). Endurance training, expression, and physiology of LDH, MCT1, and MCT4 in human skeletal muscle. *Am. J. Physiol. Endocrinol. Metab.* 278, E571–E579. doi: 10.1152/ajpendo.2000.278.4.E571

Estruel-Amades, S., Ruiz-Iglesias, P., Périz, M., Franch, À, Pérez-Cano, F. J., Camps-Bossacoma, M., et al. (2019). Changes in lymphocyte composition and functionality after intensive training and exhausting exercise in rats. *Front. Physiol.* 10:1491. doi: 10.3389/fphys.2019.01491

Foley, T. E., and Fleshner, M. (2008). Neuroplasticity of dopamine circuits after exercise: implications for central fatigue. *Neuromolecular Med.* 10, 67–80. doi: 10.1007/s12017-008-8032-3

Frayn, K. N. (1983). Calculation of substrate oxidation rates *in vivo* from gaseous exchange. *J. Appl. Physiol.* 55, 628–634. doi: 10.1152/jappl.1983.55.2.628

Gollnick, P. D., Armstrong, R. B., Saltin, B., Saubert, C. W., Sembrowich, W. L., and Shepherd, R. E. (1973). Effect of training on enzyme activity and fiber composition of human skeletal muscle. *J. Appl. Physiol.* 34, 107–111. doi: 10.1152/jappl.1973.34.1.107

Hackney, A. C. (2019). "Molecular and physiological adaptations to endurance training," in *Concurrent Aerobic and Strength Training*, eds M. Schumann and B. R. Rønnestad (Cham: Springer International Publishing), 19–34. doi: 10.1007/978-3-319-75547-2_3

Hawley, J. A., Lundby, C., Cotter, J. D., and Burke, L. M. (2018). Maximizing cellular adaptation to endurance exercise in skeletal muscle. *Cell Metab.* 27, 962–976. doi: 10.1016/j.cmet.2018.04.014

Holloszy, J. O., Rennie, M. J., Hickson, R. C., Conlee, R. K., and Hagberg, J. M. (1977). Physiological consequences of the biochemical adaptations to endurance exercise. *Ann. N. Y. Acad. Sci.* 301, 440–450. doi: 10.1111/j.1749-6632.1977.tb38220.x

Jackman, M. L., Gibala, M. J., Hultman, E., and Graham, T. E. (1997). Nutritional status affects branched-chain oxoacid dehydrogenase activity during exercise in humans. *Am. J. Physiol.* 272, E233–E238. doi: 10.1152/ajpendo.1997.272.2.E233

Jacobs, D. M., Hodgson, A. B., Randell, R. K., Mahabir-Jagessar-T, K., Garczarek, U., Jeukendrup, A. E., et al. (2014). Metabolic response to decaffeinated green tea extract during rest and moderate-intensity exercise. *J. Agric. Food Chem.* 62, 9936–9943. doi: 10.1021/jf502764r

Jacobs, I. (1986). Blood lactate. Implications for training and sports performance. *Sports Med. Auckl.* 3, 10–25. doi: 10.2165/00007256-198603010-00003

Jensen, K., Jørgensen, S., and Johansen, L. (2002). A metabolic cart for measurement of oxygen uptake during human exercise using inspiratory flow rate. *Eur. J. Appl. Physiol.* 87, 202–206. doi: 10.1007/s00421-002-0616-2

Joyner, M. J., and Coyle, E. F. (2008). Endurance exercise performance: the physiology of champions. *J. Physiol.* 586, 35–44. doi: 10.1113/jphysiol.2007.143834

Kand'ár, R., and Záková, P. (2008). Allantoin as a marker of oxidative stress in human erythrocytes. *Clin. Chem. Lab. Med.* 46, 1270–1274. doi: 10.1515/CCLM.2008.244

Karamanou, M., and Androutsos, G. (2013). Antoine-Laurent de Lavoisier (1743-1794) and the birth of respiratory physiology. *Thorax* 68, 978–979. doi: 10.1136/thoraxjnl-2013-203840

Kephart, W. C., Wachs, T. D., Thompson, R. M., Brooks Mobley, C., Fox, C. D., McDonald, J. R., et al. (2016). Ten weeks of branched-chain amino acid supplementation improves select performance and immunological variables in trained cyclists. *Amino Acids* 48, 779–789. doi: 10.1007/s00726-015-2125-8

Kim, S. H., Koh, J. H., Higashida, K., Jung, S. R., Holloszy, J. O., and Han, D.-H. (2015). PGC-1α mediates a rapid, exercise-induced downregulation of glycogenolysis in rat skeletal muscle. *J. Physiol.* 593, 635–643. doi: 10.1113/jphysiol.2014.283820

Knab, A. M., Nieman, D. C., Gillitt, N. D., Shanely, R. A., Cialdella-Kam, L., Henson, D. A., et al. (2013). Effects of a flavonoid-rich juice on inflammation, oxidative stress, and immunity in elite swimmers: a metabolomics-based approach. *Int. J. Sport Nutr. Exerc. Metab.* 23, 150–160. doi: 10.1123/ijsnem.23.2.150

Lempert, C. (1959). The chemistry of the glycocyamidines. *Chem. Rev.* 59, 667–736. doi: 10.1021/cr50028a005

Little, J. P., Safdar, A., Wilkin, G. P., Tarnopolsky, M. A., and Gibala, M. J. (2010). A practical model of low-volume high-intensity interval training induces mitochondrial biogenesis in human skeletal muscle: potential mechanisms. *J. Physiol.* 588, 1011–1022. doi: 10.1113/jphysiol.2009.181743

Lutz, H. U., and Bogdanova, A. (2013). Mechanisms tagging senescent red blood cells for clearance in healthy humans. *Front. Physiol.* 4:387. doi: 10.3389/fphys.2013.00387

Mahoney, D. J., Parise, G., Melov, S., Safdar, A., and Tarnopolsky, M. A. (2005). Analysis of global mRNA expression in human skeletal muscle during recovery from endurance exercise. *FASEB J. Off. Publ. Fed. Am. Soc. Exp. Biol.* 19, 1498–1500. doi: 10.1096/fj.04-3149fje

Mairbäurl, H. (2013). Red blood cells in sports: effects of exercise and training on oxygen supply by red blood cells. *Front. Physiol.* 4:332. doi: 10.3389/fphys.2013.00332

Maker, H. S., Weiss, C., Silides, D. J., and Cohen, G. (1981). Coupling of dopamine oxidation (Monoamine Oxidase Activity) to glutathione oxidation via the generation of hydrogen peroxide in rat brain homogenates. *J. Neurochem.* 36, 589–593. doi: 10.1111/j.1471-4159.1981.tb01631.x

Makrecka-Kuka, M., Sevostjanovs, E., Vilks, K., Volska, K., Antone, U., Kuka, J., et al. (2017). Plasma acylcarnitine concentrations reflect the acylcarnitine profile in cardiac tissues. *Sci. Rep.* 7:17528. doi: 10.1038/s41598-017-17797-x

McDermott, J. C., and Bonen, A. (1993). Endurance training increases skeletal muscle lactate transport. *Acta Physiol. Scand.* 147, 323–327. doi: 10.1111/j.1748-1716.1993.tb09505.x

Meeusen, R., and De Meirleir, K. (1995). Exercise and brain neurotransmission. *Sports Med. Auckl.* 20, 160–188. doi: 10.2165/00007256-199520030-00004

Melamud, E., Vastag, L., and Rabinowitz, J. D. (2010). Metabolomic analysis and visualization engine for LC-MS data. *Anal. Chem.* 82, 9818–9826. doi: 10.1021/ac1021166

Murphy, M. B. (2000). Dopamine: a role in the pathogenesis and treatment of hypertension. *J. Hum. Hypertens.* 14(Suppl. 1), S47–S50. doi: 10.1038/sj.jhh.1000987

Nemkov, T., Hansen, K. C., and D'Alessandro, A. (2017). A three-minute method for high-throughput quantitative metabolomics and quantitative tracing experiments of central carbon and nitrogen pathways. *Rapid Commun. Mass Spectrom.* 31, 663–673. doi: 10.1002/rcm.7834

Nemkov, T., Hansen, K. C., Dumont, L. J., and D'Alessandro, A. (2015). Metabolomics in transfusion medicine. *Transfusion* 56, 980–993. doi: 10.1111/trf.13442

Nemkov, T., Reisz, J. A., Xia, Y., Zimring, J. C., and D'Alessandro, A. (2018). Red blood cells as an organ? How deep omics characterization of the most abundant cell in the human body highlights other systemic metabolic functions beyond oxygen transport. *Expert Rev. Proteomics* 15, 855–864. doi: 10.1080/14789450.2018.1531710

Perry, C. G. R., Lally, J., Holloway, G. P., Heigenhauser, G. J. F., Bonen, A., and Spriet, L. L. (2010). Repeated transient mRNA bursts precede increases in transcriptional and mitochondrial proteins during training in human skeletal muscle. *J. Physiol.* 588, 4795–4810. doi: 10.1113/jphysiol.2010.199448

Phillips, S. M., Atkinson, S. A., Tarnopolsky, M. A., and MacDougall, J. D. (1993). Gender differences in leucine kinetics and nitrogen balance in endurance athletes. *J. Appl. Physiol.* 75, 2134–2141. doi: 10.1152/jappl.1993.75.5.2134

Reisz, J. A., Zheng, C., D'Alessandro, A., and Nemkov, T. (2019). Untargeted and semi-targeted lipid analysis of biological samples using mass spectrometry-based metabolomics. *Methods Mol. Biol.* 1978, 121–135. doi: 10.1007/978-1-4939-9236-2_8

Roberson, P. A., Haun, C. T., Mobley, C. B., Romero, M. A., Mumford, P. W., Martin, J. S., et al. (2018). Skeletal muscle amino acid transporter and BCAT2 expression prior to and following interval running or resistance exercise in mode-specific trained males. *Amino Acids* 50, 961–965. doi: 10.1007/s00726-018-2570-2

Rundqvist, H. C., Lilja, M. R., Rooyackers, O., Odrzywol, K., Murray, J. T., Esbjörnsson, M., et al. (2013). Nutrient ingestion increased mTOR signaling, but not hVps34 activity in human skeletal muscle after sprint exercise. *Physiol. Rep.* 1:e00076. doi: 10.1002/phy2.76

Sakaguchi, C. A., Nieman, D. C., Signini, E. F., Abreu, R. M., and Catai, A. M. (2019). Metabolomics-based studies assessing exercise-induced alterations of the human metabolome: a systematic review. *Metabolites* 9:E164. doi: 10.3390/metabo9080164

San-Millán, I., and Brooks, G. A. (2018). Assessment of metabolic flexibility by means of measuring blood lactate, fat, and carbohydrate oxidation responses to exercise in professional endurance athletes and less-fit individuals. *Sports Med. Auckl.* 48, 467–479. doi: 10.1007/s40279-017-0751-x

San-Millán, I., González-Haro, C., and Sagasti, M. (2009). Physiological differences between road cyclists of different categories. a new approach.: 733. *Med. Sci. Sports Exerc.* 41, 64–65. doi: 10.1249/01.mss.0000353467.61975.ae

San-Millán, I., Julian, C. G., Matarazzo, C., Martinez, J., and Brooks, G. A. (2020). Is lactate an oncometabolite? Evidence supporting a role for lactate in the regulation of transcriptional activity of cancer-related genes in mcf7 breast cancer cells. *Front. Oncol.* 9:1536. doi: 10.3389/fonc.2019.01536

Saunders, M. J., Luden, N. D., and Herrick, J. E. (2007). Consumption of an oral carbohydrate-protein gel improves cycling endurance and prevents postexercise muscle damage. *J. Strength Cond. Res.* 21, 678–684. doi: 10.1519/R-20506.1

Schrimpe-Rutledge, A. C., Codreanu, S. G., Sherrod, S. D., and McLean, J. A. (2016). Untargeted metabolomics strategies—challenges and emerging directions. *J. Am. Soc. Mass Spectrom.* 27, 1897–1905. doi: 10.1007/s13361-016-1469-y

Sender, R., Fuchs, S., and Milo, R. (2016). Revised estimates for the number of human and bacteria cells in the body. *PLoS Biol.* 14:e1002533. doi: 10.1371/journal.pbio.1002533

Shephard, R. J. (1984). Tests of maximum oxygen intake a critical review. *Sports Med.* 1, 99–124. doi: 10.2165/00007256-198401020-00002

Shi, Y., Shi, H., Nieman, D. C., Hu, Q., Yang, L., Liu, T., et al. (2019). Lactic acid accumulation during exhaustive exercise impairs release of neutrophil extracellular traps in mice. *Front. Physiol.* 10:709. doi: 10.3389/fphys.2019.00709

Sigal, R. J., Fisher, S., Halter, J. B., Vranic, M., and Marliss, E. B. (1996). The roles of catecholamines in glucoregulation in intense exercise as defined by the islet cell clamp technique. *Diabetes Metab. Res. Rev.* 45, 148–156. doi: 10.2337/diab.45.2.148

Steinhauser, M. L., Olenchock, B. A., O'Keefe, J., Lun, M., Pierce, K. A., Lee, H., et al. (2018). The circulating metabolome of human starvation. *JCI Insight* 3:e121434. doi: 10.1172/jci.insight.121434

Stepto, N. K., Benziane, B., Wadley, G. D., Chibalin, A. V., Canny, B. J., Eynon, N., et al. (2012). Short-term intensified cycle training alters acute and chronic responses of PGC1α and Cytochrome C oxidase IV to exercise in human skeletal muscle. *PLoS One* 7:e53080. doi: 10.1371/journal.pone.0053080

Terrados, N., Melichna, J., Sylvn, C., Jansson, E., and Kaijser, L. (1988). Effects of training at simulated altitude on performance and muscle metabolic capacity in competitive road cyclists. *Eur. J. Appl. Physiol.* 57, 203–209. doi: 10.1007/BF00640664

Thiele, I., Swainston, N., Fleming, R. M. T., Hoppe, A., Sahoo, S., Aurich, M. K., et al. (2013). A community-driven global reconstruction of human metabolism. *Nat. Biotechnol.* 31, 419–425. doi: 10.1038/nbt.2488

Tumilty, L., Davison, G., Beckmann, M., and Thatcher, R. (2011). Oral tyrosine supplementation improves exercise capacity in the heat. *Eur. J. Appl. Physiol.* 111, 2941–2950. doi: 10.1007/s00421-011-1921-4

Uchida, M., Horii, N., Hasegawa, N., Fujie, S., Oyanagi, E., Yano, H., et al. (2019). Gene expression profiles for macrophage in tissues in response to different exercise training protocols in senescence mice. *Front. Sports Act. Living* 1:50. doi: 10.3389/fspor.2019.00050

Underwood, E. A. (1944). Lavoisier and the history of respiration. *Proc. R. Soc. Med.* 37, 247–262. doi: 10.1177/003591574403700603

Virgili, F., Maiani, G., Zahoor, Z. H., Ciarapica, D., Raguzzini, A., and Ferro-Luzzi, A. (1994). Relationship between fat-free mass and urinary excretion of creatinine and 3-methylhistidine in adult humans. *J. Appl. Physiol.* 76, 1946–1950. doi: 10.1152/jappl.1994.76.5.1946

Wahrenberg, H., Bolinder, J., and Arner, P. (1991). Adrenergic regulation of lipolysis in human fat cells during exercise. *Eur. J. Clin. Invest.* 21, 534–541. doi: 10.1111/j.1365-2362.1991.tb01406.x

Watson, P., Enever, S., Page, A., Stockwell, J., and Maughan, R. J. (2012). Tyrosine supplementation does not influence the capacity to perform prolonged exercise in a warm environment. *Int. J. Sport Nutr. Exerc. Metab.* 22, 363–373. doi: 10.1123/ijsnem.22.5.363

Webster, M. J. (1998). Physiological and performance responses to supplementation with thiamin and pantothenic acid derivatives. *Eur. J. Appl. Physiol.* 77, 486–491. doi: 10.1007/s004210050364

Williams, M. H. (1989). Vitamin supplementation and athletic performance. *Int. J. Vitam. Nutr. Res. Suppl. Int. Z. Vitam. Ernahrungsforschung Suppl.* 30, 163–191.

Xu, G., Hansen, J. S., Zhao, X. J., Chen, S., Hoene, M., Wang, X. L., et al. (2016). Liver and muscle contribute differently to the plasma acylcarnitine pool during fasting and exercise in humans. *J. Clin. Endocrinol. Metab.* 101, 5044–5052. doi: 10.1210/jc.2016-1859

Yoshida, T., Prudent, M., and D'Alessandro, A. (2019). Red blood cell storage lesion: causes and potential clinical consequences. *Blood Transfus.* 17, 27–52. doi: 10.2450/2019.0217-18

Zhang, D., Tang, Z., Huang, H., Zhou, G., Cui, C., Weng, Y., et al. (2019). Metabolic regulation of gene expression by histone lactylation. *Nature* 574, 575–580. doi: 10.1038/s41586-019-1678-1

Zhang, J., Wang, Y. T., Miller, J. H., Day, M. M., Munger, J. C., and Brookes, P. S. (2018). Accumulation of succinate in cardiac ischemia primarily occurs via canonical krebs cycle activity. *Cell Rep.* 23, 2617–2628. doi: 10.1016/j.celrep. 2018.04.104

Zhong, G., Yang, X., Jiang, X., Kumar, A., Long, H., Xie, J., et al. (2019). Dopamine-melanin nanoparticles scavenge reactive oxygen and nitrogen species and activate autophagy for osteoarthritis therapy. *Nanoscale* 11, 11605–11616. doi: 10.1039/C9NR03060C

Zouhal, H., Jacob, C., Delamarche, P., and Gratas-Delamarche, A. (2008). Catecholamines and the effects of exercise, training and gender. *Sports Med.* 38, 401–423. doi: 10.2165/00007256-200838050-00004

13

Micro-Raman Spectroscopy Analysis of Optically Trapped Erythrocytes in Jaundice

Sanu Susan Jacob[1], Aseefhali Bankapur[2], Surekha Barkur[2], Mahendra Acharya[2], Santhosh Chidangil[2], Pragna Rao[3], Asha Kamath[4], R. Vani Lakshmi[4], Prathap M. Baby[5] and Raghavendra K. Rao[1]*

[1] Department of Physiology, Kasturba Medical College-Manipal, Manipal Academy of Higher Education, Manipal, India, [2] Department of Atomic and Molecular Physics, Centre of Excellence for Biophotonics, Manipal Academy of Higher Education, Manipal, India, [3] Department of Biochemistry, Kasturba Medical College-Manipal, Manipal Academy of Higher Education, Manipal, India, [4] Department of Data Science, Prasanna School of Public Health, Manipal Academy of Higher Education, Manipal, India, [5] Department of Physiology, Melaka Manipal Medical College, Manipal Academy of Higher Education, Manipal, India

***Correspondence:**
Sanu Susan Jacob
sanu.susan@manipal.edu

Derangements in bilirubin metabolism and/or dysfunctions in the hepato-biliary system lead to the unhealthy buildup of bilirubin in blood, resulting in jaundice. During the course of this disorder, circulating red cells are invariably subjected to toxic effects of serum bilirubin and an array of inflammatory compounds. This study aimed to investigate the vibrational spectroscopy of live red cells in jaundice using micro-Raman spectroscopy combined with optical-trap. Red cells from blood samples of healthy volunteers and patients with jaundice were optically immobilized and micro-Raman probed using a 785 nm diode laser. Raman signatures from red cells in jaundice exhibited significant variations from the normal and the spectral-markers were obtained from multivariate analytical methods. This research gives insightful views on how different pathologies can act as "stress-milieus" for red cells in circulation, possibly impeding their normal functions and also exasperating anemia. Raman spectroscopy, an emerging bio-analytical technique, is sensitive in detecting molecular-conformations *in situ*, at cellular-levels and in real-time. This study could pave way in understanding fundamental red cell behavior in different diseases by analyzing Raman markers.

Keywords: red blood cells, jaundice, Raman spectroscopy, bilirubin, hemoglobin

INTRODUCTION

Bilirubin is a yellow pigment produced in the body, primarily as a consequent-by-product of red blood cell (RBC) catabolism (Adamson and Longo, 2012). Senile RBCs at the end of their 120 days' sojourn in circulation, are physiologically detected and hemolysed by the reticulo-endothelial system. This process occurs chiefly in the spleen, causing the release of hemoglobin (Hb), the principle RBC component (Korolnek and Hamza, 2015). When the tetra-pyrrolic moiety of hemoglobin, called heme, breaks down, it generates plasma-insoluble unconjugated bilirubin (UCB) (Schiff et al., 2011). To render it plasma-soluble, UCB is reversibly-bound with plasma albumin and is transported to the liver to undergo a process called conjugation. In the liver, UCB is conjugated with glucuronic acid (Levitt and Levitt, 2014). Plasma-soluble conjugated bilirubin (CB) is then released through the bile into the intestine, from where it is excreted and eliminated from the body through the bowels and bladder in the form of stercobilinogen (Erlinger et al., 2014).

In any circumstances of increased hemolysis, hepatic dysfunction or even biliary obstruction, there ensues a build-up of bilirubin in blood, resulting in a condition termed hyperbilirubinemia, which is clinical jaundice (Fargo et al., 2017). Normal serum bilirubin levels are not known to pose any implications for general health, rather, is found to be valuable for possessing antioxidant properties in the vasculature (Park, 2016). However, uncontrolled and/or untreated hyperbilirubinemia is toxic and can cause serious deleterious effects that require immediate and appropriate life-saving treatments (Hansen, 2001). Currently, there exist numerous reports that describe the lethal effects of jaundice, but most studies are directed toward its effects on the nervous system (Shapiro, 2003; Shi et al., 2019). There exist fewer reports of bilirubin toxicity on blood components. It has been recognized that bilirubin liberated from RBC destruction, if present in larger than physiological quantities in circulation, can trigger toxic effects on other circulating RBCs. This can result in augmenting the production of more bilirubin, the whole course transitioning into a potentially vicious cycle (Lang et al., 2015). In addition to bilirubin, levels of a multitude of inflammatory markers and acute phase reactants escalate in the serum, such as endotoxins, tumor necrosis factor-α, IL-6, and C-reactive protein (Padillo et al., 2001). These systemic inflammatory cytokines are themselves known to reduce the life-span of RBCs, through a programmed process called eryptosis (Pretorius, 2018).

The focus of this study was to investigate the chemometrics of the Raman spectra of RBCs from blood samples of patients with jaundice, keeping the cells as alive as possible. This required the utility of a tool that demands only the bare minimal sample-preparation procedures, reagents or fixatives, which could themselves be potential sources of cell-injury or cell-modification (Sato et al., 2018). Keeping these viewpoints into consideration, we made use of a home-built micro-Raman spectroscopy system coupled with laser-tweezers to investigate the biochemical characteristics of RBCs (Bankapur et al., 2010). Raman spectroscopy has been gaining popularity in its application in biomedicine as this single-cell technique is tremendously sensitive to very subtle molecular changes within the cell, records vibrations of molecular bonds in its original state instantaneously and does not employ the usage of any chemical markers (Kumamoto et al., 2018). This research aimed to comprehend the molecular changes of RBCs, in a "jaundice" environment.

MATERIALS AND METHODS

Ethics Statement

This study was approved (IEC 02/2002) by the Institutional Ethics Committee of Kasturba Medical College, Manipal

Academy of Higher Education, Manipal and all experimental procedures conformed with the Ethical Committee guidelines. Written informed consent was obtained from all the volunteers. Blood samples of confirmed hyperbilirubinemic patients were collected from the laboratory of Clinical Biochemistry. The requirement for informed consent was waived concerning the acquisition of patients' samples.

Blood Sample Collection and RBC Preparation

Four milliliter anticoagulated blood samples were collected in EDTA-containing BD Vacutainers from healthy volunteers ($n = 10$). Blood samples of hyperbilirubinemic patients ($n = 28$) in EDTA were collected from the laboratory of Clinical Biochemistry. Whole anticoagulated blood sample was centrifuged (Labinet Spectrafuge 7M) at 5000 rpm for 5 minutes. Following centrifugation, the plasma and buffy-coat layers were aspirated off using a micropipette. One microliter hematocrit (Hct) was pipetted out and suspended in 2 milliliter phosphate-buffered saline (PBS)-filled Eppendorf tube to obtain a suspension of live RBCs. PBS was procured from Sigma Aldrich, India.

One microliter of RBC suspension in PBS was pipetted out from the Eppendorf tube onto the custom-made sample chamber. A total of 379 raw spectra were recorded from 379 individual RBCs, of which 147 spectra were recorded from the blood samples of healthy volunteers and 232 from that of patients with jaundice.

Experimental Set-Up

The illustration of the single-beam micro-Raman spectrometer coupled with optical tweezers has been shown in **Figure 1**. The system comprised of an inverted microscope (Nikon Eclipse Ti-U, Japan) coupled with a 785 nm wavelength-emitting diode laser (Starbright Diode Laser, Denmark) and a spectrograph (Horiba Jobin Yvon iHR320). The expanded laser-beam overfills the back aperture of a 100X, 1.3NA microscope objective (Nikon, Plan Fluor) to create a sharp focal-spot at the sample plane. Single RBCs suspended in PBS were optically-trapped under this focal-spot and simultaneously Raman-excited. The scattered light from the individual RBCs was collected from the very same objective and fed into the spectrograph using f-matching optics. The spectrograph was equipped with a 1200 groves/mm grating and a liquid nitrogen-cooled charged couple device (Symphony CCD-1024x256-OPEN-1LS) detector, with 1024×256 pixels. An optical edge-filter (Razor edge LP02-785RU-25, Semrock, United States) was placed just in front of the spectrograph for the removal of Rayleigh scattered radiations. The comprehensive description of the experimental set-up has been published in earlier studies (Bankapur et al., 2010, 2014).

Data Acquisition

Utilizing the above setup, RBCs were optically immobilized and probed using a 785 nm diode laser with an excitation-power of 10mW. RBCs were positioned at the focus

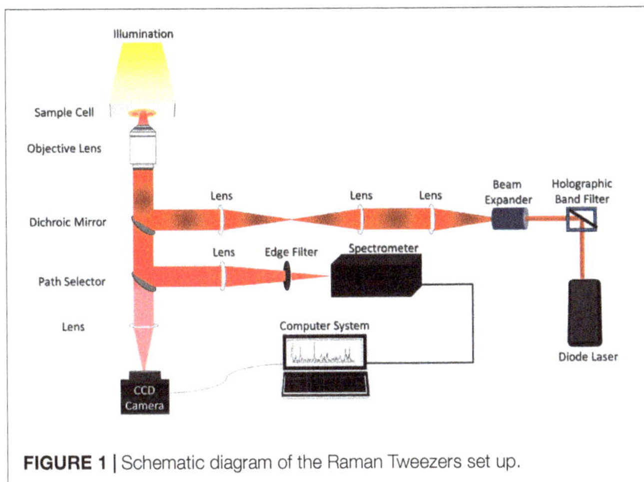

FIGURE 1 | Schematic diagram of the Raman Tweezers set up.

of the laser beam by adjusting the manual stage. Five accumulations were recorded per RBC with the exposure/acquisition time of 60 seconds. Raman spectra of RBCs were procured within the spectral range of $\sim$600 to 1,800 cm^{-1}. This spin-sensitive-region is observed to deliver maximum information on the "oxygenation-state" of Hb (De Luca et al., 2008).

Data Pre-processing

Origin (OriginLab Corp., Northampton, MA, United States), MATLAB (MATLAB® 7.0) and GRAMS (Grams/AI, PLS Plus IQ) software were employed for the pre-processing of the accrued RBC spectra. 379 raw spectra that included those from the healthy, as well as the jaundice groups, were pre-processed.

The pre-processing steps involved the exclusion of spurious noises/cosmic-spikes, spectral smoothening, baseline-correction, and vector-normalization. The data-points that corresponded to cosmic-spikes were deleted using "Origin." Using GRAMS, the raw spectra were smoothened out by second-order polynomial Savitzky-Golay moving-average-technique. Spectral normalization was accomplished by 2-norm standard-vector normalization (Barkur et al., 2015) using GRAMS. Multiple spectral baseline-correction was effected by employing asymmetric least-squares (AsLS) fitting approach proposed by Eilers (2003).

Statistical Analysis

Multivariate statistical analyses were carried out using GRAMS-AI and CRAN R 3.6.1 (©2019 The R Foundation for Statistical Computing) packages. The intensity of bands in 600–1800 cm^{-1} range of the Raman spectra was analyzed.

The goal was to choose seamless statistical tools for the classification of acquired data, based on variations in spectral features. In this study, for effective categorization, we had performed partial least squares based discriminant analysis (PLS-DA), a supervised diagnostic model, to sharpen variations between the generated principal components (PCs) using Grams IQ spectroscopy software. The PCs were rotated such that a maximum separation among classes was obtained. Additionally,

PLS-DA helped to identify the Raman signatures responsible for classification. This was followed by a non-parametric variant of Multivariate ANalysis Of VAriance (MANOVA) to re-affirm the statistically significant difference in the mean value of the measurements captured at different wavelengths across the healthy and jaundice samples. Subsequently, a paired Wilcoxon *post-hoc* test with Bonferonni correction was used to facilitate multiple comparisons. The statistical significance was defined as a $p \leq 0.05$.

RESULTS

Raman Spectra of RBCs From Healthy and Jaundice Blood Samples

The pre-processed spectra were first overlaid together separately for each group (**Supplementary Figure S1**). This exhibited the overall trend of the Raman peaks for the healthy and jaundice groups separately. **Figure 2** gives the overlaid averaged Raman spectra of both the healthy and jaundice groups. The spectral bands were assigned as was described in earlier studies. Since RBCs comprised predominantly of Hb protein, spectral data were dominated primarily by the Raman signatures from heme. There also appeared to be contributions from other organic components as well, such as aromatic amino acids, amide bonds and –CH/-CH$_2$ side-chains of globular proteins.

The averaged spectra exhibited multiple differences in terms of both intensity changes as well as frequency shifts. RBCs from the jaundice group displayed increase in intensity of Raman frequencies at 975, 1,049, 1,123, 1,165, 1,175, 1,212, 1,248, 1,258, 1,264, 1,370, 1,375, 1,389, 1,460, and 1,771 cm^{-1} and a decrease in intensity at wavelengths 752, 788, 999, 1,025, 1,080, 1,223, 1524, 1,545, 1,563, 1,602, 1,617, and 1,636 cm^{-1} when compared to those from RBCs of the healthy group. Additionally, the averaged Raman spectrum in the disease group exhibited a new peak at 663 cm^{-1} and a 2–3 cm^{-1} frequency shift in 975 and 1,248 cm^{-1} bands. The detailed vibrational assignments of the same have been summarized in **Table 1** and **Figure 3** illustrates the detailed chemical structure of heme.

Multivariate PLS-DA and Factor Analysis Approach

Figure 4A displays the two-dimensional plot of the first two PCs: PC$_1$ and PC$_2$, and **Figure 4B** gives the 3-dimensional plot of the first three PCs: PC$_1$, PC$_2$, and PC$_3$. Both scores-plots (generated by Grams IQ) exhibited a tight clustering of RBCs from the healthy group, with negligible variance. However, RBCs from the jaundice samples were scattered farther away from those of the healthy group to form a different and wider-distributed group. The variations in the PCs appeared significantly different for classification as healthy and "stressed" due to jaundice. The variance captured in this approach, taking into account all the wavenumbers in the spectra between the groups, was 31.08%.

Subsequently, we performed factor load analysis, taking into consideration the first three factors (**Figure 5**). It was observed

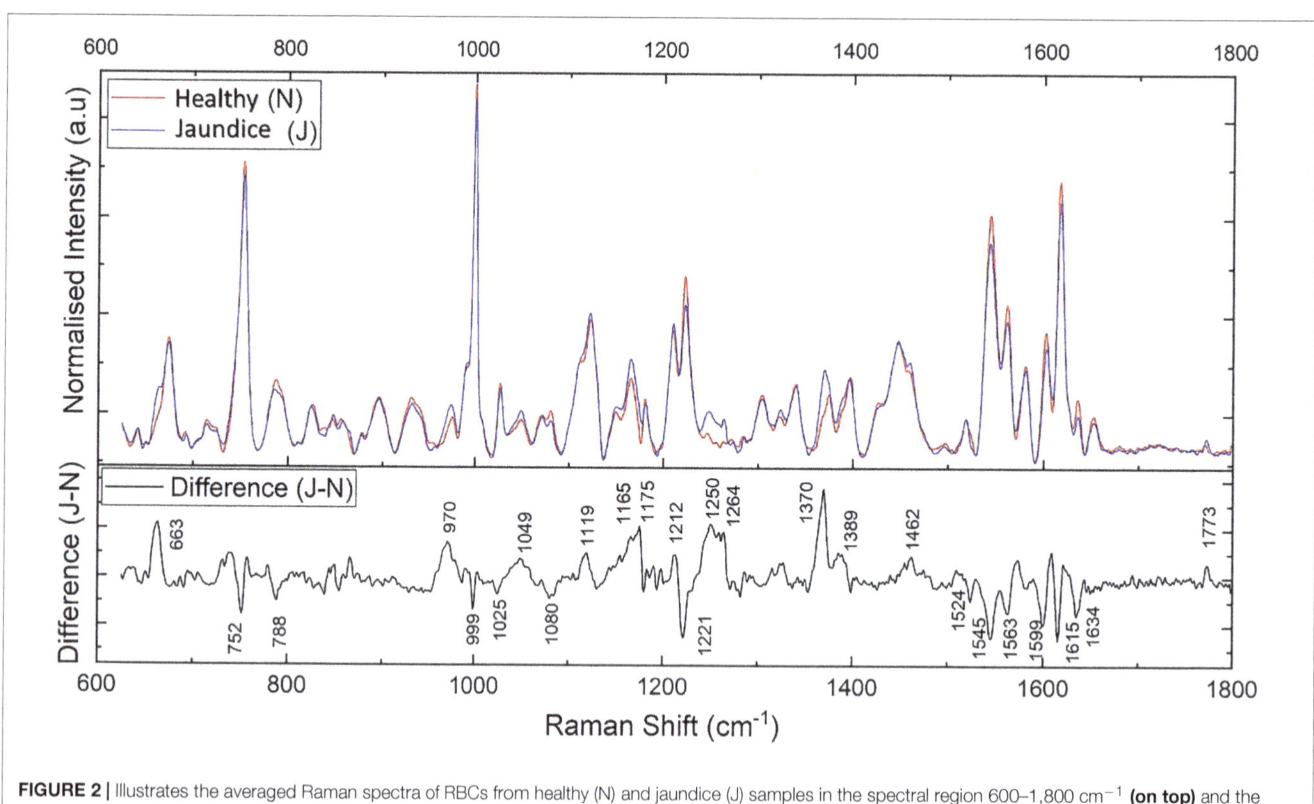

FIGURE 2 | Illustrates the averaged Raman spectra of RBCs from healthy (N) and jaundice (J) samples in the spectral region 600–1,800 cm^{-1} **(on top)** and the difference (J–N) between the averaged spectra between the healthy and jaundice blood samples **(below)**.

that all the Raman peaks that showed positive peaks in Factor-1 loading allied to their corresponding peaks of increased intensities in **Figure 2**. Similarly, all the Raman signals that showed negative peaks in Factor-1 loading allied with their corresponding peaks with decreased intensities in **Figure 2**.

Factors 2 and 3 picked up the next level variations in the spectra, chiefly of Raman peaks at 752, 999, 1,025, and 1,223 cm^{-1}.

RBC Raman Markers Detected in Jaundice

Averaged spectra, PLS-DA and the factor analysis extracted the Raman frequencies that exhibited substantial variations between RBCs from healthy and jaundice blood samples. The observed variations and their assignments have been described below under the four specific band regions, viz., core size/spin-state marker region, pyrrole-ring stretching region, methine C-H deformation region and the low wavenumber region (Wood et al., 2001). It has to be noted that the Raman peaks were assigned based on already existing data from previously conducted studies.

Core Size or Spin State Marker Band: 1,500–1,650 cm^{-1}

This specific band region comprises Raman signatures that were scattered from the C-C bonds in the porphyrin ring and rely on the spin-state of the iron atom (Wood and McNaughton,

2002). It is an important spectral range to explore, as it describes the "oxygenation" (oxy) state of Hb (Lukose et al., 2019). The Raman peaks at 1,524, 1,545, 1,563, 1,602, 1,617, and 1,636 cm^{-1} exhibited a decrease in their intensities in the jaundice samples when compared to their corresponding peaks from the healthy samples. This was an indication that Hb in jaundice was largely present in their "deoxygenated" (deoxy) state, with less bound oxygen (Rao et al., 2009). When iron conforms from the low-spin oxy-state to the deoxy-state, the iron is displaced a few nanometers out of the porphyrin plane as a result of electron withdrawal (Wood et al., 2001; Burchett et al., 2017). This "switch" becomes reflected in the vibrational modes of the porphyrin rings and marks the changes in the above-mentioned peaks. There appeared to be a significant reduction in the intensity of the peak at 1,636 cm^{-1}, a phenomenon attributed to porphyrin-doming (Rao et al., 2009). This peak is considered as the Raman-marker characterizing oxygen-concentration. An intensity decrease of this peak indicates a decline in the oxy-configuration of heme (Luo et al., 2015).

Pyrrole-Ring Stretching Band: 1,300–1,400 cm^{-1}

This Raman band region signifies the "oxidation state" of iron in the pyrrole ring and comprises heme-aggregation bands (Ong et al., 1999). In jaundice, there appeared to be a prominent increase in the intensity of the heme-aggregation Raman peak at 1,370 cm^{-1} (Wood et al., 2005) accompanied by enhanced peaks at 1,375 and 1,389 cm^{-1}. The latter peaks are designated to

TABLE 1 | Assignments of the significantly different Raman peaks of RBCs of healthy and jaundice samples.

Healthy (cm^{-1})	Jaundice (cm^{-1})	Changes	Assignments	P-value
–	663	*New Peak*	p:C-S str (gauche)	<0.00001
752	752	↓	pyrrole ring def (out-of-plane), Tryptophan	<0.00001
788	788	↓	pyrrole ring breathing	<0.00001
975	973	↑ and Shift	$\nu(C_c$–$C_d)$	<0.00001
999	999	↓	Phenylalanine	<0.00001
1,025	1,025	↓	Phenylalanine	<0.00001
1,049	1,049	↑	$\nu(O=O)$, $\delta(=C_bH_2)_{asym}$	<0.00001
1,080	1,080	↓	p: C-N str	<0.00001
1,122	1,123	↑	$\nu(C_\beta$–$C_1)_{sym}$	<0.00001
1,165	1,165	↑	Pyrrole half-ring vibration	<0.00001
1,175	1,175	↑	Pyrrole half-ring vibration	<0.00001
1,212	1,212	↑	Methine C_m–H def	<0.00001
1,223	1,223	↓	Methine C_m–H def	0.003
1,246	1,248	↑ & Shift	p:Amide III (disordered)	<0.00001
1,258	1,258	↑	p:Amide III (disordered)	<0.00001
1,264	1,264	↑	p:Amide III (disordered)	<0.00001
1,370	1,368, 1,375	↑	Pyrrole half vibration	<0.00001
1,389	1,389	↑	Pyrrole half vibration	<0.00001
1,460	1,460	↑	$\delta(=C_bH_2)_{sym}$, p:$\delta(CH_2)$	<0.00001
1,524	1,524	↓	$\nu(C_\beta$–$C_\beta)$	<0.00001
1,545	1,545	↓	$\nu(C_\beta$–$C_\beta)$	<0.00001
1,563	1,563	↓	$\nu(C_\beta$–$C_\beta)$	<0.00001
1,602	1,602	↓	$\nu(C_\alpha$–$C_m)_{asym}$, $\nu(C_a=C_b)_{venyl}$	<0.00001
1,617	1,617	↓	$\nu(C_\alpha$–$C_m)_{asym}$, $\nu(C_a=C_b)_{venyl}$	<0.00001
1,636	1,636	↓	$\nu(C_\alpha$–$C_m)_{asym}$	<0.00001
1,771	1,771	↑	p: C=O str	0.07

p, protein; sy, symmetric; asym, antisymmetric; str, stretch; ν and δ, in-plane modes; str, stretching, def, deformation. RBCs from healthy volunteers and jaundice patients suspended in PBS were optically-trapped and excited with a 785 nm diode laser and their Raman spectra were captured. The given table lists the frequencies of Raman signals that had displayed increased, decreased and shifted peaks. The assignments of the Raman peaks were carried out based on previous studies.

pyrrole deformation/breathing modes (Dybas et al., 2018). This indicated that RBCs in the "jaundice environment" produced heme-aggregates within themselves. There appeared to be a shift in the 1,370 cm^{-1} (oxy state) peak to 1,375 cm^{-1} (deoxy state) in jaundice, ascertaining lesser oxy states in jaundice (Lukose et al., 2019).

Methine C-H Deformation Band: 1,200–1,300 cm^{-1}

This is the spectral zone assigned to methine C-H bonds in heme molecules. These bonds are sensitive to minute changes in Hb conformations and are governed by porphyrin-iron associations. The high sensitivity of these bonds has been described to be because of their juxtaposition with the protein subunits (Wood et al., 2001). Thus, Hb conformational-variations alter the deformation angle of C-H vibrations instantaneously (Wood et al., 2007). In this study, the wavenumbers that displayed enhanced intensity in jaundice were the peaks at 1,212, 1,248, and 1,264 cm^{-1} which were also accompanied by a fall in intensity at the peak 1,223 cm^{-1}, consequent to changes in the methine C-H deformation regions. We also report a shift in the wavenumber 1,246 cm^{-1} to heme aggregation marker 1,248 cm^{-1} in jaundice (Lemler et al., 2014).

Low Wavenumber Band: 600–1,200 cm^{-1}

This spectral region encompasses the bands for pyrrole-breathing, pyrrole-deformation modes and RBC membrane stability. In jaundice, there was a shift of the Raman peak at 973 to 975 cm^{-1} with a significant increase in its intensity along with a shouldering of 675 cm^{-1} pyrrole-deformation band with a new peak at 663 cm^{-1}. These occurrences indicate heme-aggregation, as a consequence of protein denaturation, under "stressful" environments (Wood et al., 2005; Lukose et al., 2019). The 752 cm^{-1} porphyrin breathing mode peak, a crucial spectral-marker that confers the integrity of Hb (Deng et al., 2005), shows reduced intensity in the jaundice group. The peak 788 cm^{-1}, a deoxy Raman marker, has taken a noticeable decrease in its intensity in jaundice samples (Barkur et al., 2017). The 999 and 1,025 cm^{-1} Raman peaks, originating from phenylalanine, exhibited a reduction in their strengths, signifying the breakdown of both Hb (Wu et al., 2011; Raj et al., 2013) as well as cell membranes (Lippert et al., 1975; Goheen et al., 1993). Peak 1,122 cm^{-1}, emanating from the RBC membrane, is caused by the trans conformation vibrations of the C-C frame, the height of which has intensified and so does the peak at 1,080 cm^{-1}, signifying diminished membrane deformability (Lukose et al., 2019). Changes in intensity at 1,165

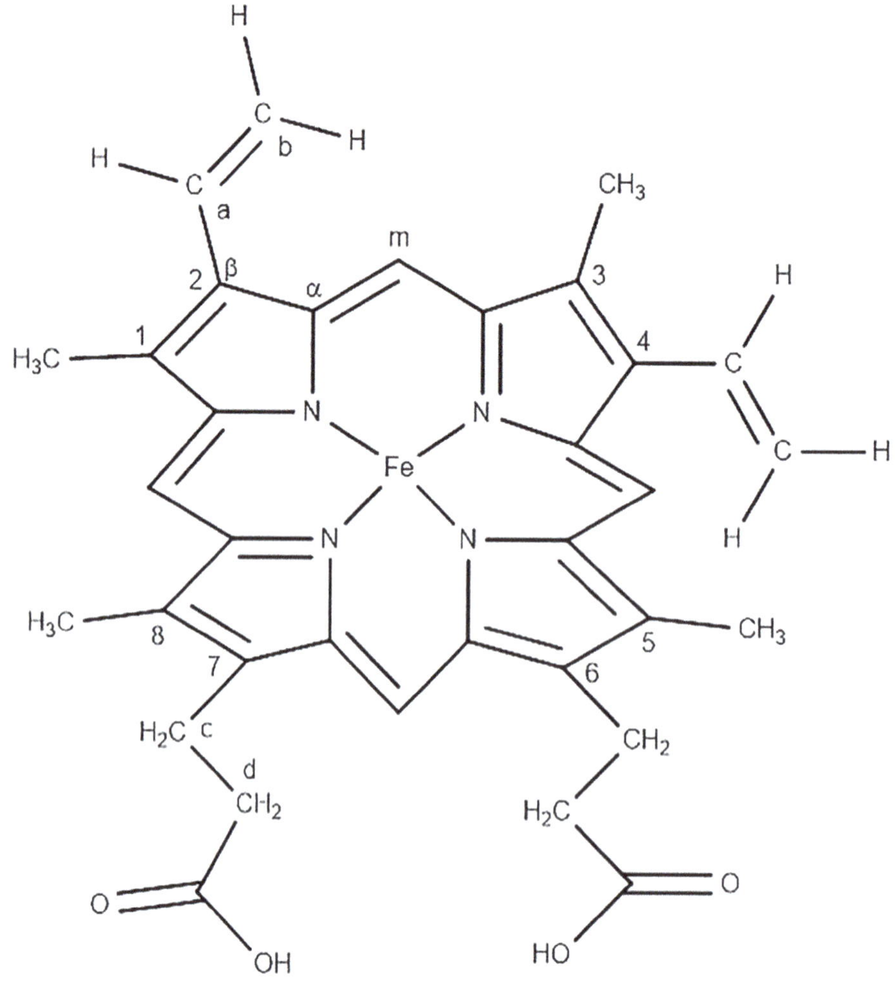

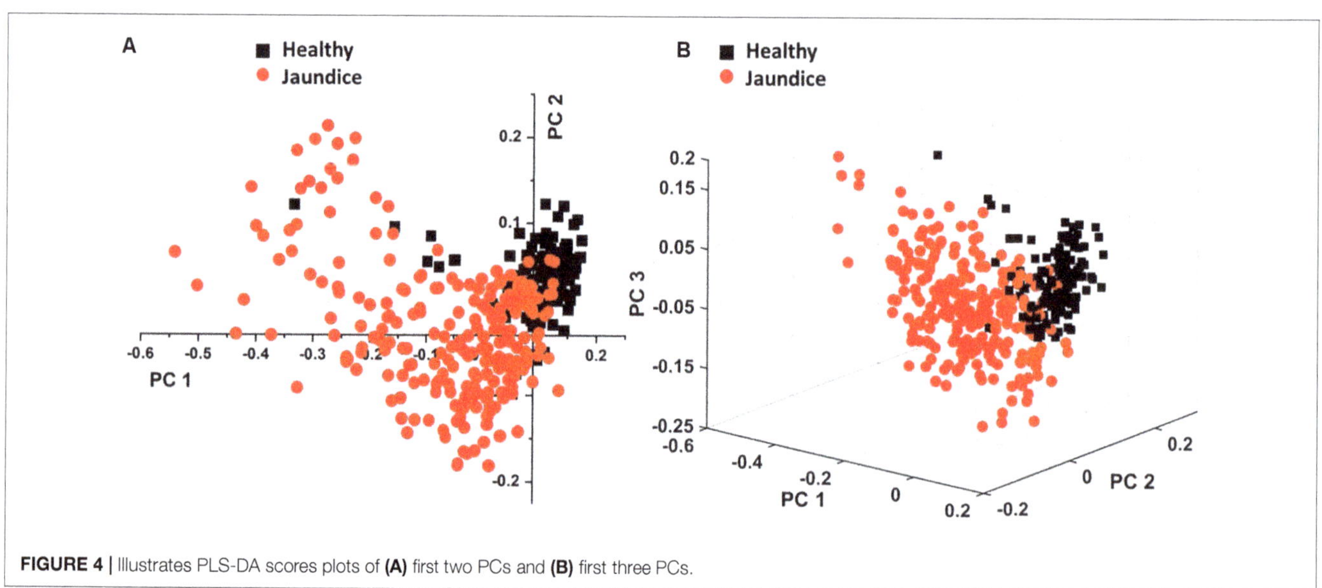

FIGURE 3 | Chemical structure of "heme" in RBC displaying the Fe-protoporphyrin-IX with four pyrrole rings, four methyl (CH_3) groups, two propionate side chains and two vinyl ($-CH = CH_2$) groups.

FIGURE 4 | Illustrates PLS-DA scores plots of **(A)** first two PCs and **(B)** first three PCs.

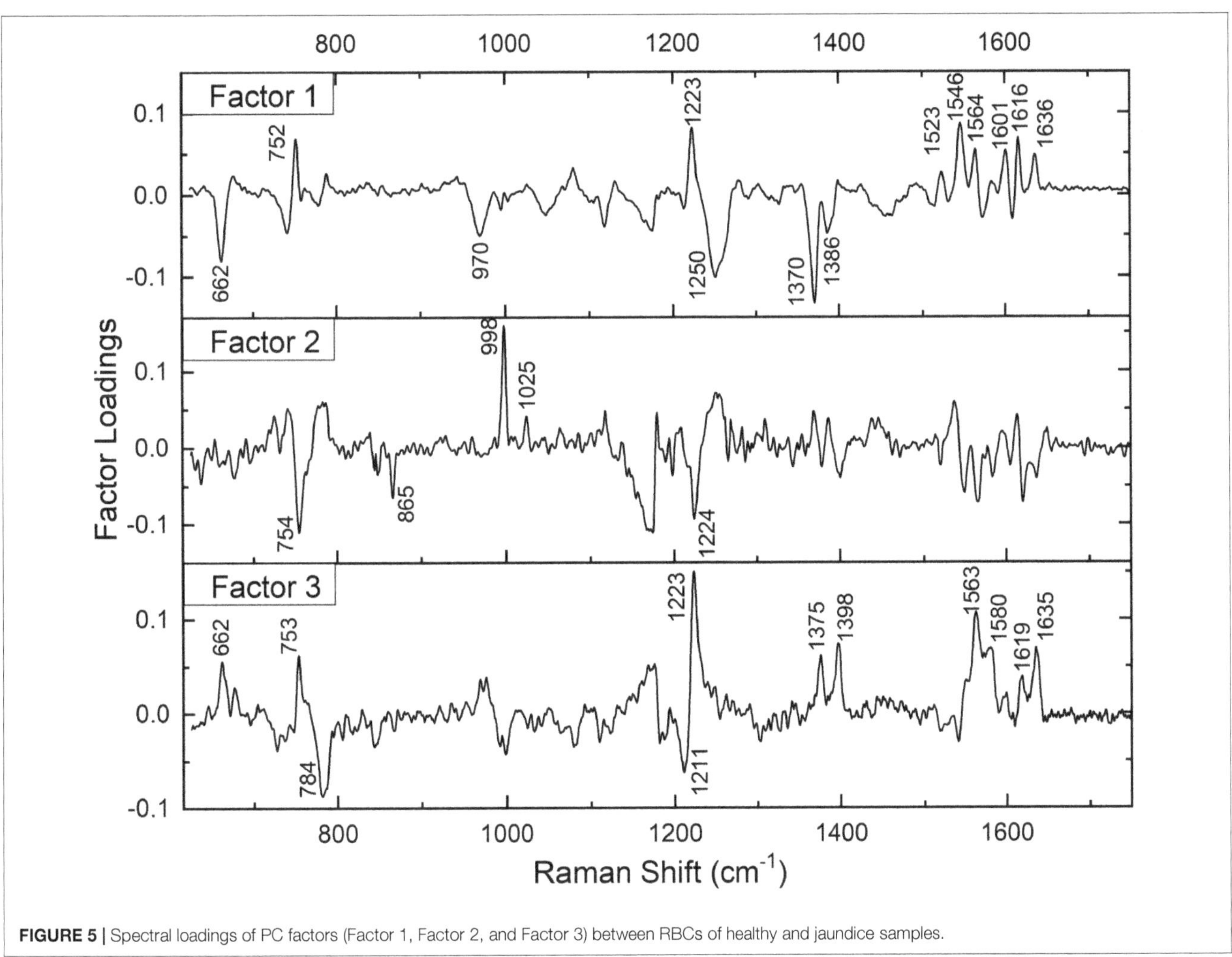

FIGURE 5 | Spectral loadings of PC factors (Factor 1, Factor 2, and Factor 3) between RBCs of healthy and jaundice samples.

and 1,175 cm^{-1} attributes to deformation in pyrrole ring of Hb (Gautam et al., 2018).

PCA of the 26 Detected Wavenumber-Bands

In total, there were 26 Raman bands (**Table 1**) detected by the spectral analysis approach that displayed a substantial difference between the healthy and jaundice samples. Accordingly, PCA was conducted again, taking into consideration the intensity-scores of just these 26 Raman peaks. The analysis captured a variance of 60.54% for the first three PCs, much higher than that obtained (31.08%) when the whole region of the spectra was considered.

Non-parametric MANOVA Approach to Raman Spectral Markers

We then proceeded to explore whether there existed a significant difference in the mean intensity-scores across the two groups for the 26 Raman bands identified in the spectral analysis by a MANOVA-based approach using CRAN 3.6.1. The packages utilized for this analysis were heplots (Fox et al.,

2018), npmv (Wood et al., 2007), and MVN (Korkmaz et al., 2014). The advantage of implementing this method was that it made use of dependencies between the wavelengths and enabled joint acceptance or rejection of the hypothesis. When we tested the assumptions of MANOVA, it was observed that the assumptions of multivariate-normality as well as the equality of covariance-matrices that were associated with the two groups were violated. Hence, we made use of a non-parametric variant of MANOVA. Since the data represented intensity-scores across different Raman peaks corresponding to two populations, the McKeon approx. for Lawley Hotelling Test ($p < 0.00001$) was considered. The results indicated a significant difference between the average values of the intensity-scores across the healthy and jaundice samples. As a *post-hoc* analysis, paired Wilcoxon test with Bonferonni corrections was implemented to identify the wavelengths that accounted for significant differences in the intensity-scores across the two populations. The p-values that were derived are presented in **Table 1**.

Excluding Raman peak at 1,223 cm^{-1} ($p = 0.003$), that emanated from the methine C-H deformation band and 1,771 cm^{-1} ($p = 0.07$) from the ν(C = O) peak of proteins, the intensity scores of all the other Raman bands detected by spectral analysis exhibited a highly significant difference ($p < 0.00001$) between the healthy and jaundice samples.

Therefore, the "vitality" or "normality" of RBCs can be deduced by just analyzing these specific Raman peaks, rather than assessing all the RBC Raman bands across the spectrum. The variations in these peaks convey information regarding deformations/distortions in both cytoplasmic Hb as well as the RBC membrane. This study consequently informs us that under pathologies relating to jaundice, RBCs in circulation, although looking "morphologically" healthy, may not always be "chemometrically" healthy.

DISCUSSION

Jaundice (derived from the French word "jaune" which means yellow) is the yellowish discoloration of the sclera, skin and mucous membrane, caused by the build-up of serum bilirubin, above 2 mg/dL, and nearly always signifies the existence of an underlying infection related to the derangement of bilirubin metabolism. Its persistence is associated with a substantial increase in morbidity and mortality (Anand and Garg, 2015). Our body produces close to 4 mg/kg bilirubin every day (Schiff et al., 2011). Normally, the total serum bilirubin level never exceeds 1 mg/dL, of which UCB is less than 0.8 mg/dL and CB levels is less than 0.3 mg/dL (Zieve et al., 1951; Werner et al., 1970). There exists a myriad of clinical pathologies that results in jaundice. Unconjugated hyperbilirubinemia results primarily due to excessive hemolysis, while hyperbilirubinemia caused by excessive CB, develops in hepatocellular diseases and obstructions of the biliary system. In the latter cases, there can be a rise in both forms of bilirubin (Fargo et al., 2017). In this study, we have included patients diagnosed with hyperbilirubinemia, irrespective of its cause.

Hyperbilirubinemia can be toxic to circulating RBCs, causing morphological and metabolic impairments, eventually leading to further hemolysis, aggravating hyperbilirubinemia and intensifying anemia (Lang et al., 2015). Bilirubin toxicity on RBCs has been found to occur in a concentration and temperature-dependent manner (Brito et al., 2000). During the course of the disease, jaundice is accompanied by elevated levels of several enzymes and inflammatory-cytokines in serum (Padillo et al., 2001). This adds up to the already existing distress on healthy RBCs (Pretorius, 2018). Taking these facets into account, we attempted to examine the biochemical status of live RBCs in jaundice employing Raman spectroscopy. Since this analytical technique is both reagent and fixation-free, it rules out any concerns regarding possible cell modifications or damage caused by the analytical processes itself (Sato et al., 2018). Because RBCs are delicate and extremely sensitive cells, it was very important to ensure utmost care in the management of samples during the procedures that might themselves lead to chemometric variations, hemolysis or even eryptosis. We had also ensured to conduct the experiments immediately on sample-procurement.

Our results showed distinct variations in the biochemical signatures of RBCs in jaundice, when compared to the normal.

Since RBCs are simple organelle-less cells with a highly deformable cell membrane that holds abundant cytoplasmic Hb, the obtained Raman spectral features were derived from these two sources. The overlaid averaged Raman spectra from the groups displayed Raman peaks that had their intensities altered and a few peaks that had undergone a shift (**Figure 2** and **Table 1**). It is to be noted that the Raman peaks were assigned to specific biochemical components of the RBC in accordance with earlier studies and in this study we report how these Raman signatures altered in clinical jaundice. Spectral analysis and the various multivariate analytical methods employed in this study indicate chemometric modifications of RBCs in jaundice, when compared to RBCs of the healthy group. The variations (section "RBC Raman Markers Detected in Jaundice") were noted in Raman peaks that represented the "oxygenation status" of Hb. The heme in jaundice exhibited largely a "deoxy" status when compared to the more "oxy" states in the healthy population. This was accompanied by a decline in the frequencies signifying "oxygen saturation" in the jaundice samples. Heme-aggregation-Raman-markers were manifested in jaundice along with a fall in peaks that depicted Hb stability. The spectra also exhibited alterations in Raman peaks portraying RBC membrane stability. The non-parametric MANOVA method served as an aid in the identification of the Raman signatures that could distinctly ($p < 0.00001$) capture the significant differences in the intensity scores across healthy and jaundice populations (**Table 1**). PCA of these identified Raman signatures captured better variance between the groups.

These results exhibit enough evidence to establish that in jaundice, intact and healthy-appearing RBCs in circulation, can experience biochemical alterations similar to those seen when under distress. The probable reasons could be the hostile milieu created by the disease itself such as hyperbilirubinemia, presence of pathogens and/or products related to the immune response that alters the physiological state of blood, thereby experiencing a shortened life-span. The intention of this study was to report the Raman spectroscopy features of RBCs in jaundice and analyze the generated findings, irrespective of the disease or it's cause. An attempt was made to correlate Raman spectral changes with serum bilirubin levels in jaundice. But the results with this group yielded less correlation (**Supplementary Figure S2** and **Supplementary Table S1**). Further exploratory studies are required to investigate the correlation of bilirubin levels and other systemic factors with RBC Raman spectral variations in specific pathologies.

CONCLUSION

It is challenging to unearth the exact biochemical status of RBCs, as existent *in-vivo*, considering their extremely delicate and sensitive nature to even minuscule changes in its environment. Therefore, on a routine basis, RBC pathology is clinically assessed by a collective assemblage of tests that encompasses the estimation of its counts, Hb and bilirubin levels and analysis using peripheral smears. Unlike routine assays or even other high-end analytical techniques such as flow cytometry, Raman spectroscopy does not demand labels and is unique by being more sensitive and specific. The Raman-scattered-light from a particular substance characterizes its own exclusive "chemical-fingerprint." In recent times this tool has been increasingly employed in hematological research for elucidating structural information from hemes, demonstrating the existence of characteristic molecular signatures from them under different conditions of stress such as oxidation and energy depletion, disorders such as thalassemia and diabetes, and diseases such as malaria and carcinomas. This study is merely a commencement to understand the chemometric behavior of RBCs in jaundice and needs to be extrapolated further to comprehend its characteristics in specific diseases/conditions that lead to it.

AUTHOR CONTRIBUTIONS

SJ, AB, PR, and SC conceived and designed the study. SJ and AB performed the experiments, interpreted the results, and wrote the manuscript. SB and MA contributed to performing the experiments. RR and PB contributed to the conception and design of the project and the interpretation of results. AK and RL contributed to the statistical analysis, interpretation of results, and wrote the manuscript. All authors contributed to the article and approved the submitted version.

ACKNOWLEDGMENTS

We acknowledge the support contributed by the Clinical Biochemistry of Kasturba Hospital-Manipal and also extend special thanks to Dr. Alex Joseph, Associate Professor in Pharmaceutical Chemistry, Manipal College of Pharmaceutical Sciences, Manipal for his contribution in creating figures in ChemDraw and Dr. Prakash P.Y., Associate Professor, Department of Microbiology, Kasturba Medical College-Manipal for his guidance in manuscript review.

REFERENCES

Adamson, J., and Longo, D. (2012). *Anemia and Polycythemia. Harrison's Principles of Internal Medicine*, 18th Edn. New York, NY: The McGraw-Hill Companies.

Anand, A. C., and Garg, H. K. (2015). Approach to clinical syndrome of jaundice and encephalopathy in tropics. *J. Clin. Expl. Hepatol.* 5(Suppl. 1), S116–S130. doi: 10.1016/j.jceh.2014.05.007

Bankapur, A., Barkur, S., Chidangil, S., and Mathur, D. (2014). A micro-Raman study of live, single red blood cells (RBCs) treated with AgNO3 nanoparticles. *PLoS One* 9:e103493. doi: 10.1371/journal.pone.0103493

Bankapur, A., Zachariah, E., Chidangil, S., Valiathan, M., and Mathur, D. (2010). Raman tweezers spectroscopy of live, single red and white blood cells. *PLoS One* 4:e10427. doi: 10.1371/journal.pone.0010427

Barkur, S., Bankapur, A., Chidangil, S., and Mathur, D. (2017). Effect of infrared light on live blood cells: role of β-carotene. *J. Photochem. Photobiol. B Biol.* 171, 104–116. doi: 10.1016/j.jphotobiol.2017.04.034

Barkur, S., Bankapur, A., Pradhan, M., Chidangil, S., Mathur, D., and Ladiwala, U. (2015). Probing differentiation in cancer cell lines by single-cell micro-Raman spectroscopy. *J. Biomed. Opt.* 20:085001. doi: 10.1117/1.JBO.20.8.085001

Brito, M. A., Silva, R., Tiribelli, C., and Brites, D. (2000). Assessment of bilirubin toxicity to erythrocytes. Implication in neonatal jaundice management. *Eur. J. Clin. Invest.* 30, 239–247. doi: 10.1046/j.1365-2362.2000.00612.x

Burchett, W. W., Ellis, A. R., Harrar, S. W., and Bathke, A. C. (2017). Nonparametric Inference for multivariate data: the R package npmv. *J. Stat. Softw.* 76, 1–18. doi: 10.18637/jss.v076.i04

De Luca, A. C., Rusciano, G., Ciancia, R., Martinelli, V., Pesce, G., Rotoli, B., et al. (2008). Spectroscopical and mechanical characterization of normal and thalassemic red blood cells by Raman Tweezers. *Opt. Express* 16, 7943–7957. doi: 10.1364/oe.16.007943

Deng, J. L., Wei, Q., Zhang, M. H., Wang, Y. Z., and Li, Y. Q. (2005). Study of the effect of alcohol on single human red blood cells using near-infrared laser tweezers Raman spectroscopy. *J. Raman Spectrosc.* 36, 257–261. doi: 10.1002/jrs.1301

Dybas, J., Grosicki, M., Baranska, M., and Marzec, K. M. (2018). Raman imaging of heme metabolism in situ in macrophages and Kupffer cells. *Analyst* 143, 3489–3498. doi: 10.1039/c8an00282g

Eilers, P. H. (2003). A perfect smoother. *Anal. Chem.* 75, 3631–3636. doi: 10.1021/ac034173t

Erlinger, S., Arias, I. M., and Dhumeaux, D. (2014). Inherited disorders of bilirubin transport and conjugation: new insights into molecular mechanisms and consequences. *Gastroenterology.* 146, 1625–1638. doi: 10.1053/j.gastro.2014.03.047

Fargo, M. V., Grogan, S. P., and Saguil, A. (2017). Evaluation of jaundice in adults. *Am. Fam. Physician* 95, 164–168.

Fox, J., Friendly, M., and Monett, G. (2018). *heplots: Visualizing Tests in Multivariate Linear Models. R package version 1.3-5.*

Gautam, R., Oh, J. Y., Marques, M. B., Dluhy, R. A., and Patel, R. P. (2018). Characterization of storage-induced red blood cell hemolysis using raman spectroscopy. *Lab. Med.* 49, 298–310. doi: 10.1093/labmed/lmy018

Goheen, S. C., Lis, L. J., Kucuk, O., Westerman, M. P., and Kauffman, J. W. (1993). Compositional dependence of spectral features in the Raman spectra of erythrocyte membranes. *J. Raman Spectrosc.* 24, 275–279. doi: 10.1002/jrs.1250240503

Hansen, T. W. J. (2001). Bilirubin brain toxicity. *Perinatology* 21(Suppl. 1), S48–S51. doi: 10.1038/sj.jp.7210634

Korkmaz, S., Goksuluk, D., and Zararsiz, G. (2014). MVN: an R package for assessing multivariate normality. *R J.* 6, 151–162. doi: 10.32614/rj-2014-031

Korolnek, T., and Hamza, I. (2015). Macrophages and iron trafficking at the birth and death of red cells. *Blood* 125, 2893–2897. doi: 10.1182/blood-2014-12-567776

Kumamoto, Y., Harada, Y., Takamatsu, T., and Tanaka, H. (2018). Label-free molecular imaging and analysis by raman spectroscopy. *Acta Histochem. Cytochem.* 51, 101–110. doi: 10.1267/ahc.18019

Lang, E., Gatidis, S., Freise, N. F., Bock, H., Kubitz, R., Lauermann, C., et al. (2015). Conjugated bilirubin triggers anemia by inducing erythrocyte death. *Hepatology* 61, 275–284. doi: 10.1002/hep.27338

Lemler, P., Premasiri, W. R., DelMonaco, A., and Ziegler, L. D. (2014). NIR Raman spectra of whole human blood: effects of laser-induced and in vitro hemoglobin denaturation. *Anal. Bioanal. Chem.* 406, 193–200. doi: 10.1007/s00216-013-7427-7

Levitt, D. G., and Levitt, M. D. (2014). Quantitative assessment of the multiple processes responsible for bilirubin homeostasis in health and disease. *Clin. Exp. Gastroenterol.* 7, 307–328. doi: 10.2147/CEG.S64283

Lippert, J. L., Gorczyca, L. E., and Meiklejohn, G. (1975). A laser Raman spectroscopic investigation of phospholipid and protein configurations in hemoglobin-free erythrocyte ghosts. *Biochim. Biophys. Acta* 382, 51–57. doi: 10.1016/0005-2736(75)90371-5

Lukose, J., Mithun, N., Mohan, G., Shastry, S., and Chidangil, S. (2019). Normal saline-induced deoxygenation of red blood cells probed by optical tweezers combined with the micro-Raman technique. *RSC Adv.* 9, 7878–7884. doi: 10.1039/c8ra10061f

Luo, M., Huang, Y. Y., Su, B. C., Shi, Y. F., Zhang, H., and Ye, X. D. (2015). Study on RBC Oxygen-Carrying Function with the Incubation Time. *Guang Pu Xue Yu Guang Pu Fen Xi* 35, 3350–3355.

Ong, C. W., Shen, Z. X., Ang, K. K. H., Kara, U. A. K., and Tang, S. H. (1999). Resonance Raman microspectroscopy of normal erythrocytes and plasmodium berghei-infected erythrocytes. *Appl. Spectrosc.* 53, 1097–1101. doi: 10.1366/0003702991947874

Padillo, F. J., Andicoberry, B., Muntane, J., Lozano, J. M., Miño, G., Sitges-Serra, A., et al. (2001). Cytokines and acute-phase response markers derangements in patients with obstructive jaundice. *Hepato Gastroenterol.* 48, 378–381.

Park, T. H. (2016). Bilirubin: a promising antioxidant for ischaemia/reperfusion injury. *Int. Wound J.* 13:1040. doi: 10.1111/iwj.12362

Pretorius, E. (2018). Erythrocyte deformability and eryptosis during inflammation, and impaired blood rheology. *Clin. Hemorheol. Microcirc.* 69(Suppl. 1), 1–6. doi: 10.3233/CH-189205

Raj, S., Wojdyla, M., and Petrov, D. (2013). Studying single red blood cells under a tunable external force by combining passive microrheology with Raman spectroscopy. *Cell Biochem. Biophys.* 65, 347–361. doi: 10.1007/s12013-012-9439-x

Rao, S., Bálint, S., Cossins, B., Guallar, V., and Petrov, D. (2009). Raman study of mechanically induced oxygenation state transition of red blood cells using optical tweezers. *Biophys. J.* 96, 209–216. doi: 10.1529/biophysj.108.139097

Sato, H., Ishigaki, M., Taketani, A., and Andriana, B. B. (2018). Raman spectroscopy and its use for live cell and tissue analysis. *Biomed. Spectrosc. Imaging* 7, 97–104. doi: 10.3233/BSI-180184

Schiff, E. R., Maddrey, W. C., and Sorrell, M. F. (2011). *Schiff's Diseases of the Liver*, 11th Edn. Hoboken, GB: Wiley-Blackwell.

Shapiro, S. M. (2003). M Bilirubin toxicity in the developing nervous system. *Rev. Pediatr. Neurol.* 29, 410–421. doi: 10.1016/j.pediatrneurol.2003.09.011

Shi, H. S., Lai, K., Yin, X. L., Liang, M., Ye, H. B., Shi, H. B., et al. (2019). Ca2+-dependent recruitment of voltage-gated sodium channels underlies bilirubin-induced overexcitation and neurotoxicity. *Cell Death Dis.* 10:774. doi: 10.1038/s41419-019-1979-1971

Werner, M., Tolls, R. E., Hultin, J. V., and Mellecker, J. (1970). Influence of sex and age on the normal range of eleven serum constituents. *Z. Klin. Chem. Klin. Biochem.* 8, 105–115. doi: 10.1515/cclm.1970.8.2.105

Wood, B. R., Caspers, P., Puppels, G. J., Pandiancherri, S., and McNaughton, D. (2007). Resonance Raman spectroscopy of red blood cells using near-infrared laser excitation. *Anal. Bioanal. Chem.* 387, 1691–1703. doi: 10.1007/s00216-006-0881-8

Wood, B. R., Hammer, L., Davis, L., and McNaughton, D. (2005). Raman microspectroscopy and imaging provides insights into heme aggregation and denaturation within human erythrocytes. *J. Biomed. Opt.* 10:014005. doi: 10.1117/1.1854678

Wood, B. R., and McNaughton, D. (2002). Micro-Raman characterization of high- and low-spin heme moieties within single living erythrocytes. *Biopolymers* 67, 259–262. doi: 10.1002/bip.10120

Wood, B. R., Tait, B., and McNaughton, D. (2001). Micro-Raman characterisation of the R to T state transition of haemoglobin within a single living erythrocyte. *Biochim. Biophys. Acta* 1539, 58–70. doi: 10.1016/s0167-4889(01)00089-1

Wu, H., Volponi, J. V., Oliver, A. E., Parikh, A. N., Simmons, B. A., and Singh, S. (2011). In vivo lipidomics using single-cell Raman spectroscopy. *Proc. Natl. Acad. Sci. U.S.A.* 108, 3809–3814. doi: 10.1073/pnas.1009043108

Zieve, L., Hill, E., Hanson, M., Falcone, A. B., and Watson, C. J. (1951). Normal and abnormal variations and clinical significance of the one-minute and total serum bilirubin determinations. *J. Lab. Clin. Med.* 38, 446–469.

14

Recent Advances in the Treatment of Sickle Cell Disease

*Gabriel Salinas Cisneros[1,2] and Swee L. Thein[1]**

[1] *Sickle Cell Branch, National Heart Lung and Blood Institute, National Institutes of Health, Bethesda, MD, United States,*
[2] *Division of Hematology and Oncology, Children's National Medical Center, Washington, DC, United States*

Correspondence:
Swee L. Thein
sl.thein@nih.gov

Sickle cell anemia (SCA) was first described in the Western literature more than 100 years ago. Elucidation of its molecular basis prompted numerous biochemical and genetic studies that have contributed to a better understanding of its pathophysiology. Unfortunately, the translation of such knowledge into developing treatments has been disproportionately slow and elusive. In the last 10 years, discovery of *BCL11A*, a major γ-globin gene repressor, has led to a better understanding of the switch from fetal to adult hemoglobin and a resurgence of efforts on exploring pharmacological and genetic/genomic approaches for reactivating fetal hemoglobin as possible therapeutic options. Alongside therapeutic reactivation of fetal hemoglobin, further understanding of stem cell transplantation and mixed chimerism as well as gene editing, and genomics have yielded very encouraging outcomes. Other advances have contributed to the FDA approval of three new medications in 2017 and 2019 for management of sickle cell disease, with several other drugs currently under development. In this review, we will focus on the most important advances in the last decade.

Keywords: sickle cell disease, anti-sickling agents, gene editing, gene therapy, hemoglobinopathies

INTRODUCTION

Sickle cell disease (SCD) is an inherited blood disorder that first appeared in the Western literature in 1910 when Dr. James Herrick described a case of severe malaise and anemia in a 20-year-old dental student from Grenada (Herrick, 1910). On examining his blood smear, he noticed many bizarrely shaped red blood cells, leading him to surmise that "…the cause of the disease may be some unrecognized change in the red corpuscle itself" (Herrick, 2014). More than 100 years later we recognize that the change in the red corpuscle is caused by a single base substitution in β-globin, and that the disease is not just present in the United States (US), but prevalent in regions where malaria was historically endemic, including sub-Saharan Africa, India, the Middle East, and the Mediterranean (Williams and Thein, 2018). Presence of SCD in the non-malarial regions is related to the recent migration patterns.

Currently, an estimated 300,000 affected babies are born each year, more than 80% of whom are in Africa. Due to recent population migrations, increasing numbers of individuals affected by SCD are encountered in countries that are not historically endemic for malaria, such as the US. It is estimated that 100,000 Americans are affected with SCD, the majority of whom are of African descent (Hassell, 2010, 2016). The numbers affected with SCD are predicted to increase exponentially; Piel et al. (2013) estimated that between 2010 and 2050, the overall number of births affected by SCD will be 14,242,000; human migration and further globalization will continue to expand SCD throughout the world in the coming decades. While 75% or more of newborns with SCD in sub-Saharan Africa do not make their fifth birthday (McGann, 2014), in medium- to well-resourced countries almost all of affected babies can now expect to live to adulthood but overall survival still lags behind that of a non-SCD person by 20–30 years (Telfer et al., 2007; Quinn et al., 2010; Elmariah et al., 2014; Gardner et al., 2016; Serjeant et al., 2018). Despite these global prevalence figures, and the fact that SCD is by far the largest public health concern among the hemoglobinopathies, it was not until 2006 when the World Health Organization (WHO) recognized SCD as a global public health problem[1].

In 1949, Linus Pauling showed that an abnormal protein (hemoglobin S, HbS) was the cause of sickle cell anemia (SCA), making SCD the first molecular disease and motivating an enormous amount of scientific and medical research. Because of its genetic simplicity, SCA has been used to illustrate many of the advances in molecular genetics such as detection of a DNA mutation by restriction fragment enzyme analysis, and was used as proof of principle for the polymerase chain reaction (PCR) that we now take for granted (Wilson et al., 1982; Saiki et al., 1985).

In the last 50 years, tremendous progress has been made in understanding the pathophysiology and pathobiological complexities of SCD, but developing treatments has been disproportionately slow and elusive; a history of Perils and Progress, so succinctly summarized by Wailoo (2017). We are confident that in the next 30 years, the therapeutic landscape for SCD will change due to a combination of recent advancements in genetics and genomics, an increase in the number of competing clinical trials, and also an increased awareness from the funding bodies, in particular the NIH, USA.

Here, after a brief review of the pathophysiology, we will focus on the advances in treatment of SCD that have occurred in the last 10 years and that have reached phase 2/3 of clinical trials (**Figure 1**).

PATHOPHYSIOLOGY OF SICKLE CELL DISEASE

Sickle cell disease is caused by an abnormal HbS ($\alpha_2\beta^S_2$) in which glutamic acid at position 6 of the β-globin chain of hemoglobin is changed to valine. Goldstein et al. (1963) showed that this amino acid substitution arose from a single base change (A > T) at codon 6 (*rs334*). The genetic causes of SCD include homozygosity for the *rs334* mutation (HbSS, commonly referred as SCA) and compound heterozygosity between *rs334* and mutations that lead to either other structural variants of β-globin (such as HbC, causing HbSC) or reduced levels of β-globin production as in β-thalassemia (causing HbS/β-thalassemia). In patients of African ancestry, HbSS is the most common cause of SCD (65–70%), followed by HbSC (about 30%), with HbS/β-thalassemia being responsible for most of the rest (Steinberg et al., 2001). SCA in which the intracellular concentration of HbS is almost 100%, is by far the most severe and well described (Brittenham et al., 1985). The majority of the therapeutic developments and interventions have focused on this genotype, which is also

the focus of this review, although they also impact the other SCD genotypes.

The fundamental event that underlies the complex pathophysiology and multi-systemic consequences of SCD is the polymerization of HbS that occurs under low oxygen tension (**Figure 2**). Polymerization of the de-oxygenated HbS alters the structure and function of the red blood cells (RBCs). These damaged (typically sickled shaped) RBCs are not only less flexible compared to normal RBCs, but also highly adhesive. Repeated cycles of sickling and unsickling shortens the lifespan of the damaged sickle RBCs to about 1/6th that of normal RBCs (Bunn, 1997; Hebbel, 2011). The outcome is the occlusion of blood vessels in almost every organ of the body and chronic hemolytic anemia, the two hallmarks of the disease, that result in recurrent episodic acute clinical events, of which acute pain is the most common, and accumulative organ damage. Acute sickle pain is so severe that it is often referred to as "vaso-occlusive sickle crisis" or VOC.

These events trigger a cascade of pro-inflammatory activity setting off multiple pathophysiological factors that also involve neutrophils, platelets, and vascular endothelium (Sundd et al., 2019). The continual release of cell-free hemoglobin from hemolysis depletes hemopexin and haptoglobin, a consequence of which is the reduced bioavailability of nitric oxide (NO), and vascular endothelial dysfunction that underlies the chronic organ damage in SCD pathology.

The sickle red blood cells do not just interact with the vascular endothelium but trigger activation of neutrophils, monocytes and platelets. During steady-state, patients with SCD have above normal values of neutrophils, monocytes and platelets which further increase during acute events (Villagra et al., 2007). Neutrophilia has been consistently correlated with SCD severity (Ohene-Frempong et al., 1998; Miller et al., 2000); neutrophils play a central role in vaso-occlusion through their interactions with both erythrocytes and endothelium upregulating expression of cytoadhesion molecules such as P- and E-selectins, current therapeutic targets (Zhang et al., 2016).

Platelets, when activated, form aggregates with erythrocytes, monocytes, and neutrophils both in patients and in murine models (Wun et al., 1997; Zhang et al., 2016). As with neutrophils, it appears that platelet aggregation is dependent on P-selectin. As part of this constant inflammatory state, the coagulation cascade is also hyperactivated in SCD. The repeated interaction between RBCs and endothelium promote expression of pro-adhesive and procoagulant proteins evidenced by increased levels of plasma coagulation factors, tissue factor (TF) and interactions between monocyte-endothelium, platelet-neutrophil and platelet-RBC. Patients with SCD have increased rates of venous and arterial thrombotic events (Brunson et al., 2017).

Unraveling these pathophysiological targets has provided insights on clinical trials on anti-platelet and anti-adhesion agents, as well as anti-coagulation factors for the prevention of acute VOC pain in SCD (Telen, 2016; Nasimuzzaman and Malik, 2019; Telen et al., 2019). A case in point is the development of an anti-P-selection molecule (Crizanlizumab) for treatment of sickle VOC, recently approved by the FDA in November 2019 and marketed as Adakveo®.

---

[1] https://apps.who.int/iris/handle/10665/20890

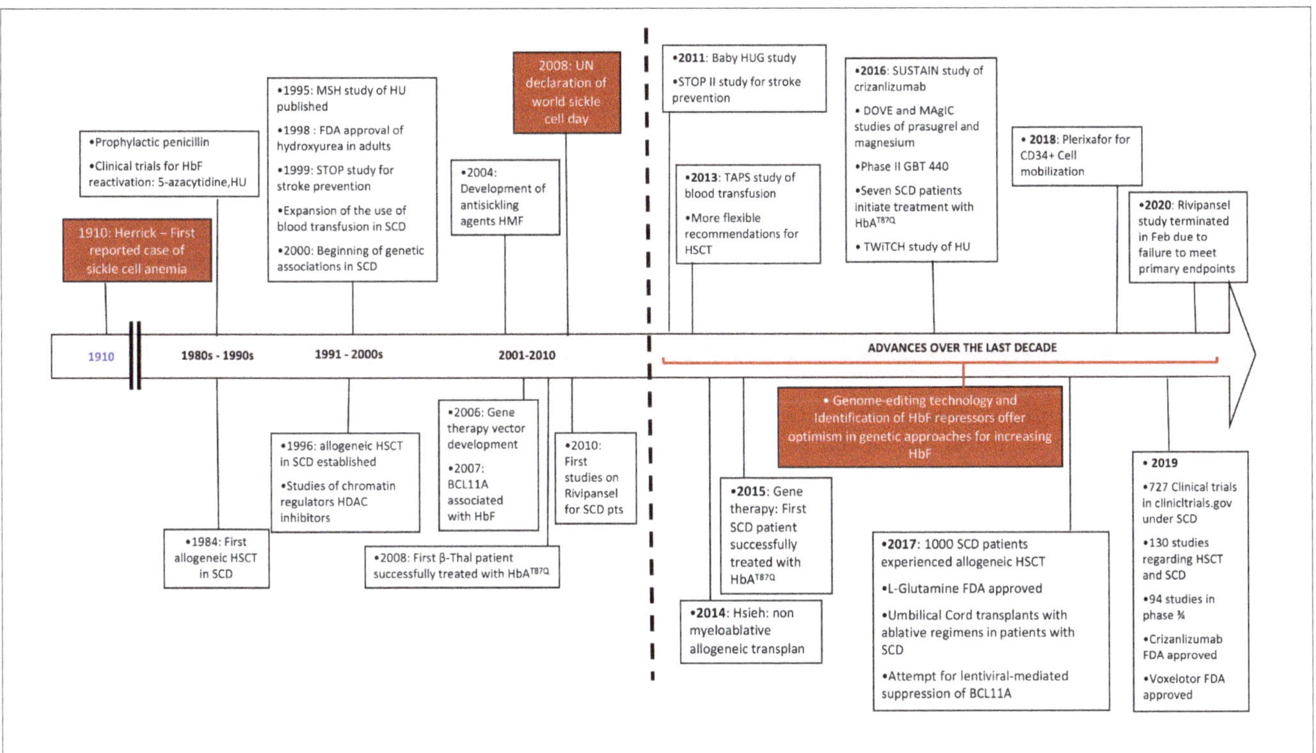

FIGURE 1 | Timeline review of historic events since the diagnosis of sickle cell disease with an emphasis over the last decade. SCD, sickle cell disease; HSCT, hematopoietic stem cell transplant; HU, hydroxyurea.

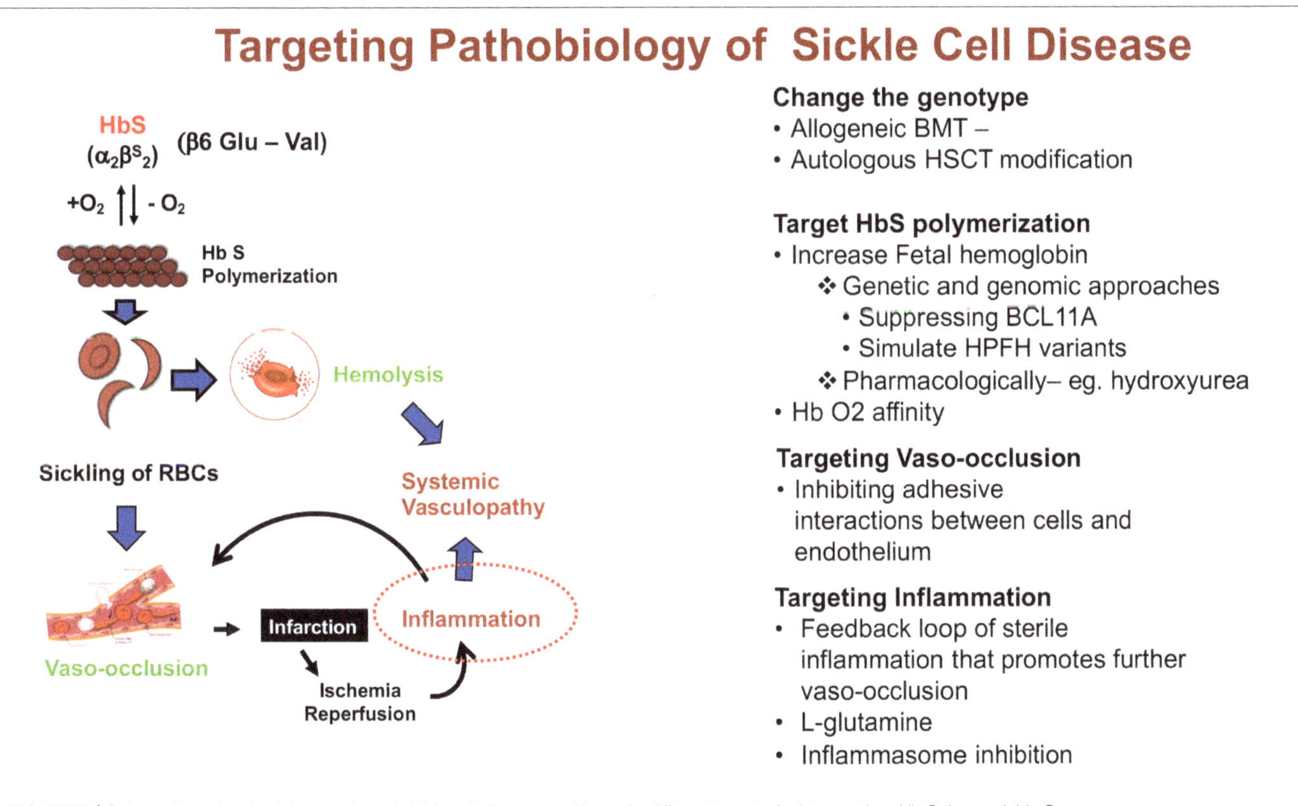

FIGURE 2 | Schematic pathophysiology review of sickle cell disease and its main different targets for intervention. Hb S, hemoglobin S.

New therapeutic approaches that use drugs to ameliorate the downstream sequelae of HbS polymerization have not proved to be as effective as hydroxyurea (HU) which has an "anti-sickling" effect via induction of fetal hemoglobin (HbF, α2γ2) (Ware and Aygun, 2009). Other effects of HU include improvement of RBC hydration, reduction of neutrophil count, reduction of leucocyte adhesion, and reduction of pro-inflammatory markers, all of which add to the clinical efficacy of HU. In addition, HU also acts as NO donor, promoting vasodilation (Cokic et al., 2003). Increasing HbF is highly effective because it dilutes the intracellular HbS concentration, thereby increasing the delay time to HbS polymerization (Eaton and Bunn, 2017); in addition to which, the γ-chains also have an inhibitory effect on the polymerization process. Hydroxyurea, however, is only partially successful because the increase in fetal hemoglobin is uneven and not present in all cells. Nonetheless, the well-established clinical efficacy of HbF increase, substantiated by numerous clinical and epidemiological studies, has motivated both pharmacological and genetic approaches to induce HbF (Nevitt et al., 2017).

A more detailed understanding of the switch from fetal to adult hemoglobin, and identification of transcriptional regulators such as BCL11A, aided by the developments in genetic and genomic platforms, provide hope that genomic-based approaches for therapeutic reactivation of HbF may soon be possible (Vinjamur et al., 2018). In the meanwhile, a gene addition approach that infects the patient's stem cells with a virus expressing an anti-sickling β-globin variant, T87Q, shows great promise (Negre et al., 2016; Ribeil et al., 2017). The most successful "curative" approach so far, is transplantation with stem cells from an immunologically matched sibling but this is severely limited by the lack of availability of matched donors (Walters et al., 1996a; Gluckman et al., 2017).

Parallel to the new medications being developed blood transfusions with normal red blood cells, remain an effective and increasing therapeutic option for managing and preventing SCD complications, but this strategy has limitations (not uniformly accessible, accompanied by risks of alloimmunization, hemolytic transfusion reactions and transfusional iron overload). Blood transfusion improves the oxygen-carrying capacity and improves microvascular perfusion by decreasing the HbS percentage. A major complication of blood transfusion is hemolytic transfusion reactions that occur primarily in RBC alloimmunized patients and SCD patients, in particular, are at high risk because of the mismatch in donor pool (predominantly Northern European descent) while SCD patients are predominantly of African descent (Vichinsky et al., 1990; Thein et al., 2020). Limiting blood from ethnic-matched donors has reduced but did not eliminate alloimmunization (Chou et al., 2013), and a major cause is the mismatch between serologic Rh phenotype and *RHD* or *RHCE* genotype due to variant *RH* alleles in a large proportion of the individuals (Chou et al., 2013). *RH* genotyping in addition to serologic typing may be required to identify the most compatible RBCs and recent studies have shown that a prospective rather than reactive (after appearance of allo-antibodies) genotyping approach may be feasible (Chou et al., 2018, 2020; Hendrickson and Tormey, 2018). Until prospective genotyping of RBC antigens become a practical feasibility, as

a prevention, many blood transfusion centers have adopted extended red cell phenotyping, including ABO, Rh, Kell, Kidd, Duffy, and S and s antigens, and some centers have also adopted molecular genotyping for red blood cell phenotype prediction using microarray chips (e.g., the PreciseType HEA BeadChip assay). It should be noted that, while blood transfusion remains an important therapeutic option in SCD, evidence for its role in management of acute or chronic complications is lacking except for prevention of primary and secondary strokes (Howard, 2016). Supportive evidence for the role of preoperative transfusion in patients with HbSS or HbS/β⁰-thalassemia was demonstrated in the Transfusion Alternatives Preoperatively in Sickle Cell disease (TAPS) study (Howard et al., 2013).

Insight on the pathophysiology of SCD (**Figure 2**) has allowed different targets for interventions in patients with SCD summarized under four categories of its pathobiology – (1). Modifying the genotype, (2). Targeting HbS polymerization, (3). Targeting vasocclusion, and (4). Targeting inflammation.

Understanding of the kinetics of HbS polymerization suggest that there are many ways to inhibit HbS polymerization (Eaton and Bunn, 2017) other than induction of HbF (**Table 1**). One approach is to increase oxygen affinity of the hemoglobin molecule, an example is Oxbryta^{TM} (Voxelotor/GBT440) (Vichinsky et al., 2019) that was recently approved by the FDA in November 2019, making this the second anti-sickling agent.

One of the biggest challenges in managing SCD is the clinical complexity and extreme variable clinical course that cannot be explained by the specific disease genotype. Patients with identical sickle genotype still display extreme clinical course; both acquired and inherited factors contribute to this clinical complexity of SCD (Gardner and Thein, 2016). Although laboratory prognostic factors (HbF, hemoglobin, reticulocyte count, leukocytosis) and clinical phenotypes (such as stroke/TIA, acute chest syndrome/pulmonary hypertension, avascular necrosis, kidney injury, or skin ulcers) have been described and analyzed, classifying disease severity remains complex and should be assessed individually. Prediction of disease severity and clinical course of SCD has been the topic of many reviews and, to date there is no clear algorithm using genetic and/or imaging, and/or laboratory markers that can reliably predict mortality risk in SCD (Quinn, 2016).

CURRENT ADVANCES IN THERAPY

(1) Modifying the Patient's Genotype

Modifying the patient's genotype via hemopoietic stem cell transplantation (HSCT) was first reported to be performed over 30 years ago in an 8-year-old child who had SCD (HbSS) with frequent VOCs; she subsequently developed acute myeloid leukemia. The patient received HSCT for the acute myeloid leukemia from an HLA-matched sister who was a carrier for HbS (HbAS). She was cured of her leukemia and at the same time, her sickle cell complications also resolved (Johnson et al., 1984; Johnson, 1985). Until then, HSCT had not been considered as a therapeutic option for SCD. This successful HSCT demonstrated

TABLE 1 | Current advances on therapy for sickle cell disease.

Changing the genotype		
(1) Allogeneic stem cell transplant	Myeloablative regimens (MAC), reduced intensity regimens (RIC), and non-myeloablative regimens (NMA)	50 clinical trials listed in ClinicalTrials.gov
(2) Autologous transplant		10 clinical trials listed in ClinicalTrials.gov
	a) Gene therapy	
	Lentiviral strategies (NCT02247843, NCT02140554, NCT02186418)	
	Inducing fetal hemoglobin	Downregulation of *BCL11*A (NCT03282656) Globin chromatin structure manipulation Downregulating beta^s globin expression
	b) Gene editing	
	Using zinc finger nucleosomes (ZFN), transcription activator-like effector nucleases (TALENs), CRISPR/Cas9 techniques (NCT03745287)	Downregulation of *BCL11*A Reactivation of HbF by HPFH mutations Globin gene repair
Hemoglobin S polymerization	Hydroxyurea (FDA approved)	Ribonucleotide diphosphate reductase inhibitor
	LBH589/Panobinostat (NCT01245179)	Pan histone deacetylase inhibitor
	Voxelotor/GBT440 (NCT03036813) (FDA approved)	α-Globin reversible binding
	Decitabine/THU (NCT01685515)	DNMT1 inhibition
	Sanguinate (NCT02411708)	Targeting carbon monoxide delivery
	IMR-687 (NCT04053803)	Phosphodiesterase 9 inhibitor
Vasocclusion	L-Glutamine (FDA approved)	Increase NADH and NAD redox potential
	Crizanlizumab (NCT03264989) (FDA approved)	P-selectin inhibitor
	Heparinoids: Sevuparin (NCT02515838)	P-selectin and L-selectin inhibitor
	Poloxamer and Vepoloxamer	Nonionic block copolymer surfactant
Inflammation	Prasugrel, ticagrelor (NCT02482298)	P2Y2 inhibitors
	Intravenous immunoglobulin (NCT01783691)	Effects on neutrophils and monocytes activation
	Simvastatin (NCT03599609)	Vascular endothelium
	Rivaroxaban (NCT02072668)	Anti factor Xa
	N-Acetylcysteine (NCT01800526)	Oxidative stress reduction

HbF, hemoglobin F; HPFH, hereditary persistence of fetal hemoglobin; THU, tetrahydrouridine; DNMT1, DNA methyltransferase type 1.

that reversal of SCD could be achieved without complete reversal of the hematological phenotype to HbAA, and paved the way for bone marrow transplant (BMT) as a curative option for children with severe SCD (Walters et al., 1996b).

The conclusion was that, as long as stable mixed hemopoietic chimerism after BMT can be achieved, patients can be cured of their SCD without complete replacement of their bone marrow (Walters et al., 2001).

Allogeneic Bone Marrow Transplant

Hematopoietic stem cell transplant (HSCT) has now become an important therapeutic option for patients with SCD. Currently there are about 35 clinical trials at ClinicalTrials.gov studying allogeneic BMT in patients with SCD. As described by Walters et al. (2010), HSCT can establish donor-derived erythropoiesis, but even more importantly, can stabilize or even restore function in affected organs of patients with SCD when performed in time.

Between 1986 and 2013, 1,000 patients received HLA-identical matched sibling donor (MSD) HSCTs (Gluckman et al., 2017). The outcomes for both children and adults were excellent, demonstrating 93% overall survival. Eighty seven percent of the patients received myeloablative chemotherapy (MAC) and the

rest (13%) received reduced intensity chemotherapy (RIC). It is important to note that patients 16 years or older had worse overall survival (95% vs. 81% p = 0.001) and a higher probability of graft versus host disease (GVHD)-free survival (77% vs. 86% p = 0.001). These results should encourage physicians to provide early referrals to SCD patients for transplant evaluation so that the donor search can be started in a timely matter (Gluckman et al., 2017).

Although myeloablative conditioning has achieved high rates of overall and event free survival, the conditioning is too toxic for adult patients with pre-existing organ dysfunction. Reversal of the sickle hematology without complete replacement of the patient's bone marrow led to the development of less intense conditioning regimens expanding allogeneic transplantation in adult patients, who otherwise would not be able to tolerate the intense myeloablative conditioning. Donors could be HbAA or HbAS, and in order to reverse the sickle hematological genotype, the myeloid donor chimerism has to be >20% (Fitzhugh et al., 2017).

In an international, multicenter study, 59 patients had MSD HSCT, of which 50 survived and were cured of SCD. Of the nine patients that had a negative outcome, five had graft rejection and four intracranial hemorrhage. Thirteen patients developed mixed chimerism. Of those patients that developed mixed chimerism, there was no GVHD or disease recurrence/graft rejection. Patients with stable mixed chimerism did not have worse outcomes related to complications of SCD. Hsieh et al. (2009) developed a protocol for non-myeloablative HSCT with low dose total body radiation, alemtuzumab, and sirolimus. In the initial 10 patients with SCD, nine had long-term, stable, mixed donor chimerism and reversal of their sickle cell phenotype (Hsieh et al., 2009). An updated report showed that 87% of the 30 patients had long-term stable donor engraftment without acute or chronic graft-versus-host disease (Clinical trials [NCT00061568]) (Walters et al., 2001; Hsieh et al., 2014). More recent data reported at least 95% cure rate in 234 children and young adults (<30 years) with SCA after MSD with no increased mortality compared to SCA itself and better quality of life. The data also showed that myeloablative HSCT can be a safe option for patients <15 years old if a MSD is available unless there is a clear and strong recommendation not to undergo transplant (Bernaudin et al., 2020).

However, in the US, less than 15% of patients with SCD have HLA- matched siblings as donors, but a promising alternative donor source is haplo-identical family members. Studies are now underway in several centers to find a balance of conditioning regime that provides adequate immunosuppression without rejection and minimal GVHD (Joseph et al., 2018). Matched unrelated donors (MUD) have shown promising results in patients with thalassemia major and are currently being evaluated in patients with SCD (Fitzhugh et al., 2014). One of the main limitations, unfortunately, is the low probability of finding suitable donors for African and African American populations as per the National Marrow Donor Program and so, not sufficient MUD transplants have been completed in patients with SCD. HLA-haploidentical HSCT following RIC has been reported to show promising results with prolonged and stable engraftment, but for both unrelated umbilical cord blood (UCB) and haploidentical HSCT, rejection remains a major obstacle in the context of RIC (Bolanos-Meade et al., 2012; Angelucci et al., 2014; Fitzhugh et al., 2014; Saraf et al., 2018; Bolanos-Meade et al., 2019).

Although encouraging options with promising results in clinical trials, acute and chronic GVHD remain major complications which can be life threatening and have severe effects on quality of life. Multiple factors affect the development of GVHD in patients undergoing transplant, including the source of the stem cells, the intensity of immunosuppression in the conditioning regime (dose of anti-thymoglobulin) and the mismatch status of the donor to the recipient (Shenoy, 2013; Inamoto et al., 2016; Bernaudin et al., 2020).

Acute GVHD remains a concern in patients receiving mismatched donor transplants but UCB continues to show reduced rates of chronic GVHD (Kamani et al., 2012). Reduced-intensity conditioning regimens have also been studied in related and unrelated HSCT, and while a suitable option for patients with a matched sibling, patients with unrelated donor should be made aware of the not-so-favorable short and long-term outcomes (Guilcher et al., 2018).

As new transplant modalities emerge with less transplant related mortality, better immunomodulators to prevent GVHD are being developed and graft rejection has become less frequent and accepted indications for HSCT have become less restrictive (**Table 2**). Nonetheless, clinicians continue to have reservation toward transplant and tend to delay the referral to a HSCT specialist because of concerns for GVHD, mortality/morbidity related to transplant itself and the risk of graft rejection, which has not been eliminated completely (Leonard and Tisdale, 2018). An ongoing clinical trial will compare 2-year overall survival and outcomes related to SCD in patients that undergo transplant compared with current standard of care (ClinicalTrail.gov Identifier: NCT02766465).

In allogeneic transplant, the source of hematopoietic stem cells (HSCs) is from a donor (matched sibling, haplo-identical family members, UCB or MUD). Allogeneic BMT using HSCs from the latter 3 donor sources are still risky; and donor availability presents a huge limitation. These limitations can be overcome by autologous transplant, in which the patient receives his own cells after being modified by gene therapy.

TABLE 2 | Indications for HSCT balanced with donor availability: Risk/benefit ratio considerations.

Matched sibling donor	Matched unrelated donor or minimally mismatched good quality cord product	Mismatched marrow donor, haploidentical donor
• Stroke	• Stroke	• Recurrent stroke despite adequate chronic transfusion therapy
• Elevated TCD velocity	• Elevated TCD velocity	• Inability to tolerate supportive care though strongly indicated, e.g., red cell alloimmunization, severe VOC and inability to take hydroxyurea
• Acute chest syndrome	• Recurrent acute chest syndrome despite supportive care	
• VOC	• Recurrent severe VOC despite supportive care	
• Pulmonary Hypertension/tricuspid regurgitation jet velocity.2.5 m/s	• Red cell alloimmunization despite intervention plus established indication for chronic transfusion therapy	
• Osteonecrosis/AVN	• Pulmonary hypertension	
• Red cell alloimmunization		
• Silent stroke specially with cognitive impairment		
• Recurrent priapism		
• Sickle nephropathy		

HSCT, hematopoietic stem cell transplantation; AVN, avascular necrosis; TCD, transcranial doppler; VOC, vaso-occlusive crisis.

Autologous Hematopoietic Stem Cell Transplant Modification: Gene Editing or Gene Therapy

Genetically engineered autologous cells eliminate the need to find a HSCT donor, and thus available to all patients. Since these are the patient's own stem cells, there is no need for immunosuppression, thus eliminating the risks of GVHD and immune-mediated graft rejection (Esrick and Bauer, 2018; Orkin and Bauer, 2019).

Sickle cell disease patients represent a special and complicated population for this therapy for two major reasons. First, patients that undergo autologous stem cell transplant require collection of hematopoietic stem cells (CD34+) and the traditional method of collection is a bone marrow harvest done by a specialist but in patients with SCD this process yields CD34+ cells with suboptimal quantity and quality requiring multiple harvests, each harvesting procedure increasing the risk of triggering acute pain crisis. Second, the current gold standard procedure for cell mobilization is with granulocyte-colony stimulating factor (G-CSF) but this is contraindicated in patients with SCD due to risk of causing complications such as pain crisis, acute chest syndrome, and even death, from the increased white cell counts.

Recently, great advances have been made in using an alternative approach for harvesting CD34+ cells using Plerixafor. Plerixafor acts by reversibly blocking the binding between chemokine CXC-receptor 4 (CXCR4) and the stromal cell derived factor-1α triggering the mobilization of progenitor cells into the peripheral blood. It allows peripheral mobilization of stem cells by releasing CD34+ cells from the bone marrow niches, without the massive increase in white blood cells. Its development has been crucial in optimization of CD34+ collection in patients with SCD. Results have shown appropriate mobilization of CD34+ cells 6 h after a single dose of Plerixafor and are of higher quality and purity, decreasing the need for multiple bone marrow harvests and the associated stress/pain. Associated with hyper-transfusion therapy, it has become the preferred way of marrow stimulation to yield appropriate hematopoietic stem/progenitor cells in patients with SCD (Boulad et al., 2018; Esrick et al., 2018; Hsieh and Tisdale, 2018; Lagresle-Peyrou et al., 2018).

The genetic defect in the sickle HSPCs can be corrected via several approaches.

(A) Gene addition using lentiviral vector-based strategies

(a) Anti- or non-sickling strategies: Several gene therapies based on gene addition using viral vectors to carry therapeutic genes in HSCs are being actively developed with curative purposes. Gene addition strategies that have reached clinical trials include a promising one where the patient's stem cells are infected with a lentivirus expressing an anti-sickling β-globin variant, T87Q. The unique feature of this vector is that the amino acid substitution (β^{A-T87Q}) allows for high performance liquid chromatography (HPLC) monitoring of the transgene globin levels in the patient's cells (Cavazzana-Calvo et al., 2010). The first SCD patient who received this Bluebird vector (protocol HGB-205) was reported in 2017; engraftment was stable with no sickle cell crises reported at 15 months of follow up (Ribeil et al., 2017), with further undergoing studies (ClinicalTrials.gov Identifier: NCT02140554, NCT03282656). Other approaches to anti-sickling gene therapy in erythroid-specific lentiviral vectors include utilizing a β-globin gene with three specific point mutations that confer anti-sickling properties (ClinicalTrials.gov Identifier: NCT02247843) or the introduction of a γ-globin coding sequence in a β-globin gene to increase HbF levels and decrease HbS (ClinicalTrials.gov Identifier: NCT02186418) (Cavazzana et al., 2017). Thus far, the most promising of these LV vectors is the one utilizing anti-sickling β-globin variant, T87Q.

(b) Hb F induction: The well-established efficacy of increasing HbF has motivated both pharmacological and genetic approaches to HbF induction.

A gene addition approach that is already in clinical trials (ClinicalTrials.gov Identifier: NCT03282656) utilizes a lentiviral mediated erythroid specific short hairpin RNA (shRNA) for BCL11A. This shRNA is modified to target the specific gene and downregulate its expression (Brendel et al., 2016). As of December 2018, three adults have been enrolled, utilizing plerixafor mobilized HSC, all three patients showed prompt neutrophil engraftment, and at 2 months follow up, the average HbF was 30% (ASH abstract #1023 – 2018 ASH conference). Other lentiviral therapies using zinc-finger nucleases (ZFN) directed against the γ-globin promoter have been proposed. This would force an interacting loop between the LCR and γ-globin which would reactivate γ-globin production, increasing HbF and decreasing HbS production at the same time. These lentiviral-based approaches still need preclinical *in vivo* studies to address safety and specificity before they can be considered in human patients (Breda et al., 2016; Orkin and Bauer, 2019).

Viral vectors, such as lentivirus, are a great tool for gene therapy but these results underscore the need to develop gene transfer protocols that ensure efficient and consistent delivery of the therapeutic globin gene cargo to HSC. Their major limitations include:

(1) Their immunogenicity which can create an inflammatory response in the donor which can lead to degeneration of the transduced tissue, (2) they can produce non-specific toxins, (3) due to the semi-random integration to the genome, there is a theoretical risk of insertional mutagenesis, (4) they have limitations of transgenic capacity size. An additional challenge in SCD is the ability to maintain a persistent myeloid donor chimerism of >20% to prevent return of SCD symptoms (Fitzhugh et al., 2017). Due to these limitations, long-term monitoring of patients to evaluate both safety and efficacy is necessary. Until now, over the last decade of clinical trials, no genotoxicity secondary to LV vectors has been reported but the main challenge has been to keep the myeloid donor chimerism above the 20% threshold (Nayerossadat et al., 2012).

(B) Gene editing

Gene-editing corrects a specific defective DNA in its native location. SCD with its simple single base change presents a very attractive prototype. Over the last couple of decades, there has been a spectacular growth of such strategies, setting the scene for developing therapies that could precisely genetically correct a single base mutation in patient with SCD. These strategies include ZFNs, transcription activator-like effector nucleases (TALENs) and the clustered regularly interspaced short palindromic repeat (CRISPR)-associated nuclease Cas9 approach which is the most advanced of the three. The CRISPR-Cas9 technology typically make a double-stranded break (DSB) in a particular genomic sequence directed to that site by a guide RNA. The most common method of DSB repair is non-homologous end joining, often resulting in gene disruption or knockout. This strategy is currently being tested in a clinical trial (ClinicalTrials.gov Identifier: NCT03745287) in which the patient's own *BCL11A* gene (a major inhibitor of γ-globin gene expression) is disrupted to induce HbF expression. BCL11A also has roles in lymphoid and neurological development but gene-editing for SCD exploits the erythroid-specific enhancers in intron 2 of the gene (Bauer et al., 2013; Brendel et al., 2016). CRISPR-Cas9 technology is also being explored to mimic the rare, genetic variants that promote expression of the γ-globin genes as in hereditary persistence of fetal hemoglobin (Traxler et al., 2016; Wienert et al., 2018). Disrupting the putative binding sites for γ-globin repressors like BCL11A to induce HbF production will be an attractive therapeutic strategy for both β-thalassemic and SCD patients (Masuda et al., 2016; Liu et al., 2018; Martyn et al., 2018). The ultimate challenge, however, is to genetically correct the mutation, a single nucleotide change in the codon of the globin gene from GAG to GTG, by providing a homology template with the correct sequence at the sixth codon. Although this has been completed in preclinical studies, current techniques do not allow for specific transversion mutations like those required to cure SCD in humans (Dever et al., 2016; Orkin and Bauer, 2019). The enormous selective advantage of red blood cells with normal hemoglobin or anti-sickling hemoglobin predicts that genetic modification of a proportion of HSCs (estimated 10–20%) may suffice as a one-off treatment (Fitzhugh et al., 2017). Before gene therapy can become a reality, however, many hurdles need to be overcome; genetically manipulated HSCs need to be able to retain long-term repopulating potential; pre-transplant conditioning is toxic and needs to be modified to reduce the morbidity. A clinical trial exploring antibody-mediated non-chemotherapy conditioning is being evaluated in patients with severe combined immunodeficiency, in an attempt to reduce the exposure to chemotherapy and its toxicities is currently recruiting patients (ClincialTrials.gov Identifier: NCT02963064). Further understanding of this technology could represent a new option for patients with SCD.

Although different gene strategies have reached clinical trials showing promising results they remain in early phases of development and allogeneic HSCT remain the only curative treatment modality for SCD. For the majority of patients without a MSD, haploidentical HSCT with recent promising data of improved overall survival presents an alternative for curative therapy. Multiple gene therapy strategies utilizing patient's own stem cells, are also being pursued, but this has the disadvantage of myeloablative conditioning (Leonard et al., 2020).

In addition to great advances in HSCT and gene therapy, new pharmacological anti-sickling approaches have developed.

(2) Targeting Hemoglobin S Polymerization

Approaches targeting HbS polymerization presents a very attractive strategy as this "puts out the fire" rather than dealing with the sequelae of the sickling event (Eaton and Bunn, 2017). HbF has long been known to have a major beneficial effect in SCD – increased intracellular HbF not only dilutes the intracellular HbS concentration but inhibits sickling as the mixed hybrid tetramers do not partake in HbS polymerization. Hydroxyurea (HU) works via induction of fetal hemoglobin (HbF, α2γ2) synthesis, but hydroxyurea is only partially successful as the increase in HbF is uneven and not equally present in all the red blood cells (Ware, 2015). Nonetheless, use of HU therapy in SCD has expanded substantially in recent years. Follow on studies include demontration of its efficacy and safety in the pediatric population (BABY HUG) (Wang et al., 2011), the Transcranial doppler with Transfusion Changing to Hydroxyurea Study (TWiTCH) that showed HU was comparable to blood transfusions for primary stroke prevention (Ware et al., 2016) although the Stroke with Transfusion Changing to Hydroxyurea study (SWiTCH) concluded that HU is not comparable to blood transfusion in secondary stroke prevention (Ware et al., 2011).

More recently, two clinical studies have shown that HU is relatively safe in Sub Saharan Africa, a setting with high infectious disease and SCD burden. Hydroxyurea has been shown to not only decrease complications from SCD such as VOC, acute chest syndrome, frequency of transfusions, death and infections – including malaria but also to be a feasible approach in these under-resourced countries (Opoka et al., 2017; Tshilolo et al., 2019).

Despite having a significant impact in patients with SCD, there are still multiple unanswered questions regarding HU. Its mechanism of action has not been fully understood and its impact on HbF will decrease over time. Older patients become more sensitive to the dosage and they require frequent blood tests and readjustment of their dose. Regardless of the advances, there is no clear evidence of the long-term effect of hydroxyurea in preventing end organ damage (Nevitt et al., 2017; Luzzatto and Makani, 2019). There is also conflicting evidence of the effects of HU on male fertility (DeBaun, 2014). Chronic complications of SCD such as recurrent episodes of priapism, asymptomatic testicular infarctions and primary hypogonadism have been described as potential etiologies of low fertility in male SCD patients. Studies in transgenic SCD mice showed that SCD itself was associated with inhibition of spermatogenesis and primary hypogonadism but when compared to HU (25 mg/kg/day), testicular volume was lower in those mice with SCD exposed to HU, inferring lower spermatogenesis. Berthaut et al. (2008) measured the semen quality of 4 patients with SCA at baseline and 4 years after starting hydroxyurea. In three of four patients the spermatozoan concentration continued to drop while patients were taking the medication and did not return to baseline after discontinuing HU (Berthaut et al., 2008). Although the evidence is limited, full disclosure regarding implications on male fertility should be given to patients and families in order to make an informed decision before starting HU (Jones et al., 2009).

Other than HU, other pharmacological options to increase HbF are still experimental undergoing clinical trials. Molecular studies on γ-globin identified regulatory elements in the gene expression and subsequent HbF production. Such molecules; histone deacetylase (HDAC), DNA methyltransferase 1 (DNMT1), BCL11A and SOX6 modifying HbF expression have been explored as possible therapeutic options.

One of the proposed mechanisms for HU effect on HbF is stimulation of cyclic guanosine monophosphate (cGMP). Phosphodiesterase 9 (PDE9) is a specific enzyme in charge of degrading cGMP and is highly present in neutrophils and RBCs of patients with SCD. A novel, potent and selective PDE9 inhibitor (IMR-687) has been shown to increase levels of cGMP and HbF without signs of myelosuppression in cell lines of patients with SCD. An open-label extension to a previous phase 2a study is ongoing in adults with SCD (ClinicalTrials.gov Identifier: NCT04053803) (McArthur et al., 2019).

Panobinostat is a pan HDAC inhibitor currently being studied in adult patients with SCD as a phase 1 study (ClinicalTrials.gov Identifier: NCT01245179). *In vitro* analysis of human erythroid progenitor cells that underwent shRNA knockdown of HDAC1 or HDAC2 genes resulted in increased levels of γ-globin but without altering cellular proliferation of the cell cycle phase.

Associated with HU, HDAC gene inhibition produced a more pronounced increase of γ-globin and HbF (Esrick et al., 2015).

DNA Methyltransferase 1 is involved in the shutting down of γ-globin gene after birth and its subsequent production. DNA methylransferase inhibitor 5-azacytidine was one of the chemotherapeutic agents used to reactivate HbF but it was quickly abandoned due to its toxicity and carcinogenicity. Decitabine, an analog of 5-azacytidine, is also a potent DNMT1 inhibitor with a more favorable safety profile but decitabine is rapidly deaminated and inactivated by cytosine deaminase, if taken orally. To overcome this limitation, a clinical study combines decitabine and tetrahydrouridine (THU), a cytosine deaminase inhibitor, as a therapeutic strategy for inducing HbF (ClinicalTrials.gov Identifier: NCT01685515). In a phase 1 study, Molokie et al. (2017) showed that the inhibition of DNMT1 led to appropriate blood levels of decitabine that were safe and induced a large increase in fetal hemoglobin in healthy red blood cells. These agents did not induce cytoreduction, but increased platelets count that can potentially trigger vaso-occlusion in SCD patients (Molokie et al., 2017).

Voxelotor (Oxbryta/GBT440) binds specifically to the N-terminus of the alpha subunit of HbS to stabilize the oxygenated hemoglobin state (Strader et al., 2019), thus reducing the predisposition to sickling. Voxelotor (Oxbryta/GBT440) was approved by the FDA in November 2019 for the treatment of SCD in adults and pediatric patients 12 years of age and older. The HOPE study showed an increase in hemoglobin levels and reduced markers of hemolysis in 274 patients with HbS that were randomly assigned to receive the study drug versus placebo. These findings have not correlated with reduced episodes of pain crisis and/or end organ damage. Agents that shift Hb oxygen affinity present some concerns of potential negative effects as the bound oxygen cannot be off loaded in tissues with high oxygen requirements, particularly concerning in a disease characterized by decreased oxygen delivery (Hebbel and Hedlund, 2018; Thompson, 2019). These concerns are being addressed in a current phase 3, double-blind, randomized, placebo-controlled, multicenter study of Voxelotor (ClinicalTrials.gov Identifier: NCT03036813) (Vichinsky et al., 2019).

Dehydration of the RBC appears to be closely controlled by the efflux of potassium through 2 specific pathways; one is the potassium chloride cotransport and the other, calcium-activated potassium efflux (Gardos channel). Senicapoc blocks the Gardos channels, thus preventing dehydration of the red cells. Preclinical and phase 1/2 showed that inhibition of potassium flow through the Gardos channel increased Hb levels and decreased hemolysis (ClinicalTrials.gov Identifier: NCT00040677). A phase 3 study was terminated for lack of efficacy (ClinicalTrials.gov Identifier: NCT00294541) (Ataga et al., 2008; Ataga and Stocker, 2009).

N-Methyl D-aspartate receptors (NMDARs) are non-selective calcium channels present in erythroid precursors and circulating RBCs and have been shown to be abnormally increased in RBCs of patients with SCD (Hanggi et al., 2014). These channels are closely related with RBC hydration that affects the intracellular HbS concentration and thereby HbS polymerization and sickling of RBCs. Memantine is a NMDAR inhibitor which has shown to improve hydration of RBCs of patients with SCD *in vitro* and to reduce sickling in the setting of deoxygenation. It is being explored in an ongoing phase 2 clinical trial (ClinicalTrials.gov Identifier: NCT03247218).

Sanguinate which is a bovine PEGylated hemoglobin product attempts to block polymerization by targeting carbon monoxide (CO) delivery. By binding to HbS polymers, CO enhances their melting and minimize their persistence in peripheral blood. However, this equilibrium is based on high concentrations of CO. A phase 1/2 single-blind, randomized, placebo-controlled study of this agent in the management of pain crisis has been carried out but no results have yet been posted (ClinicalTrials.gov Identifier: NCT02411708).

(3) Targeting Vasocclusion

Increased expression and activation of normally inactive erythroid adhesion molecules promote cytoadherence of sickle RBCs to the endothelium accompanied by platelets and leukocytes. Activated leukocytes and platelets further increase the risk to develop VOC (Nasimuzzaman and Malik, 2019; Sundd et al., 2019; Telen et al., 2019).

Previous *in vitro* studies had demonstrated that glutamine depletion contributed to red blood cell membrane damage and adhesion. Uptake of L-glutamine uptake is markedly increased in patients with SCD, primarily to increase the total intracellular NAD level (Morris et al., 2008). In a phase 3 study, L-glutamine demonstrated a 25% reduction in the median number of pain crisis, 30% less hospitalizations and reduced acute chest episodes in children and adults with SCD with or without HU over a 48-week period. There were 36% drop-out rate in the glutamine arm and 24% in the placebo control arm from unknown reasons. L-Glutamine appears to significantly increase NADH and NAD redox potential and decrease endothelial adhesion, but its mechanism remains still unknown and there are concerns regarding its use in patients with renal impairment, a common sickle-related complication (Quinn, 2018). In July 2017, the pharmacological grade of L-glutamine (Endari) was approved by the FDA for use in patients with SCD, 5 years or older (Niihara et al., 2018). Of note, L-glutamine has not been approved by the European Medicines Agency for treating SCD.

In the future it could be a useful combination therapy with HU (Minniti, 2018) but uptake among patients is still low, one of the reasons is the unpleasant taste. There are potentially less expensive pharmaceutical formulations of L-glutamine available off the counter, but purity of the effective agents in these compounds have not been validated.

As the endothelium emerge as a key factor in the constant activation of adhesion molecules in sickle RBCs, these adhesion molecules present a very attractive therapeutic target. Selectins, which are present in endothelial cells and are the initial step toward a firm adhesion between RBCs and the endothelium, have been further studied and targeted as possible therapeutic approaches.

Crizanlizumab is a monoclonal antibody to P-selectin and its mechanism of action is to block the adhesion of activated erythrocytes, neutrophils and platelets. In a phase 2, multicenter, randomized, placebo controlled double blind study, crizanlizumab with or without hydroxyurea (SUSTAIN study) (ClinicalTrails.gov Identifier: NCT01895361) showed that patients on the treatment arm had significantly lower rate of sickle-related pain crises compared to placebo with a lower incidence of adverse events - 10% of patients suffered from moderate side effects while one patient suffered from an intracranial bleed during treatment with this drug that could also interfere with platelet function via its effects on selectins (Ataga et al., 2017). *Post hoc* analyses showed that more patients were VOC event-free in the crizanlizumab arm than in the placebo arm, and that crizanlizumab also significantly increased time-to-first VOC compared to the placebo (Kutlar et al., 2019). A phase 3 interventional, multicenter, randomized, double-blind clinical trial is ongoing to assess safety and efficacy of crinalizumab with or without hydroxyurea in patients with SCD and history of VOC (ClinicalTrials.gov Identifier: NCT03814716). In November 2019, the US Food and Drug Administration approved crizanlizumab-tmca (ADAKVEO, Novartis) to reduce the frequency of VOC in adults and pediatric patients aged 16 years and older with SCD.

Rivipansel is a pan-selectin inhibitor with its strongest activity against E-selectin. In a multicenter, randomized, double—blind, placebo—controlled phase 2 study (ClinicalTrails.gov Identifier: NCT01119833), Rivipansel showed clinical and meaningful reductions in multiple measures of VOC compared with those receiving standard of care treatment (Telen et al., 2015). A phase 3 study (Identifier: NCT02187003) to evaluate the efficacy and safety of rivipansel in the treatment of VOC in hospitalized patients with SCD was terminated (posted on ClinicalTrials.gov February 20, 2020) based on failure of the primary study (NCT02433158) to meet the study efficacy endpoints of time to readiness-for-discharge.

In a phase 1, dose-escalation study propranolol showed it significantly reduced epinephrine-stimulated sickle RBCs adhesion. A phase 2 study (NCT01077921) showed decrease in adhesion molecules such as E-selectin and P-selectin but results were not statistically significant and no clinical endpoints were discussed (De Castro et al., 2012).

Due to their P-selectin mediated adhesion inhibition properties, heparinoids have been additionally investigated with interesting results. Sevuparin, a heparin derivate polysaccharide that has shown to bind to P— and L—selectins, thrombospondin, fibronectin and von Willebrand factor, all of which are thought to contribute to vasocclusion in SCD. It has been reported to inhibit sickle RBC adhesion to the endothelial cells and to reduce tumor necrosis factor-induced vasocclusion. It is currently being tested in a phase 2 clinical trial, placebo controlled, to study its efficacy and safety in patients with SCD during VOC (ClinicalTrials.gov Identifier: NCT02515838) (Telen et al., 2016). Other heparinoids such as Dalteparin showed incomplete evidence to support or refute its effectiveness in the management of patients with SCD. There are ongoing trials (ClinicalTrials.gov Identifier: NCT02098993) to assess the feasibility of unfractionated heparin in patients with SCD admitted with pain crisis. Well-designed studies are still needed to clarify its role in the management of patients with SCD and to assess the safety of this approach (van Zuuren and Fedorowicz, 2015).

Poloxamer 188 is a non-ionic block copolymer surfactant thought to seal stable defects in the microvasculature leading to an improvement in blood flow and decreasing blood viscosity. Although its mechanism is not well understood, a randomized, double-blind, placebo-controlled trial showed that it decreased the duration of sickle crisis by 8 h compared to placebo (133 h vs. 141 h, $p = 0.04$) and more patients receiving the medication reported crisis resolution (52% vs. 37%, $p = 0.02$) (Orringer et al., 2001). In an early phase 2 study, one patient receiving the medication developed renal dysfunction due to presence of low molecular weight substances and a purified version was designed (Adams-Graves et al., 1997). Vepoloxamer, a purified form of Poloxamer 188 with multi mechanistic properties, was believed to improve RBC adhesion, membrane fragility and organ damage. Unfortunately, a phase 3 study failed to reduce the mean duration of VOC in patients with SCD compared to placebo (Adams-Graves et al., 1997).

(4) Targeting Inflammation

Continual background inflammation contributes to organ damage in patients with SCD. Persistent activation of platelets, neutrophils, monocytes, endothelium, and coagulation factors are key participants in this vicious cycle. Different therapeutic approaches have been proposed to assess the impact in patients with SCD (Nasimuzzaman and Malik, 2019; Sundd et al., 2019; Telen et al., 2019).

Intravenous immunoglobulin (IVIG) and statins have been studied for their anti-inflammatory effects on neutrophils and monocyte adhesion. Patients on statin demonstrated a decrease in C-reactive protein, soluble ICAM1, soluble E-selectin and vascular endothelial growth. Simvastatin was found to reduce adhesion of white blood cells and in combination with hydroxyurea, was found to decrease the number of pain crisis and markers of inflammation (Hoppe et al., 2017). Currently,

FIGURE 3 | The different therapeutic approaches for sickle cell disease and their mechanisms and current status in clinical trials. Orange: targeting hemoglobin S polymerization; gray: targeting vasocclusion; light blue: targeting inflammation and green: modification of the genotype. shRNA, short hairpin RNA; Hb S, hemoglobin S; Hb F, hemoglobin F; PDE9, phosphodiesterase 9. *FDA approved July 2017; **FDA approved November 2019; ***Terminated in February 20, 2020 due to failure to meet primary endpoints.

there is an active clinical trial to assess the effect of simvastatin on central nervous system vasculature in patients with SCD (ClinicalTrials.gov Identifier: NCT03599609).

N-Acetylcysteine (NAC) commonly used in respiratory conditions has also been tested for patients with SCD. In a phase 2 study, NAC proved to inhibit dense cell formation and restored glutathione levels toward normal. The decrease in irreversible sickling of RBCs was not statistically significant but a downward trend was observed (Pace et al., 2003; Nur et al., 2012). Further studies have shown decreased red cell membrane expression of phosphatidylserine which seems to reflect overall reduced oxidative stress. To better assess its clinical effect in patients with SCD, a pilot study, currently enrolling with invitation is studying its effect in redox and RBC function during VOC (ClinicalTrials.gov Identifier: NCT01800526).

Aberrant activation of the coagulation cascade, abnormal excess of TF on the endothelial wall and high plasma levels of different coagulation factors drive increased thrombin and fibrin production leading to further inflammation and risk of VOC (Sundd et al., 2019). In a SCD mouse model, factor Xa, TF, and thrombin differentially contributed to vascular inflammation (Sparkenbaugh and Pawlinski, 2013). Factor Xa inhibition demonstrated a decrease in vascular inflammation as assessed by the lower interleukin 6 levels. Although thrombin had no effect on interleukin 6, it was a significant factor for neutrophil infiltration and further inflammation (Sparkenbaugh et al., 2014). A retrospective analysis of rivaroxaban, a factor Xa inhibitor, demonstrated non-inferiority with regard to thrombosis compared to warfarin with the advantage of less outpatient visits and monitoring (Bhat and Han, 2017). Currently, a two-treatment phase clinical trial with rivaroxaban on the pathology of SCD has been completed but results are pending (ClinicalTrials.gov Identifier: NCT02072668). Patients with SCD have increased platelet levels at baseline that are further increased during acute VOC. Platelet activation triggers further leukocyte activation and promote RBC adhesion to an exposed endothelium (Conran and Belcher, 2018) setting off a vicious cycle of adhesion events. Antiplatelet therapy with Clopidogrel in patients with SCD, unfortunately, were disappointing. New, third generation P2Y12 inhibitors such as ticagrelor and prasugrel have also been studied in patients with SCD. Prasugrel showed appropriate levels of anti-platelet aggregation compared to healthy patients in *ex vivo* studies, and was well tolerated by patients, but on a 24-month follow up, patients on the treatment arm failed to show reduction in the frequency of VOC (Heeney et al., 2016; Conran and Rees, 2017). Ticagrelor, in a phase 2b study, was well tolerated, but failed to show effect in the frequency of VOC (Kanter et al., 2019) (ClinicalTrials.gov identifier: NCT02482298). Previous studies have also showed that aspirin as an anticoagulant therapy did not provide benefit over placebo, although it is used as an analgesic in many parts of Africa (Sins et al., 2017).

In patients with SCD, continual lysis of RBCs activates the inflammasome triggering the release of multiple cytokines, including IL-1β (Awojoodu et al., 2014). Canakinumab is a humanized monoclonal antibody that targets interleukin 1-β (IL-1β), and thus potentially could be useful in mitigating some of the inflammation in SCD. Canakinumab was shown to be well tolerated and not associated with major side effects in pediatric and young adult patients (Rees, 2019). A clinical trial to assess its efficacy, safety and tolerability is ongoing in the pediatric population (ClinicalTrials.gov Identifier: NCT02961218).

CONCLUSION

In the last 30 years, there has been a revolution in the medical sciences, and SCD because of its genetic simplicity, has been at the forefront of the numerous scientific discoveries. Tremendous progress has been made in understanding its pathophysiology and pathobiological complexities, but developing treatments, has been disproportionately slow and elusive. However, after a century of neglect, going back to basics offers hope for translating these insights into better therapeutic options – pharmacological and genetic – and for finding curative genetic options for SCD (**Figure 3**). Although frequent in the US, SCD is far more prevalent in Africa where patients have less access to resources, medical treatment and facilities and the consequences of the disease are devastating. As we move forward, we have to continue focus our therapeutic approaches so that they can be accessed by those that suffer the most.

AUTHOR CONTRIBUTIONS

GSC and ST wrote and revised the manuscript.

REFERENCES

Adams-Graves, P., Kedar, A., Koshy, M., Steinberg, M., Veith, R., Ward, D., et al. (1997). RheothRx (poloxamer 188) injection for the acute painful episode of sickle cell disease: a pilot study. *Blood* 90, 2041–2046.

Angelucci, E., Matthes-Martin, S., Baronciani, D., Bernaudin, F., Bonanomi, S., Cappellini, M. D., et al. (2014). Hematopoietic stem cell transplantation in thalassemia major and sickle cell disease: indications and management recommendations from an international expert panel. *Haematologica* 99, 811–820. doi: 10.3324/haematol.2013.099747

Ataga, K. I., Kutlar, A., Kanter, J., Liles, D., Cancado, R., Friedrisch, J., et al. (2017). Crizanlizumab for the prevention of pain crises in sickle cell disease. *N. Engl. J. Med.* 376, 429–439. doi: 10.1056/NEJMoa1611770

Ataga, K. I., Smith, W. R., De Castro, L. M., Swerdlow, P., Saunthararajah, Y., Castro, O., et al. (2008). Efficacy and safety of the Gardos channel blocker, senicapoc (ICA-17043), in patients with sickle cell anemia. *Blood* 111, 3991–3997. doi: 10.1182/blood-2007-08-110098

Ataga, K. I., and Stocker, J. (2009). Senicapoc (ICA-17043): a potential therapy for the prevention and treatment of hemolysis-associated complications in sickle cell anemia. *Expert Opin. Investig. Drugs* 18, 231–239. doi: 10.1517/13543780802708011

Awojoodu, A. O., Keegan, P. M., Lane, A. R., Zhang, Y., Lynch, K. R., Platt, M. O., et al. (2014). Acid sphingomyelinase is activated in sickle cell erythrocytes and contributes to inflammatory microparticle generation in SCD. *Blood* 124, 1941–1950. doi: 10.1182/blood-2014-01-543652

Bauer, D. E., Kamran, S. C., Lessard, S., Xu, J., Fujiwara, Y., Lin, C., et al. (2013). An erythroid enhancer of BCL11A subject to genetic variation determines fetal hemoglobin level. *Science* 342, 253–257. doi: 10.1126/science.1242088

Bernaudin, F., Dalle, J. H., Bories, D., de Latour, R. P., Robin, M., Bertrand, Y., et al. (2020). Long-term event-free survival, chimerism and fertility outcomes in 234 patients with sickle-cell anemia younger than 30 years after myeloablative conditioning and matched-sibling transplantation in France. *Haematologica* 105, 91–101. doi: 10.3324/haematol.2018.213207

Berthaut, I., Guignedoux, G., Kirsch-Noir, F., de Larouziere, V., Ravel, C., Bachir, D., et al. (2008). Influence of sickle cell disease and treatment with hydroxyurea on sperm parameters and fertility of human males. *Haematologica* 93, 988–993. doi: 10.3324/haematol.11515

Bhat, S., and Han, J. (2017). Outcomes of rivaroxaban use in patients with sickle cell disease. *Ann. Pharmacother.* 51, 357–358. doi: 10.1177/1060028016681129

Bolanos-Meade, J., Cooke, K. R., Gamper, C. J., Ali, S. A., Ambinder, R. F., Borrello, I. M., et al. (2019). *Lancet Haematol.* 6, e183–e193.

Bolanos-Meade, J., Fuchs, E. J., Luznik, L., Lanzkron, S. M., Gamper, C. J., Jones, R. J., et al. (2012). HLA-haploidentical bone marrow transplantation with post-transplant cyclophosphamide expands the donor pool for patients with sickle cell disease. *Blood* 120, 4285–4291. doi: 10.1182/blood-2012-07-438408

Boulad, F., Shore, T., van Besien, K., Minniti, C., Barbu-Stevanovic, M., Fedus, S. W., et al. (2018). Safety and efficacy of plerixafor dose escalation for the mobilization of CD34(+) hematopoietic progenitor cells in patients with sickle cell disease: interim results. *Haematologica* 103, 770–777. doi: 10.3324/haematol.2017.187047

Breda, L., Motta, I., Lourenco, S., Gemmo, C., Deng, W., Rupon, J. W., et al. (2016). Forced chromatin looping raises fetal hemoglobin in adult sickle cells to higher levels than pharmacologic inducers. *Blood* 128, 1139–1143. doi: 10.1182/blood-2016-01-691089

Brendel, C., Guda, S., Renella, R., Bauer, D. E., Canver, M. C., Kim, Y. J., et al. (2016). Lineage-specific BCL11A knockdown circumvents toxicities and reverses sickle phenotype. *J. Clin. Invest.* 126, 3868–3878. doi: 10.1172/JCI87885

Brittenham, G. M., Schechter, A. N., and Noguchi, C. T. (1985). Hemoglobin S polymerization: primary determinant of the hemolytic and clinical severity of the sickling syndromes. *Blood* 65, 183–189.

Brunson, A., Lei, A., Rosenberg, A. S., White, R. H., Keegan, T., and Wun, T. (2017). Increased incidence of VTE in sickle cell disease patients: risk factors, recurrence and impact on mortality. *Br. J. Haematol.* 178, 319–326. doi: 10.1111/bjh.14655

Bunn, H. F. (1997). Pathogenesis and treatment of sickle cell disease. *N. Engl. J. Med.* 337, 762–769.

Cavazzana, M., Antoniani, C., and Miccio, A. (2017). Gene therapy for beta-hemoglobinopathies. *Mol. Ther.* 25, 1142–1154. doi: 10.1016/j.ymthe.2017.03.024

Cavazzana-Calvo, M., Payen, E., Negre, O., Wang, G., Hehir, K., Fusil, F., et al. (2010). Transfusion independence and HMGA2 activation after gene therapy of human beta-thalassaemia. *Nature* 467, 318–322. doi: 10.1038/nature09328

Chou, S. T., Alsawas, M., Fasano, R. M., Field, J. J., Hendrickson, J. E., Howard, J., et al. (2020). American society of hematology 2020 guidelines for sickle cell disease: transfusion support. *Blood Adv.* 4, 327–355. doi: 10.1182/bloodadvances.2019001143

Chou, S. T., Evans, P., Vege, S., Coleman, S. L., Friedman, D. F., Keller, M., et al. (2018). RH genotype matching for transfusion support in sickle cell disease. *Blood* 132, 1198–1207. doi: 10.1182/blood-2018-05-851360

Chou, S. T., Jackson, T., Vege, S., Smith-Whitley, K., Friedman, D. F., and Westhoff, C. M. (2013). High prevalence of red blood cell alloimmunization in sickle cell disease despite transfusion from RH-matched minority donors. *Blood* 122, 1062–1071. doi: 10.1182/blood-2013-03-490623

Cokic, V. P., Smith, R. D., Beleslin-Cokic, B. B., Njoroge, J. M., Miller, J. L., Gladwin, M. T., et al. (2003). Hydroxyurea induces fetal hemoglobin by the nitric oxide-dependent activation of soluble guanylyl cyclase. *J. Clin. Invest.* 111, 231–239. doi: 10.1172/JCI16672

Conran, N., and Belcher, J. D. (2018). Inflammation in sickle cell disease. *Clin. Hemorheol. Microcirc.* 68, 263–299. doi: 10.3233/CH-189012

Conran, N., and Rees, D. C. (2017). Prasugrel hydrochloride for the treatment of sickle cell disease. *Expert Opin. Investig. Drugs* 26, 865–872. doi: 10.1080/13543784.2017.1335710

De Castro, L. M., Zennadi, R., Jonassaint, J. C., Batchvarova, M., and Telen, M. J. (2012). Effect of propranolol as antiadhesive therapy in sickle cell disease. *Clin. Transl. Sci.* 5, 437–444. doi: 10.1111/cts.12005

DeBaun, M. R. (2014). Hydroxyurea therapy contributes to infertility in adult men with sickle cell disease: a review. *Expert Rev. Hematol.* 7, 767–773. doi: 10.1586/17474086.2014.959922

Dever, D. P., Bak, R. O., Reinisch, A., Camarena, J., Washington, G., Nicolas, C. E., et al. (2016). CRISPR/Cas9 beta-globin gene targeting in human haematopoietic stem cells. *Nature* 539, 384–389. doi: 10.1038/nature20134

Eaton, W. A., and Bunn, H. F. (2017). Treating sickle cell disease by targeting HbS polymerization. *Blood* 129, 2719–2726. doi: 10.1182/blood-2017-02-765891

Elmariah, H., Garrett, M. E., De Castro, L. M., Jonassaint, J. C., Ataga, K. I., Eckman, J. R., et al. (2014). Factors associated with survival in a contemporary adult sickle cell disease cohort. *Am. J. Hematol.* 89, 530–535. doi: 10.1002/ajh.23683

Esrick, E. B., and Bauer, D. E. (2018). Genetic therapies for sickle cell disease. *Semin. Hematol.* 55, 76–86. doi: 10.1053/j.seminhematol.2018.04.014

Esrick, E. B., Manis, J. P., Daley, H., Baricordi, C., Trebeden-Negre, H., Pierciey, F. J., et al. (2018). Successful hematopoietic stem cell mobilization and apheresis collection using plerixafor alone in sickle cell patients. *Blood Adv.* 2, 2505–2512. doi: 10.1182/bloodadvances.2018016725

Esrick, E. B., McConkey, M., Lin, K., Frisbee, A., and Ebert, B. L. (2015). Inactivation of HDAC1 or HDAC2 induces gamma globin expression without altering cell cycle or proliferation. *Am. J. Hematol.* 90, 624–628. doi: 10.1002/ajh.24019

Fitzhugh, C. D., Abraham, A. A., Tisdale, J. F., and Hsieh, M. M. (2014). Hematopoietic stem cell transplantation for patients with sickle cell disease: progress and future directions. *Hematol. Oncol. Clin. N. Am.* 28, 1171–1185. doi: 10.1016/j.hoc.2014.08.014

Fitzhugh, C. D., Cordes, S., Taylor, T., Coles, W., Roskom, K., Link, M., et al. (2017). At least 20% donor myeloid chimerism is necessary to reverse the sickle phenotype after allogeneic HSCT. *Blood* 130, 1946–1948. doi: 10.1182/blood-2017-03-772392

Gardner, K., Douiri, A., Drasar, E., Allman, M., Mwirigi, A., Awogbade, M., et al. (2016). Survival in adults with sickle cell disease in a high-income setting. *Blood* 128, 1436–1438. doi: 10.1182/blood-2016-05-716910

Gardner, K., and Thein, S. L. (2016). "Genetic factors modifying sickle cell disease severity," in *Sickle Cell Anemia - From Basic Science to Clinical Practice*, eds F. F. Costa and N. Conran (Cham: Springer International), 371–397.

Gluckman, E., Cappelli, B., Bernaudin, F., Labopin, M., Volt, F., Carreras, J., et al. (2017). Sickle cell disease: an international survey of results of HLA-identical sibling hematopoietic stem cell transplantation. *Blood* 129, 1548–1556. doi: 10.1182/blood-2016-10-745711

Goldstein, J., Konigsberg, W., and Hill, R. J. (1963). The structure of human hemoglobin: VI. The sequence of amino acids in the tryptic peptides of the β chain. *J. Biol. Chem.* 238, 2016–2027.

Guilcher, G. M. T., Truong, T. H., Saraf, S. L., Joseph, J. J., Rondelli, D., and Hsieh, M. M. (2018). Curative therapies: allogeneic hematopoietic cell transplantation from matched related donors using myeloablative, reduced intensity, and nonmyeloablative conditioning in sickle cell disease. *Semin. Hematol.* 55, 87–93. doi: 10.1053/j.seminhematol.2018.04.011

Hanggi, P., Makhro, A., Gassmann, M., Schmugge, M., Goede, J. S., Speer, O., et al. (2014). Red blood cells of sickle cell disease patients exhibit abnormally high abundance of N-methyl D-aspartate receptors mediating excessive calcium uptake. *Br. J. Haematol.* 167, 252–264. doi: 10.1111/bjh.13028

Hassell, K. L. (2010). Population estimates of sickle cell disease in the U.S. *Am. J. Prev. Med.* 38, S512–S521. doi: 10.1016/j.amepre.2009.12.022

Hassell, K. L. (2016). Sickle cell disease: a continued call to action. *Am. J. Prev. Med.* 51, S1–S2. doi: 10.1016/j.amepre.2015.11.002

Hebbel, R. P. (2011). Reconstructing sickle cell disease: a data-based analysis of the "hyperhemolysis paradigm" for pulmonary hypertension from the perspective of evidence-based medicine. *Am. J. Hematol.* 86, 123–154. doi: 10.1002/ajh.21952

Hebbel, R. P., and Hedlund, B. E. (2018). Sickle hemoglobin oxygen affinity-shifting strategies have unequal cerebrovascular risks. *Am. J. Hematol.* 93, 321–325. doi: 10.1002/ajh.24975

Heeney, M. M., Hoppe, C. C., Abboud, M. R., Inusa, B., Kanter, J., Ogutu, B., et al. (2016). A multinational trial of prasugrel for sickle cell vaso-occlusive events. *N. Engl. J. Med.* 374, 625–635. doi: 10.1056/NEJMoa1512021

Hendrickson, J. E., and Tormey, C. A. (2018). Rhesus pieces: genotype matching of RBCs. *Blood* 132, 1091–1093. doi: 10.1182/blood-2018-07-865634

Herrick, J. B. (1910). Peculiar elongated and sickle-shaped red blood corpuscles in a case of severe anemia. *Arch. Intern. Med.* 6, 517–521. doi: 10.1001/jama.2014.11011

Herrick, J. B. (2014). Peculiar elongated and sickle-shaped red blood corpuscles in a case of severe anemia. *JAMA* 312:1063. doi: 10.1001/jama.2014.11011

Hoppe, C., Jacob, E., Styles, L., Kuypers, F., Larkin, S., and Vichinsky, E. (2017). Simvastatin reduces vaso-occlusive pain in sickle cell anaemia: a pilot efficacy trial. *Br. J. Haematol.* 177, 620–629. doi: 10.1111/bjh.14580

Howard, J. (2016). Sickle cell disease: when and how to transfuse. *Hematol. Am. Soc. Hematol. Educ. Program* 2016, 625–631. doi: 10.1182/asheducation-2016.1.625

Howard, J., Malfroy, M., Llewelyn, C., Choo, L., Hodge, R., Johnson, T., et al. (2013). The transfusion alternatives preoperatively in sickle cell disease (TAPS) study: a randomised, controlled, multicentre clinical trial. *Lancet* 381, 930–938. doi: 10.1016/S0140-6736(12)61726-7

Hsieh, M. M., Fitzhugh, C. D., Weitzel, R. P., Link, M. E., Coles, W. A., Zhao, X., et al. (2014). Nonmyeloablative HLA-matched sibling allogeneic hematopoietic stem cell transplantation for severe sickle cell phenotype. *JAMA* 312, 48–56. doi: 10.1001/jama.2014.7192

Hsieh, M. M., Kang, E. M., Fitzhugh, C. D., Link, M. B., Bolan, C. D., Kurlander, R., et al. (2009). Allogeneic hematopoietic stem-cell transplantation for sickle cell disease. *N. Engl. J. Med.* 361, 2309–2317. doi: 10.1056/NEJMoa0904971

Hsieh, M. M., and Tisdale, J. F. (2018). Hematopoietic stem cell mobilization with plerixafor in sickle cell disease. *Haematologica* 103, 749–750. doi: 10.3324/haematol.2018.190876

Inamoto, Y., Kimura, F., Kanda, J., Sugita, J., Ikegame, K., Nakasone, H., et al. (2016). Comparison of graft-versus-host disease-free, relapse-free survival according to a variety of graft sources: antithymocyte globulin and single cord blood provide favorable outcomes in some subgroups. *Haematologica* 101, 1592–1602. doi: 10.3324/haematol.2016.149427

Johnson, F. L. (1985). Bone marrow transplantation in the treatment of sickle cell anemia. *Am. J. Pediatr. Hematol. Oncol.* 7, 254–257.

Johnson, F. L., Look, A. T., Gockerman, J., Ruggiero, M. R., Dalla-Pozza, L., and Billings, F. T. (1984). Bone-marrow transplantation in a patient with sickle-cell anemia. *N. Engl. J. Med.* 311, 780–783. doi: 10.1056/NEJM198409203111207

Jones, K. M., Niaz, M. S., Brooks, C. M., Roberson, S. I., Aguinaga, M. P., Hills, E. R., et al. (2009). Adverse effects of a clinically relevant dose of hydroxyurea used for the treatment of sickle cell disease on male fertility endpoints. *Int. J. Environ. Res. Public Health* 6, 1124–1144. doi: 10.3390/ijerph6031124

Joseph, J. J., Abraham, A. A., and Fitzhugh, C. D. (2018). When there is no match, the game is not over: alternative donor options for hematopoietic stem cell transplantation in sickle cell disease. *Semin. Hematol.* 55, 94–101. doi: 10.1053/j.seminhematol.2018.04.013

Kamani, N. R., Walters, M. C., Carter, S., Aquino, V., Brochstein, J. A., Chaudhury, S., et al. (2012). Unrelated donor cord blood transplantation for children with severe sickle cell disease: results of one cohort from the phase II study from the blood and marrow transplant clinical trials network (BMT CTN). *Biol. Blood Marrow Transplant.* 18, 1265–1272. doi: 10.1016/j.bbmt.2012.01.019

Kanter, J., Abboud, M. R., Kaya, B., Nduba, V., Amilon, C., Gottfridsson, C., et al. (2019). Ticagrelor does not impact patient-reported pain in young adults with sickle cell disease: a multicentre, randomised phase IIB study. *Br. J. Haematol.* 184, 269–278. doi: 10.1111/bjh.15646

Kutlar, A., Kanter, J., Liles, D. K., Alvarez, O. A., Cancado, R. D., Friedrisch, J. R., et al. (2019). Effect of crizanlizumab on pain crises in subgroups of patients with sickle cell disease: a SUSTAIN study analysis. *Am. J. Hematol.* 94, 55–61. doi: 10.1002/ajh.25308

Lagresle-Peyrou, C., Lefrere, F., Magrin, E., Ribeil, J. A., Romano, O., Weber, L., et al. (2018). Plerixafor enables safe, rapid, efficient mobilization of hematopoietic stem cells in sickle cell disease patients after exchange transfusion. *Haematologica* 103, 778–786. doi: 10.3324/haematol.2017.184788

Leonard, A., Tisdale, J., and Abraham, A. (2020). Curative options for sickle cell disease: haploidentical stem cell transplantation or gene therapy? *Br. J. Haematol.* doi: 10.1111/bjh.16437 [Epub ahead of print].

Leonard, A., and Tisdale, J. F. (2018). Stem cell transplantation in sickle cell disease: therapeutic potential and challenges faced. *Expert Rev. Hematol.* 11, 547–565. doi: 10.1080/17474086.2018.1486703

Liu, N., Hargreaves, V. V., Zhu, Q., Kurland, J. V., Hong, J., Kim, W., et al. (2018). Direct promoter repression by BCL11A controls the fetal to adult hemoglobin switch. *Cell* 173, 430–42.e17. doi: 10.1016/j.cell.2018.03.016

Luzzatto, L., and Makani, J. (2019). Hydroxyurea - an essential medicine for sickle cell disease in Africa. *N. Engl. J. Med.* 380, 187–189. doi: 10.1056/NEJMe1814706

Martyn, G. E., Wienert, B., Yang, L., Shah, M., Norton, L. J., Burdach, J., et al. (2018). Natural regulatory mutations elevate the fetal globin gene via disruption of BCL11A or ZBTB7A binding. *Nat. Genet.* 50, 498–503. doi: 10.1038/s41588-018-0085-0

Masuda, T., Wang, X., Maeda, M., Canver, M. C., Sher, F., Funnell, A. P., et al. (2016). Transcription factors LRF and BCL11A independently repress expression of fetal hemoglobin. *Science* 351, 285–289. doi: 10.1126/science.aad3312

McArthur, J. G., Svenstrup, N., Chen, C., Fricot, A., Carvalho, C., Nguyen, J., et al. (2019). A novel, highly potent and selective phosphodiesterase-9 inhibitor for the treatment of sickle cell disease. *Haematologica* 105, 623–631. doi: 10.3324/haematol.2018.213462

McGann, P. T. (2014). Sickle cell anemia: an underappreciated and unaddressed contributor to global childhood mortality. *J. Pediatr.* 165, 18–22. doi: 10.1016/j.jpeds.2014.01.070

Miller, S. T., Sleeper, L. A., Pegelow, C. H., Enos, L. E., Wang, W. C., Weiner, S. J., et al. (2000). Prediction of adverse outcomes in children with sickle cell disease. *N. Engl. J. Med.* 342, 83–89. doi: 10.1056/NEJM200005253422114

Minniti, C. P. (2018). l-glutamine and the dawn of combination therapy for sickle cell disease. *N. Engl. J. Med.* 379, 292–294. doi: 10.1056/NEJMe1800976

Molokie, R., Lavelle, D., Gowhari, M., Pacini, M., Krauz, L., Hassan, J., et al. (2017). Oral tetrahydrouridine and decitabine for non-cytotoxic epigenetic gene regulation in sickle cell disease: a randomized phase 1 study. *PLoS Med.* 14:e1002382. doi: 10.1371/journal.pmed.1002382

Morris, C. R., Suh, J. H., Hagar, W., Larkin, S., Bland, D. A., Steinberg, M. H., et al. (2008). Erythrocyte glutamine depletion, altered redox environment, and pulmonary hypertension in sickle cell disease. *Blood* 111, 402–410. doi: 10.1182/blood-2007-04-081703

Nasimuzzaman, M., and Malik, P. (2019). Role of the coagulation system in the pathogenesis of sickle cell disease. *Blood Adv.* 3, 3170–3180. doi: 10.1182/bloodadvances.2019000193

Nayerossadat, N., Maedeh, T., and Ali, P. A. (2012). Viral and nonviral delivery systems for gene delivery. *Adv. Biomed. Res.* 1:27. doi: 10.4103/2277-9175.98152

Negre, O., Eggimann, A. V., Beuzard, Y., Ribeil, J. A., Bourget, P., Borwornpinyo, S., et al. (2016). Gene therapy of the beta-hemoglobinopathies by lentiviral transfer of the beta(A(T87Q))-globin gene. *Hum. Gene Ther.* 27, 148–165. doi: 10.1089/hum.2016.007

Nevitt, S. J., Jones, A. P., and Howard, J. (2017). Hydroxyurea (hydroxycarbamide) for sickle cell disease. *Cochrane Database. Syst. Rev.* 4:CD002202. doi: 10.1002/14651858.CD002202.pub2

Niihara, Y., Miller, S. T., Kanter, J., Lanzkron, S., Smith, W. R., Hsu, L. L., et al. (2018). A phase 3 trial of l-glutamine in sickle cell disease. *N. Engl. J. Med.* 379, 226–235. doi: 10.1056/NEJMoa1715971

Nur, E., Brandjes, D. P., Teerlink, T., Otten, H. M., Oude Elferink, R. P., Muskiet, F., et al. (2012). N-acetylcysteine reduces oxidative stress in sickle cell patients. *Ann. Hematol.* 91, 1097–1105. doi: 10.1007/s00277-011-1404-z

Ohene-Frempong, K., Weiner, S. J., Sleeper, L. A., Miller, S. T., Embury, S., Moohr, J. W., et al. (1998). Cerebrovascular accidents in sickle cell disease: rates and risk factors. *Blood* 91, 288–294.

Opoka, R. O., Ndugwa, C. M., Latham, T. S., Lane, A., Hume, H. A., Kasirye, P., et al. (2017). Novel use of hydroxyurea in an African region with malaria (NOHARM): a trial for children with sickle cell anemia. *Blood* 130, 2585–2593. doi: 10.1182/blood-2017-06-788935

Orkin, S. H., and Bauer, D. E. (2019). Emerging genetic therapy for sickle cell disease. *Annu. Rev. Med.* 70, 257–271. doi: 10.1146/annurev-med-041817-125507

Orringer, E. P., Casella, J. F., Ataga, K., Koshy, M., Adams-Graves, P., Luchtman-Jones, L., et al. (2001). Purified poloxamer 188 for treatment of acute vaso-occlusive crisis of sickle cell disease: a randomized controlled trial. *JAMA* 286, 2099–2106. doi: 10.1001/jama.286.17.2099

Pace, B. S., Shartava, A., Pack-Mabien, A., Mulekar, M., Ardia, A., and Goodman, S. R. (2003). Effects of N-acetylcysteine on dense cell formation in sickle cell disease. *Am. J. Hematol.* 73, 26–32. doi: 10.1002/ajh.10321

Piel, F. B., Hay, S. I., Gupta, S., Weatherall, D. J., and Williams, T. N. (2013). Global burden of sickle cell anaemia in children under five, 2010-2050: modelling based on demographics, excess mortality, and interventions. *PLoS Med.* 10:e1001484. doi: 10.1371/journal.pmed.1001484

Quinn, C. T. (2016). Minireview: clinical severity in sickle cell disease: the challenges of definition and prognostication. *Exp. Biol. Med.* 241, 679–688. doi: 10.1177/1535370216640385

Quinn, C. T. (2018). l-Glutamine for sickle cell anemia: more questions than answers. *Blood* 132, 689–693. doi: 10.1182/blood-2018-03-834440

Quinn, C. T., Rogers, Z. R., McCavit, T. L., and Buchanan, G. R. (2010). Improved survival of children and adolescents with sickle cell disease. *Blood* 115, 3447–3452. doi: 10.1182/blood-2009-07-233700

Rees, D. C. (2019). Double-blind, randomized study of canakinumab treatment in pediatric and young adult patients with sickle cell anemia. *Blood* 134(Suppl._1):615.

Ribeil, J. A., Hacein-Bey-Abina, S., Payen, E., Magnani, A., Semeraro, M., Magrin, E., et al. (2017). Gene therapy in a patient with sickle cell disease. *N. Engl. J. Med.* 376, 848–855.

Saiki, R. K., Scharf, S., Faloona, F., Mullis, K. B., Horn, G. T., Erlich, H. A., et al. (1985). Enzymatic amplification of b-globin genomic sequences and restriction site analysis for diagnosis of sickle cell anaemia. *Science* 230, 1350–1354.

Saraf, S. L., Oh, A. L., Patel, P. R., Sweiss, K., Koshy, M., Campbell-Lee, S., et al. (2018). Haploidentical peripheral blood stem cell transplantation demonstrates stable engraftment in adults with sickle cell disease. *Biol. Blood Marrow Transplant.* 24, 1759–1765. doi: 10.1016/j.bbmt.2018.03.031

Serjeant, G. R., Chin, N., Asnani, M. R., Serjeant, B. E., Mason, K. P., Hambleton, I. R., et al. (2018). Causes of death and early life determinants of survival in homozygous sickle cell disease: the Jamaican cohort study from birth. *PLoS One* 13:e0192710. doi: 10.1371/journal.pone.0192710

Shenoy, S. (2013). Hematopoietic stem-cell transplantation for sickle cell disease: current evidence and opinions. *Ther. Adv. Hematol.* 4, 335–344. doi: 10.1177/2040620713483063

Sins, J. W. R., Mager, D. J., Davis, S., Biemond, B. J., and Fijnvandraat, K. (2017). Pharmacotherapeutical strategies in the prevention of acute, vaso-occlusive pain in sickle cell disease: a systematic review. *Blood Adv.* 1, 1598–1616. doi: 10.1182/bloodadvances.2017007211

Sparkenbaugh, E., Chantrathammachart, P., Mickelson, J., van Ryn, J., Hebbel, R. P., Monroe, D. M., et al. (2014). Differential contribution of FXa and thrombin to vascular inflammation in a mouse model of sickle cell disease. *Blood* 123, 1747–1756. doi: 10.1182/blood-2013-08-523936

Sparkenbaugh, E., and Pawlinski, R. (2013). Interplay between coagulation and vascular inflammation in sickle cell disease. *Br. J. Haematol.* 162, 3–14. doi: 10.1111/bjh.12336

Steinberg, M. H., Forget, B. G., Higgs, D. R., and Nagel, R. L. (2001). *Disorders of Hemoglobin: Genetics, Pathophysiology, and Clinical Management*, 1st Edn. Cambridge: Cambridge University Press.

Strader, M. B., Liang, H., Meng, F., Harper, J., Ostrowski, D. A., Henry, E. R., et al. (2019). Interactions of an anti-sickling drug with hemoglobin in red blood cells from a patient with sickle cell anemia. *Bioconjug. Chem.* 30, 568–571. doi: 10.1021/acs.bioconjchem.9b00130

Sundd, P., Gladwin, M. T., and Novelli, E. M. (2019). Pathophysiology of sickle cell disease. *Annu. Rev. Pathol.* 14, 263–292.

Telen, M. J. (2016). Beyond hydroxyurea: new and old drugs in the pipeline for sickle cell disease. *Blood* 127, 810–819. doi: 10.1182/blood-2015-09-618553

Telen, M. J., Batchvarova, M., Shan, S., Bovee-Geurts, P. H., Zennadi, R., Leitgeb, A., et al. (2016). Sevuparin binds to multiple adhesive ligands and reduces sickle red blood cell-induced vaso-occlusion. *Br. J. Haematol.* 175, 935–948. doi: 10.1111/bjh.14303

Telen, M. J., Malik, P., and Vercellotti, G. M. (2019). Therapeutic strategies for sickle cell disease: towards a multi-agent approach. *Nat. Rev. Drug Discov.* 18, 139–158. doi: 10.1038/s41573-018-0003-2

Telen, M. J., Wun, T., McCavit, T. L., De Castro, L. M., Krishnamurti, L., Lanzkron, S., et al. (2015). Randomized phase 2 study of GMI-1070 in SCD: reduction in time to resolution of vaso-occlusive events and decreased opioid use. *Blood* 125, 2656–2664. doi: 10.1182/blood-2014-06-583351

Telfer, P., Coen, P., Chakravorty, S., Wilkey, O., Evans, J., Newell, H., et al. (2007). Clinical outcomes in children with sickle cell disease living in England: a neonatal cohort in East London. *Haematologica* 92, 905–912. doi: 10.3324/haematol.10937

Thein, S. L., Pirenne, F., Fasano, R. M., Habibi, A., Bartolucci, P., Chonat, S., et al. (2020). Hemolytic transfusion reactions in sickle cell disease: underappreciated and potentially fatal. *Haematologica* 105, 539–544. doi: 10.3324/haematol.2019.224709

Thompson, A. A. (2019). Targeted agent for sickle cell disease - changing the protein but not the gene. *N. Engl. J. Med.* 381, 579–580. doi: 10.1056/NEJMe1906771

Traxler, E. A., Yao, Y., Wang, Y. D., Woodard, K. J., Kurita, R., Nakamura, Y., et al. (2016). A genome-editing strategy to treat beta-hemoglobinopathies that recapitulates a mutation associated with a benign genetic condition. *Nat. Med.* 22, 987–990. doi: 10.1038/nm.4170

Tshilolo, L., Tomlinson, G., Williams, T. N., Santos, B., Olupot-Olupot, P., Lane, A., et al. (2019). Hydroxyurea for children with sickle cell anemia in sub-saharan Africa. *N. Engl. J. Med.* 380, 121–131. doi: 10.1056/NEJMoa1813598

van Zuuren, E. J., and Fedorowicz, Z. (2015). Low-molecular-weight heparins for managing vaso-occlusive crises in people with sickle cell disease. *Cochrane Database Syst. Rev.* 6:CD010155. doi: 10.1002/14651858.CD010155.pub3

Vichinsky, E., Hoppe, C. C., Ataga, K. I., Ware, R. E., Nduba, V., El-Beshlawy, A., et al. (2019). A phase 3 randomized trial of voxelotor in sickle cell disease. *N. Engl. J. Med.* 381, 509–519. doi: 10.1056/NEJMoa1903212

Vichinsky, E. P., Earles, A., Johnson, R. A., Hoag, M. S., Williams, A., and Lubin, B. (1990). Alloimmunization in sickle cell anemia and transfusion of racially unmatched blood. *N. Engl. J. Med.* 322, 1617–1621. doi: 10.1056/NEJM199006073222301

Villagra, J., Shiva, S., Hunter, L. A., Machado, R. F., Gladwin, M. T., and Kato, G. J. (2007). Platelet activation in patients with sickle disease, hemolysis-associated pulmonary hypertension, and nitric oxide scavenging by cell-free hemoglobin. *Blood* 110, 2166–2172. doi: 10.1182/blood-2006-12-061697

Vinjamur, D. S., Bauer, D. E., and Orkin, S. H. (2018). Recent progress in understanding and manipulating haemoglobin switching for the haemoglobinopathies. *Br. J. Haematol.* 180, 630–643. doi: 10.1111/bjh.15038

Wailoo, K. (2017). Sickle cell disease - a history of progress and peril. *N. Engl. J. Med.* 376, 805–807. doi: 10.1056/NEJMp1700101

Walters, M. C., Hardy, K., Edwards, S., Adamkiewicz, T., Barkovich, J., Bernaudin, F., et al. (2010). Pulmonary, gonadal, and central nervous system status after bone marrow transplantation for sickle cell disease. *Biol. Blood Marrow Transplant.* 16, 263–272. doi: 10.1016/j.bbmt.2009.10.005

Walters, M. C., Patience, M., Leisenring, W., Eckman, J. R., Buchanan, G. R., Rogers, Z. R., et al. (1996a). Barriers to bone marrow transplantation for sickle cell anemia. *Biol. Blood Marrow Transplant.* 2, 100–104.

Walters, M. C., Patience, M., Leisenring, W., Eckman, J. R., Scott, J. P., Mentzer, W. C., et al. (1996b). Bone marrow transplantation for sickle cell disease. *N. Engl. J. Med.* 335, 369–376. doi: 10.1002/ajh.24995

Walters, M. C., Patience, M., Leisenring, W., Rogers, Z. R., Aquino, V. M., Buchanan, G. R., et al. (2001). Stable mixed hematopoietic chimerism after bone marrow transplantation for sickle cell anemia. *Biol. Blood Marrow Transplant.* 7, 665–673. doi: 10.1053/bbmt.2001.v7.pm11787529

Wang, W. C., Ware, R. E., Miller, S. T., Iyer, R. V., Casella, J. F., Minniti, C. P., et al. (2011). Hydroxycarbamide in very young children with sickle-cell anaemia: a multicentre, randomised, controlled trial (BABY HUG). *Lancet* 377, 1663–1672. doi: 10.1016/S0140-6736(11)60355-3

Ware, R. E. (2015). Optimizing hydroxyurea therapy for sickle cell anemia. *Hematol. Am. Soc. Hematol. Educ. Program* 2015, 436–443. doi: 10.1182/asheducation-2015.1.436

Ware, R. E., and Aygun, B. (2009). Advances in the use of hydroxyurea. *Hematol. Am. Soc. Hematol. Educ. Program* 2009, 62–69.

Ware, R. E., Davis, B. R., Schultz, W. H., Brown, R. C., Aygun, B., Sarnaik, S., et al. (2016). Hydroxycarbamide versus chronic transfusion for maintenance of transcranial doppler flow velocities in children with sickle cell anaemia-TCD with transfusions changing to hydroxyurea (TWiTCH): a multicentre, open-label, phase 3, non-inferiority trial. *Lancet* 387, 661–670. doi: 10.1016/S0140-6736(15)01041-7

Ware, R. E., Schultz, W. H., Yovetich, N., Mortier, N. A., Alvarez, O., Hilliard, L., et al. (2011). Stroke with transfusions changing to hydroxyurea (SWiTCH): a phase III randomized clinical trial for treatment of children with sickle cell anemia, stroke, and iron overload. *Pediatr. Blood Cancer* 57, 1011–1017. doi: 10.1002/pbc.23145

Wienert, B., Martyn, G. E., Funnell, A. P. W., Quinlan, K. G. R., and Crossley, M. (2018). Wake-up sleepy gene: reactivating fetal globin for beta-hemoglobinopathies. *Trends Genet.* 34, 927–940. doi: 10.1016/j.tig.2018.09.004

Williams, T. N., and Thein, S. L. (2018). Sickle cell anemia and its phenotypes. *Annu. Rev. Genomics Hum. Genet.* 19, 113–147. doi: 10.1146/annurev-genom-083117-021320

Wilson, J. T., Milner, P. F., Summer, M. E., Nallaseth, F. S., Fadel, H. E., Reindollar, R. H., et al. (1982). Use of restriction endonucleases for mapping the allele for beta s-globin. *Proc. Natl. Acad. Sci. U.S.A.* 79, 3628–3631. doi: 10.1073/pnas.79.11.3628

Wun, T., Paglieroni, T., Tablin, F., Welborn, J., Nelson, K., and Cheung, A. (1997). Platelet activation and platelet-erythrocyte aggregates in patients with sickle cell anemia. *J. Lab. Clin. Med.* 129, 507–516. doi: 10.1016/s0022-2143(97)90005-6

Zhang, D., Xu, C., Manwani, D., and Frenette, P. S. (2016). Neutrophils, platelets, and inflammatory pathways at the nexus of sickle cell disease pathophysiology. *Blood* 127, 801–809. doi: 10.1182/blood-2015-09-618538

15

Usefulness of NGS for Diagnosis of Dominant Beta-Thalassemia and Unstable Hemoglobinopathies in Five Clinical Cases

Valeria Rizzuto[1,2,3], Tamara T. Koopmann[4], Adoración Blanco-Álvarez[5],
Barbara Tazón-Vega[5], Amira Idrizovic[1], Cristina Díaz de Heredia[6], Rafael Del Orbe[7],
Miriam Vara Pampliega[7], Pablo Velasco[6], David Beneitez[8], Gijs W. E. Santen[4],
Quinten Waisfisz[9], Mariet Elting[9], Frans J. W. Smiers[10], Anne J. de Pagter[10],
Jean-Louis H. Kerkhoffs[11], Cornelis L. Harteveld[4] and Maria del Mar Mañú-Pereira[1]*

[1] Translational Research in Child and Adolescent Cancer – Rare Anemia Disorders Research Laboratory, Vall d'Hebron Research Institute, ERN-EuroBloodNet Member, Barcelona, Spain, [2] Josep Carreras Leukaemia Research Institute, Badalona, Spain, [3] Department of Medicine, Universitat de Barcelona, Barcelona, Spain, [4] Department of Clinical Genetics, Leiden University Medical Center, ERN-EuroBloodNet Member, Leiden, Netherlands, [5] Hematologic Molecular Genetics Unit, Hematology Department, Hospital Universitari Vall d'Hebron, ERN-EuroBloodNet Member, Barcelona, Spain, [6] Oncohematologic Pediatrics Department, Hospital Universitari Vall d'Hebron, ERN-EuroBloodNet Member, Barcelona, Spain, [7] Hematology Department, Hospital Universitario Cruces, Barakaldo, Spain, [8] Red Blood Cell Disorders Unit, Hematology Department, Hospital Universitari Vall d'Hebron, ERN-EuroBloodNet Member, Barcelona, Spain, [9] Department of Clinical Genetics, VU Medical Center, Amsterdam, Netherlands, [10] Department of Pediatric Hematology, Leiden University Medical Center, Leiden, Netherlands, [11] Department of Hematology, HAGA City Hospital, The Hague, Netherlands

*Correspondence:
Maria del Mar Mañú-Pereira
mar.manu@vhir.org

Unstable hemoglobinopathies (UHs) are rare anemia disorders (RADs) characterized by abnormal hemoglobin (Hb) variants with decreased stability. UHs are therefore easily precipitating, causing hemolysis and, in some cases, leading to dominant beta-thalassemia (dBTHAL). The clinical picture of UHs is highly heterogeneous, inheritance pattern is dominant, instead of recessive as in more prevalent major Hb syndromes, and may occur *de novo*. Most cases of UHs are not detected by conventional testing, therefore diagnosis requires a high index of suspicion of the treating physician. Here, we highlight the importance of next generation sequencing (NGS) methodologies for the diagnosis of patients with dBTHAL and other less severe UH variants. We present five unrelated clinical cases referred with chronic hemolytic anemia, three of them with severe blood transfusion dependent anemia. Targeted NGS analysis was performed in three cases while whole exome sequencing (WES) analysis was performed in two cases. Five different UH variants were identified correlating with patients' clinical manifestations. Four variants were related to the beta-globin gene (Hb Bristol—Alesha, Hb Debrousse, Hb Zunyi, and the novel Hb Mokum) meanwhile one case was caused by a mutation in the alpha-globin gene leading to Hb Evans. Inclusion of alpha and beta-globin genes in routine NGS approaches for RADs has to be considered to improve diagnosis' efficiency of RAD due to UHs. Reducing misdiagnoses and underdiagnoses of UH variants, especially of the severe forms leading to dBTHAL would also facilitate the early start of intensive or curative treatments for these patients.

Keywords: unstable hemoglobinopathies, dominant beta-thalassemia, next generation sequencing, whole exome sequencing, rare anemia disorders

INTRODUCTION

Beta-thalassemia major (BTHAL) is a well-known life-threatening condition characterized by severe transfusion-dependent anemia. BTHAL is an autosomal recessive disorder presenting with high frequencies in populations from the Mediterranean area. Currently, up to 257 genetic variants in the beta-globin gene (*HBB*) have been identified as BTHAL disease-causing, leading to a total or partial reduction of beta-globin chain synthesis. The clinical severity of BTHAL is related to the extent of imbalance between the alpha and non-alpha-globin chains, while clinical management consists of regular life-long red blood cell (RBC) transfusions and iron chelation therapy. At present, the only definitive cure is bone marrow transplant (Efremov, 2007; Galanello and Origa, 2010). Both BTHAL patients and carriers are usually easily diagnosed through routine laboratory tests. However, there is an ultra-rare condition overlapping BTHAL clinical manifestations known as dominant beta-thalassemia (dBTHAL), which is caused by the presence of certain unstable (UH) or hyper unstable (HUH) hemoglobinopathies.

UHs are a group of congenital disorders caused by mutations in globin genes leading to destabilization of hemoglobin (Hb) molecules as a consequence of (a) amino acid substitutions within the heme pocket, (b) disruption of secondary structure, (c) substitution in the hydrophobic interior of the subunit, (d) amino acid deletions, and (e) elongation of the subunit. Thus, altering any of the steps in globin processing, including subunit folding, heme interaction, dimerization, or tetramerization (Bunn and Forget, 1986). These abnormal Hb variants undergo rapid denaturation followed by precipitation, leading to the formation of Heinz bodies, which cause hemolysis of RBCs. Clinical manifestations may vary from asymptomatic to severely affected forms. Treatment is mainly symptomatic and based on transfusion requirements as for BTHAL (Steinberg et al., 2009; Thom et al., 2013).

UHs are dominantly inherited with a significant rate of *de novo* mutations. They generally do not separate from normal Hb using standard methods. Thus, diagnosis of dBTHAL can be challenging since it requires a high index of suspicion and the diagnosis may be delayed for years hampering the access to timely treatment interventions.

The study we present herein confirms the relevance of including globin genes in next generation sequencing (NGS) approaches for the diagnosis of rare anemia disorders (RADs), especially for cases with no family history in which the anemia is not easily explained.

PATIENTS AND METHODS

Clinical Reports

Here we present five clinical cases diagnosed with UH after NGS analysis. Clinical data and laboratory findings are shown in **Table 1**.

The first case is a male pediatric patient referred with severe chronic blood dependent anemia since he was 4-month-old,

asthenia, jaundice, and short stature. No family history of hemolytic anemia. Examination of blood smear revealed polychromasia, anisopoikilocytosis, basophil stippling, Cabot rings, schistocytes, and spherocytes. Separation and quantification of Hb fractions did not reveal any extraordinary peak and showed normal values for HbA_2 and HbF. At 5-year-old he underwent splenectomy. After the surgery, Heinz bodies were present (**Figures 1**, **2**) and isopropanol stability test, performed according to standard methodology, appeared positive (**Figure 3**). Family studies in both parents were strictly normal, including evaluation of Hb fractions. Enzyme activity assays, EMA-binding test, and osmotic gradient ectacytometry (LoRRca MaxSis) were performed to rule out hemolytic anemia due to RBC defects other than hemoglobinopathy. Results, although not strictly normal, did not reveal any RBC defect. However, they should be taken with caution since the patient was intensively transfused. Genetic analysis was performed on *PKLR* and *G6PD* genes failing to reveal any disease-causing mutation.

The second case is a female adult patient with mild chronic compensated hemolysis referred for diagnosis when she was 20 years old. The father also presented with mild compensated hemolysis. No further examinations were performed before referral. Although the presence of extravascular hemolysis, the examination of blood smear was not informative. Separation and quantification of Hb fractions did not reveal any extra peaks and Heinz body and stability tests were normal. Further laboratory tests were performed to rule out hemolytic anemia due to RBC enzyme and membrane defects, including enzyme activity assays, EMA-binding test, and osmotic gradient ectacytometry (LoRRca MaxSis). All of them showed normal values.

The third case is a male adult patient. He presented with several episodes of hemolytic crises during childhood requiring blood transfusion on two occasions. He underwent splenectomy at the age of 25-year-old. The patient was diagnosed with hereditary spherocytosis (HS) following a previous HS diagnosis of his mother and the absence of abnormal Hb peaks by conventional electrophoresis.

Patients who underwent splenectomy neither clinically improved nor presented complications as pulmonary hypertension, thrombosis or increased hemolysis during 10-year follow-up.

The last two cases are two unrelated children who presented with macrocephaly and severe congenital anemia. The parents of both patients had no family history for abnormal Hb or thalassemia and had normal hematological features. Therefore, conventional testing for abnormal Hb was not performed. All siblings were unaffected.

The first of these two unrelated children is a male patient presenting with large head circumference and hepatosplenomegaly. Congenital dyserythropoietic anemia was suspected. However, no genetic analysis was performed for confirmation. He underwent successfully bone marrow transplant at the age of 4.

The second child is a female patient presenting with frontal bossing, macrocephaly, and severe anemia at the age of 2. Congenital dyserythropoietic anemia was suspected. Therefore, genetic analysis of *CDAN1* and *SEC23B* genes was

TABLE 1 | Overview on clinical and genetic data of the five reported clinical cases.

Parameters	Case 1	Case 2	Case 3	Case 4	Case 5
Gender/Age	**Male/Pediatric**	**Female/Adult**	**Male/Adult**	**Male/Pediatric**	**Female/Pediatric**
Hb (120–170 g/L)	70–80	119	141	82	79
MCV (80–100 fL)	110–115	97.8	102.3	83	Not done
MCHC (27–33.5 g/dL)	28	32.1	30.8	Not done	Not done
Reticulocyte count (50–100 · 10⁹/L)	900	293	331	810	Not done
Reticulocyte count (%)	34	7.71	7.39	Not done	Not done
Lactate dehydrogenase-LDH (U/L)	4,500–5,000	243	145	186	259
Hb Fractions	Normal	Normal	Normal	Not done	Not done
Heinz bodies	Positive	Negative	Positive	Not done	Not done
Stability test	Positive	Negative	Positive	Not done	Not done
Age of onset (months)	4	Unknown	Unknown	Unknown	2
Family history	No family history	Father presents mild compensated hemolysis	Mother diagnosed with hereditary spherocytosis	No family history	No family history
Transfusion need	8 U/Year	No	2 times	Multiple	Multiple
Splenectomy	Yes (5 y)	No	Yes (25 y)	No	No
Stem cell transplant (age years)	No	No	No	Yes (4 y)	Yes (3 y)
Genotype	*HBB*c.202G > A (p.Val67Met)	*HBA1*c.187G > A (p.Val62Met)	*HBB*c.290T > C (p.Leu96Pro)	*HBB*c.442T > C (p.Ter147Glnext*21)	*HBB*c.442T > A (p.Ter147Lysext*21)
Hb variant name	Hb Bristol-Alesha	Hb Evans	Hb Debrousse	Hb Zunyi	Hb Mokum

Performed after the diagnosis of UH.

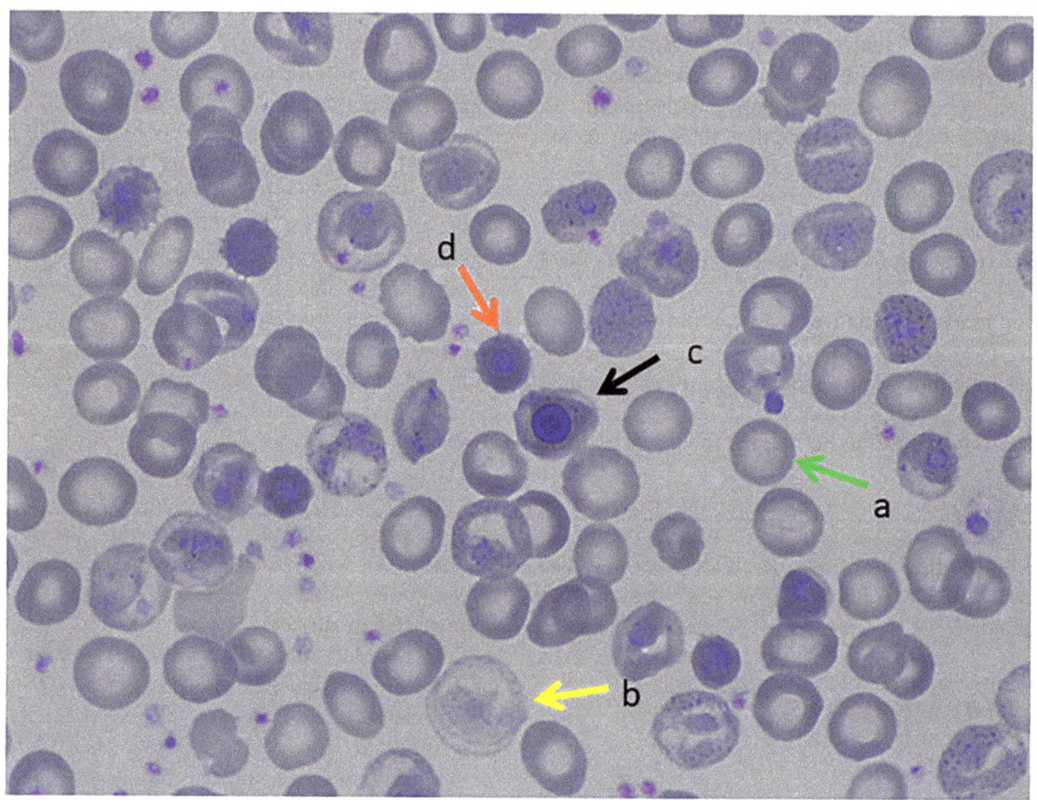

FIGURE 1 | Peripheral Blood Smear, May Grunwald Giemsa Stain. **(a)** Transfused red blood cells, **(b)** non-transfused red blood cells with hemoglobinization abnormalities, **(c)** orthochromatic erythroblast, **(d)** erythrocitary inclusions that correspond to Heinz bodies.

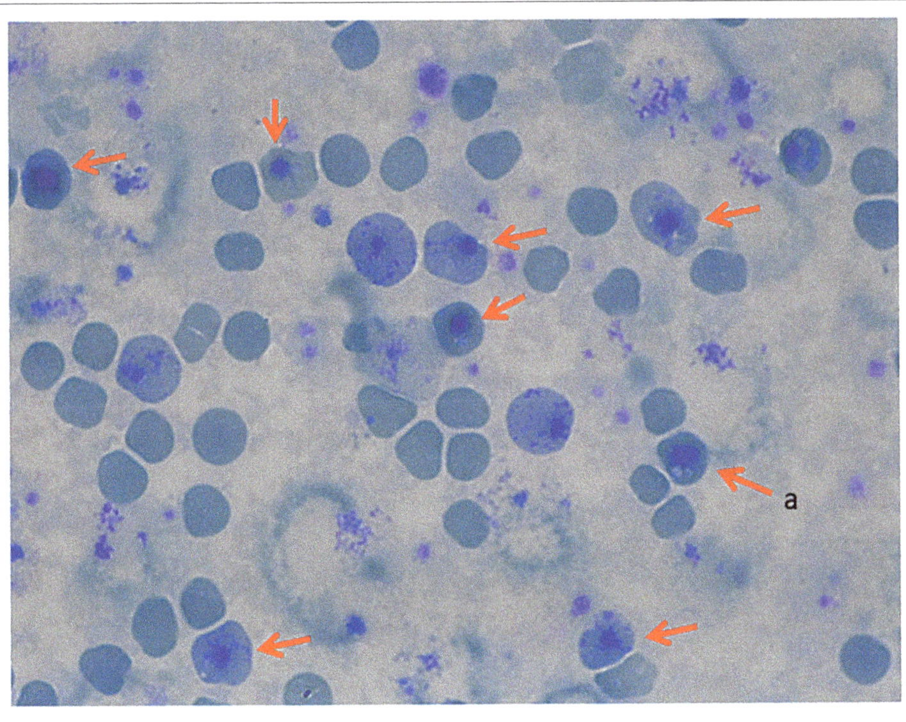

FIGURE 2 | Peripheral Blood Smear, Brilliant Cresyl Blue Stain. **(a)** Heinz bodies.

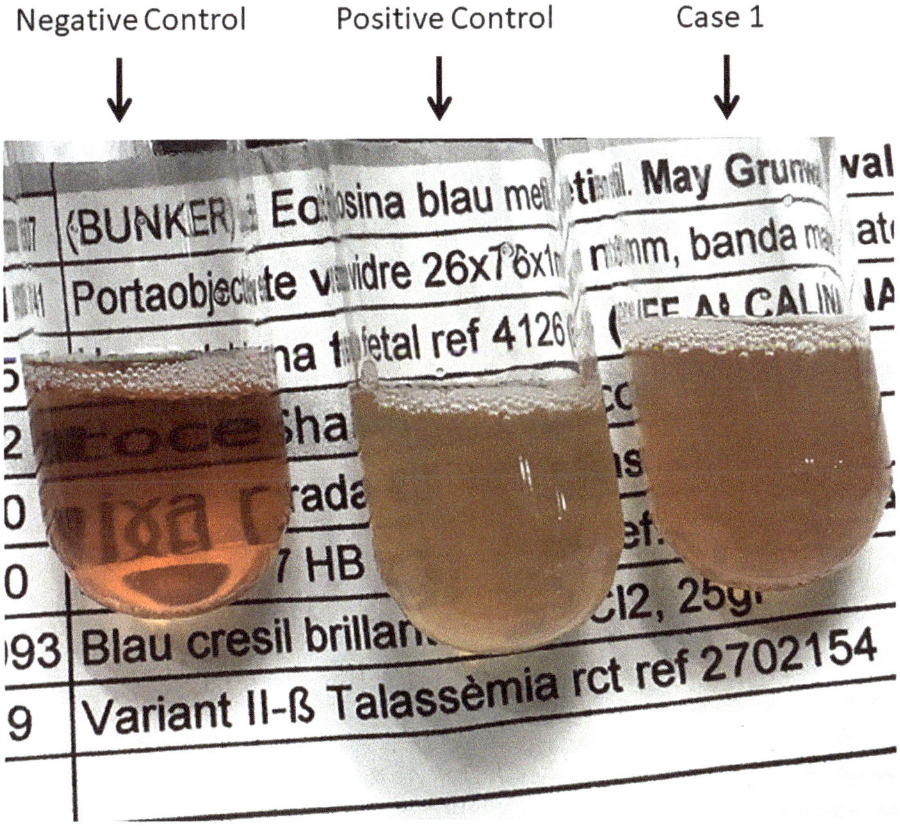

FIGURE 3 | Isopropanol Test_01. Negative control (Hb AS), Positive control (Hb F), and Case 1.

TABLE 2 | List of genes included in the t-NGS approach.

Symbol	Phenotype MIM number	Gene/Locus MIM number	Category	Description
ADA	102700	608958	Enzymopathy	Adenosine deaminase
AK1	103000	103000	Enzymopathy	Adenylate kinase 1
ALDOA	611881	103850	Enzymopathy	Aldolase, fructose-bisphosphate a
ANK1	616089	612641	Membranopathy	Ankyrin 1
ATRX	301040	300032	Alpha-thalassemia myelodysplasia syndrome, somatic; Alpha-thalassemia/mental retardation syndrome; Mental retardation-hypotonic facies syndrome, X-linked	Helicase 2, x-linked
BPGM	222800	613896	Erythrocytosis and methemoglobinemia due to enzyme alteration	Bisphosphoglycerate mutase
C15orf41	615631	615626	Congenital dyserythropoietic anemia	Chromosome 15 open reading frame 41
CDAN1	224120	224120	Congenital dyserythropoietic anemia	Codanin 1
CYB5R3	250800	613213	Methemoglobinemia, type I; Methemoglobinemia, type II	Cytochrome b5 reductase 3
EPB41	611804	130500	Membranopathy	Erythrocyte membrane protein band 4.1
EPB42	612690	177070	Membranopathy	Erythrocyte membrane protein band 4.2
EPO	617907	133170	Erythropoiesis modulator	Erythropoietin
EPOR	133100	133171	Erythropoiesis modulator	Erythropoietin receptor
G6PD	300908	305900	Enzymopathy	Glucose-6-phosphate dehydrogenase
GAPDH	*	138400	Enzymopathy	Glyceraldehyde-3-phosphate dehydrogenase
GATA1	300835	305371	Congenital dyserythropoietic anemia	Gata binding protein 1 (globin transcription factor 1)
GCLC	230450	606857	Enzymopathy	Glutamate-cysteine ligase, catalytic subunit
GPI	613470	172400	Enzymopathy	Glucose-6-phosphate isomerase
GSR	618660	138300	Enzymopathy	Glutathione reductase
GSS	266130	601002	Enzymopathy	Glutathione synthetase
GYPC	616089	110750	Membranopathy	Glycophorin c (gerbich blood group)
HBA1	617981	141800	Hemoglobinopathy	Hemoglobin–alpha locus 1
HBA2	617981	141850	Hemoglobinopathy	Hemoglobin–alpha locus 2
HBB	617980	141900	Hemoglobinopathy	Hemoglobin subunit beta
HBD	*	142000	Thalassemia due to Hb Lepore; Thalassemia, delta-	Hemoglobin–delta locus
HBG1	141900	141749	Fetal hemoglobin quantitative trait locus 1	Hemoglobin, gamma a
HBG2	613977	142250	Cyanosis, transient neonatal; Fetal hemoglobin quantitative trait locus 1	Hemoglobin, gamma g
HK1	235700	142600	Enzymopathy	Hexokinase 1
KCNN4	616689	602754	Membranopathy	Potassium channel, calcium activated intermediate/small conductance subfamily n alpha, member 4
KIF23	*	605064	Congenital dyserythropoietic anemia	Kinesin family member 23
KLF1	613673	600599	Congenital dyserythropoietic anemia	Kruppel-like factor 1 (erythroid)
NT5C3A	266120	606224	Enzymopathy	5′-nucleotidase, cytosolic iiia
PFKL	*	171860	Hemolytic anemia due to phosphofructokinase deficiency	Phosphofructokinase, liver type
PFKM	232800	610681	Enzymopathy	Phosphofructokinase, muscle
PGD	*	172200	Enzymopathy	6-phosphogluconate dehydrogenase, erythrocyte
PGK1	300653	311800	Enzymopathy	Phosphoglycerate kinase 1
PIEZO1	616089	611184	Membranopathy	Piezo-type mechanosensitive ion channel component 1
PKLR	266200	609712	Enzymopathy	Pyruvate kinase, liver and rbc
RHAG	185000	180297	Membranopathy	rh-associated glycoprotein
SEC23B	224100	610512	Congenital dyserythropoietic anemia	Sec23 homolog b, copii coat complex component

(Continued)

TABLE 2 | Continued

Symbol	Phenotype MIM number	Gene/Locus MIM number	Category	Description
SLC2A1	606777	138140	Membranopathy	Solute carrier family 2 (facilitated glucose transporter), member 1
SLC4A1	612653	109270	Membranopathy	Solute carrier family 4 (anion exchanger), member 1 (diego blood group)
SPTA1	130600	182860	Membranopathy	Spectrin alpha, erythrocytic 1
SPTB	616649	182870	Membranopathy	Spectrin beta, erythrocytic
TPI1	615512	190450	Enzymopathy	Triosephosphate isomerase 1
UGT1A1	237900	191740	Gilbert syndrome	udp glucuronosyltransferase 1 family, polypeptide a1

Not available.

performed not revealing any disease-causing mutation. She underwent successfully bone marrow transplant when she was almost 3-year-old.

In all cases, RAD due to Hb variant was not suspected mainly due to the fact that parents did not present family history of RADs, except for case 3, RBC parameters were found to be normal and abnormal Hb fractions were absent when analyzed. Therefore, genetic testing was performed for genes associated with RADs other than globin genes, failing to show a conclusive diagnosis.

Genetics Analysis

Written informed consent was obtained from cases or legal guardian. Targeted NGS (t-NGS) analysis was performed in cases 1, 2, and 3 while whole exome sequencing (WES) analysis was performed in cases 4 and 5. For all the patients, genomic DNA was extracted from peripheral blood. For patients who underwent bone marrow transplant, DNA samples were previously stored.

The designed t-NGS panel covered 46 genes described as disease causing for RADs, including genes responsible for membrane disorders, enzyme defects, congenital dyserytrhopoietic anemia and the *HBA1/HBA2* and *HBB* genes responsible for alpha and beta-globin chains, respectively. The full list of genes included is shown in **Table 2**. Exon and exon/intron boundaries were capture using a NimbleGen SeqCap EZ HyperCap (Roche) solution-based capture system followed by next generation sequencing on the MySeq (Illumina) with 150 bp paired-end reads. For the bioinformatics analysis, alignment to the hg38 genome was performed with BWA-MEM (Li H. 203 arXIV:1303.3997v2) and detection of changes with GATK[1]. Obtained variants were filtered and annotated based on variant effect, coverage (>30) and MAF (>0.05). Resulting variants were assessed for technique pitfalls through IGV. The nomenclature used was the recommended by HGVS[2]. Finally, disease-causing variants were prioritized based on inheritance pattern and VarSome[3] for previous evidence as disease causing mutations or predictions score information. Variants were reported according to American College of Medical Genetics (ACMG) guidelines.

For case 4, WES was performed in a trio approach (patient and both parents). Libraries were prepared using the Kapa HTP kit (Illumina, San Diego, CA, United States) and capture was performed using the SeqCap EZ Human Exome Library v3.0 (Roche NimbleGen Madison, WI, United States). Sequencing was done on an Illumina HiSeq2500 HTv4 (Illumina, San Diego, CA, United States) with paired-end 125-bp reads. Read alignment to hg19 and variant calling were done with a pipeline based on BWA-MEM0.7 and GATK 3.3.0. The median coverage of the captured target region was at least 98×. Variant annotation and prioritizing were done using Cartagenia Bench Lab NGS (Agilent Technologies). Variants located outside the exons and intron/exon boundaries and variants with a minor allele frequency (MAF) of >1% in control databases, including dbSNP137[4], 1000 Genomes Project (phase 3)[5], and Exome Variant Server (EVS), NHLBI Exome Sequencing Project National Heart, Lung, and Blood Institute GO Exome Sequencing Project (ESP6500 release)[6] and in-house exome controls were excluded. Variants that fitted with a *de novo* or recessive mode of inheritance were further prioritized based on literature, predicted (deleterious) effects on protein function by e.g., truncating the protein, affecting splicing, amino acid change, and evolutionary conservation.

For case 5, WES was performed in a trio approach (patient and both parents). Exomes were captured using the Agilent SureSelectXT Human All Exon v5 (Agilent, Santa Clara, CA, United States) accompanied by Illumina paired-end sequencing on the HiSeq2000 (Illumina, San Diego, CA, United States). The in-house sequence analysis pipeline Modular GATK-Based Variant Calling Pipeline (MAGPIE) (LUMC Sequencing Analysis Support Core, LUMC) was used to call the SNVs/indels. LOVDplus (Leiden Genome Technology Center, LUMC, Leiden) was used for interpretation of variants.

RESULTS

Genetic variants in globin genes responsible for UH or HUH were found in all five cases as shown in **Table 1**. All variants were confirmed by Sanger sequencing.

In case 1, variant *HBB*:c.202G > A (p.Val67Met) was found in exon 2 in the heterozygous state. This *HBB* variant is known as Hb Bristol-Alesha, a UH associated with moderate-severe hemolytic anemia. The variant was not found in the parents, suggesting a *de novo* variant in the patient.

[1]https://software.broadinstitute.org/gatk/

[2]http://www.hgvs.org

[3]https://varsome.com

[4]http://www.ncbi.nlm.nih.gov/projects/SNP

[5]http://www.internationalgenome.org/

[6]http://evs.gs.washington.edu/EVS/

In case 2, variant *HBA1*:c.187G > A (p.Val62Met) was found in exon 2 in the heterozygous state. This *HBA1* variant is known as Hb Evans and is associated wild with mild hemolytic anemia and classified as UH. Parents were not sequenced.

In Case 3, variant *HBB*:c.290T > C (p.Leu96Pro) was found in in exon 2 in the heterozygous state. This *HBB* variant is known as Hb Debrousse and is described as a moderate UH. Parents were not sequenced. Nevertheless, antecedents of hemolytic anemia are present in the mother, suggesting a dominant inheritance pattern.

In cases 4 and 5, two missense stop-loss mutations at position 422 of the *HBB* gene were found. The first variant *HBB*:c.442T > C (p.Ter147Glnext*21), found in case 4, is known as Hb Zunyi, while the second variant *HBB*:c.442T > A (p.Ter147Lysnext*21), found in case 5, constitutes a novel variant which was called Hb Mokum. Both variants cause the loss of a stop codon and elongation of the translated beta-globin chain of 21 amino acids due to a new stop codon in the 3′ untranslated region (3′UTR) of the *HBB* gene. The variants were not found in the parents suggesting *de novo* variants in the patients.

According to the ACMG guidelines, all the variants were classified as pathogenic (Richards et al., 2015).

DISCUSSION

We highlight the importance of including globin genes in the NGS analysis of RAD for enabling the diagnosis of UH. We present five clinical cases affected with RAD due to UH variants, four are related to the beta-globin gene (Hb Bristol—Alesha, Hb Debrousse, Hb Zunyi, and the novel Hb Mokum), meanwhile, one is related to the alpha-globin gene (Hb Evans). The use of NGS has been crucial for the final conclusive diagnosis.

The severity of RADs due to UHs depends on the mutation's impact on protein stability and consequently on the degree of hemolysis and inefficient erythropoiesis. Patients' RBCs typically display abnormal but unspecific morphology with microcytosis, hypochromia, moderate to severe anisopoikilocytosis, basophilic stippling, and inclusions that may become particularly prominent following splenectomy (Steinberg et al., 2009; Kent et al., 2014). UHs are commonly inherited in a dominant way or presented as *de novo*, although there are some examples of recessive inheritance leading to mild phenotypes. According to results obtained through the HbVar Query page (dated 14th January 2020), 1,534 Hb variants have been described so far due to mutations on either *HBA1*/*HBA2* or the *HBB* genes. Up to 251 variants (16.4%) are classified as UH or HUH based on heat or isopropanol stability tests and/or low Hb abundancy (Giardine et al., 2007, 2014). It is worthy to highlight that all the HUH variants involving the *HBB* gene reported positive stability tests, meanwhile in most of the HUH involving the alpha-globin genes, hyper instability has been only deduced from low abundance. This must be cautiously taken since mutations in alpha-globin genes are lower expressed (<25%) than in beta-globin gene due to the existence of duplicated alpha-globin genes, *HBA1* and *HBA2*, especially in mutations involving the *HBA2* gene, as it encodes a 2-3–fold higher level of mRNA than *HBA1* (Liebhaberts et al., 1986). Thus, the beta-globin gene is the first option to investigate for disease-causing mutations leading to RADs due to UHs/HUHs especially in cases with moderate to severe phenotypes.

Interestingly, Hb Bristol-Alesha is classified as a UH variant, not as a HUH as we expected based on the severity of the patient's clinical picture. The change to methionine at position 67 of the beta-globin chain alters the hydrophobic heme pocket causing the instability of the protein (Kano et al., 2004). As described in previous clinical reports, at physical examination, splenomegaly and jaundice may be found. Iron overload and gallstones may develop due to the rapid turnover of RBCs.

Hb Debrousse, reported twice in literature, is a UH characterized by well-compensated chronic hemolytic anemia due to its high oxygen affinity. Hb Debrousse is caused by leucine to proline substitution at position 96 involving the hydrophobic environment of the proximal side of the heme. In the previously reported cases, Hb Debrousse discovery was possible after a Parvovirus B19 infection that caused a hemolytic crisis (Lacan et al., 1996). Indeed, since affected patients show a chronic well-compensated hemolytic anemia, the diagnosis of such a variant is unlikely until the globin genes are investigated. Such a study is usually performed only when some complications occur.

Hb Zunyi was recently reported for the first time as a *de novo* mutation in a Chinese child with severe anemia requiring blood transfusion, malnutrition, growth delay, splenomegaly and hepatomegaly (Su et al., 2019). In the study herein, we identified both Hb Zunyi and Hb Mokum as *de novo* mutations in the heterozygous state. Hb Zunyi and the novel Hb Mokum are stop-loss mutations at position 442 in *HBB*, resulting in an elongated beta-globin chain leading to HUHs. The extra amino acids in the elongated beta-globin chain (169 a.a.) are probably affecting its helical sequence, interfering with its tertiary structure and causing an unstable tetramer. Frameshift mutations in the *HBB* gene, resulting in the elongated beta-globin chain, have been described before but resulted in shorter beta-chains (max. 157 aa.) and milder phenotypes than the mutations described here (Su et al., 2019).

Finally, Hb Evans is classified as UH. It is consequence of a valine to methionine substitution at position 62 of the alpha2-globin chain encoding gene *HBA2*. Hb Evans has been reported in patients presenting with mild hemolytic anemia that was getting worse particularly in case of stress (Wilson et al., 1989).

The standard tests to detect abnormal anemias are High Precision Liquid Chromatography (HPLC) or conventional or capillary electrophoresis (CE). However, UHs/HUHs do not normally appear in the peak-patterns or appear as small peaks that may be mistaken for degradation products. In three of the five UH cases reported here, extra peaks were not detected. More confined methods are Heinz Bodies test or stability tests as isopropanol precipitation or heat stability tests, which are affordable screening techniques for UHs/HUHs variants. In the patient with Hb Bristol-Alesha, Heinz bodies were detected and heat stability test was positive only after splenectomy (**Figures 1**, **2**), while in the other patients, Heinz bodies and heat stability test were not performed. Genetic analysis of globin genes should be performed for diagnosis confirmation. Inclusion of *HBA1/HBA2* and *HBB* in NGS approaches will facilitate timely conclusive diagnosis. as a screening tool for hemolytic anemias will assist in reaching a definitive diagnosis sooner.

Furthermore, the occurrence of *de novo* mutations causing UHs/HUHs should also be considered in the analysis of genetic variants.

The usefulness of NGS in improving the diagnosis of RADs has already been demonstrated in several studies as well as its relevance in new gene discovery (Shang et al., 2017; Duez et al., 2018). In the case of overlapping phenotypes, which frustrate proper diagnosis, the use of NGS may be beneficial for ultra-rare RADs. In a recent publication, 36% of patients initially diagnosed with congenital dyserytrhopoietic anemia, received a final diagnosis of pyruvate kinase deficiency after NGS analysis (Russo et al., 2018). Nevertheless, in the majority of the t-NGS panels reported, globin genes are not included, since globin genes are quite short and molecular diagnosis of most common Hb disorders, such as sickle cell disease (SCD) and thalassemia syndromes, is well-established through Sanger sequencing and GAP-PCR/MLPA. Therefore, dBTHAL disorders due to UH/HUH may also benefit from NGS approaches for RADs by including globin genes, as presented herein.

Current literature on dBTHAL and UH/HUH variants is mainly composed of retrospective case reports, which makes evidenced-based management of this RAD unlikely. In addition, to benefit from the most adequate management it is necessary to achieve a diagnosis as early as possible. In conclusion, this study confirms the importance of NGS as a fundamental tool to early identify and treat UH/HUH in patients with RAD without an established diagnosis after standard methodologies.

Future challenges include a better understanding of disease characteristics and management, and consideration of bone marrow transplant as a curative option. Therefore, we encourage that these patients are referred to expert Units in referral centers for enabling basic and clinical research taking advantage of the already established European Reference Networks for rare hematological disorders, ERN-EuroBloodNet.

AUTHOR CONTRIBUTIONS

VR, MM-P, CLH, TK, and DB wrote the manuscript. All authors critically revised the manuscript.

ACKNOWLEDGMENTS

This work was generated within the European Reference Network on Rare Hematological Diseases (ERN-EuroBloodNet, FPA 739541).

REFERENCES

Bunn, H. F., and Forget, B. G. (1986). *Hemoglobin: Molecular, Genetic and Clinical Aspects*. Philadelphia, PA: W. B. Saunders Company, doi: 10.1016/0092-8674(87)90069-9

Duez, J., Carucci, M., Garcia-Barbazan, I., Corral, M., Perez, O., Luis Presa, J., et al. (2018). High-throughput microsphiltration to assess red blood cell deformability and screen for malaria transmission-blocking drugs. *Nat. Protoc.* 13, 1362–1376. doi: 10.1038/nprot.2018.035

Efremov, G. D. (2007). Dominantly inherited β-Thalassemia. *Hemoglobin* 31, 193–207. doi: 10.1080/03630260701290092

Galanello, R., and Origa, R. (2010). Beta-Thalassemia. *Orphanet J. Rare Dis.* 5:11. doi: 10.1186/1750-1172-5-11

Giardine, B., Borg, J., Viennas, E., Pavlidis, C., Moradkhani, K., Joly, P., et al. (2014). Updates of the hbvar database of human hemoglobin variants and Thalassemia Mutations. *Nucleic Acids Res.* 42, 1063–1069. doi: 10.1093/nar/gkt911

Giardine, B., van Baal, S., Kaimakis, P., Riemer, C., Miller, W., Samara, M., et al. (2007). HbVar database of human hemoglobin variants and thalassemia mutations: 2007 update. *Hum. Mutat.* 28:206. doi: 10.1002/humu.9479

Kano, G., Morimoto, A., Hibi, S., Tokuda, C., Todo, S., and Sugimoto, T. (2004). Hb Bristol-Alesha presenting Thalassemia-Type hyperunstable hemoglobinopathy. *Int. J. Hematol.* 80, 410–415. doi: 10.1532/IJH97.04048

Kent, M. W., Oliveira, J. L., Hoyer, J. D., Swanson, K. C., Kluge, M. L., Dawson, D. B., et al. (2014). Hb grand junction (HBB: C.348-349delinsG; P.His117IlefsX42): a new hyperunstable hemoglobin variant. *Hemoglobin* 38, 8–12. doi: 10.3109/03630269.2013.853672

Lacan, P., Kister, J., Francina, A., Souillet, G., Galactéros, F., Delaunay, J., et al. (1996). Hemoglobin debrousse (B96[FG3]Leu → Pro): a new unstable hemoglobin with twofold increased oxygen affinity. *Am. J. Hematol.* 51, 276–281. doi: 10.1002/(SICI)1096-8652(199604)51:4<276::AID-AJH5<3.0.CO;2-T

Liebhaberts, S. A., Cash, F. E., Ballad, S. K., and Human Gene Expression (1986). Human A-Globin gene expression. The dominant role of the Alpha 2-Locus in MRNA and protein synthesis. *J. Biol. Chem.* 261, 15327–15333.

Richards, S., Aziz, S., Bale, S., Bick, D., Das, S., Acmg Laboratory Quality Assurance Committee, et al. (2015). Standards and guidelines standards and guidelines for the interpretation of sequence variants: a joint consensus recommendation of the American College of Medical Genetics and Genomics and the Association for Molecular Pathology. *Genet. Med.* 17, 405–424. doi: 10.1038/gim.2015.30

Russo, R., Manna, F., Gambale, A., Marra, R., Rosato, B. E., Caforio, P., et al. (2018). Multi-Gene panel testing improves diagnosis and management of patients with Hereditary Anemias. *Am. J. Hematol.* 93, 672–682. doi: 10.1002/ajh.25058

Shang, X., Peng, Z., Ye, Y., Asan, Zhang, X., Chen, Y., et al. (2017). Rapid targeted next-generation sequencing platform for molecular screening and clinical genotyping in subjects with hemoglobinopathies. *EBioMedicine* 23, 150–159. doi: 10.1016/j.ebiom.2017.08.015

Steinberg, M. H., Forget, B. G., Higgs, D. R., and Weatherall, D. J. (2009). *Disorders of Hemoglobin: Genetics, Pathophysiology, and Clinical Management, Second Edition*, Vol. 94. Cambridge: Cambridge University Press, i–iv. doi: 10.1017/CBO9780511596582

Su, Q., Chen, S., Wu, L., Tian, R., Yang, X., Huang, X., et al. (2019). Severe thalassemia caused by Hb Zunyi [B147(HC3)Stop→Gln; HBB: C.442T>C)] on the β-Globin gene. *Hemoglobin* 43, 7–11. doi: 10.1080/03630269.2019.1582430

Thom, C. S., Dickson, C. F., Gell, D. A., and Weiss, M. J. (2013). Hemoglobin variants: biochemical properties and clinical correlates. *Cold Spring Harb. Perspect. Med.* 3, 1–22. doi: 10.1101/cshperspect.a011858

Wilson, J. B., Webber, B. B., Kutlar, A., Reese, A. L., Mckie, V. C., Lutcher, C. L., et al. (1989). Hb evans or A262(E11)Val→metβ2; an unstable hemoglobin causing a mild hemolytic anemia. *Hemoglobin* 13, 557–566. doi: 10.3109/03630268908993106

16

Theoretical Bases for the Role of Red Blood Cell Shape in the Regulation of its Volume

*Saša Svetina[1,2]**

[1] *Institute of Biophysics, Faculty of Medicine, University of Ljubljana, Ljubljana, Slovenia,* [2] *Jožef Stefan Institute, Ljubljana, Slovenia*

***Correspondence:**
Saša Svetina
sasa.svetina@mf.uni-lj.si

The red blood cell (RBC) membrane contains a mechanosensitive cation channel Piezo1 that is involved in RBC volume homeostasis. In a recent model of the mechanism of its action it was proposed that Piezo1 cation permeability responds to changes of the RBC shape. The aim here is to review in a descriptive manner different previous studies of RBC behavior that formed the basis for this proposal. These studies include the interpretation of RBC and vesicle shapes based on the minimization of membrane bending energy, the analyses of various consequences of compositional and structural features of RBC membrane, in particular of its membrane skeleton and its integral membrane proteins, and the modeling of the establishment of RBC volume. The proposed model of Piezo1 action is critically evaluated, and a perspective presented for solving some remaining experimental and theoretical problems. Part of the discussion is devoted to the usefulness of theoretical modeling in studies of the behavior of cell systems in general.

Keywords: Piezo1, Gárdos channel, mechanosensitivity, spectrin skeleton, curvature dependent protein–membrane interaction, cell to cell variability, osmotic fragility, negative feedback loop

INTRODUCTION

The red blood cell (RBC) shape is, basically, assumed to depend on the cohesion and mechanical stability of its membrane (Mohandas and Chasis, 1993) and its volume to depend on the harmonized action of several different membrane pumps and channels that define the content of cytoplasm cations (Hoffmann et al., 2009). It is therefore considered that RBC shape and volume attain their physiological states independently of each other. The discovery that the RBC membrane includes a mechanosensitive channel, Piezo1, that has an effect on RBC dehydration (Murthy et al., 2017), indicated that RBC volume may also depend on membrane mechanics. Piezo1 acts through the activation of Gárdos channels by Ca^{++} ions that enter the cell when it is open (Cahalan et al., 2015). Recently we proposed a theoretical model in which it was postulated that Piezo1 cation permeability depends on an RBC discoid shape (Svetina et al., 2019). The model revealed the existence of a negative feedback loop that interrelates this shape with the RBC content of potassium ions and, thus, also with its volume. At the Monte Verita RBC meeting I reported about how predictions of the model were verified by utilizing the concepts developed in studies on RBC cell to cell variability (Svetina, 1982, 2017; Svetina et al., 2003). However, the model is a combination of these and several other concepts, together with views expressed previously in different theoretical studies on RBC shape and volume behavior. The present review will include the topics of these studies. This review is also motivated by the fact that the organizers of the Monte Verita meeting asked some senior participants to disseminate to newcomers to the field their research experiences. In this sense it will be rather subjective and thus largely concentrated on the work of our research group. The model discussed here is an example of the research approach by which, on the basis of theoretical analyses and exploitation of existing experimental data, it is possible to make predictions about the behavior of a treated system, thus providing new ideas as to how to advance the corresponding inquiries (Goldstein, 2018). We shall therefore discuss also some general capabilities of theoretical approaches in studies of cell processes.

To understand a given cell process it is necessary to identify the structural elements responsible and to provide a description of the mode of their operation. The corresponding theoretical studies are aimed at obtaining their structure–function relationship in a quantitative manner. This task is, in general, difficult, since cells are complex. The only way to make progress is frequently by analyses of mathematical models. In modeling it is usually necessary first to identify the structural level that is proper for the description of different aspects and for the function of a treated physiological process, and then to reveal its essential features on the basis of the simplest possible system. Models are, as a rule, built on the basis of a set of assumptions that can then be tested experimentally. When these assumptions are found to be correct, and it is thus possible to obtain model predictions by exact either analytical or numerical calculations, a model becomes a theory. The modeling approach should be distinguished from the use of mathematics in the analysis of experimental results and from simulations where, on the basis of the already established theory, the system's behavior can be described mathematically in an exact manner. When modeling the behavior of whole cells it is advantageous to study those that are simple. RBCs, although composed of several thousand different molecules and ions, are, in some aspects, extremely simple. Basically, they are constituted by a concentrated hemoglobin solution enclosed by an essentially smooth membrane. Moreover, they also have a well-defined main function of carrying respiratory gasses. Therefore, and because of its availability, the RBC served, and still serves, as an ideal system for developing the principles of modeling structure–function relationships in cell systems in general (Lux, 2016).

This review will be focused on the bases on which we recently developed a model of the role of Piezo1 in the regulation of the RBC volume (Svetina et al., 2019). The aim is to help build a more thorough critical view on this model. The model was formed on the basis of several RBC and other research directions. It illustrates a circuitous nature of modeling approaches: the past theoretical studies on RBC shape have opened up some other research topics which have turned out to be relevant to studies of the regulation of RBC volume after the identification of the mechanosensitive protein Piezo1 (Coste et al., 2010) and the elucidation of

its role in hereditary xerocytosis (Zarychanski et al., 2012). Briefly, some years ago we examined the possible osmotic states of RBC in dependence on the permeability state of its membrane (Brumen et al., 1979, 1981). In another study we presented (Svetina et al., 1982) a theoretical counterpart of the earlier proposed bilayer couple hypothesis of RBC shape transformation (Sheetz and Singer, 1974). This led us to formulate a general theory of shapes of vesicular objects with flexible membranes (Svetina and Žekš, 1989). This theory predicted that, among the possible stable shapes, some exhibit polar symmetry. We proposed that such shapes could serve as a mechanical origin of cell polarity, and also speculated that this could have been realized through curvature dependent interaction between membrane inclusions such as channels and pumps and the surrounding membrane (Svetina and Žekš, 1990; Svetina et al., 1990). We later derived a general phenomenological interaction term for the curvature dependent inclusion–lipid matrix interaction (Kralj-Iglič et al., 1999), and formulated the procedure for treating the mutual effects of the shape of a vesicular object and the lateral distribution of membrane inclusions on each other (Kralj-Iglič et al., 1996; Božič et al., 2006). These results gained significance because, in the meantime, several membrane proteins had been disclosed that were characterized by their membrane sensing and curvature forming capabilities (McMahon and Gallop, 2005; Zimmerberg and Kozlov, 2006). The curved Piezo1 structure (Ge et al., 2015; Guo and MacKinnon, 2017; Saotome et al., 2018; Zhao et al., 2018) indicates that it affects the shape of the surrounding membrane. A possible role of curvature dependent protein–membrane interaction in the process of mechanosensitivity has also been indicated (Svetina, 2015). The described broad modeling background thus seemed to be well suited also for analyzing different possible modes of Piezo1 operation in the regulation of RBC volume.

The review is organized as follows. The treated model (Svetina et al., 2019) will be described and commented in the last section (see section "Model of the Effect of RBC Discocyte Shape on RBC Volume and its Outlook"). The two intermediate sections will describe the model background. Section "The Mechanical and Thermodynamic Bases of RBC Shape and Deformability" deals with the RBC shape and deformability. In its first subsection it will be described how the initial theoretical studies of RBC shapes in which it was assumed that its membrane is laterally homogeneous led to a general theory of vesicular objects with flexible membranes. In the second subsection it will be shown how the difference between the predictions of this theory and the behavior of RBCs helps to understand the role of RBC membrane skeleton. Section "RBC Volume and Related Aspects of the Variability of RBC Population" will deal with the models of the regulation of RBC volume. Special attention will be devoted to the aspects of RBC population variability. It will then be shown how the results described in two previous sections can be combined in the model of the effect of Piezo1 on RBC volume. A critical review of this model will be given and some suggestions presented for the necessary future work. Throughout the review, the emphasis will be on the development of concepts, therefore it will be mostly presented in a descriptive manner. The corresponding equations and their derivation can be found in the cited literature.

THE MECHANICAL AND THERMODYNAMIC BASES OF RBC SHAPE AND DEFORMABILITY

The function of RBC as the carrier of respiratory gasses led to its adoption throughout evolution of numerous specific mechanical and thermodynamic properties. In the absence of external forces, the normal RBCs of most vertebrates assume the shape of a disk that involves, at its poles, the presence of symmetrically indented dimples. RBCs are deformable, e.g., under microcirculatory flow conditions, at sufficiently high shear stress, deform into rolling stomatocytes and, finally, adopt polylobed shapes (Lanotte et al., 2016). An important factor that allows for these shape transformations is that RBC occupies only about 60% of the volume that a cell could at a given area of its membrane. This property of RBCs is conveniently quantified in terms of the reduced volume (v) defined as the ratio between the RBC volume (V) and the volume of the sphere with the same membrane area (A):

$$v = 3V/4\pi R_s^3 \qquad (1)$$

where $R_s = (A/4\pi)^{1/2}$. The mechanism for the establishment of RBC volume will be dealt with in section "RBC Volume and Related Aspects of the Variability of RBC Population." Here our focus is on how, at a given value of v, RBC shape and its deformation depend on the mechanical properties of its membrane. RBC membrane is composed of a lipid bilayer occupied densely by membrane integral proteins, and the underlying membrane skeleton, a two-dimensional pseudo-hexagonal network with actin based protein complexes as nodes and spectrin tetramers as bonds. The bilayer and the skeleton are linked by chemical bonds between spectrin and integral membrane proteins band3 and glycophorin C, via ankirin and actin complexes, respectively (Mohandas and Gallagher, 2008; Lux, 2016). The RBC membrane differs from those of most other eukaryotic cells in that it has no cytoplasm reservoirs and has therefore a smooth appearance. Because of a relatively large value of the compressibility modulus of the bilayer it is, laterally, practically incompressible. RBC mechanical behavior depends crucially on the characteristic of its membrane that its three layers, the two leaflets of the bilayer and the underlying skeleton, can slide, one over the other. The bilayer resists bending of the membrane and the skeleton to exhibit shear deformation. The first reasonable models of RBC shape behavior were based on the assumption that their shapes correspond to the minimum of membrane bending energy. In the subsequent subsection it will be revealed how, out of these models, a theory of shapes of simple vesicular objects such as phospholipid vesicles developed, and how the stability of shapes depends on the elastic properties of multilayered membranes. In the second subsection it will be shown how the comparison between the predictions of this theory and

the behavior of RBC helps the different roles of its membrane skeleton to be understood.

Interpretation of RBC and Vesicle Shapes on the Basis of Membrane Bending

Red blood cell shape has been treated by assuming its membrane to be a single, thin, laterally homogeneous mechanical entity. RBC membrane can, at its reduced volume v of about 0.6, take up an infinite number of shapes, exhibiting different values of the total membrane bending energy (W_b) that can be for symmetrical bilayer obtained by the integral of the square of the mean membrane curvature ($H = (C_1 + C_2)/2$ where C_1 and C_2 are the principal membrane curvatures) over the whole membrane area expressed as

$$W_b = 2k_c \int H^2 dA \qquad (2)$$

with k_c membrane bending constant. In general membrane bending energy involves also a contribution due to Gaussian curvature ($K = C_1 C_2$) (Helfrich, 1973). However, because the integral of K over the membrane area is for a given membrane topology constant this term will be in further discussions here ignored. Canham (1970) looked for the minimum of W_b (Eq. 2) and found that, at $v = 0.6$, the shape is a discoid. He assumed that the membrane has zero energy when it is flat and took into consideration that the membrane has no lateral shear, i.e., that it behaves laterally as a two-dimensional liquid. Helfrich (1973) generalized the expression for membrane bending energy by assuming that the membrane may have, due to transmembrane asymmetry, zero energy when it is bent to its spontaneous curvature (C_0). Deuling and Helfrich (1976), by applying their "spontaneous curvature model," obtained by minimizing the Helfrich's (1973) expression of membrane bending energy at given reduced volumes and reduced values of the spontaneous curvature ($R_s C_0$) beside the discocyte also several other shapes including cup shaped stomatocytes. At about the same time Sheetz and Singer (1974) introduced the "bilayer couple hypothesis" based on the evidence that RBCs change shape under conditions of asymmetric changes of the areas of the outer and inner leaflets of the membrane bilayer. They showed that, by adding a drug (chlorpromazine) that intercalates into the inner monolayer of the RBC bilayer, the discocyte transforms into a cup shape (stomatocyte) whereas drugs intercalating into the outer layer cause shape transformation into a spiculated echinocyte. In a theoretical treatment of the bilayer couple hypothesis, RBC shapes (which were defined in terms of the finite number of geometrical parameters) were obtained by minimization of the membrane bending energy at a fixed difference between the areas of the outer and inner leaflets of the bilayer (Svetina et al., 1982). It was implied that this area difference (ΔA) constitutes a convenient single parameter whose continuous decrease causes the shape to be transformed from discocyte to stomatocyte in a continuous manner. This result was confirmed and, also, further explored by an exact variational procedure for minimizing membrane bending energy under the constraints of constant membrane area, cell volume, and area difference (Svetina and Žekš, 1989). While bilayer couple hypothesis represents for some aspects of RBC shape transformations a useful workable model, it also turned out to be a strict theory for shapes of simple vesicular objects defined as a liquid interior enclosed by a flexible membrane. For students of RBC shape behavior and deformability it is useful to be familiar with the basic results of this theory because knowledge of its predictions may help to distinguish which aspects of RBC behavior depend on properties of its bilayer and which on its other structural features.

The bilayer couple theory (Svetina and Žekš, 1989) predicts that vesicle shapes depend on only two geometrical parameters, the reduced volume v and the reduced area difference Δa (defined as the ratio between ΔA and its value for the sphere which is $8\pi h R_s$ with h the distance between the neutral surfaces of the bilayer leaflets and R_s, as already defined, the radius of the sphere). The geometrical meaning of Δa is that it is also equal to the integral of the reduced mean membrane curvature ($R_s H$) over the membrane surface. Vesicle shapes can be grouped into classes that occupy different parts of the $v - \Delta a$ (or $\Delta a - v$ as used in Seifert et al., 1991) shape phase diagram. The shapes belong to a given class if they have the same symmetry and if, by continuously changing Δa and/or v, they are changing continuously. The sense of such shape classification is illustrated in **Figures 1A–C**. There are two types of shape class boundaries. One type comprises shapes obtained by variational search of the extreme values of the reduced volume v at a fixed value of the reduced area difference Δa (**Figure 1A**). They are composed of spheres or spherical parts with only two possible values of their radius (Svetina and Žekš, 1983, 1989). For example, lines 1 and 6 are boundaries of shape class to which belongs the discocyte (located in the minimum of the bending energy curve "S" in **Figure 1C**). All these shapes are axisymmetric and involve equatorial mirror symmetry. The second type of shape class boundaries are symmetry breaking lines (lines 9 to 12 in **Figure 1B**). For example, the class of cup (stomatocyte) shapes is, on one side, bounded by the limiting shape (line 5 in **Figure 1A**) and, on the other side, by the symmetry breaking line (line 9 in **Figure 1B**) that connects the points at which the equatorial mirror symmetry of disk shapes breaks down (shown by an arrow in **Figure 1C**) at all reduced volumes. Notably, the class of non-axisymmetric (ellipsoidal) shapes is bounded by symmetry breaking lines on both of its sides (lines 10 and 11 in **Figure 1B**) at which disk and cigar shapes, respectively, break down their axial symmetry (Heinrich et al., 1993). As demonstrated by curves A and S in **Figure 1C**, classes overlap. Only the shape with the lowest bending energy is stable. **Figure 1B** shows which shapes are stable within the presented central part of the $v - \Delta a$ shape phase diagram. Red point and triangle in **Figures 1A–C** indicate where in the $v - \Delta a$ shape phase diagram are located the discocyte and typical stomatocyte, respectively. The significance of the bilayer couple theory is that it predicts all possible shapes of vesicular objects with laterally homogeneous membranes. If a vesicle shape differs from any of these shapes it means that there are external forces acting on it (Svetina and Žekš, 1996) or that its membrane is laterally inhomogeneous (Božič et al., 2006).

The described predictions of the bilayer couple model are strictly only valid if the two equally composed leaflets of a bilayer are incompressible. In reality they are compressible and therefore it has to be taken into account that, in general, in a given shape, they might be deformed differently, for example in that the area of one is extended and of the other compressed. In such cases the reduced area difference (Δa) differs from analogously defined equilibrium (preferred) area difference (Δa_0) which corresponds to the situation where leaflets are neither extended nor compressed. The bilayer thus exhibits, in addition to the already defined bending energy (Eq. 2), also the non-local bending energy (W_k) (Evans, 1980), termed also as area difference elastic term (Miao et al., 1994), expressed in its reduced form ($w_k = W_k/8\pi k_c$) as

$$w_k = k_r(\Delta a - \Delta a_0)^2/2 \qquad (3)$$

where k_r is the non-local bending constant. The derivation and consequences of non-local bending energy were comprehensively reviewed in Svetina and Žekš (2014). Briefly, in the generalized bilayer couple model, vesicle shapes correspond to the minimum of the sum of the bending (Helfrich, 1973) and non-local bending energies. The shape equation to be solved is the same as in the limit of the strict bilayer couple model so the shapes obtained are the same. However, not all of them are necessarily stable. Why it is so is demonstrated by **Figure 1D**. The minimization of the sum of the two bending energies with respect to Δa gives rise to the requirement

$$\partial w_b/\partial \Delta a = -(k_r/k_c)(\Delta a - \Delta a_0) \qquad (4)$$

The solution of Eq. 4 for its unknown Δa can, for a given value of Δa_0, be obtained graphically as a point on the graph of **Figure 1D** where a dashed curve (right hand side of Eq. 4) crosses one (either S or A) of the $\partial w_b/\partial \Delta a$ curves (left hand side of Eq. 4). The number of solutions of Eq. 4 at given Δa_0 depends on the slope of dashed curves that is proportional to the ratio k_r/k_c. There is only one solution if this slope is steeper than that of the largest derivative by Δa of the function $\partial w_b/\partial \Delta a$ of asymmetrical shapes (A) which is at the symmetry breaking point (**Figure 1C**). A vesicle can thus attain all possible shapes predicted by the strict bilayer couple model. At values of k_r/k_c that are smaller than above defined critical value of $\partial w_b/\partial \Delta a$ there are, for some values of Δa_0 (e.g., 0.82 in **Figure 1D**), three solutions of Eq. 4. The shape at the middle value of Δa is not stable. Consequently there is, e.g., at continuously decreasing value of Δa_0, a discontinuous shape transformation from the Δa at a cross-section of a dotted line with the curve S to the smaller Δa at which this line crosses the curve A. The shapes of the strict bilayer couple model that correspond to the intermediate Δa values are not stable. Possible stable shapes of the generalized bilayer couple model are thus defined by the 3-dimensional $v - \Delta a - k_r/k_c$ shape phase diagram. The example of the cross-section of this diagram is for $v = 0.85$ shown in Figure 12 of Svetina and Žekš (1996). The slope of the $\partial w_b/\partial \Delta a$ curve at the symmetry breaking point is at $v = 0.6$ close to -3. The estimated ratio k_r/k_c for RBC membrane is about 2 (Hwang and Waugh, 1997) which indicates that the discocyte–stomatocyte transition is discontinuous. The described reasoning can be generalized straightforwardly to cover also the effects of transmembrane asymmetry characterized by membrane spontaneous curvature

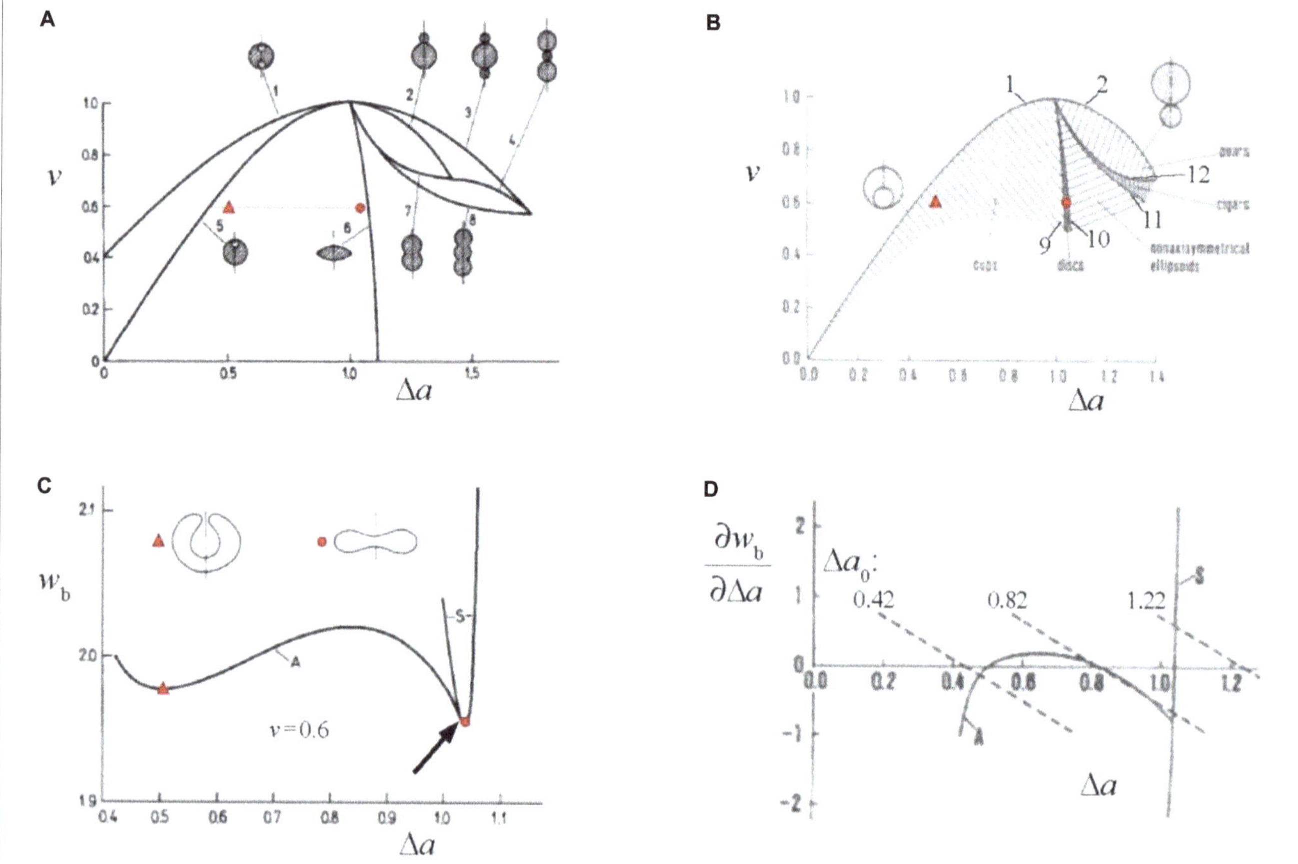

FIGURE 1 | Demonstration of basic features of the bilayer couple theories of vesicle shapes [**(A–C)** of "strict" and **(D)** of "generalized"]. **(A)** $v(\Delta a)$ dependences of limiting shapes which correspond to the extreme values of v at a given value of Δa (lines 1 to 8) plotted in the $v - \Delta a$ shape phase diagram (Svetina and Žekš, 1989). Examples of corresponding shape cross-section are also shown. The dotted horizontal line at $v = 0.6$ indicates the range of Δa values for which minimal membrane bending energy is presented in **(C)**. Red point indicates the phase diagram location of a discocyte and red triangle of a characteristic stomatocyte. **(B)** Stable vesicle shapes in the central part of the $v - \Delta a$ shape phase diagram. Lines 9–12 are symmetry breaking lines. Marked areas represent the parts of the shape phase diagram where there are no contacts between different regions of the membrane. The meaning of red points is the same as in **(A)**. **(C)** Membrane bending energy w_b (defined relative to the bending energy of the sphere which is $8\pi k_c$) calculated at $v = 0.6$ for the Δa values indicated by the dotted line in **(A)**. Line S shows bending energy of disk shapes and line A of the cup shapes. The arrow indicates the point of the symmetry breaking of the disk shape as predicted by Svetina and Žekš (1989). Shown are also contours of cross-sections of a discocyte (red point) and a stomatocyte (red triangle) and it is indicated which are their Δa values. **(D)** The dependence of the partial derivatives of the bending energies S and A presented in **(C)** by the area difference Δa. The dotted lines present the right hand side of Eq. 4 for the indicated values of Δa_0 and $k_r/k_c = 3$ (reprinted with permission from Svetina, 1998).

C_0. C_0 and ΔA_0, in spite of having different physical background affect the shapes of vesicular objects with bilayer membranes in a similar manner. The stationary shapes obtained by solving the shape equation are the same as in the strict bilayer couple model if for the reduced equilibrium area difference is taken an effective one defined as

$$\Delta a_{0,eff} = \Delta a_0 + c_0 k_c / 2k_r \qquad (5)$$

It has to be noted that in this case the region of stable shapes in the generalized shape phase diagram $v - \Delta a_{0,eff} - k_r/k_c$ depends on the relative contribution to $\Delta a_{0,eff}$ of Δa_0 and c_0. It is because the energy term due to Δa_0 (Eq. 3) involves Δa^2 whereas the energy term due to c_0 is a linear function of Δa. Therefore the discontinuous transition indicated in **Figure 1D** occurs at shifted Δa values, such that at increasing the relative contribution of c_0,

the region of stable shapes is diminishing. The limit $k_r/k_c = 0$ represents the spontaneous curvature model of Deuling and Helfrich (1976). A full description of which shapes are stable in this limit was presented by Seifert et al. (1991) (reviewed in Seifert, 1997). The spontaneous curvature model also applies in the case that due to transmembrane lipid transport the bilayer relaxes into the state with $\Delta a_0 = \Delta a$. The relaxation time for this process was estimated to be 8 min (Raphael and Waugh, 1996) or even much less (Svetina et al., 1998).

Effects of Compositional and Structural Features of the RBC Membrane

Red blood cell membrane is, compared to phospholipid membranes, complex. Its bilayer part is crowded with integral membrane proteins such as band3 that is involved in RBC's function of carrying carbon dioxide, different pumps and channels that take care of the establishment of RBC volume, and many other proteins serving in its protection (Mohandas and Gallagher, 2008). As already noted, it contains, on its cytoplasmic side, a spectrin based membrane skeleton which is the main element that accounts for how RBC shape behavior differs from that of simple phospholipid vesicles. In this respect we here discuss skeleton shear elasticity, its role in the formation of RBC shape, and its possible effects on the lateral distribution of integral membrane proteins and their lateral diffusion.

Red blood cell membrane exhibits shear elasticity. Because its bilayer part can be considered as two-dimensional liquid, the shear elasticity can be ascribed solely to its membrane skeleton which is a two-dimensional pseudo-hexagonal network of spectrin tetramers as bonds and acting filaments as nodes. To understand the skeleton behavior it is crucial to realize that RBC membrane deformation may cause an alteration of local skeleton densities while the density of the lipids remains the same, as was observed by measuring skeleton lateral distribution in RBC partially aspirated into the micropipette (Discher et al., 1994; Discher and Mohandas, 1996; **Figure 2A**). These results imply that skeleton nodes shift their position relative to the bilayer and that the bonds deform elastically. This is possible because the bilayer integral proteins to which the skeleton is anchored can move laterally in the plane of the bilayer. The observed changes of skeleton density indicate that the extension ratios (λ_i, the ratio between the final length and the initial length of the deformed skeleton material in the i-th direction) may reach a value of about 3, which corresponds to a fully extended spectrin tetramer of length ~ 200 nm. Skeleton deformation may be described by shear and area compressibility deformational modes (Mohandas and Evans, 1994). The area compressibility elastic modulus of the RBC skeleton is estimated at 26 μN/m (Svetina et al., 2016), which is about four orders of magnitude less than that of the whole RBC membrane determined to be 0.29 N/m (Evans et al., 1976). Consequently, the deformed skeleton redistributes over the RBC membrane at its practically constant total area A. Different parts of RBC skeleton are kept together by non-covalent bonds which can break and reform, indicating the possibility that it is plastic. The undeformed state of the skeleton may thus depend on the RBC's history and is in general not well defined.

Theoretical modeling of the RBC skeleton is developing in several different directions (e.g., Discher et al., 1998; Fedosov et al., 2010; Peng et al., 2010; Svetina et al., 2016). The present discussion will focus on the type of models aimed at making a distinction between skeleton deformation due to the change of the shape of the cell membrane and that due to its mechanical properties. In this respect Mukhopadhyay et al. (2002) applied in their model of the RBC membrane the concept of a mapping function defined in terms of the dependence of the original position of a skeleton element on its position in the deformed state. They used this concept in their studies of the effect of the skeleton on RBC shapes (see below). For axisymmetric shapes it is possible to express the mapping function as $s_0(s)$ where s_0 is the arc-length distance from the cell pole to a given contour point of the original skeleton state and s the corresponding distance of the deformed state. Mathematically,

the dependence $s_0(s)$ is sought at which the sum of the skeleton energy and the membrane bending energy is minimal. We treated a simplified version of this problem by studying cases in which the deformed shape is defined by rigid walls and therefore there is no need to consider the bending energy (Svetina et al., 2016). Examples of such deformations are RBC aspiration into medium sized (with respect to the RBC size) (**Figure 2A**) and narrow (~ 1 μm) pipettes. Following Discher et al. (1998) we also assumed that the main contribution to the skeleton energy is the energy of spectrin bonds. Instead of using for bond energy a more realistic flexible chain model (Discher et al., 1998), we described this energy by a harmonic potential which is a good approximation at sufficiently small deformations (Figure 8 in Svetina et al., 2016). The energy of the deformed skeleton was determined in the mean field approximation. In this simple model the skeleton deformation is the same for any value of the bond strength. This means that the most important factor for the observed change in skeleton lateral distribution is the changed cell geometry. For axisymmetric shapes it is, to some extent, possible to reason about the effect of changed geometry in qualitative terms. When a patch of the skeleton that is at a distance r_0 from the axis moves in the deformed state to the distance r from the axis, its extension along the parallels is $\lambda_p = r/r_0$. The corresponding compression (in case that $r < r_0$) exerts a tendency to make λ_m larger than 1. The local magnitudes of the extensions along meridians are restricted by the requirement that the total skeleton area is constant. The effect of this requirement cannot be visualized so clearly; however, it can still be concluded that the skeleton prevents shape changes with significant changes of the distances of the membrane from the axis. In **Figure 2B**, as an example, is shown by this model predicted deformation of RBC discocyte aspirated into a medium sized pipette.

Red blood cell shape behavior differs qualitatively from that of phospholipid vesicles in the region of the $v - \Delta a$ shape phase diagram, where prolates are the typical equilibrium shapes of vesicles with simple membranes (e.g., dumb-bells and pears) with their limiting shapes involving external buds (**Figures 1A,B**). In contrast, RBC shapes are, in the respective Δa region, echinocytic (reviewed in Svetina et al., 2004). Mukhopadhyay et al. (2002) studied the formation of echinocytes on the basis of continuum mechanics by searching for the minimum of the sum of the local and non-local bending energies of the bilayer, and the stretching and shear elastic energies of the membrane skeleton. The bending energy of the spiculated membrane is larger than that of the dumb-bell shape. However, the echinocyte can be understood to be more stable since the deformation of the skeleton into the geometry of a dumb-bell would, because of large changes of the distances from the axis, require a much larger increase of the skeleton energy than that for its transformation into a quasi-spherical echinocyte (neglecting that some skeleton albeit with decreased density is also present in its spicules). The formation of echinocytes has a physiological advantage by preventing the occurrence of shapes with large external buds that form in simple vesicles at increased values of Δa (c.f. line 3 in **Figure 1A**). The probable pinching off of such buds in

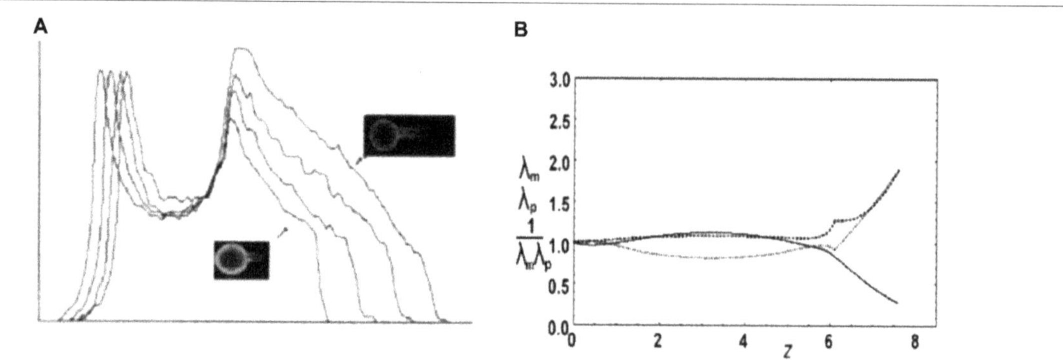

FIGURE 2 | Deformation of RBC skeleton. **(A)** Experimental evidence for the deformation of the membrane skeleton when RBC ghost is aspirated into a medium size pipette ($R_p \approx 2\ \mu$m) (from Discher et al., 1994; reprinted with permission from AAAS). The skeleton density profiles along the projection of four aspirated RBC ghosts are shown, obtained by measuring fluorescein-phalloidin-labeled actin. Relevant for the present discussion is the decrease of the intensity along the aspirated part of the ghosts (arrows). **(B)** Skeleton extension ratios along parallels λ_p (dashed) and along meridians λ_m (dots) and the density $1/\lambda_p\lambda_m$ (relative to its mean value; full line) calculated according to the described model (Svetina et al., 2016) for the RBC aspirated into medium size pipette **(A)** (reprinted with permission from Švelc and Svetina, 2012). The pipette is on the right. The initial shape with, presumably, homogeneous skeleton distribution is a discocyte.

the turbulent blood flow would increase RBC reduced volume considerably, thus diminishing its deformability. It should be noted that when RBC changes its shape from discocyte to stomatocyte, e.g., due to possible decrease of membrane Δa_0, the skeleton is not deformed so much because these two shapes are both oblate and the distances of skeleton elements from the axis in them do not differ appreciably. Therefore, the behavior of RBC shape in the "oblate" region of the $v - \Delta a$ shape phase diagram does not differ essentially from the behavior of simple phospholipid vesicles. In cases where the buds are internal (c.f. lines 1 and 5 in **Figure 1A**) there is also no danger that they would be pinched off.

Another physiological role of the RBC membrane skeleton is that of the prevention of formation of risky budded shapes due to lateral segregation of its integral proteins. Membrane embedded proteins interact with the surrounding membrane when their intrinsic principal curvatures differ from those of the membrane at their location. For example, when the drastically curved protein Piezo1 (Guo and MacKinnon, 2017) is embedded in a flat phospholipid membrane it is predicted that it will cause the membrane in its surroundings to form a kind of dome-shaped invagination (**Figure 3A**, from Haselwandter and MacKinnon, 2018). It is possible to treat such a protein–membrane interaction in terms of a phenomenological expression that takes into account the fact that there is a mismatch between the intrinsic principal curvatures of membrane inclusion (e.g., a protein) and those of the membrane. The general expression for the corresponding energy term, in the limit of a rigid inclusion surface, is conveniently written as (Kralj-Iglič et al., 1999)

$$W_{curv,j} = \frac{\kappa_j}{2}\left(H - H_{P,j}\right)^2$$

$$+ \frac{\kappa_j^*}{2}\left[\Delta H^2 - 2\Delta H \Delta H_{P,j}\cos\left(2\omega_j\right) + \Delta H_{P,j}^2\right] \quad (6)$$

where $H_{P,j} = (C_{1,P,j} + C_{2,P,j})/2$ is the mean principal intrinsic curvature of the transmembrane part of the inclusion and

$\Delta H_{P,j} = (C_{1,P,j} - C_{2,P,j})/2$ is a measure of the difference between its two principal curvatures. κ_j and κ_j^* are independent interaction constants. The angle ω_j defines the mutual orientation of the coordinate systems of the intrinsic principal curvatures of the inclusion and the principal curvatures of the membrane. One consequence of such interaction term is curvature sensing, meaning that mobile membrane proteins, due to curvature dependent interaction energy term, accumulate in membrane regions where this mismatch is small and are depleted from regions where it is large. For example, it is reasonable to expect that it is more probable for the Piezo1, due to its curved structure, to reside in regions of RBC discocyte poles (dimples) than on its equator. The second possible consequence of the curvature dependent protein–membrane interaction is its effect on shape which, for a membrane with mobile proteins, corresponds to the minimum of the sum of their distributional free energy and the bending energy of the membrane (Božič et al., 2006). The effect of inclusions on RBC shape depends on their number. Using a model study (Svetina et al., 1996), it was shown that, for the discocyte, the minimum system's free energy depends on the number of inclusions in a much stronger manner than in that for a budded shape (**Figure 3B**). If this number is sufficiently small the bending energy prevails and the stable shape remains to be the disk. However, above a certain critical value, the budded shape has a lower free energy. RBC membrane contains about 10^6 band3 proteins which could constitute a potential danger for the stability of the RBC if most of them were not linked to the skeleton.

Red blood cell membrane proteins that are not linked to the skeleton can, upon the deformation, redistribute over the membrane with a time constant that depends on their diffusion coefficient. The latter can be smaller than in a vesicle because of the corralling effect of the spectrin skeleton (Tomishige et al., 1998). A drastic reduction of the diffusion constant can be expected for Piezo1 because the membrane indentation that it causes (**Figure 3A**). It just about fits into the triangle formed

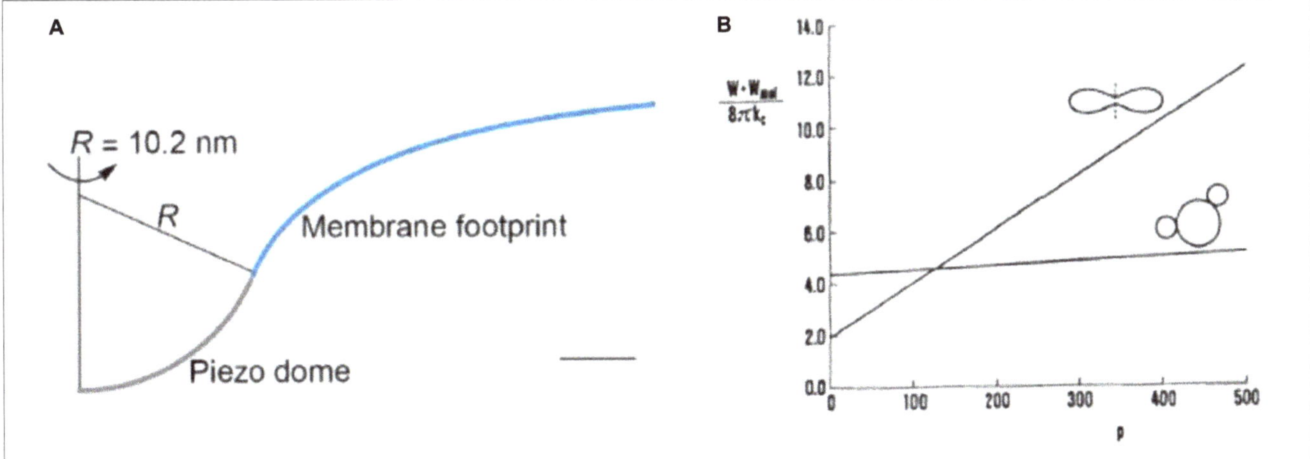

FIGURE 3 | Illustrations of effects of protein–membrane interaction. **(A)** The shape of the Piezo1 membrane footprint shown as the cross-section of the mid-bilayer surface and its intersection with the Piezo1 dome; scale bar: 4 nm (reprinted from Haselwandter and MacKinnon, 2018). **(B)** The sum of the membrane bending energy and the distributional free energy of mobile membrane inclusions calculated for the disk shape and for the shape that involves two buds for a cell with the reduced volume 0.6 in the dependence of the number of inclusions p (in some reduced units) (reprinted with permission from Svetina et al., 1996).

by the three spectrin tetramers, with each of them having in the RBC's resting state the length of about 70 nm.

RBC VOLUME AND RELATED ASPECTS OF THE VARIABILITY OF RBC POPULATION

Red blood cell membrane is, as those of most mammalian cells, well permeable for water. Therefore the RBC's water content, and thus also its volume, depend on its content of osmotically active substances and on the external tonicity (Hoffmann et al., 2009). Hemoglobin, the main protein constituent of the RBC cytoplasm, cannot cross its membrane, so therefore the RBC is under osmotic stress. In RBCs of many species, their reduced volume is, by virtue of the active pump-leak system for monovalent cations K^+ and Na^+, nevertheless kept at about 0.6. These cations are pumped by Na^+/K^+- ATPase which actively expels three sodium ions and takes in two potassium ions (Tosteson and Hoffman, 1960; Kay and Blaustein, 2019). The consequent higher cell concentration of K^+ and lower concentration of Na^+, both relative to their concentration in the environment, cause fluxes of these two cations in the direction of their concentration gradients. The pumping and leaking of K^+ and Na^+ eventually leads to the stationary volume level. There are several channels/transporters involved in the passive leakage of these two cations. The results of many studies of their action and of the data on cation pumping made it possible to formulate realistic mathematical modeling of RBC volume regulation (Lew and Bookchin, 1986; Armstrong, 2003; Ataullakhanov et al., 2009). K^+ and Na^+ attain their stationary value in a time scale which is several orders of magnitude larger than that of water and also of univalent anions Cl^- and HCO_3^- which can thus be treated at correspondingly short time scales as being

in quasi-equilibrium between the inside and outside solutions. Because hemoglobin is charged, this equilibrium can be described by models that involve a version of the Donnan equilibrium in which cations do not exchange (Brumen et al., 1979; Freedman and Hoffman, 1979).

The issue here is the extension of already established models of RBC volume regulation that take into consideration the role of Piezo1 and Gárdos channels (Svetina et al., 2019). The predictions of the proposed model were supported by studies on properties of RBC population cell to cell variability. RBCs, in otherwise homogeneous RBC population, are known to be variable with respect to many of their measurable parameters such as cell volume, membrane area, hemoglobin content, density, etc. RBCs vary because of the variable properties of their precursors and because, throughout their lifespan, they are releasing nanovesicles (Westerman and Porter, 2016). In general, cells of the same kind are presumably organized in an identical manner, meaning that their state is defined by the same physical-chemical processes. On the basis of this assumption it can be deduced that, if there are some strict algebraic relations between the parameters that define the state of a single cell, there are also relationships between the parameters that measure the variability of these parameters, e.g., coefficients of variation and correlation coefficients. This notion has been confirmed by analysis of relations between standard deviations and correlation coefficients extracted from different single cell measurements of RBC volume, membrane area, density, and hemoglobin and cation contents (Svetina, 1982). Moreover, this analysis also revealed the possibility that an RBC involves a strict relationship, of at that time unknown origin, between membrane area and hemoglobin and cation contents. The analogous conclusion was later, by a different approach, obtained by Lew et al. (1995). The basic source for such relation was realized to be the strong correlation between RBC volume and membrane area. Simultaneous single cell determinations of these two cell

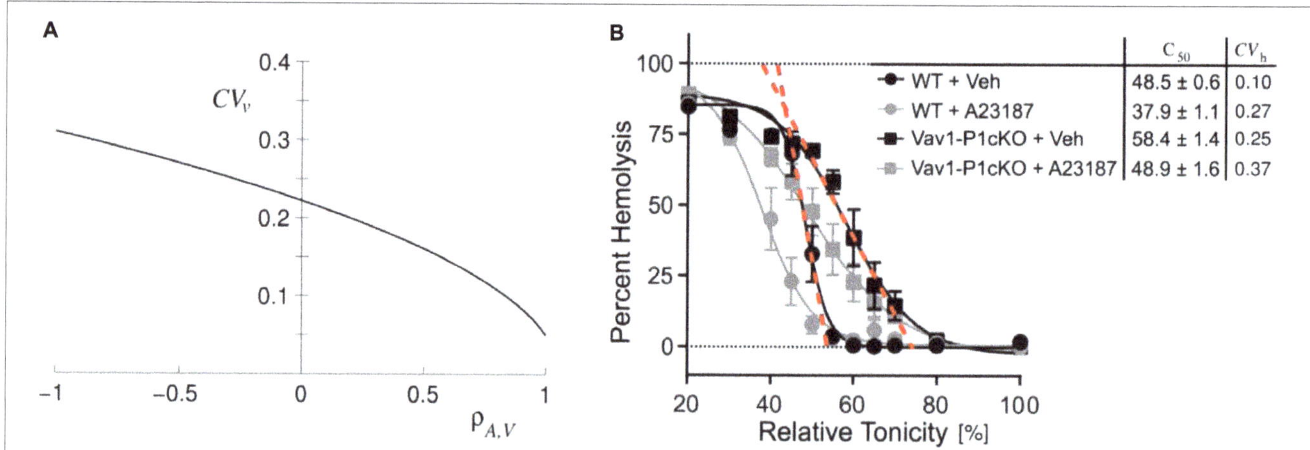

FIGURE 4 | Illustrations of consequences of the correlation between RBC area (*A*) and volume (*V*). **(A)** Coefficient of variation of RBC reduced volume (*CV*$_v$) in dependence on the correlation coefficient $\rho_{A,V}$ obtained for the values of coefficients of variations of RBC volume and membrane area to be 0.12 and 0.13, respectively (reprinted with permission from Svetina et al., 2019). *CV*$_v$ at $\rho_{A,V}$ = 0.96 is about 0.06. **(B)** Evidence for the role of Piezo1 based regulation of RBC volume on the mouse RBC *V − A* correlation. Cahalan et al. (2015) measured osmotic fragility of normal (WT) and of Piezo1 knocked out (Vav1-PicKO) mouse RBC in the absence (+Veh) and presence (+A23187) of Ca^{2+} ionophore A23187 (reprinted with permission from Svetina et al., 2019). In Figure 5 of Svetina et al. (2019) we added to Figure 3D of Cahalan et al. (2015) the column of the corresponding coefficients of variation (*CV*$_h$) obtained from steepness of osmotic fragility curves as indicated by red dashed lines.

parameters yielded for the corresponding correlation coefficient a value of $\rho_{A,V} \sim 0.97$ (Canham and Burton, 1968) and of ~ 0.96 (Gifford et al., 2003). The strong correlation between *V* and *A* is reflected in the fact that the coefficient of variation of the reduced volume is only about one half of the coefficients of variation of *V* and *A* (**Figure 4A**). It also explains the narrowness of the coefficient of variation for the hemolytic osmotic pressure presented in Figure 2 of Lew et al. (1995). With respect to the modeling of the regulation of the RBC volume it can be concluded that one of the model parameters should be membrane area. Recent evidence indicates that the process responsible for *V − A* correlation involves Piezo1 (Cahalan et al., 2015). Namely, in Piezo1 knockout mice, and in the case where Piezo1 action of opening Gárdos channels was overruled by the Ca^{2+} ionophore A23187, this correlation was lost (**Figure 4B**).

MODEL OF THE EFFECT OF RBC DISCOCYTE SHAPE ON RBC VOLUME AND ITS OUTLOOK

The fact that RBC dehydration in hereditary xerocytosis can be caused by malfunctioning of a mechanosensitive protein Piezo1 indicates that RBC volume may also depend on mechanical properties of RBC membrane. Piezo1 system appeared to represent a relatively independent module of the otherwise complex regulation of RBC volume, and could thus be considered as an ideal candidate for application of the modeling approach. In the model under consideration (Svetina et al., 2019) we chose, for its elements that are crucial for the action of Piezo1, the sensing of the membrane curvature through curvature dependent inclusion–membrane interaction (Svetina et al., 1990; Kralj-Iglič et al., 1996, 1999), and the assumption that the mechanical

properties of an RBC membrane affect its volume through the dependence of Piezo1 Ca^{++} permeability on RBC discocyte shape. These ideas were supported by curved structure of the Piezo1 trimer, evidenced by its structural studies (Ge et al., 2015; Guo and MacKinnon, 2017; Saotome et al., 2018; Zhao et al., 2018) and, indirectly, also by the altered RBC membrane cation permeability when stressing the cells by a distorting device (Kuchel and Shishmarev, 2017) or by a flow through a capillary constriction (Cinar et al., 2015; Danielczok et al., 2017). Here we shall first outline which of the topics presented in the previous two sections formed the basis of this model. Then we shall evaluate its outcome, define its deficiencies and indicate some model implications to improve the understanding of RBC Piezo1 action by further experimental and theoretical studies.

The essential ingredients of the proposed mechanism of the effect of Piezo1 on RBC volume are schematically represented by the cause-effect links shown in **Figure 5**. These links represent either experimental evidence or the results of theory or modeling. The upper dashed link indicates that RBC discocyte shape, according to theories described in the subsection "Interpretation of RBC and Vesicle Shapes on the Basis of Membrane Bending," depends on RBC reduced volume. The lower dashed line link represents Eq.1. The link between RBC content of K$^+$ and RBC volume is in a broad sense the consequence of the fact that RBC volume is established through osmotic equilibrium with the surrounding solution and that thus depends on the level of its cytoplasm cations. As discussed in section "RBC Volume and Related Aspects of the Variability of RBC Population" the regulation of cell cation content operates on the basis of active and passive membrane cation permeabilities (Hoffmann et al., 2009). The model concentrates on the homeostasis of K$^+$. The preceding two links are thus based on the experiments of Cahalan et al. (2015) who have shown that Piezo1 channels act by the

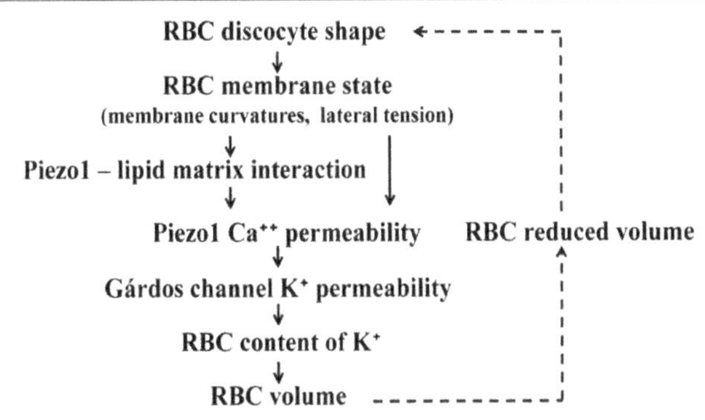

FIGURE 5 | Schematic presentation of processes involved in the effect of RBC discocyte shape on RBC volume. The meanings of the links are described in the text. Because the volume affects the shape (dashed links) the described system as a whole represents a closed regulatory loop. It is indicated that Piezo1 Ca^{++} permeability can be affected either by membrane curvature or membrane lateral tension.

way that, when they are temporarily transformed into their open conformation, allow the influx calcium ions which activate the potassium specific Gárdos channels. The enhanced leakage of potassium ions then causes a decrease of their cell content and consequent loss of water. The Piezo1–Gárdos channel system is thus considered as a complement to the mechanism of RBC volume regulation based on the balance between influx and efflux of cations K$^+$ and Na$^+$. In the model it is assumed that, due to active Ca^{++} efflux, the Gárdos channels are on the average open only part of the time and that it is therefore reasonable to express membrane permeability coefficient for K$^+$ (P_K) as the sum two contributions

$$P_K = P_{K,0} + f_G P_{K,G} \qquad (7)$$

where $P_{K,G}$ is RBC K$^+$ permeability of its Gárdos channels, f_G the average fraction of them that are open, and $P_{K,0}$ the potassium permeability of its other K$^+$ channels. Due to osmotic equilibrium between its interior and exterior (see section "RBC Volume and Related Aspects of the Variability of RBC Population"), RBC volume is at larger values of f_G smaller. In the treated model we derived a relationship between f_G and the reduced volume v in which appeared as model parameters the ratio $P_{K,G}/P_{K,0}$, the relative amount of other RBC cytoplasm ingredients that cannot penetrate the membrane, and the reduced volume at $f_G = 0$. The crucial task of the model was to reveal a plausible mechanism for the effect of RBC shape on the fraction of time that Piezo1 channels are open. In the model it was proposed that there is another relationship between f_G and v based on the dependence of Piezo1 cation permeability on RBC shape. This relationship is represented in **Figure 5** by the links that relate RBC discocyte shape and Piezo1 Ca^{++} permeability. The theory described in sub-section "Interpretation of RBC and Vesicle Shapes on the Basis of Membrane Bending" makes it possible to determine reduced mean membrane curvature (h) at each point on the membrane and its dependence on the reduced volume v. In the model (Svetina et al., 2019) it was shown

that its value in RBC poles can be well represented by a linear function

$$h_{\text{pole}} = h_{\text{pole,r}} + \beta_{\text{pole}}(v - v_{\text{r}}) \qquad (8)$$

where $h_{\text{pole,r}}$ is the reduced mean curvature at an arbitrarily chosen reference reduced volume v_{r}. The value of the coefficient β_{pole} is 4.0. It was then taken into account that due to Piezo1 intrinsic curvature and its interaction with the membrane (Eq. 6), its molecules would tend to concentrate in the regions of RBC poles. On the basis of the assumption of that open Piezo1 conformation is less curved than its closed conformation it follows that, at the decrease of v, the probability that Piezo1 is closed increases. The parameters that defined thus obtained increasing function $f_G(v)$ are a combination of parameters that appear in Eqs. 6 and 8. Due to thus obtained relationships between f_G and v it is possible to express these two parameters in terms of other RBC structural parameters. On the basis of assuming the curvature dependent Piezo1–lipid matrix interaction it has been thus established that the system operates as a negative feedback regulatory loop between the average of the fraction of open Gárdos channels (f_G) and the RBC reduced volume (v).

The described model was meant primarily to serve as the proof of principle for Piezo1 based regulation of RBC volume. Therefore it involves many simplifications of the real system. For example, it was restricted to K$^+$ homeostasis and did not take into consideration possible concomitant changes of RBC Na$^+$ content; the fraction of open Piezo1 channels was calculated as if they would all be located at the RBC poles; it was assumed that there are only two relevant Piezo1 conformations, etc. However, some of the model predictions are general in that they do not depend on its specific features. The main outcome of the effect of the RBC discoid shape on its volume is that it implies the existence of a closed regulatory loop for RBC volume regulation. The Piezo1–Gárdos channel system can be considered as a complement to the mechanism for the regulation of cell volume that operates on the basis of active and passive membrane cation permeabilities (Hoffmann et al., 2009). Within the mechanism of RBC volume regulation based on the balance between influx and efflux of cations K$^+$ and Na$^+$, Piezo1 acts at the level of K$^+$ efflux. The membrane permeability coefficient for K$^+$ involves a contribution that depends on the RBC reduced volume v and thus also on the membrane area A. The consequence of the regulation of v is the strong V–A correlation. Such correlation has been observed by simultaneous measurements of V and A in the RBC population (Canham and Burton, 1968; Gifford et al., 2003) and is evidenced by the steepness of the osmotic fragility curve (**Figure 4B**). Confirmation of this prediction of the model lies in the fact that the V–A correlation is lost, either in the absence of Piezo1 or by the application of the Ca^{2+} ionophore A23187 which overrules its action (Cahalan et al., 2015). The regulation of the RBC reduced volume is physiologically important because RBC, during the process of its aging, constantly releases nanovesicles (Westerman and Porter, 2016).

In this type of vesiculation the loss of area is more significant than the loss of volume and, therefore, without this regulation the cells would lose their deformability because of the consequent increase of their reduced volume.

The model presented here points to the involvement of the RBC discoid shape in the fine regulation of its volume in a rather consistent manner. However, there are still many unanswered questions that require further experimentation. One such concern is whether the response of Piezo1 to change of RBC shape is due to change of membrane curvature or to the change of membrane lateral tension (**Figure 5**). For example, the theory of vesicle shapes predicts that the lateral tension is negative and that its absolute value at lowering the v increases (Svetina and Žekš, 1989). Changes of the lateral tension thus act in the appropriate direction but their effect was estimated to be small (Svetina et al., 2019). It could, however, be enhanced by the action of myosin motors (Smith et al., 2018). The dilemma about possible role of lateral tension could be resolved by experimental determination of the distribution of Piezo1 channels over the RBC membrane. Because mean principal membrane curvature is most negative in the RBC dimples, the analogously curved Piezo1 (Guo and MacKinnon, 2017) would preferentially reside in this region. With regard to the question as to whether the described mechanism of fine regulation of RBC volume also functions *in vivo* it should be realized that RBCs spend only about 50% of their time in veins, in which hydrodynamic conditions allow them to establish their discocyte shape. Namely, freely movable, membrane embedded proteins have diffusion coefficients that would cause them to equilibrate along the whole RBC surface in about 1 min. It would thus be important to determine whether the Piezo1 lateral diffusion coefficient is, for one reason or another, sufficiently smaller. It could be smaller because the RBC membrane is crowded (with more than 10^4 proteins per 1 μm^2) or because the Piezo1 molecule modifies the shape of the surrounding membrane to the size (Haselwandter and MacKinnon, 2018) that fits well into the area of the triangle of the structural unit of the hexagonal network of the spectrin skeleton. With regard to the Piezo1 oligomeric homo-trimer structure it can be noted that, in the case where its subunits can, independently, have two different conformation, it may attain at least four different structures, of which two have axial symmetry and two not. It still has to be established which of these structures corresponds to the Piezo1 open state. In the analyses of the variations

within RBC population it has been assumed that there is no variation in membrane areal density of different RBC membrane proteins. In the context of the presented model it would be of particular interest to determine whether there are differences in the variability parameters of pumps and channels that are involved in the regulation of RBC volume.

There are also many aspects of the proposed model that require further theoretical modeling. For example, there is the question as to what is causing, in the $A–V$ scatter plot, the remaining cell to cell variability. It could be ascribed to RBC variability with respect to its hemoglobin content (Svetina, 1982), but also to cell to cell variability of the ratio between the permeabilities of Gárdos channel and other potassium channels. In the model presented here it was taken that, for a given fixed membrane shape, the inclusions redistribute due to their interaction with the surrounding membrane. However, in general the effect is mutual: due to the curvature dependent interaction of Piezo1 molecules with the surrounding membrane, the RBC shape may also change (Božič et al., 2006). The consequent coupling between the RBC shape and the conformational state of Piezo1 molecules could cause oscillatory non-stationary behavior of the treated system. One has to be aware also of possible second order factors such as membrane lateral inhomogeneity (Hoffman, 2019). It still needs to be established as to the nature of the physical basis for the Piezo1–lipid matrix interaction. It could be based on the perturbed energy of the lipid bilayer (Haselwandter and MacKinnon, 2018) but may also involve specific interactions between a protein and its surrounding molecules, e.g., the curvature dependence of the number of hydrogen bonds that Piezo1 forms with surrounding lipids. Curvature dependent Piezo1–lipid matrix interaction may also involve energy terms due to Piezo1 intrinsic elasticity (Lin et al., 2019).

AUTHOR CONTRIBUTIONS

The author confirms being the sole contributor of this work and has approved it for publication.

ACKNOWLEDGMENTS

The author thanks Prof. Roger H. Pain for critical reading of the manuscript.

REFERENCES

Armstrong, C. M. (2003). The Na/K pump, Cl ion, and osmotic stabilization of cells. *Proc. Natl. Acad. Sci. U.S.A.* 100, 6257–6262. doi: 10.1073/pnas.0931278100

Ataullakhanov, F. I., Korunova, N. O., Spiridonov, I. S., Pivovarov, I. O., Kalyagina, N. V., and Martinov, M. V. (2009). How erythrocyte volume is regulated, or what mathematical models can and cannot do for biology. *Biochem. Mosc. Suppl. Ser. A Membr. Cell Biol.* 3, 101–115. doi: 10.1134/s1990747809020019

Božič, B., Kralj-Iglič, V., and Svetina, S. (2006). Coupling between vesicle shape and lateral distribution of mobile membrane inclusions. *Phys. Rev. E* 73:041915.

Brumen, M., Glaser, R., and Svetina, S. (1979). Osmotic states of red blood cells. *Bioelectrochem. Bioenerget.* 6, 227–241. doi: 10.1016/0302-4598(79)87010-5

Brumen, M., Glaser, R., and Svetina, S. (1981). Study of the red blood cell osmotic behaviour in the "pump-leak" model. *Period. Biol.* 83, 151–153.

Cahalan, S. M., Lukacs, V., Ranade, S. S., Chien, S., Bandell, M., and Patapoutian, A. (2015). Piezo1 links mechanical forces to red blood cell volume. *eLife* 4:e07370. doi: 10.7554/eLife.07370

Canham, P. B. (1970). Minimum energy of bending as a possible explanation of biconcave shape of human red blood cell. *J. Theor. Biol.* 26, 61–81. doi: 10.1016/s0022-5193(70)80032-7

Canham, P. B., and Burton, A. C. (1968). Distribution of size and shape in populations of normal human red cells. *Circ. Res.* 22, 405–422. doi: 10.1161/01.res.22.3.405

Cinar, E., Zhou, S., DeCourcey, J., Wang, Y., Waugh, R. E., and Wan, J. (2015). Piezo1 regulates mechanotransductive release of ATP from human RBCs. *Proc. Natl. Acad. Sci. U.S.A.* 112, 11783–11788. doi: 10.1073/pnas.1507309112

Coste, B., Mathur, J., Schmidt, M., Earley, T. J., Ranade, S., Petrus, M. J., et al. (2010). Piezo1 and Piezo2 are essential components of distinct mechanically activated cation channels. *Science* 330, 55–60. doi: 10.1126/science.1193270

Danielczok, J. G., Terriac, E., Hertz, L., Petkova-Kirova, P., Lautenschläger, F., Laschke, M. W., et al. (2017). Red blood cell passage of small capillaries is associated with transient Ca2+-mediated adaptations. *Front. Physiol.* 8:979. doi: 10.3389/fphys.2017.00979

Deuling, H. J., and Helfrich, W. (1976). Red blood cell shapes as explained on the basis of curvature elasticity. *Biophys. J.* 71, 861–868. doi: 10.1016/s0006-3495(76)85736-0

Discher, D. E., Boal, D. H., and Boey, S. K. (1998). Simulations of the erythrocyte skeleton at large deformation. II. Micropipette aspiration. *Biophys. J.* 75, 1584–1597. doi: 10.1016/s0006-3495(98)74076-7

Discher, D. E., and Mohandas, N. (1996). Kinematics of red cell aspiration by fluorescence-imaged microdeformation. *Biophys. J.* 71, 1680–1694. doi: 10.1016/s0006-3495(96)79424-9

Discher, D. E., Mohandas, N., and Evans, E. A. (1994). Molecular maps of red cell deformation: hidden elasticity and in situ connectivity. *Science* 266, 1032–1035. doi: 10.1126/science.7973655

Evans, E. A. (1980). Minimum energy analysis of membrane deformation applied to pipet aspiration and surface adhesion of red blood cells. *Biophys. J.* 30, 265–284. doi: 10.1016/s0006-3495(80)85093-4

Evans, E. A., Waugh, R., and Melnik, L. (1976). Elastic area compressibility modulus of red cell membrane. *Biophys. J.* 16, 585–595. doi: 10.1016/s0006-3495(76)85713-x

Fedosov, D. A., Caswell, B., and Karniadakis, G. E. (2010). Systematic coarse-graining of spectrin-level red blood cell models. *Comput. Methods Appl. Mech. Eng.* 199, 1937–1948. doi: 10.1016/j.cma.2010.02.001

Freedman, J. C., and Hoffman, J. F. (1979). Ionic and osmotic equilibria of human red blood cells treated by nystatin. *J. Gen. Physiol.* 74, 157–185. doi: 10.1085/jgp.74.2.157

Ge, J., Li, W., Zhao, Q., Chen, M., Zhi, P., Li, R., et al. (2015). Architecture of the mammalian mechanosensitive Piezo1 channel. *Nature* 527, 64–69. doi: 10.1038/nature15247

Gifford, S. C., Frank, M. G., Derganc, J., Gabel, C., Austin, R. H., Yoshida, T., et al. (2003). Parallel microchannel-based measurements of individual erythrocyte areas and volumes. *Biophys. J.* 84, 623–633. doi: 10.1016/s0006-3495(03)74882-6

Goldstein, R. E. (2018). Are theoretical results 'results'? *eLife* 7:e40018.

Guo, Y. R., and MacKinnon, R. (2017). Structure-based membrane dome mechanism for Piezo mechanosensitivity. *eLife* 6:e33660. doi: 10.7554/eLife.33660

Haselwandter, C. A., and MacKinnon, R. (2018). Piezo's membrane footprint and its contribution to mechanosensitivity. *eLife* 7:e41968. doi: 10.7554/eLife.41968

Heinrich, V., Svetina, S., and Žekš, B. (1993). Nonaxisymmetric vesicle shapes in a generalized bilayer-couple model and the transition between oblate and prolate axisymmetric shapes. *Phys. Rev. E* 48, 3112–3123. doi: 10.1103/physreve.48.3112

Helfrich, W. (1973). Elastic properties of lipid bilayers: theory and possible experiments. *Z. Naturforsch.* 28c, 693–703. doi: 10.1515/znc-1973-11-1209

Hoffman, J. F. (2019). Reflections on the crooked timber of red blood cell physiology. *Blood Cells Mol. Dis.* 79:102354. doi: 10.1016/j.bcmd.2019.102354

Hoffmann, E. K., Lambert, I. H., and Pedersen, S. F. (2009). Physiology of cell volume regulation in vertebrates. *Physiol. Rev.* 89, 193–277. doi: 10.1152/physrev.00037.2007

Hwang, W. C., and Waugh, R. E. (1997). Energy of dissociation of lipid bilayer from the membrane skeleton of red blood cells. *Biophys. J.* 72, 2669–2678. doi: 10.1016/s0006-3495(97)78910-0

Kay, A. R., and Blaustein, M. P. (2019). Evolution of our understanding of cell volume regulation by the pump-leak mechanism. *J. Gen. Physiol.* 151, 407–416. doi: 10.1085/jgp.201812274

Kralj-Iglič, V., Heinrich, V., Svetina, S., and Žekš, B. (1999). Free energy of closed membrane with anisotropic inclusions. *Eur. Phys. J. B* 10, 5–8. doi: 10.1007/s100510050822

Kralj-Iglič, V., Svetina, S., and Žekš, B. (1996). Shapes of bilayer vesicles with membrane embedded molecules. *Eur. Biophys. J.* 24, 311–321. doi: 10.1007/bf00180372

Kuchel, P. W., and Shishmarev, D. (2017). Accelerating metabolism and transmembrane cation flux by distorting red blood cells. *Sci. Adv.* 3:eaao1016. doi: 10.1126/sciadv.aao1016

Lanotte, L., Mauer, J., Mendez, S., Fedosov, D. A., Fromental, J.-M., Claveria, V., et al. (2016). Red cells' dynamic morphologies govern blood shear thinning under microcirculatory flow conditions. *Proc. Natl. Acad. Sci. U.S.A.* 113, 13289–13294. doi: 10.1073/pnas.1608074113

Lew, V. L., and Bookchin, R. M. (1986). Volume, pH, and ion-content regulation in human red cells: analysis of transient behavior with an integrated model. *J. Membr. Biol.* 92, 57–74. doi: 10.1007/bf01869016

Lew, V. L., Raftos, J. E., Sorette, M., Bookchin, R. M., and Mohandas, N. (1995). Generation of normal human red cell volume, hemoglobin content, and membrane area distributions by "birth" or regulation? *Blood* 86, 334–341. doi: 10.1182/blood.v86.1.334.bloodjournal861334

Lin, Y. C., Guo, R., Miyagi, A., Levring, J., MacKinnon, R., and Scheuring, S. (2019). Force-induced conformational changes in PIEZO1. *Nature* 92, 57–74.

Lux, S. E. (2016). Anatomy of the red cell membrane skeleton: unanswered questions. *Blood* 127, 187–199. doi: 10.1182/blood-2014-12-512772

McMahon, H. T., and Gallop, J. L. (2005). Membrane curvature and mechanisms of dynamic cell membrane remodelling. *Nature* 438, 590–596. doi: 10.1038/nature04396

Miao, L., Seifert, U., Wortis, M., and Döbereiner, H.-G. (1994). Budding transitions of fluid-bilayer vesicles: the effect of area-difference elasticity. *Phys. Rev. E* 49, 5389–5407. doi: 10.1103/physreve.49.5389

Mohandas, N., and Chasis, J. A. (1993). Red blood cell deformability, membrane material properties and shape: regulation by transmembrane, skeletal and cytosolic proteins and lipids. *Semin. Hematol.* 30, 171–192.

Mohandas, N., and Evans, E. (1994). Mechanical properties of the red cell membrane in relation to molecular structure and genetic effects. *Ann. Rev. Biophys. Biomol. Struct.* 23, 787–818. doi: 10.1146/annurev.bb.23.060194.004035

Mohandas, N., and Gallagher, P. G. (2008). Red cell membrane: past, present, and future. *Blood* 112, 3939–3947. doi: 10.1182/blood-2008-07-161166

Mukhopadhyay, R., Lim, G. H. W., and Wortis, M. (2002). Echinocyte shapes: bending, stretching, and shear determine spicule shape and spacing. *Biophys. J.* 82, 1756–1772. doi: 10.1016/s0006-3495(02)75527-6

Murthy, S. E., Dubin, A. E., and Patapoutian, A. (2017). Piezos thrive under pressure: mechanically activated ion channels in health and disease. *Nat. Rev. Mol. Cell Biol.* 18, 771–783. doi: 10.1038/nrm.2017.92

Peng, Z., Asaro, R. J., and Zhu, Q. (2010). Multiscale simulation of erythrocyte membranes. *Phys. Rev. E* 81:031904.

Raphael, R. M., and Waugh, R. E. (1996). Accelerated interleaflet transport of phosphatidylcholine molecules in membranes under deformation. *Biophys. J.* 82, 1756–1772.

Saotome, K., Murthy, S. E., Kefauver, J. M., Whitwam, T., Patapoutian, A., and Ward, A. B. (2018). Structure of the mechanically activated ion channel Piezo1. *Nature* 554, 481–486. doi: 10.1038/nature25453

Seifert, U. (1997). Configurations of fluid membranes and vesicles. *Adv. Phys.* 46, 13–137. doi: 10.1080/00018739700101488

Seifert, U., Berndl, K., and Lipowsky, R. (1991). Shape transformations of vesicles - phase-diagram for spontaneous-curvature and bilayer-coupling models. *Phys. Rev. A* 44, 1182–1202. doi: 10.1103/physreva.44.1182

Sheetz, M. P., and Singer, S. J. (1974). Biological-membranes as bilayer couples - molecular mechanism of drug-erythrocyte interactions. *Proc. Natl. Acad. Sci. U.S.A.* 71, 4457–4461. doi: 10.1073/pnas.71.11.4457

Smith, A. S., Nowak, R. B., Zhou, S., Giannetto, M., Gokhin, D. S., Papoin, J., et al. (2018). Myosin IIA interacts with the spectrin-actin membrane skeleton to control red blood cell membrane curvature and deformability. *Proc. Natl. Acad. Sci. U.S.A.* 115, E4377–E4385.

Švelc, T., and Svetina, S. (2012). Stress-free state of the red blood cell membrane and the deformation of its skeleton. *Cell. Mol. Biol. Lett.* 17, 217–227. doi: 10.2478/s11658-012-0005-8

Svetina, S. (1982). Relations among variations in human red cell volume, density, membrane area, hemoglobin content and cation content. *J. Theor. Biol.* 95, 123–134. doi: 10.1016/0022-5193(82)90291-0

Svetina, S. (1998). Skeleton –bilayer interaction and the shape of red blood cells. *Cell. Mol. Biol. Lett.* 3, 449–463.

Svetina, S. (2015). Curvature–dependent protein–lipid bilayer interaction and cell mechanosensitivity. *Eur. Biophys. J.* 44, 513–519. doi: 10.1007/s00249-015-1046-5

Svetina, S. (2017). Investigating cell functioning by theoretical analysis of cell-to-cell variability. *Eur. Biophys. J.* 46, 739–748. doi: 10.1007/s00249-017-1258-y

Svetina, S., Brumen, M., Gros, M., Vrhovec, S., and Žnidarčič, T. (2003). "On the variation of parameters that characterize the state of a physiological system. Red blood cells as an example," in *Simulations in Biomedicine V*, eds Z. M. Arnež, C. A. Brebbia, F. Solina, and V. Stankovski (Southampton: WITT Press), 3–14.

Svetina, S., Iglič, A., Kralj-Iglič, V., and Žekš, B. (1996). Cytoskeleton and red cell shape. *Cell. Mol. Biol. Lett.* 1, 67–78.

Svetina, S., Kokot, G., Švelc Kebe, T., Žekš, B., and Waugh, R. E. (2016). A novel strain energy relationship for red blood cell membrane skeleton based on spectrin stiffness and its application to micropipette deformation. *Biomech. Model. Mechanobiol.* 15, 745–758. doi: 10.1007/s10237-015-0721-x

Svetina, S., Kralj-Iglič, V., and Žekš, B. (1990). "Cell shape and lateral distribution of mobile membrane constituents," in *Biophysics of Membrane Transport: Tenth School on Biophysics of Membrane Transport, Part II*, eds J. Kuczera and S. Przestalski (Wrocław: Agricultural University of Wrocław), 139–155.

Svetina, S., Kuzman, D., Waugh, R. E., Ziherl, P., and Žekš, B. (2004). The cooperative role of membrane skeleton and bilayer in the mechanical behaviour of red blood cells. *Bioelectrochemistry* 62, 107–113. doi: 10.1016/j.bioelechem. 2003.08.002

Svetina, S., Ottova-Leitmannová, A., and Glaser, R. (1982). Membrane bending energy relation to bilayer couples concept of red blood cell shape transformations. *J. Theor. Biol.* 94, 13–23. doi: 10.1016/0022-5193(82)90327-7

Svetina, S., Švelc Kebe, T., and Božič, B. (2019). A model of Piezo1 based regulation of red blood cell volume. *Biophys. J.* 116, 151–164. doi: 10.1016/j.bpj.2018.11. 3130

Svetina, S., and Žekš, B. (1983). Bilayer couple hypothesis of red-cell shape transformations and osmotic hemolysis. *Biomed. Biochim. Acta* 42, S86–S90.

Svetina, S., and Žekš, B. (1989). Membrane bending energy and shape determination of phospholipid vesicles and red blood cells. *Eur. Biophys. J.* 17, 101–111. doi: 10.1007/bf00257107

Svetina, S., and Žekš, B. (1990). The mechanical behavior of cell membranes as a possible physical origin of cell polarity. *J. Theor. Biol.* 146, 115–122. doi: 10.1016/s0022-5193(05)80047-5

Svetina, S., and Žekš, B. (1996). "Elastic properties of closed bilayer membranes and the shapes of giant phospholipid vesicles," in *Handbook of Nonmedical Applications of Liposomes*, Vol. I, eds D. D. Lasic and Y. Barenholz (Boca Raton, FL: CRC), 13–42.

Svetina, S., and Žekš, B. (2014). Nonlocal membrane bending: a reflection, the facts and its relevance. *Adv. Coll. Interface Sci.* 208, 189–196. doi: 10.1016/j.cis.2014. 01.010

Svetina, S., Žekš, B., Waugh, R. E., and Raphael, R. M. (1998). Theoretical analysis of the effect of the transbilayer movement of phospholipid molecules on the dynamic behavior of a microtube pulled out of an aspirated vesicle. *Eur. Biophys. J.* 27, 197–209. doi: 10.1007/s0024900 50126

Tomishige, M., Sako, Y., and Kusumi, A. (1998). Regulation mechanism of the lateral diffusion of band 3 in erythrocyte membranes by the membrane skeleton. *J. Cell Biol.* 142, 989–1000. doi: 10.1083/jcb.142.4.989

Tosteson, D. C., and Hoffman, J. F. (1960). Regulation of cell volume by active cation transport in high and low potassium sheep red cells. *J. Gen. Physiol.* 44, 169–196.

Westerman, M., and Porter, J. B. (2016). Red blood cell-derived microparticles: an overview. *Blood Cells Mol. Dis.* 59, 134–139. doi: 10.1016/j.bcmd.2016.04.003

Zarychanski, R., Schulz, V. P., Brett, L. H., Maksimova, Y., Houston, D. S., Smith, J., et al. (2012). Mutations in the mechanotransduction protein PIEZO1 are associated with hereditary xerocytosis. *Blood* 120, 1908–1915. doi: 10.1182/ blood-2012-04-422253

Zhao, Q., Zhou, H., Chi, S., Wang, Y., Wang, J., Geng, J., et al. (2018). Structure and mechanogating mechanism of the Piezo1 channel. *Nature* 554, 487–492. doi: 10.1038/nature25743

Zimmerberg, J., and Kozlov, M. M. (2006). How proteins produce cellular membrane curvature. *Nat. Rev. Mol. Cell Biol.* 7, 9–19.

17

Mechanical Signature of Red Blood Cells Flowing Out of a Microfluidic Constriction is Impacted by Membrane Elasticity, Cell Surface-to-Volume Ratio and Diseases

Magalie Faivre[1], Céline Renoux[2,3,4], Amel Bessaa[2,3], Lydie Da Costa[3,5,6,7], Philippe Joly[2,3,4], Alexandra Gauthier[2,3,8] and Philippe Connes[2,3,9]*

[1] Université de Lyon, Institut des Nanotechnologies de Lyon INL-UMR 5270 CNRS, Université Lyon 1, Villeurbanne, France, [2] Laboratoire Interuniversitaire de Biologie de la Motricité (LIBM) EA7424, Equipe "Biologie Vasculaire et du Globule Rouge", UCBL1, Villeurbanne, France, [3] Laboratoire d'Excellence (Labex) GR-Ex, Paris, France, [4] Biochimie des Pathologies Érythrocytaires, Centre de Biologie et de Pathologie Est, HCL, Bron, France, [5] AP-HP, Service d'Hématologie Biologique, Hôpital Robert-Debré, Paris, France, [6] Université Paris Diderot, Université Sorbonne, Paris Cité, Paris, France, [7] INSERM U1149, CRI, Faculté de Médecine Bichat-Claude Bernard, Paris, France, [8] Institut d'Hématologie et d'Oncologie Pédiatrique (IHOP), Hospices Civils de Lyon, Lyon, France, [9] Institut Universitaire de France, Paris, France

***Correspondence:**
Magalie Faivre
magalie.faivre@univ-lyon1.fr

Despite the fact that Red Blood Cells (RBCs) have been intensively studied in the past 50 years to characterize mechanical phenotypes associated with both healthy and pathological states, only ektacytometry (i.e., laser diffractometry) is currently used by hematologists to screen for RBC membrane disorders. Therefore, the development of new diagnostic tools able to perform analysis at the scale of a single cell, over a statistically relevant population, would provide important complementary information. But these new diagnostic tools would have to be able to discriminate between different disorders causing a change in RBCs mechanical properties. We evaluated the mechanical response of artificially rigidified RBCs flowing through a microfluidic constriction. The geometry consists in a 50 μm wide channel with a succession of 14 tooth-like patterns, each composed of a 5 μm wide and 10 μm long constriction, associated with a 25 μm wide and 10 μm long enlargement. RBCs deformability was altered using two chemical treatments, known to affect RBCs membrane surface area and membrane deformability, lysolecithine (LPC) and diamide, respectively. Differences between samples were highlighted by the representation of the inverse of the shape recovery time $(1/\tau_r)$, versus the extension at the exit of the constriction, D_{out}. The results demonstrate that our approach is able to provide a direct signature of RBCs membrane composition and architecture, as it allows discriminating the effect of changes in RBCs membrane surface area from changes in RBCs membrane deformability. Finally, in order to evaluate the potential of our microsystem to detect pathological cells, we have performed preliminary experiments on patients with Hereditary Spherocytosis (HS) or Sickle Cell Anemia (SCA).

Keywords: mechanical phenotype, microfluidics, red blood cells, pathologies, chemical treatment

INTRODUCTION

Red Blood Cells (RBCs) membrane possesses a unique structure responsible for their remarkable ability to deform to flow through the small capillaries of the microcirculation. RBCs have an elastic 2D mesh-like spectrin cytoskeleton anchored to the internal side of a lipid bilayer (Mohandas and Evans, 1994). The membrane of RBCs encloses a cytoplasm made of hemoglobin which viscosity is 10 mPa.s at 25°C. The membrane and the absence of nucleus in the internal media, plays a key role in the regulation of RBCs deformability.

Under normal conditions, the membrane deforms at constant surface area and exhibits a viscoelastic behavior. Once the cell surface area excess has been unfolded, further extension of the membrane is governed by the lipid bilayer, which tends to resist area expansion. The shear resistance of the membrane is directly related to the density of spectrin and thus, to the cytoskeleton molecular structure (Waught and Agre, 1988). RBCs have been extensively studied in the past 50 years in order to characterize mechanical phenotypes associated with both healthy and pathological states.

Various diseases such as malaria (Shelby et al., 2003; Suresh et al., 2005; Mauritz et al., 2010) diabetes (Buys et al., 2003) Sickle Cell Anemia (Ballas and Mohandas, 2004; Maciaszek and Lykotrafitis, 2011; Iragorri et al., 2018) (SCA) or Hereditary Spherocytosis (Waught and Agre, 1988) (HS) are associated with variation of RBCs deformability. Although conventional techniques allowing the quantification of cellular mechanical properties, (Suresh, 2007) such as atomic force microscopy (AFM), micropipette aspiration and optical tweezers are well-established, they present throughput too low (of the order of several tens of cells per day) to be envisaged as routine diagnostic tools. Currently only osmotic gradient ektacytometry (osmoscan), which consists in following the behavior of a suspension of RBCs sheared into a Couette system across an osmotic gradient, is used by hematologists to screen for RBC membrane disorders (Mohandas et al., 1980; Da Costa et al., 2013, 2016). However, up to now, they cannot provide any quantitative information such as parasitemia for malaria or distribution of cell populations which would provide valuable insights for SCA or HS (Dondorp et al., 1997).

Due to a match between cell sizes and typical dimensions accessible by microfabrication methods, microfluidic technologies propose attractive solutions for the study of cellular mechanics at the single cell level, while being compatible with high throughput (Antia et al., 2008; Hauck et al., 2010). Previous studies have demonstrated the use of various microfluidic geometries (Tomaiuolo, 2014) to detect alteration of RBCs deformability. Very different readouts have been used, such as deformation index, cell flowing velocity, transit time, relaxation time, etc. For example, Tsukada et al. (2001) have shown that the shape of cells flowing in micro-capillaries according to their velocity can be used to discriminate diabetic from healthy RBCs. Shelby et al. (2003) have shown that the ability of malaria infected RBCs to flow through microfluidic constrictions could be related to the stage of maturation of the parasite inside the host cells. Zheng et al. (2013) combined the flow of RBCs in a geometric

constriction and impedance measurement to differentiate adult from neonatal RBCs. Guo et al. (2012) have measured the cortical tension associated with healthy, and *Plasmodium falciparum* infected RBCs, using funnel channels. Differences in velocities of RBCs flowing through geometrical constrictions have been used to discriminate healthy and diamide treated RBCs (Vilas Boas et al., 2018) or healthy and malaria infected RBCs (Bow et al., 2011). Faustino et al. (2014) took advantage of the strong extensional flow associated with a microchannel implementing a hyperbolic constriction, to differentiate RBCs from RBCs in contact with tumoral cells and healthy RBCs from RBCs of End-Stage Kidney Disease (ESKD) patients (Faustino et al., 2019). Tomaiuolo et al. (2011) have used the behavior of RBCs flowing through converging constrictions in order to evaluate cell membrane viscoelastic properties. Several papers have focused on the measurement of RBCs relaxation time (Tomaiuolo and Guido, 2011; Braunmüller et al., 2012; Prado et al., 2015) in microfluidics – i.e., the time necessary for the cell to recover its discocyte-like shape after cessation of the flow.

Although many research teams have demonstrated the ability to detect modification of molecular structure, (Forsyth et al., 2010; Hansen et al., 2011; Guo et al., 2012) to the best of our knowledge, none of them have tried to evaluate the specificity of their measurement by demonstrating their ability to discriminate between several diseases inducing an overall stiffening of the cells.

While many researches focused on the flow of RBCs through a microfluidic geometric constriction, we explored for the first time the maximum deformation of the cells being stretched by the sudden extension of the channel, in relation to its shape recovery time. We report the effect of two chemical treatments known to affect RBCs membrane surface area or membrane deformability on the dynamical behavior of RBCs flowing out a microfluidic constriction. We evaluated whether the response of the cells at the exit of the constriction was sensitive enough to discriminate between both effects (i.e., excess surface area and membrane elasticity). Finally, preliminary results highlighting the mechanical responses of RBCs in few patients with HS and SCA are discussed, thus addressing the specificity of our approach for potential diagnosis applications.

MATERIALS AND METHODS

Blood Samples
Healthy Samples

Healthy blood samples were collected in EDTA tubes. Mechanically impaired RBCs were obtained by treating healthy RBCs with one of the following molecules: lysolecithin (LPC) or diamide (Clark et al., 1983). Before and after treatment, RBCs were washed twice with phosphate buffered saline (PBS) 1X (Biosolve chemicals BV, Netherlands) and then re-suspended at a concentration of 19×10^6 RBCs/mL – corresponding to a hematocrit (Ht) of 0.17% – in PBS 1X. This dilution was necessary in order to avoid the flow of several cells simultaneously in the microchannel. Finally, after centrifugation,

RBCs were re-suspended in dextran solutions (Sigma-Aldrich, Saint-Louis, MO, United States) (Dextran from *Leuconostoc* spp., $M_w = 2 \times 10^6$ g/mol was used) at 0.9 mg/mL of PBS 1X, to avoid RBCs sedimentation in the reservoir and guaranty the injection of a homogeneous concentration of cells during the time of an experiment while maintaining the low Ht condition. Dextran solutions were always filtered at 0.2 μm on the day of use. The use of dextran solutions also increased the hydrodynamic stress undergone by the cells in the channel. Viscosity, pH and osmolarity of the solutions were verified to be 31.5 mPa.s, 7.4 and 300 mOsmol, respectively.

Cell Surface and *S/V* Reduction (LPC Treatment)

Washed RBCs were incubated for 5 min at room temperature with LPC (Sigma-Aldrich, Saint-Louis, MO, United States) (LPC from egg yolk with $M_w = 505$ g/mol was used) at final concentrations ranging from 0 to 1.0 μmol/mL of cells, according to the protocol reported by Clark et al. (1983) LPC changes discocyte RBCs to type III echinocytes, induces membrane vesiculation and loss of membrane, resulting in a reduction of the surface to volume ratio (noted *S/V*).

Membrane Deformability Reduction (Diamide Treatment)

As previously described,30,31 washed RBCs were incubated with diamide (Sigma-Aldrich, Saint-Louis, MO, United-States) at final concentrations ranging from 0 to 1.0 mmol/L for 1 h at 37°C. Diamide induces the formation of disulfide bonds between spectrin proteins and increases the shear modulus of RBCs (Fischer et al., 1978; Safeukui et al., 2012).

Pathological Samples

Blood samples from 3 healthy individuals, 3 patients with HS and 2 SCA patients were collected in EDTA tubes. The protocol was approved by the "Hospices Civils de Lyon – CPP Est" Ethics Committee (L14-127).

Device Fabrication

Microfluidic channels in polydimethylsiloxane Sylgard 184 (PDMS) were manufactured using standard soft photolithographic techniques (Duffy et al., 2004) and sealed on glass via oxygen plasma treatment (Harrick Plasma, Ithaca, NY, United States). The geometry consisted in a 50 μm wide and 10 μm high channel, in which a succession of 15 tooth-like patterns have been implemented as illustrated in **Figure 1**. Each teeth-like pattern was composed of a 5 μm wide and 10 μm

long constriction, associated with a 25 μm wide and 10 μm long enlargement. This width oscillation has been repeated over 290 μm and was chosen for its ability to significantly center the RBCs at the exit of the last constriction.

Video-Microscopy

Polyethylene (PE 20) tubes (Harvard Apparatus, Holliston, MA, United States) connected the blood reservoirs to the inlet hole in the device and the outlet hole to the trash reservoir. RBC suspensions were injected in the microsystems by the flow control system MFCS^{TM}-EZ (Fluigent, Paris, France) at a pressure of 200 mbar. Video-microscopic recordings of the cell behavior were performed with an inverted phase contrast microscope (Leica DMI 4000B, Germany) with a 40× magnification and a high speed camera (Mikrotron EoSens MC1362, Germany). The microscope was equipped with an environmental chamber (Ibidi, Martinsried, Germany) thus allowing the experiments to be done at 37°C.

Image Analysis

Post processing of the movies was performed using a self-edited Matlab code to study cell dynamics and deformation. Briefly, on each image, after background subtraction, allowing to get rid of the microchannel walls, RBCs were automatically detected and cells contour was fitted with an elliptical shape. The position of the ellipse center of mass, the length of both axes along (*x*-direction) and perpendicular to (*y*-direction) the flow direction (2*a* and 2*b*, respectively, see inset **Figure 1**), were measured, allowing the calculation of the deformation index *D* defined as $D = (2a-2b)/(2a + 2b)$. Currently, the routine takes a couple of minutes (2–5 min) per cell on a regular computer, because the process is not fully automatized yet, which account for the quite low number of cells investigated (~60 cells/sample). However, this step could be optimized in the future to reach nearly real-time analysis as reported by Deng and Chung (2016) on similar analysis with a throughput of 2000 cells/s.

For each condition, roughly 50 cells were analyzed. Results are presented as box-and-whisker plots. A non-parametric ANOVA test for independent measurements was used to compare the different readouts before and after chemical treatments. *Post hoc* comparisons were performed using Fisher's Least Squares Difference method. A Multivariate Analysis of Variance (MANOVA) test was used to compare the different pathological samples (HS, SCA, and control samples). The significance level was defined as $p < 0.05$.

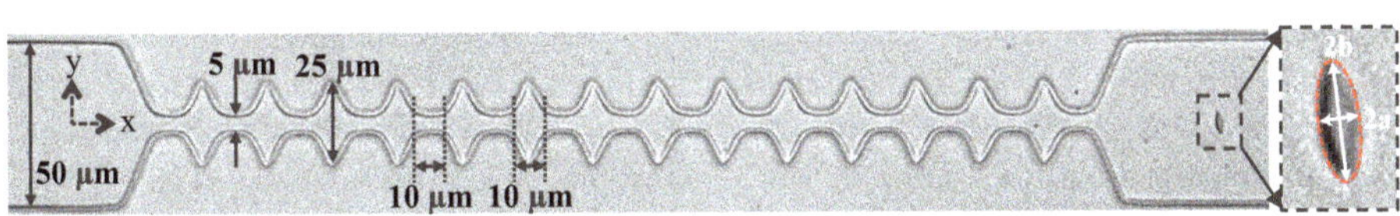

FIGURE 1 | The geometry consists in a 50 μm wide channel implementing 14 tooth-like patterns. The height of the device is 10 μm. The different dimensions are reported on the pictures. The inset shows a close-up of a RBC which contour is fitted by an ellipse (dashed red line). The two axes of the ellipse along (2a) and perpendicular (2b) to the flow direction allows the calculation of a deformation index such as $D = (2a–2b)/(2a + 2b)$.

Ektacytometry

Osmoscan experiments have also been performed on the LoRRca MaxSis® device (RR Mechatronics, Hoorn, Netherlands) to confirm the effects of the different molecules tested on RBCs, as well as to verity the presence of the specific osmoscan signatures already described in HS and SCA patients. Osmoscan consisted in the measurements of a RBC deformability or elongation index (EI) under a defined shear stress (30 Pa), at increasing osmolality from 90 to 600 mOsm/kg and at 37°C, as recommended (Nemeth et al., 2015; Da Costa et al., 2016; Zaninoni et al., 2018). Buffer viscosity was 30.4 mPa.s. Several parameters were determined: O_{min} (i.e., the osmolality at which RBC deformability value reaches a minimum in the hypotonic region of the curve), EI_{max} (i.e., the highest RBC deformability) and O_{hyper} (also called O'), which corresponds to the osmolality at half of the EI_{max} on the hypertonic region of the curve (Zheng et al., 2013). O_{min} reflects the osmotic fragility and the surface-to-volume ratio, EI_{max} depends on the membrane deformability and RBC surface area, and O_{hyper} reflects Mean Cellular Hemoglobin Concentration (MCHC) and Mean Cell Volume (MCV) and is therefore highly dependent on the hydration status of the cells (Clark et al., 1983; Da Costa et al., 2013).

RESULTS AND DISCUSSION

Typical Behavior of Healthy RBCs

Video-microscopic recordings of RBCs flowing in the microfluidic channel allowed the visualization of the cell deformation – quantified using the deformation index D – as it travels through the device. **Figure 2** presents the typical behavior of a RBC flowing in the geometry of interest. The sequence of deformation associated with the flow of a healthy RBC in the geometry is reported in **Figure 2a** and the associated deformation index D is presented in **Figure 2b**. The cell flowing in the 50 µm wide channel presents a slipper-like shape, typical of RBC in confined flow ($D \approx -0.2$). As it approached the first constriction, the cells got compressed which is traduced by its elongation along the flow direction (x-axis). Indeed, D rose until reaching a maximum value around $x = 0$. Then, it underwent a stretching along the y-axis (i.e., perpendicular to the flow direction) when entering in the enlargement, before being compressed again by the next constriction. Accordingly, D dropped until reaching a local minimum (for $x = 10$ µm), before increasing again at $x = 20$ µm. This cycle of compressing/stretching is repeated due to the tooth-like patterns as illustrated by the oscillations of D versus x. The cyclic deformation of the cell within the geometry can be characterized through the measurement of the amplitude of deformation ΔD defined as $\Delta D = D_{max} - D_{min}$, where D_{max} and D_{min} are the mean of D in the constrictions and the enlargements, respectively, as illustrated in **Figure 2b**. Finally, as it exited the last narrowing, around $x = 290$ µm, RBC got strongly stretched perpendicular to the flow direction, which is traduced by the sudden drop of D. After reaching a minimum value noted D_{out} around $x = 310$ µm, D returns slowly to its initial value and reaches a plateau corresponding to the steady slipper-like shape.

The variation of D between the last deformed state at the exit, D_{out}, and the final steady shape, corresponding to the plateau value, is represented versus time in **Figure 2c**. The experimental data were fitted using an exponential growth, hence allowing the extraction of the shape recovery time τ_r defined as the time necessary for the cell to return to a stationary shape after exiting the last constriction, while being still under hydrodynamic stress. We have measured the shape recovery time of healthy RBCs while varying different viscous stress applied, i.e., hydrodynamic parameters such as buffer viscosity and cell speed (**Supplementary Figure SI.2B**). Because RBC steady shape is reached under hydrodynamic stress rather than at rest, the recovery time of RBCs does not directly correspond to their relaxation time. Indeed the relaxation time is defined as the time necessary for the cell to adopt the resting « discocyte-like » shape after total cessation of any stress. As already reported in the literature, the relaxation time τ depends only on the intrinsic mechanical properties of the RBC membrane, i.e., $\tau = \eta_m/\mu$, where η_m is the 2D membrane viscosity and μ the membrane shear modulus. According to the literature, τ has been measured using different techniques to be in the range 100–300 ms (Hochmuth et al., 1979; Mohandas et al., 1980; Braunmüller et al., 2012). Indeed, in specific experimental conditions, recovery times were measured at $\tau_r = 129$ ms for $\eta_{out} = 1.3$ mPa.s and $V_{cell} = 170$ µm/s, which is in good agreement with values previously reported in the literature for the relaxation time (Hochmuth et al., 1979; Mohandas et al., 1980; Evans, 1989; Tomaiuolo and Guido, 2011; Braunmüller et al., 2012). But they can also reach values as low as $\tau_r = 4$ ms for $\eta_{out} = 20.3$ mPa.s and $V_{cell} = 1500$ µm/s (Amirouche et al., 2020). We could explain these results in terms of coupling between the cell and the flow. This assumption would explain why at low hydrodynamic stress – where the cell properties dominate the recovery process and where the hydrodynamics can be neglected - we can assume being in an almost static configuration, the recovery time tend toward relaxation time values. In a previous study, we reported that at fixed hydrodynamic stress, shape recovery times of RBCs can be used to discriminate between healthy and mechanically impaired RBCs (Amirouche et al., 2017). In the present paper, we aim at demonstrating that our approach is sensitive enough to discriminate between different membrane modifications.

The repetition of the teeth-like pattern tends to focus cells within the microchannel (see **Supplementary Figure SI.1** in **Supplementary Information**), hence ensuring a symmetrical deformation at the exit of the last narrowing. This prevents them from rotating and imposes that all cells are exposed to the same hydrodynamic stress, all RBCs being aligned on the same flow line. Despite this advantage, the repetition of the constriction could also be a drawback. Indeed, it has been reported in literature (Lee et al., 2004; Simmonds and Meiselman, 2016; Horobin et al., 2017; Qiang et al., 2019) that the application of a stress too high or for a too long period of time can induce a mechanical fatigue of the cells. For example, Simmonds et al. report that RBCs present impaired deformability when exposed to physiological levels of shear stress (Simmonds and Meiselman, 2016) (above 40 Pa) for 1–64 s. Other studies report

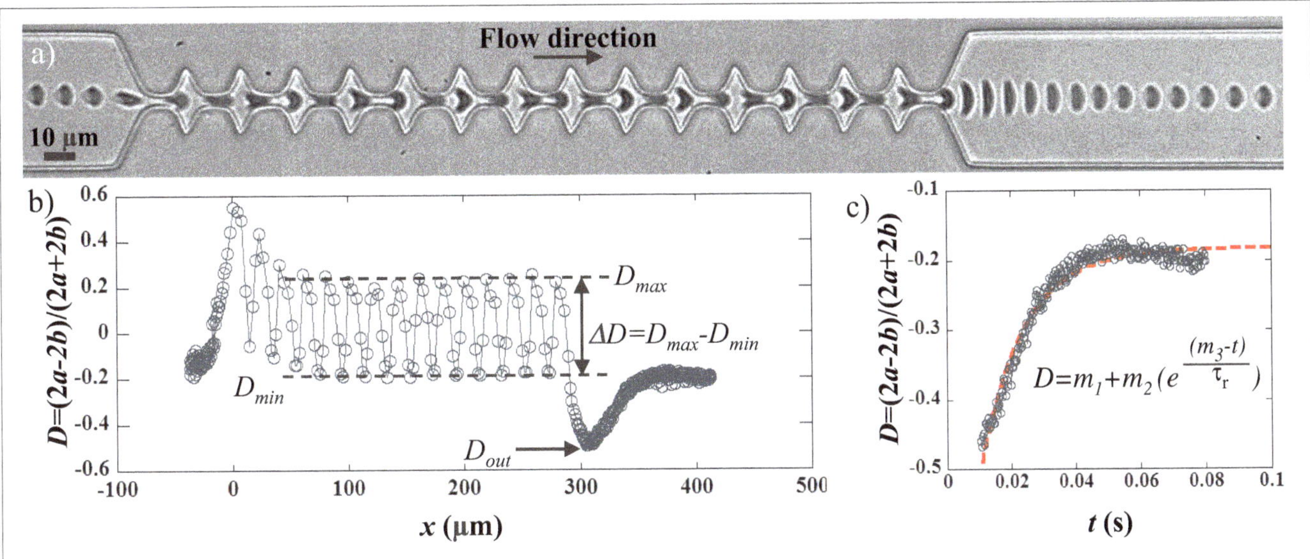

FIGURE 2 | (a) sequence of deformation of a healthy RBC flowing in the geometry at 200 mbar and **(b)** the associated variation of the deformation index D [$D = (2a-2b)/(2a + 2b)$] as a function of the position of the cell center of mass. The origin of the graph has been arbitrarily chosen at the entrance of the first narrowing. The amplitude of deformation ΔD is calculated as $\Delta D = D_{max}-D_{min}$, where D_{max} and D_{min} are the mean of D in the constrictions and the enlargements, respectively. The minimum of the curve corresponds to the large cell elongation at the exit of the last constriction, noted D_{out}. **(c)** representation of D versus time of the previous graph, where only the time window of the cell recovery is reported. The exit of the last narrowing was set to be $t = 0$. The dashed red line is an exponential fit allowing the determination of the recovery time τ_r according to the following equation $D = m_1 + m_2 \exp((m_3-t)/\tau_r)$, where m_1, m_2, and m_3 are constant to be adjusted.

that cyclic mechanical solicitations of RBCs lead to significantly greater loss of membrane deformability, compared to continuous deformation under the same maximum load and duration (Lee et al., 2004). Therefore, we have verified that the repetition of the restriction did not impact the behavior of RBCs exiting the geometry. As presented in **Supplementary Figure SI.2**, the shape recovery time of healthy RBCs was identical after exiting a single 10 µm long constriction and the 15 repetitions of a 10 µm long constriction, hence demonstrating that no fatigue was detected in our experiments.

Chemically Altered RBCs

Two chemical treatments were performed on healthy RBCs in order to affect RBC deformability. For each sample, osmotic gradient ektacytometry was performed (some results are shown on **Figure 3**) and the mechanical responses of RBCs flowing in the microfluidic system were evaluated (**Figure 4** for LPC treated RBCs, and **Figure 5** for diamide treated RBCs). LPC is known to cause a reduction in the surface area of RBCs (Mohandas et al., 1980; Clark et al., 1983). Osmotic gradient ektacytometry experiments performed on LPC treated RBCs at 1.0 µM (**Figure 3**) showed a significant rise in Omin and Ohyper, and a decrease in EImax. These findings confirm the effects of LPC on cell sphericity (decrease of surface-to-volume ratio). The increase in Ohyper could be explained by the slight increase in cell volume upon LPC treatment (MCV were measured to be 87.8 and 103.4 fL for the control and 1.0 µM LPC treated sample, respectively) as already reported (Safeukui et al., 2012). **Figure 4** reports the effect of LPC at various concentrations (from 0.25 to 1 µM) on the RBCs mechanical response while flowing in the geometry. **Figure 4a**

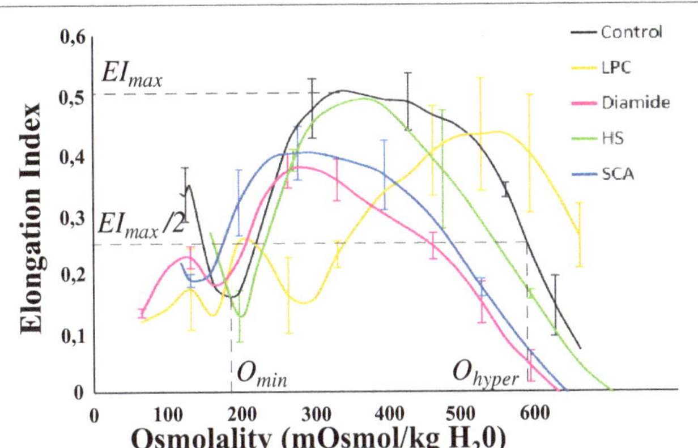

FIGURE 3 | Schematic representation of Osmoscan experiments. Several parameters can be determined on this curve: O_{min} (i.e., the osmolality at which RBC deformability value reaches a minimum in the hypotonic region of the curve), EI_{max} (i.e., the highest RBC deformability) and O_{hyper} which corresponds to the osmolality at half of the EI_{max} on the hypertonic region of the curve. LPC treated RBCs at 1.0 µM show an increase in O_{min} and O_{hyper} and a decrease in EI_{max}. Diamide treated RBCs at 0.5 mM show a decrease in EI_{max} and O_{hyper}. These HS patients have a decreased EI_{max} and O_{hyper} compared to control. O_{min}, O_{hyper} and EI_{max} are decreased in these SCA patients compared to control.

shows a typical sequence of deformation of a LPC-treated cell at 1 µM.

Lysolecithine treated RBCs behaved qualitatively similarly to healthy RBCs. They got compressed and stretched according to the width of the channel, and got elongated by the extensional flow at the exit before recovering a stationary shape. However,

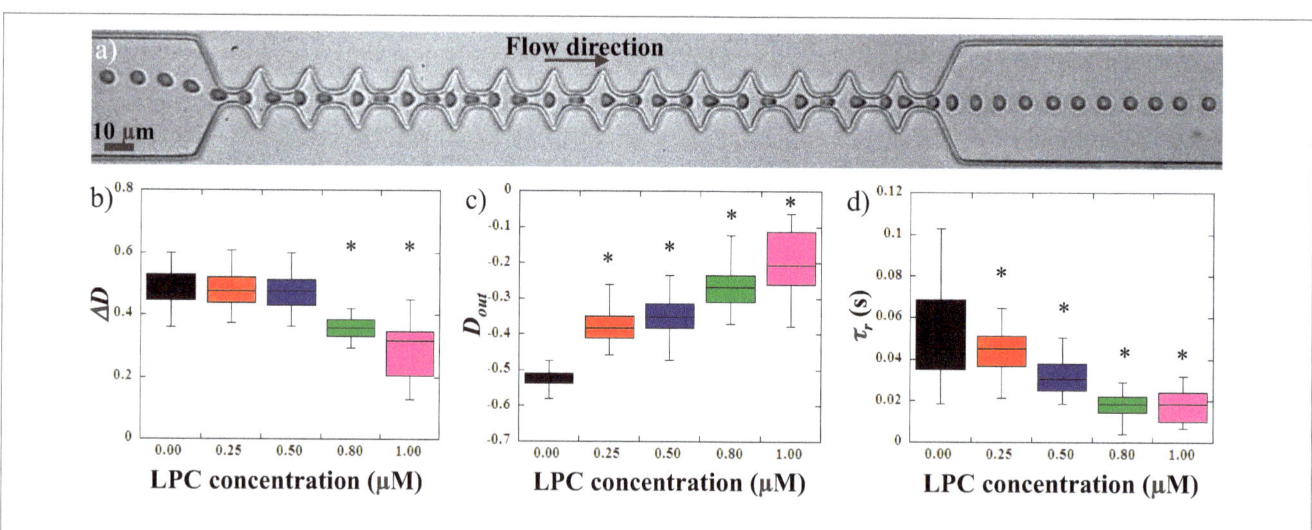

FIGURE 4 | (a) sequence of deformation of a 1 μM LPC treated RBC flowing in the geometry. Evolution of **(b)** the amplitude of deformation ΔD, **(c)** the stretching at the exit D_{out} and **(d)** the recovery time τ_r, versus LPC concentration. *Different from control sample (0 μM), $p < 0.05$.

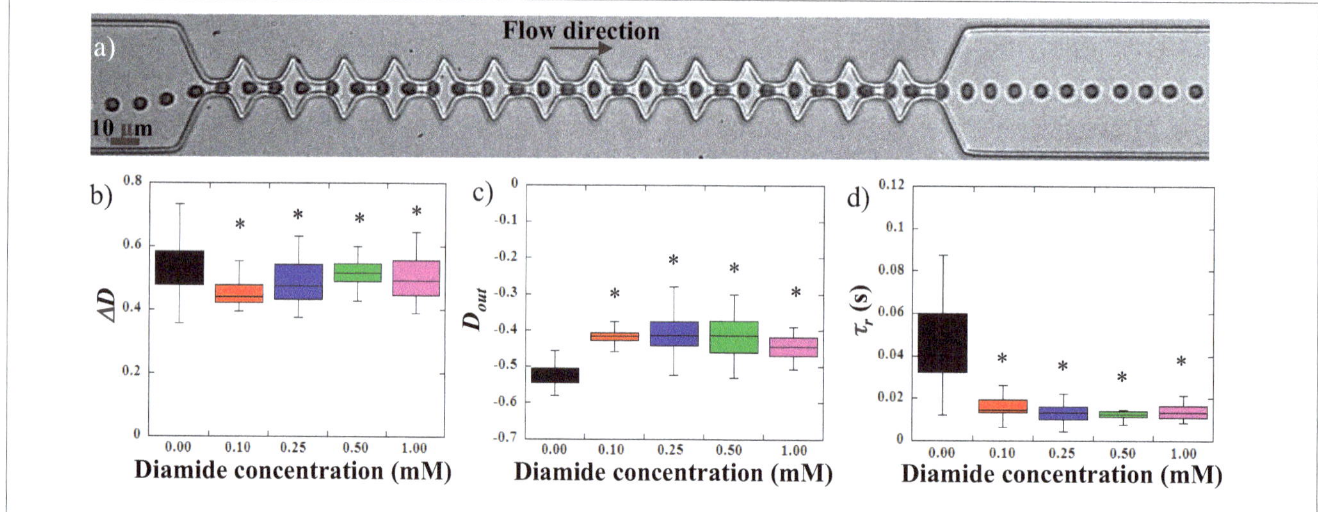

FIGURE 5 | (a) sequence of deformation of a 1 mM diamide treated RBC flowing in the geometry. Evolution of **(b)** the amplitude of deformation ΔD, **(c)** the stretching at the exit D_{out}, and **(d)** τ_r, the recovery time versus diamide concentration. *Different from control sample (0 mM), $p < 0.05$.

a reduction of the amount of deformation experienced by the cells can be visually detected on the picture while undergoing the same amount of stress. These observations are confirmed by the measurements of the amplitude of deformation, ΔD versus LPC concentration upon treatments (**Figure 4b**). LPC treatment at a concentration up to 0.5 μM had no significant impact on the amplitude of deformation of RBCs compared to healthy RBCs. However, treatment with higher concentrations led to a significant reduction of ΔD (**Figure 4b**). **Figures 3d, 4c** show the evolution of the elongation at the exit, D_{out}, and the recovery time, τ_r, respectively. The stretching at the exit, D_{out}, and recovery time, τ_r were gradually decreased upon increasing LPC concentration; hence showing that RBCs deformability drops gradually with the concentration of LPC.

Previous works used diamide to rigidify the RBCs membrane (Mohandas et al., 1980; Forsyth et al., 2010; Prado et al., 2015).

Indeed, diamide treatment (**Figure 3**) decreased EI_{max} and O_{hyper}. The increase in diamide concentration caused asymmetry in the hump of the osmotic gradient ektacytometry curve with a greater reduction in EI on the hypertonic part of the curve than on the hypotonic side. These findings confirmed the effects of diamide on membrane deformability, i.e., an increased shear modulus caused by the cross-linking between spectrins (Mohandas et al., 1980). **Figure 5** illustrates the effect of diamide treatment with concentrations ranging from 0.1 to 1.0 mM. **Figure 5a** presents the typical sequence of deformation of a diamide treated RBC at 1 mM. As for LPC treatments, RBCs deformability was clearly reduced when the cell was incubated with diamide, although the qualitative behavior of the cells was similar to that of healthy RBCs. Diamide treated cells showed a slight, yet statistically significant, drop in the amplitude of deformation ΔD, as illustrated in **Figure 5b**. Nevertheless, upon incubation

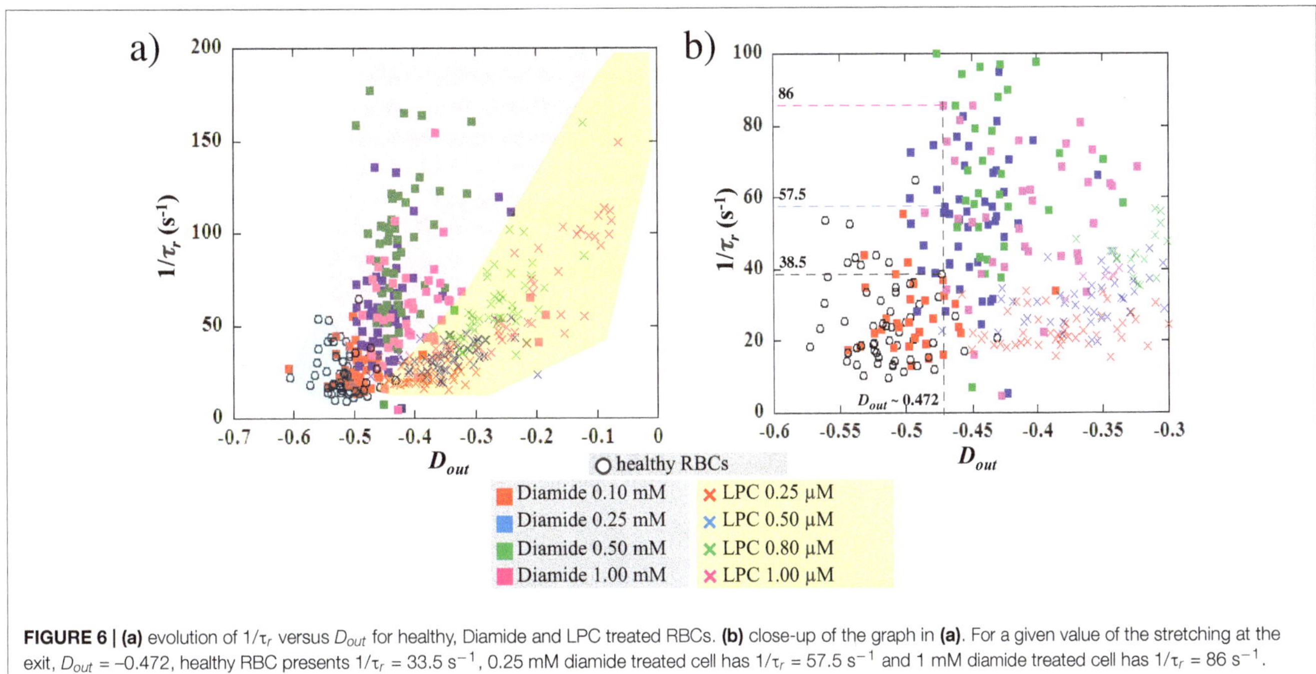

FIGURE 6 | **(a)** evolution of $1/\tau_r$ versus D_{out} for healthy, Diamide and LPC treated RBCs. **(b)** close-up of the graph in **(a)**. For a given value of the stretching at the exit, $D_{out} = -0.472$, healthy RBC presents $1/\tau_r = 33.5$ s^{-1}, 0.25 mM diamide treated cell has $1/\tau_r = 57.5$ s^{-1} and 1 mM diamide treated cell has $1/\tau_r = 86$ s^{-1}.

with diamide, elongation at the exit D_{out}, of treated RBCs was reduced (**Figure 5c**), although the measurements were not able to make a distinction between the different diamide concentrations. Diamide treatment also impacted the recovery time τ_r of RBCs as highlighted by the decrease from 0.048 s for healthy RBCs to roughly 0.013 s for diamide treated cells (**Figure 5d**).

From the results presented above, it seems that ΔD is not sensitive enough to clearly detect modifications of the cells mechanical properties. This may be explained by the fact that this parameter is highly impacted by the various off-centered initial positions of the cells when entering the zone of interest.

However, the measurements of the maximum elongation at the exit D_{out} and the RBCs recovery time τ_r can be used to discriminate healthy from chemically treated RBCs. But whether these readouts can differentiate between the different chemical concentrations is unclear. In the case of a visco-elastic object such as a RBC, the relaxation time, and therefore the recovery time, is linked to the deformed state. Indeed, τ_r decreases as D_{out} decreases, i.e., it takes less time to recover from a less elongated shape than from a more elongated one. **Figure 6** represents $1/\tau_r$ versus D_{out}, for the different concentrations of diamide and LPC as well as the healthy samples. It can be clearly observed on the figure that, while healthy RBCs mechanical signature is limited to low $1/\tau_r$ (<50 s^{-1}) values and strong elongation ($D_{out} < -0.5$), chemically rigidified RBCs present lower elongation at the exit associated with shorter recovery time. We can see on **Figure 6b**, that for a given extension at the exit $D_{out} \sim -0.472$, 3 RBCs present three different values of recovery times according to diamide concentration: $1/\tau_r = 38.5$ s^{-1} for the healthy cell ($\tau_r \sim 26$ ms), $1/\tau_r = 57.5$ s^{-1} for the RBC treated with diamide at 0.25 mM ($\tau_r \sim 17$ ms) and $1/\tau_r = 86$ s^{-1} for 1 mM diamide treated RBC ($\tau_r \sim 12$ ms). These observations suggest that cells treated with diamide (at a concentration above 0.1 mM)

exhibit a recovery time smaller than that of healthy ones for an equivalent extension at the exit D_{out}. This representation allows to discriminate easily the chemically rigidified RBCs from the healthy ones (blue zone). Moreover, it is possible to differentiate the two treatments as all the data points associated with LPC treated cells are located in a region (yellow zone) of the graph while diamide treated RBCs are located in another region (pink zone). Although both treatments lead to an increase in the overall cell rigidity, which is traduced by a lower D_{out} and a shorter τ_r, we can distinguish the effect of increased membrane stiffness from the effect of excess surface area.

The results obtained with chemically treated cells highlight that our microsystem is not only able to quantify the RBCs ability to deform, but also to provide a direct signature of the composition and architecture of their membrane. Indeed, we suspect that our microfluidic approach could be useful to differentiate and diagnose different kind of RBC disorders according to the origin of the affection (RBC membrane disorders, changes in surface-to-volume ratio, etc...).

Pathological RBCs

In order to evaluate the potential of our microsystem to detect pathological cells, we have studied the mechanical response of HS and SCA RBCs flowing out of our tooth-like pattern. Osmoscan experiments (**Figure 3**) showed a typical signature obtained for HS and SCA RBCs by ektacytometry. **Figure 7** reports the measurements of $1/\tau_r$ as a function of D_{out} for three HS patients, two SCA patients and one healthy donor. In order to help decipher if the two pathological samples are not only different from the healthy sample but also between each other, we decided to represent all pathological data on the same graph. SCA data correspond to the two different patients pooled together. In the same way, the points corresponding to the HS RBCs,

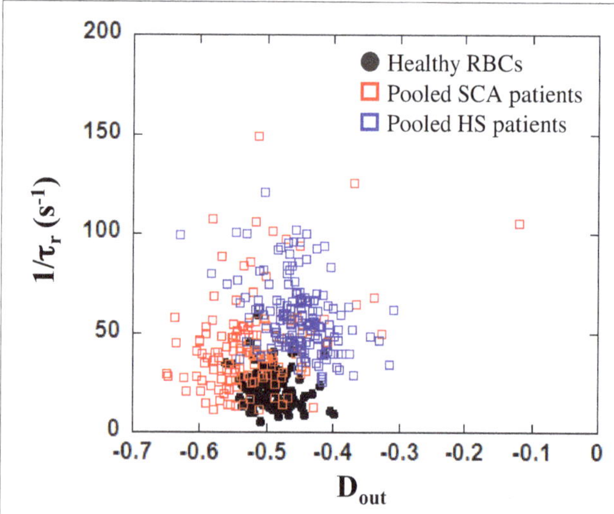

FIGURE 7 | D_{out} versus $1/\tau_r$ for healthy RBCs (solid black circles), pooled SCA patients (open red squares) and pooled HS patients (open blue squares). A MANOVA statistical analysis revealed that the RBCs distribution was significantly different between the three groups ($p < 0.05$ between HS and healthy, $p < 0.05$ between SCA and healthy, and $p < 0.05$ between HS and SCA).

were obtained by pooling the three HS patients. As illustrated on the Figure, RBCs from HS patients showed an increase of $1/\tau_r$ compared to healthy RBCs, while RBCs from SCA patients had both a decrease in D_{out} and a slight increase in $1/\tau_r$. Such increase in $1/\tau_r$ is expected because these diseases are known to increase cell stiffness (Nakashima and Beutler, 1979; Waught and Agre, 1988; Da Costa et al., 2013).

A MANOVA analysis showed that the distribution of RBCs according to D_{out} and $1/\tau_r$ was significantly different between HS, SCA and control individuals ($p < 0.05$). The morphologic signature of HS is the presence of microspherocytes (smaller spherical RBCs), which is caused by loss of RBC membrane surface area, leading to an abnormal osmotic fragility. Although one would have expected to find similar response with HS patients than with LPC treated cells, both being associated with a loss of membrane surface area, other modifications of HS RBCs may explain the difference between the LPC and HS microfluidic signatures. For instance, HS blood also contains dense RBCs (usually more than 4% of RBCs with MCHC > 41 g/dL) which may impact on the microfluidic behavior of the cells (Mohandas et al., 1980; Da Costa et al., 2016). Finally as usual, the inter-cells variability within a blood sample was higher in SCA than in HS. This discrepancy could be attributed to the presence of different sub-populations of more or less dense cells in SCA, suggesting the possibility to retrieve information about the composition of the different sub-populations. Although only few patients were tested, our preliminary results seem to highlight that our microfluidic device could be sensitive enough to make a distinction between healthy and pathological RBCs. Nevertheless, HS patients included in this study were issued from the same family, i.e., presenting the same genetic molecular modifications, and it remains unknown whether our system is able to discriminate RBCs from HS patients with different molecular modifications (ankyrin, α–spectrin, β–spectrin …). However, in this particular HS family, with the same red cell membrane molecular defect, we reproduced consistently in the three affected relatives the same microfluidic signature.

CONCLUSION

Differences between healthy and chemically treated cells were highlighted by the measurement of both the extension at the exit, D_{out}, and the shape recovery time τ_r. Although the measures were not able to distinguish between the different chemical concentrations, the representation of $1/\tau_r$ versus D_{out} membrane surface area from membrane elasticity. It is able to provide a direct signature of RBCs membrane composition and architecture.

Our preliminary results highlight that the mechanical responses of healthy and pathological RBCs are not only different but also suggest that the discrimination of both diseases could be possible, although more experiments using samples from various RBC disorders are needed. To our knowledge, this is the first attempt to evaluate the specificity of a passive microfluidic approach to perform diagnosis based on the alteration of RBC deformability, although more experiments are necessary to prove it. In addition to is diagnostic potential, further studies including large cohort of patients would be necessary to evaluate the clinical usefulness of our microfluidic approach for various RBC disorders. The single cell approach, while analyzing a statistically relevant population, could also provide complementary information (quantification of the heterogeneity of the cell population) that might be of great importance for the diagnosis and prognosis.

AUTHOR CONTRIBUTIONS

MF, CR, PJ, and PC have designed the research. MF and AB have performed the microfluidic experiments. CR and AB have performed the ektacytometry assays. LD and AG included the patients. MF, CR, and PC have analyzed the results. MF has written the main manuscript text and prepared the figures. MF, CR, LD, PJ, AG, and PC have reviewed and approved the final version of the manuscript.

ACKNOWLEDGMENTS

We thank CNRS MITI for financial support. We also thank Julie Galimand and Dr. Odile Fenneteau for their technical help and cytology expertise, respectively.

AUTHOR'S NOTE

‡ The repetition of the tooth-like pattern tends to focus cells at the exit. Such behavior has also been observed at the exit of very long straight constrictions, however, it would need a fairly longer geometry to obtain the same effect.

REFERENCES

Amirouche, A., Esteves, J., Ferrigno, R., and Faivre, M. (2020). Dual shape recovery of red blood cells flowing out of a microfluidic constriction. *Biomicrofluidics* 14:024116. doi: 10.1063/5.0005198

Amirouche, A., Ferrigno, R., and Faivre, M. (2017). Impact of channel geometry on the discrimination of mechanically impaired red blood cells in passive microfluidics. *Multidiscip. Digit. Publ. Inst. Proc.* 1:512. doi: 10.3390/proceedings1040512

Antia, M., Herricks, T., and Rathod, P. K. (2008). Microfluidic approaches to malaria pathogenesis. *Cell. Microbiol.* 10, 1968–1974. doi: 10.1111/j.1462-5822.2008.01216.x

Ballas, S. K., and Mohandas, N. (2004). Sickle red cell microrheology and sickle blood rheology. *Microcirculation* 11, 209–225. doi: 10.1080/10739680490279410

Bow, H., Pivkin, I. V., Diez-Silva, M., Goldfless, S. J., Dao, M., Niles, J. C., et al. (2011). A microfabricated deformability-based flow cytometer with application to malaria. *Lab Chip* 11, 1065–1073.

Braunmüller, S., Schmid, L., Sackmann, E., and Franke, T. (2012). Hydrodynamic deformation reveals two coupled modes/time scales of red blood cell relaxation. *Soft Matter* 8, 11240–11248.

Buys, A. V., Van Rooy, M.-J., Soma, P., Van Papendorp, D., Lipinski, B., and Pretorius, E. (2003). Changes in red blood cell membrane structure in type 2 diabetes: a scanning electron and atomic force microscopy study. *Cardiovasc. Diabetol.* 12:25. doi: 10.1186/1475-2840-12-25

Clark, M. R., Mohandas, N., and Shohet, S. B. (1983). Osmotic gradient ektacytometry: comprehensive characterization of red cell volume and surface maintenance. *Blood* 61, 899–910. doi: 10.1182/blood.v61.5.899.bloodjournal615899

Da Costa, L., Galimand, J., Fenneteau, O., and Mohandas, N. (2013). Hereditary spherocytosis, elliptocytosis, and other red cell membrane disorders. *Blood Rev.* 27, 167–178. doi: 10.1016/j.blre.2013.04.003

Da Costa, L., Suner, L., Galimand, J., Bonnel, A., Pascreau, T., Couque, N., et al. (2016). Diagnostic tool for red blood cell membrane disorders: assessment of a new generation ektacytometer. *Blood Cells Mol. Dis.* 56, 9–22. doi: 10.1016/j.bcmd.2015.09.001

Deng, Y., and Chung, A. J. (2016). "Next generation deformability cytometry: fully automated, high-throughput and near real-time cell mechanotyping," in *Proceedings of the 20th International Conference on Miniaturized Systems for Chemistry and Life Sciences*, Dublin, 148–149.

Dondorp, A. M., Angus, B. J., Hardeman, M. R., Chotivanich, K. T., Silamut, K., Ruangveerayuth, R., et al. (1997). The role of reduced red cell deformability in the pathogenesis of severe falciparum malaria and its restoration by blood transfusion. *Am. J. Trop. Med. Hyg.* 57, 507–511. doi: 10.4269/ajtmh.1997.57.507

Duffy, D. C., McDonald, J. C., Schueller, O. J. A., and Whitesides, G. M. (2004). Rapid prototyping of microfluidic systems in poly(dimethylsiloxane). *Anal. Chem.* 70, 4974–4984. doi: 10.1021/ac980656z

Evans, E. A. (1989). Structure and deformation properties of red blood cells: concepts and quantitative methods. *Methods Enzymol.* 1173, 3–35. doi: 10.1016/s0076-6879(89)73003-2

Faustino, V., Pinho, D., Yaginuma, T., Calhelha, R. C., Ferreira, I. C. F. R., and Lima, R. (2014). Extensional flow-based microfluidic device: deformability assessment of red blood cells in contact with tumor cells. *Biochip J.* 8, 42–47. doi: 10.1007/s13206-014-8107-1

Faustino, V., Rodrigues, R. O., Pinho, D., Costa, E., Santos-Silva, A., Miranda, V., et al. (2019). A microfluidic deformability assessment of pathological red blood cells flowing in a hyperbolic converging microchannel. *Micromachines* 10:645. doi: 10.3390/mi10100645

Fischer, T. M., Haest, C. W. M., Stöhr, M., Kamp, D., and Deuticke, B. (1978). Selective alteration of erythrocyte deformability by SH-reagents. Evidence for an involvement of spectrin in membrane shear elasticity. *Biochim. Biophys. Acta* 510, 270–282. doi: 10.1016/0005-2736(78)90027-5

Forsyth, A. M., Wan, J., Ristenpart, W. D., and Stone, H. A. (2010). The dynamic behavior of chemically stiffened red blood cells in microchannel flows. *Microvasc. Res.* 80, 37–43. doi: 10.1016/j.mvr.2010.03.008

Guo, Q., Reiling, S. J., Rohrbach, P., and Ma, H. (2012). Microfluidic biomechanical assay for red blood cells parasitized by *Plasmodium falciparum*. *Lab Chip* 12, 1143–1150.

Hansen, B., Pivkin, I. V., Diez-Silva, M., Goldfless, S. J., Dao, M., Jacquin, C., et al. (2011). A microfabricated deformability-based flow cytometer with application to malaria. *Lab Chip* 11, 1065–1073.

Hauck, T. S., Giri, S., Gao, Y., and Chan, W. C. W. (2010). Nanotechnology diagnostics for infectious diseases prevalent in developing countries. *Adv. Drug Deliv. Rev.* 62, 438–448. doi: 10.1016/j.addr.2009.11.015

Hochmuth, R. M., Worthy, P. R., and Evans, E. A. (1979). Red cell extensional recovery and the determination of membrane viscosity. *Biophys. J.* 26, 101–114. doi: 10.1016/s0006-3495(79)85238-8

Horobin, J. T., Sabapathy, S., and Simmonds, M. J. (2017). Repetitive supra-physiological shear stress impairs red blood cell deformability and induces hemolysis. *Artif. Organs* 41, 1017–1025. doi: 10.1111/aor.12890

Iragorri, M. A. L., El Hoss, S., Brousse, V., Lefevre, S. D., Dussiot, M., Xu, T. Y., et al. (2018). A microfluidic approach to study the effect of mechanical stress on erythrocytes in sickle cell disease. *Lab Chip* 18, 2975–2984. doi: 10.1039/c8lc00637g

Lee, S., Ahn, K. H., Lee, S. J., Sun, K., Goedhart, P. T., and Hardeman, M. R. (2004). Shear induced damage of red blood cells monitored by the decrease of their deformability. *Korea Aust. Rheol. J.* 16, 141–146.

Maciaszek, J. L., and Lykotrafitis, G. (2011). Sickle cell trait human erythrocytes are significantly stiffer than normal. *J. Biomech.* 44, 657–661. doi: 10.1016/j.jbiomech.2010.11.008

Mauritz, J. M. A., Tiffert, T., Seear, R., Lautenschlger, F., Esposito, A., Lew, V. L., et al. (2010). Detection of *Plasmodium falciparum*-infected red blood cells by optical stretching. *J. Biomed. Opt.* 15:030517. doi: 10.1117/1.3458919

Mohandas, N., Clark, M. R., Jacobs, M. S., and Shohet, S. B. (1980). Analysis of factors regulating erythrocyte deformability. *J. Clin. Invest.* 66, 563–573. doi: 10.1172/jci109888

Mohandas, N., and Evans, E. (1994). Mechanical properties of the red cell membrane in relation to molecular structure and genetic defects. *Annu. Rev. Biophys. Biomol. Struct.* 23, 787–818. doi: 10.1146/annurev.bb.23.060194.004035

Nakashima, N., and Beutler, E. (1979). Erythrocyte cellular and membrane deformability in hereditary spherocytosis. *Blood* 53, 481–485. doi: 10.1182/blood.v53.3.481.bloodjournal533481

Nemeth, N., Kiss, F., and Miszti-Blasius, K. (2015). Interpretation of osmotic gradient ektacytometry (osmoscan) data: a comparative study for methodological standards. *Scand. J. Clin. Lab. Invest.* 75, 213–222. doi: 10.3109/00365513.2014.993695

Prado, G., Farutin, A., Misbah, C., and Bureau, L. (2015). Viscoelastic transient of confined red blood cells. *Biophys. J.* 108, 2126–2136. doi: 10.1016/j.bpj.2015.03.046

Qiang, Y., Liu, J., Dao, M., Suresh, S., and Du, E. (2019). Mechanical fatigue of human red blood cells. *Proc. Natl. Acad. Ssci. U.S.A.* 116, 19828–19834. doi: 10.1073/pnas.1910336116

Safeukui, I., Buffet, P. A., Deplaine, G., Perrot, S., Brousse, V., Ndour, A., et al. (2012). Quantitative assessment of sensing and sequestration of spherocytic erythrocytes by the human spleen. *Blood* 120, 424–430. doi: 10.1182/blood-2012-01-404103

Shelby, J. P., White, J., Ganesan, K., Rathod, P. K., and Chiu, D. T. (2003). A microfluidic model for single-cell capillary obstruction by *Plasmodium falciparum*-infected erythrocytes. *Proc. Natl. Acad. Sci. U.S.A.* 100, 14618–14622. doi: 10.1073/pnas.2433968100

Simmonds, M. J., and Meiselman, H. J. (2016). Prediction of the level and duration of shear stress exposure that induces subhemolytic damage to erythrocytes. *Biorheology* 53, 237–249. doi: 10.3233/bir-16120

Suresh, S. (2007). Biomechanics and biophysics of cancer cells. *Acta Biomater.* 3, 413–438. doi: 10.1016/j.actbio.2007.04.002

Suresh, S., Spatz, J., Mills, J. P., Micoulet, A., Dao, M., Lim, C. T., et al. (2005). Connections between single-cell biomechanics and human disease states: gastrointestinal cancer and malaria. *Acta Biomater.* 1, 15–30. doi: 10.1016/j.actbio.2004.09.001

Tomaiuolo, G. (2014). Biomechanical properties of red blood cells in health and disease towards microfluidics. *Biomicrofluidics* 8:051501. doi: 10.1063/1.4895755

Tomaiuolo, G., Barra, M., Preziosi, V., Cassinese, A., Rotoli, B., and Guido, S. (2011). Microfluidics analysis of red blood cell membrane viscoelasticity. *Lab Chip* 11, 440–454.

Tomaiuolo, G., and Guido, S. (2011). Start-up shape dynamics of red blood cells in microcapillary flow. *Microvasc. Res.* 82, 35–41. doi: 10.1016/j.mvr.2011.03.004

Tsukada, K., Sekizuka, E., Oshio, C., and Minamitani, H. (2001). Direct measurement of erythrocyte deformability in diabetes mellitus with a transparent microchannel capillary model and high-speed video camera system. *Microvasc. Res.* 61, 231–239. doi: 10.1006/mvre.2001.2307

Vilas Boas, L., Faustino, V., Lima, R., Miranda, J. M., Minas, G., Veiga Fernandes, C. S., et al. (2018). Assessment of the deformability and velocity of healthy and artificially impaired red blood cells in narrow polydimethylsiloxane (PDMS) microchannels. *Micromachines* 9:384. doi: 10.3390/mi9080384

Waught, R. E., and Agre, P. (1988). Reductions of erythrocyte membrane viscoelastic coefficients reflect spectrin deficiencies in hereditary spherocytosis. *J. Clim. Invest.* 81, 133–141. doi: 10.1172/jci113284

Zaninoni, A., Fermo, E., Vercellati, C., Consonni, D., Marcello, A. P., Zanella, A., et al. (2018). Use of laser assisted optical rotational cell analyzer (LoRRca MaxSis) in the diagnosis of RBC membrane disorders, enzyme defects, and congenital dyserythropoietic anemias: a monocentric study on 202 patients. *Front. Physiol.* 9:451. doi: 10.3389/fphys.2018.00451

Zheng, Y., Nguyen, J., Wai, Y., and Sun, Y. (2013). Recent advances in microfluidic techniques for single-cell biophysical characterization. *Lab Chip* 13, 2464–2483.

18

Trends in the Development of Diagnostic Tools for Red Blood Cell-Related Diseases and Anemias

Lars Kaestner[1,2] and Paola Bianchi[3]*

[1] Theoretical Medicine and Biosciences, Medical Faculty, Saarland University, Homburg, Germany, [2] Experimental Physics, Faculty of Natural Science and Technology, Saarland University, Saarbrücken, Germany, [3] Fondazione IRCCS Ca' Granda Ospedale Maggiore Policlinico Milano, UOC Ematologia, UOS Fisiopatologia delle Anemie, Milan, Italy

***Correspondence:**
Lars Kaestner
lars_kaestner@me.com

In the recent years, the progress in genetic analysis and next-generation sequencing technologies have opened up exciting landscapes for diagnosis and study of molecular mechanisms, allowing the determination of a particular mutation for individual patients suffering from hereditary red blood cell-related diseases or anemia. However, the huge amount of data obtained makes the interpretation of the results and the identification of the pathogenetic variant responsible for the diseases sometime difficult. Moreover, there is increasing evidence that the same mutation can result in varying cellular properties and different symptoms of the disease. Even for the same patient, the phenotypic expression of the disorder can change over time. Therefore, on top of genetic analysis, there is a further request for functional tests that allow to confirm the pathogenicity of a molecular variant, possibly to predict prognosis and complications (e.g., vaso-occlusive pain crises or other thrombotic events) and, in the best case, to enable personalized theranostics (drug and/or dose) according to the disease state and progression. The mini-review will reflect recent and future directions in the development of diagnostic tools for red blood cell-related diseases and anemias. This includes point of care devices, new incarnations of well-known principles addressing physico-chemical properties, and interactions of red blood cells as well as high-tech screening equipment and mobile laboratories.

Keywords: point of care, functional screening, physico-chemical properties, mobile laboratory, sickle cell disease, personalized medication, artificial intelligence

DO WE NEED NOVEL DIAGNOSTIC TOOLS?

There is a demand for novel diagnostic assays and devices from several perspectives. (i) Since we are still facing huge economic differences across our planet, there is a need (and a market) for low-cost diagnosis of common and rare red blood cell-related diseases. This includes sickle cell disease, thalassemia, malaria, and other less common hereditary and acquired hemolytic anemias. (ii) Scientific progress and the omics era allowed to unravel new diseases. This covers so far undiagnosed red blood cell diseases as well as identifying new pathogenetic variants in previously phenomenologically defined diseases. However, the huge amount of data obtained need to be interpreted, and often variants predicted at *in silico* analysis as possibly pathogenic—the so-called variants of unknown significance (VUS)—may have no or very little impact on protein function and obviously require further diagnostic tests to assess their functional involvement in the disease. (iii) Even knowing the molecular defect in diseases does not tell us much about the severity and the current state of the disease (e.g., severity of anemia, vaso-occlusive crisis in sickle cell disease). Thus, there is a need to establish the prognosis and possible complications of particular disease states and to determine an appropriate treatment. This is not restricted to the selection or combination of particular drugs but also the dose of these drugs. The concept of personalized theranostics addresses this issue but requires new approaches to become effective.

Given the rarity and the heterogeneity of this group of disorders, the interest of pharmaceutical companies and device maufacturers in developing drugs and technological devices, respectively was limited in the past. The strong possibility not to reach a sufficient volume of requests, the approach to rare disorders, in particular the congenital ones, was scanty up to some years ago. Funding initiatives for rare diseases, especially by the European Commission (within the 7th Framework Program and Horizon 2020) boosted research in the field of rare anemias, and this will continue within the coming years (Horizon Europe). A consensus document highlighting major achievements in diagnosis and treatment of blood disorders, including rare red blood cell disorders, and identifying the greatest unmet clinical and scientific needs has been recently prepared by more than 300 experts belonging to the European Hematology Association (EHA) (Engert et al., 2016).

POINT OF CARE DIAGNOSTIC DEVICES

There are many requirements that should be taken into consideration in the development of a point of care device. It is important that the test performed will be rapid, user-friendly, easily interpretable, sensitive and specific (to avoid false negative and positive results). Another aspect that should always be considered in the development of point of care diagnostic devices is their size. They need to be transportable and in the best case being pocket size. Furthermore, such devices need to be affordable, although the threshold for the market price is different depending on the socio-economic environment of the patient(s). In particular, one of the technical developments of the past decade is in favor of such developments: the smartphones (and tablets) are ever improving mini-computers with innovative interaction interfaces that often "only" require particular sensors and the complementary software to turn into a diagnostic device. Even smartwatches (or fitness watches) already measure routinely heart rate and other health-related parameters start to become routine read-outs such as oxygen saturation (see, e.g., Garmin watch portfolio) or blood pressure. It is worthwhile to mention that most point of care devices work completely non-invasive or at least only require such small amounts of blood that can be taken by finger prick avoiding venous puncture (e.g., Pandey et al.,

2018). Examples of early developments of smartphone-based diagnostic devices in conjunction with an "App" and appropriate sensors have been described for the detection of sickle cell disease (Knowlton et al., 2015; Ung et al., 2015), although not all point of care diagnostic devices are smartphone based. Promising recent developments are the HemoTypeSC to determine the hemoglobin types by Silver Lake Research (Azusa, CA, United States) (e.g., Nankanja et al., 2019; Steele et al., 2019; Mukherjee et al., 2020) or Sickle SCAN with a similar application by BioMedomics (Morrisville, NC, United States) (Nguyen-Khoa et al., 2018). Also, a recent enrichment on the market was Q-POC by QuantuMDx (Newcastle, United Kingdom). Although initially developed to diagnose infectious diseases, it is also tested to diagnose all the different β-thalassemia mutations (Elion, 2019).

A very special test for sickle cell disease was recently developed. The test is utilizing filter paper and, thus, costs less than 0.05€ per test (Delobel et al., 2018).

A NEW GENERATION OF DEVICES PROBING FOR PHYSICO-CHEMICAL PROPERTIES OF RED BLOOD CELLS

As outlined in the introduction, although the identification of a molecular lesion is mandatory in genetic disorders to confirm the diagnosis, clinical observations reveal that the genotype/phenotype correlation is not always possible and that other genetic/epigenetic factors (other than the molecular defect) may contribute to the clinical phenotype. This is particularly evident in case of intrafamily clinical variability in presence of the same mutation, or in clinical variability in the same patient during his life (Grace et al., 2018; Bianchi et al., 2020). Therefore, diagnostic devices that address physico-chemical properties such as cellular deformability, hemolysis, and red blood cell interaction properties entered the market or are under development.

An example addressing cell deformability is the LoRRca (RR Mechatronics, Hoorn, Netherlands) (**Figure 1**). The instrument was developed and commercialized only some years ago. Initially, it was used in a few highly specialized centers, but now, it is increasingly used to routinely diagnose rare red cell disorders (Da Costa et al., 2016; Zaninoni et al., 2018). Moreover, its new oxygen scan application, measuring the relative oxygen pressure at the critical point the red blood cells start to sickle, might offer in the future new opportunities for monitoring sickling during new treatment strategies, personalized medicine, and prediction of complication in sickle cell disease (Rab et al., 2020).

A neat concept to test red blood cell stability based on a bead mill and spectral measurement of hemoglobins was developed by Blaze Medical Devices (Ann Arbor, MI, United States). Although the concept was convincing (Tarasev et al., 2016), the device never entered the market. However, measurements as a service are offered by Functional Fluidics (Detroit, MI, United States).

Yet another example is a table top device called MeCheM (mechanical and chemical modulator) that was developed by Epigem Ltd. (Redcar, United Kingdom) within the project CoMMiTMenT (Combined Molecular Microscopy for Therapy

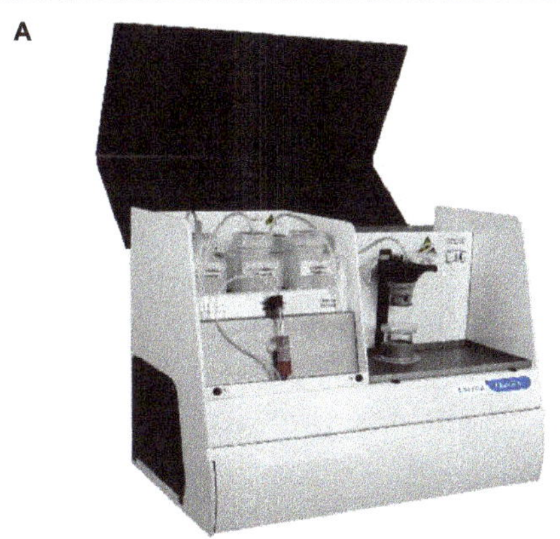

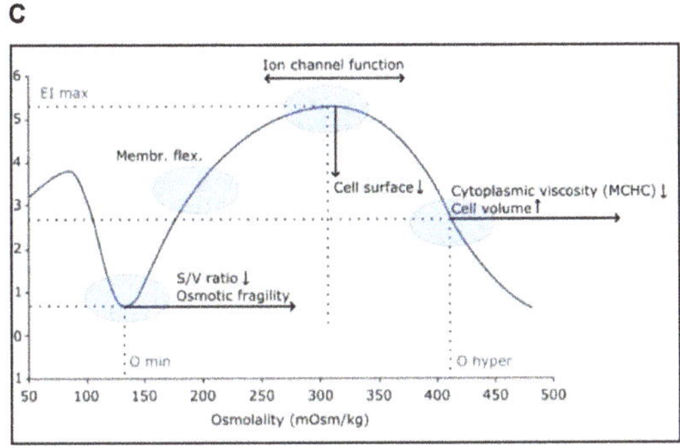

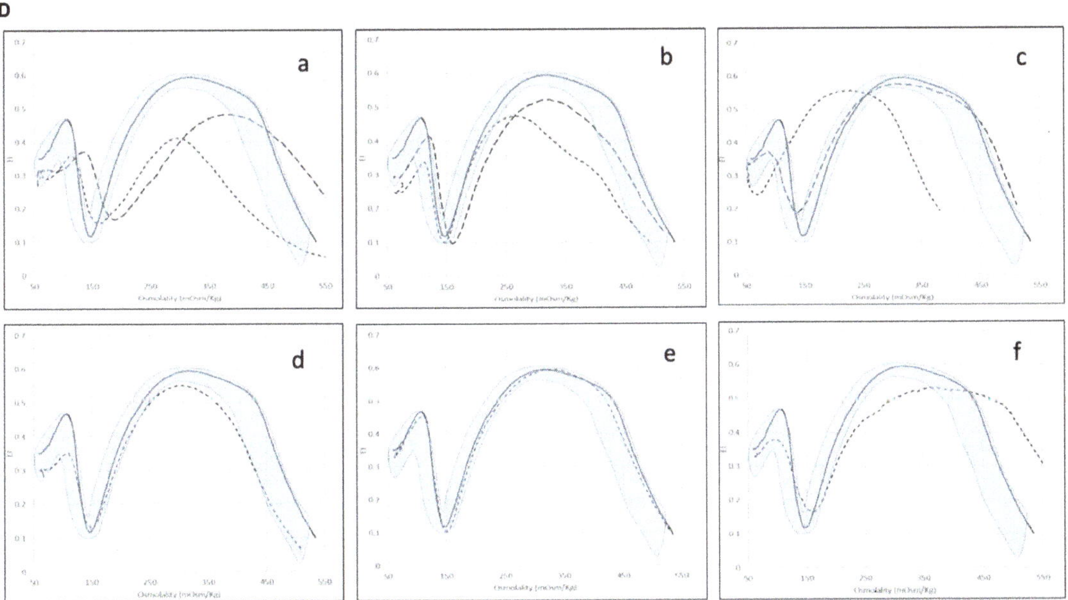

FIGURE 1 | Analysis of RBC membrane disorders and other rare haemolytic anaemias by ektacytometry analysis. **(A)** Image of the Laser Optical Rotational Red Cell Analyzer (LoRRca Maxsis RR Mechatronics, Netherlands). **(B)** List of key parameters analyzed by the instruments. **(C)** Osmoscan profile in normal subjects and parameters analyzed: the Omin value represents the 50% of the RBCs hemolysis in conventional osmotic fragility assays, reflecting mean cellular surface-to volume ratio; the Elongation Index (EI) max corresponds to the maximal deformability obtained near the isotonic osmolality and is an expression of the membrane surface; the Ohyper reflects mean cellular hydration status; the AUC correspond to the area under the curve beginning from a starting point in the hypo-osmolar region and an ending point in the hyper-osmolar region. **(D)** Examples of typical osmoscan profiles in hemolytic anemias resulting from the analysis of 202 patients affected by congenital hemolytic anemia of different etiology. Continuous line represents a daily control and shaded area the control range curve. **(a)** HS = hereditary spherocytosis, **(b)** HE = hereditary elliptocytosis, **(c)** HSt = hereditary stomatocytosis: HSt-PIEZO1 (hereditary xerocytosis) (dotted line), HSt-KCNN4 (Gardos channelopathy) (dashed line), **(d)** CDAII = congenital diserythropoietic anemia type II, **(e)** RBC enzymopathies (pyruvate kinase deficiency), **(f)** other rarer RBC enzymopathies (glucosephosphate isomerase deficiency). Panels **(A)** and **(C)** are reproduced with permission from RR Mechatronics. Panels **(D)** is reproduced from Zaninoni et al. (2018).

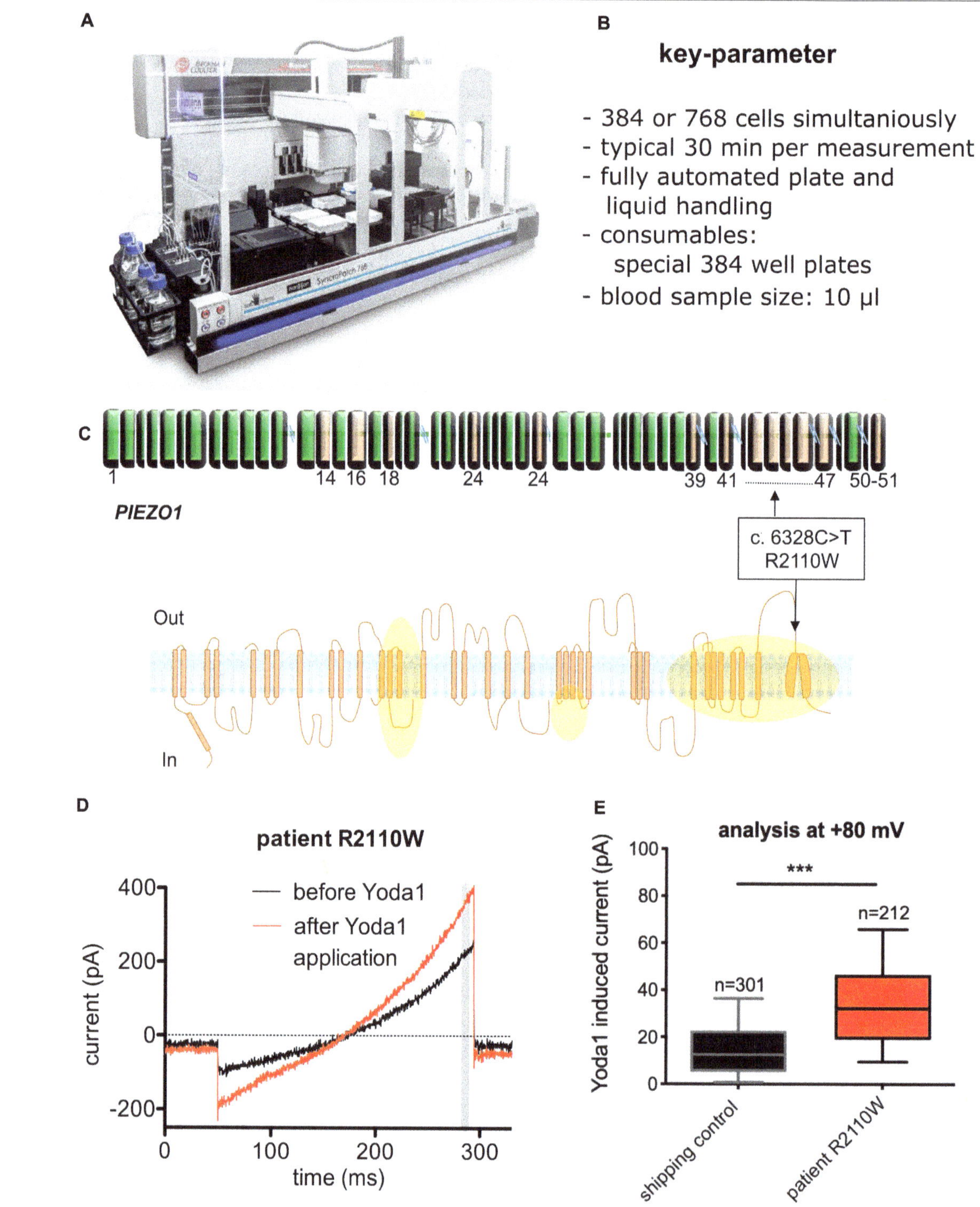

FIGURE 2 | Diagnosis of a novel *PIEZO* mutation with automated patch-clamp technology. **(A)** Image of the SyncroPatch device (Nanion Technologies, Munich, Germany). **(B)** List of key parameters of the SyncroPatch. **(C)** Illustration of a novel mutation (R2110W) of the Piezo 1 ion channel. Although detected *per se*, it was unknown if the mutation has a functional effect on the red blood cells. Orange areas represent regions affected by previously reported mutations. **(D)** Raw data traces of a red blood cell recording for illustration. Yoda1 is a specific activator of Piezo 1. The gray bar depicts the time point (= membrane potential), which was used for the statistical analysis. **(E)** Statistical analysis of all measured cells (R2110W mutation vs. control) to exemplify the functional impact of the mutation. n gives the number of successful measured and analyzed cells. **(A)** Reproduced with permission from Nanion Technologies. **(C–E)** Reproduced from Rotordam et al. (2019) with permission of the Ferrata Storti Foundation.

and Personalized Medication in Rare Anemias Treatments, funded in the European Community 7th Framework Program). It is a microfluidic device that can challenge red blood cells chemically or with functional surfaces, while red blood cells are microscopically observed. Although it is not yet on the market, it is under investigation in several hematologic laboratories within Europe and in a clinical trial testing the efficiency of Memantine for the treatment of sickle cell disease (Makhro et al., 2020).

NOVEL SCREENS BASED ON HIGH-TECH, ROBOTICS, AND ARTIFICIAL INTELLIGENCE

The opposite of point of care devices are machines or procedures that are so complicated and/or expensive that they are exclusively established in expert centers or even in specialized research laboratories. For these devices/procedures, it is sometimes hard to distinguish between the generation of new knowledge and diagnosis—at least this borderline is fuzzy. An example of such a device is an automated patch-clamp robot that in the past proofed to be useful for the investigation of red blood cells (Makhro et al., 2013; Minetti et al., 2013). An image of the device and the joint test is given in **Figure 2**. This patch-clamp robot, originally developed for pharmacological compound screening, was used for a functional diagnosis of a new variant of a mutation of the mechanosensitive ion-channel Piezo1, which is associated with hereditary xerocytosis (Rotordam et al., 2019).

Another procedure used to understand/diagnose diseases is the *in vitro* erythropoiesis, which was refined and optimized considerably also in the past decade. Although bioreactors for an automated and controlled differentiation from peripheral stem cells to erythrocytes are under development (Heshusius et al., 2020), up to date, *in vitro* erythropoiesis still requires the human resource of a scientist or technician to be performed. However, we like to emphasize that recent results showed the importance of erythropoiesis for determination of the severity of the disease (Moura et al., 2019; Caulier et al., 2020).

Artificial intelligence based on artificial neuronal networks is a concept that will enter all incarnations of diagnostic devices as long as they involve computational power. A very old and established diagnostic tool, the analysis of blood smears, is currently reinvented based on the recordings of confocal stacks, three-dimensional rendering of the cells, and classification of the cell shapes by artificial neural networks (compare Bogdanova et al., 2020 within this research topic). Whether it will indeed be possible to link the occurrence of particular cell shapes with concrete mutations still needs to be explored.

LOGISTIC CONCEPTS FOR HIGH-TECH DIAGNOSIS AND RESEARCH

Given the rarity and the heterogeneity of this group of disorders, the confinement between research and diagnostics is sometimes faint, especially for the rare or undiagnosed diseases. In the presence of very rare disorders (e.g., some rare red blood cell enzyme defects or in defects of cell volume regulation), each case seems unique and worth to be described and deeply characterized. Networking activities to recruit similar cases and to joint expertise and collaborations is utmost important in this field (e.g., Vives Corrons et al., 2014; Fermo et al., 2017; Petkova-Kirova et al., 2019). However, this requires a logistic organization of collaboration, in particular for sharing the blood samples. The most common mode is the shipment of samples. This is easy and straightforward when cells or cell extracts can be preserved, like for blood smears, chemically fixed cells for morphological investigations (Abay et al., 2019), isolated RNA for genetic investigations, frozen cells, e.g., for protein analysis, etc. However, it is much more complicated when assays are based on living cells. Some years ago, we performed a dedicated study on healthy red blood cells to mimic transportation conditions (Makhro et al., 2016). The outcome was surprising in the respect that different red blood cell parameters require different conditions in terms of anticoagulant and temperature to resemble the results of fresh red blood cells. With cells of patients, the situation can be even worse. In a recent study on cellular intracellular Ca^{2+} (Hertz et al., 2017), the effect of the transportation was bigger than the effect of the disease. In this particular study, the data could be "rescued" by normalizing to healthy transportation controls. However, we also found that in certain conditions, differences between patients and control can easily be lost during some hours *in vitro* (Rotordam et al., 2019). Taking all these indications and although shipment of blood samples is the most common and popular method of interlaboratory collaboration, it is by far not an ideal configuration. However, a much better option would be if the patients travel to the specialized laboratories. Although we recently introduced this practice in our laboratory, it does work only for a minority of patients (due to the state of the patients, their compliance, or other restrictions) and is only a kind of control that hardly can reach statistical power, especially for rare and very rare diseases. The third option would be mobile specialized laboratories. Surprisingly, this idea, so far, got stuck within discussion among researchers, presumably because appropriate funding programs are the restriction. From the technical point of view, (i) we are, in principle, able to catapult even confocal microscopes with biological samples into space (Thiel et al., 2019) and (ii) an increasing number of devices are designed for transportation. This is not restricted to the classical point of care devices mentioned above but also applies for fairly complicated machines such as flow cytometers (e.g., CyFlow Cube6, Sysmex, Germany). Therefore, a mobile laboratory down on earth for red blood cell-related diseases should be a challenging project but with high chances of success and only limited risks as it was realized for other purposes before (e.g., Weidmann et al., 2018). However, the intended project/use should define whether such a laboratory should be on wheels, on a boat, or on board an aircraft.

CONCLUSION AND OUTLOOK

Point of care, artificial intelligence, and personalized theranostocs are probably the major keywords that characterize current and future developments in diagnostic devices for red blood cell-related diseases, in general, and for rare anemias, in particular. From a conceptual point of view, genetic analysis is well established, and decrease in the size of devices and significant cost drops will increase the spread and the regular use of this diagnostic tool. In line with this, a targeted analysis of a defined (group of) protein(s) will be replaced by a full genome analysis. However, functional analysis (more and more based on individual cells) on top of gene characterization will become increasingly important. This may go far beyond the samples given above and is likely to include further miniaturized assays of well-known tests such as the investigation of cell density distributions or measurements considering filterability properties.

AUTHOR CONTRIBUTIONS

Both authors wrote and approved the manuscript.

ACKNOWLEDGMENTS

We would like to thank Prof. Jacques Elion for valuable information exchange.

REFERENCES

Abay, A., Simionato, G., Chachanidze, R., Bogdanova, A., Hertz, L., Bianchi, P., et al. (2019). Glutaraldehyde – a subtle tool in the investigation of healthy and pathologic red blood cells. *Front. Physiol.* 10:514. doi: 10.3389/fphys.2019.00514

Bianchi, P., Fermo, E., Lezon-Geyda, K., van Beers, E., Morton, D. H., Barcellini, W., et al. (2020). Genotype-phenotype correlation and molecular heterogeneity in pyruvate kinase deficiency. *Am. J. Hemtol.* 95, 472–482. doi: 10.1002/ajh.25753

Bogdanova, A., Kaestner, L., Simionato, G., Wickrema, A., and Makhro, A. (2020). Heterogeneity of red blood cells: causes and consequences. *Front. Physiol.* 11:392. doi: 10.3389/fphys.2020.00392

Caulier, A., Jankovsky, N., Demont, Y., Ouled-Haddou, H., Demagny, J., Guitton, C., et al. (2020). PIEZO1 activation delays erythroid differentiation of normal and hereditary xerocytosis-derived human progenitor cells. *Haematologica* 105, 610–622. doi: 10.3324/haematol.2019.218503

Da Costa, L., Suner, L., Galimand, J., Bonnel, A., Pascreau, T., Couque, N., et al. (2016). Diagnostic tool for red blood cell membrane disorders: assessment of a new generation ektacytometer. *Blood Cells Mol. Dis.* 56, 9–22. doi: 10.1016/j.bcmd.2015.09.001

Delobel, J., Keitel, K., Balmas-Bourloud, K., and Mlaganile, T. (2018). Harnessing the power of global health studies for sickle cell disease: validation of a rapid, open-source, paper- based screening assay in a cohort of 1103 Tanzanian children. *Blood* 132:150.

Elion, J. (2019). personal communication. UMR Inserm U1134 - Université Paris Diderot/USCP Institut National de la Transfusion Sanguine.

Engert, A., Balduini, C., Brand, A., Coiffier, B., Cordonnier, C., Döhner, H., et al. (2016). The European hematology association roadmap for european hematology research: a consensus document. *Haematologica* 101, 115–208. doi: 10.3324/haematol.2015.136739

Fermo, E., Petkova-Kirova, P., Zaninoni, A., Marcello, A. P., Makhro, A., et al. (2017). 'Gardos Channelopathy': a variant of hereditary Stomatocytosis with complex molecular regulation. *Sci. Rep.* 7:1744. doi: 10.1038/s41598-017-01591-w

Grace, R. F., Bianchi, P., van Beers, E. J., Eber, S. W., Glader, B., Yaish, H. M., et al. (2018). Clinical spectrum of pyruvate kinase deficiency: data from the pyruvate kinase deficiency natural history study. *Blood* 131, 2183–2192. doi: 10.1182/blood-2017-10-810796

Hertz, L., Huisjes, R., Llaudet-Planas, E., Petkova-Kirova, P., Makhro, A., Danielczok, J., et al. (2017). Is increased intracellular calcium in red blood cells a common component in the molecular mechanism causing anemia? *Front. Physiol.* 8:673. doi: 10.3389/fphys.2017.00673

Heshusius, S., Heideveld, E., Burger, P., Thiel-Valkhof, M., Sellink, E., Varga, E., et al. (2020). Large-scale in vitro production of red blood cells from human peripheral blood mononuclear cells. *Blood Adv.* 3, 3337–3350. doi: 10.1182/bloodadvances.2019000689

Knowlton, S. M., Sencan, I., Aytar, Y., Khoory, J., and Heeney, M. M. (2015). Sickle cell detection using a smartphone. *Sci. Rep.* 5:15022.

Makhro, A., Hanggi, P., Goede, J., Wang, J., Bruggemann, A., Gassmann, M., et al. (2013). N-methyl D-aspartate (n.d.) receptors in human erythroid precursor cells and in circulating red blood cells contribute to the intracellular calcium regulation. *Am. J. Physiol. Cell Physiol.* 305, C1123–C1138. doi: 10.1152/ajpcell.00031.2013

Makhro, A., Hegemann, I., Seiler, E., Simionato, G., Claveria, V., Bogdanov, N., et al. (2020). MemSID clinical trial: acute and long-term changes of red blood cells of sickle cell disease patients on memantine treatment. *eJHaem* doi: 10.1002/jha2.11

Makhro, A., Huisjes, R., Verhagen, L. P., Mañú-Pereira, M. M., Llaudet-Planas, E., Petkova-Kirova, P., et al. (2016). Red cell properties after different modes of blood transportation. *Front. Physiol.* 7:288. doi: 10.3389/fphys.2016.00288

Minetti, G., Egée, S., Mörsdorf, D., Steffen, P., Makhro, A., Achilli, C., et al. (2013). Red cell investigations: art and artefacts. *Blood Rev.* 27, 91–101. doi: 10.1016/j.blre.2013.02.002

Moura, P. L., Hawley, B. R., Dobbe, J. G. G., Streekstra, G. J., Rab, M. A. E., Bianchi, P., et al. (2019). PIEZO1 gain-of-function mutations delay reticulocyte maturation in hereditary xerocytosis. *Haematologica* [Epub ahead of print]. doi: 10.3324/haematol.2019.231159

Mukherjee, M. B., Colah, R. B., Mehta, P. R., Shinde, N., Jain, D., Desai, S., et al. (2020). Multicenter Evaluation of HemoTypeSC as a point-of-care sickle cell disease rapid diagnostic test for newborns and adults across India. *Am. J. Clin. Pathol.* 153, 82–87. doi: 10.1093/ajcp/aqz108

Nankanja, R., Kadhumbula, S., Tagoola, A., Geisberg, M., Serrao, E., and Balyegyusa, S. (2019). HemoTypeSC Demonstrates >99% field accuracy in a sickle cell disease screening initiative in children of Southeastern Uganda. *Am. J. Hematol.* 94, E164–E166. doi: 10.1002/ajh.25458

Nguyen-Khoa, T., Mine, L., Allaf, B., Ribeil, J. A., Remus, C., Stanislas, A., et al. (2018). Sickle SCAN™ (BioMedomics) fulfills analytical conditions for neonatal screening of sickle .cell disease. *Ann. Biol. Clin.* 76, 416–420. doi: 10.1684/abc.2018.1354

Pandey, C. M., Augustine, S., Kumar, S., Kumar, S., Nara, S., Srivastava, S., et al. (2018). Microfluidics based point-of-care diagnostics. *Biotechnol. J.* 13:1700047.

Petkova-Kirova, P., Hertz, L., Danielczok, J., Huisjes, R., Makhro, A., Bogdanova, A., et al. (2019). Red blood cell membrane conductance in hereditary haemolytic anaemias. *Front. Physiol.* 10:386. doi: 10.3389/fphys.2019.00386

Rab, M. A. E., Kanne, C. K., Bos, J., Boisson, C., van Oirschot, B. A., Nader, E., et al. (2020). Methodological aspects of the oxygenscan in sickle cell disease: a need for standardization. *Am. J. Hematol.* 95, E5–E8. doi: 10.1002/ajh.25655

Rotordam, G. M., Fermo, E., Becker, N., Barcellini, W., Brüggemann, A., Fertig, N., et al. (2019). A novel gain-of-function mutation of Piezo1 is functionally affirmed in red blood cells by high-throughput patch clamp. *Haematologica* 104:e181. doi: 10.3324/haematol.2018.201160

Steele, C., Sinski, A., Asibey, J., Hardy-Dessources, M. D., Elana, G., Brennan, C., et al. (2019). Point-of-care screening for sickle cell disease in low-resource settings: a multi-center evaluation of HemoTypeSC, a novel rapid test. *Am. J. Hematol.* 94, 39–45. doi: 10.1002/ajh.25305

Tarasev, M., Muchnik, M., Light, L., Alfano, K., and Chakraborty, S. (2016). Individual variability in response to a single sickling event for normal, sickle cell, and sickle trait erythrocytes. *Transl. Res.* 181, 96–107. doi: 10.1016/j.trsl. 2016.09.005

Thiel, C. S., Tauber, S., Lauber, B., Polzer, J., Seebacher, C., Uhl, R., et al. (2019). Rapid morphological and cytoskeletal response to microgravity in human primary macrophages. *Int. J. Mol. Sci.* 20:E2402. doi: 10.3390/ijms201 02402

Ung, R., Alapan, Y., Hasan, M. N., Romelfanger, M., He, P., Tam, A., et al. (2015). Point-of-care screening for sickle cell disease by a mobile micro-electrophoresis platform. *Blood* 126:3379.

Vives Corrons, J. L., Manu Pereira, M. D. M., Casabona, C. R., Nicolas, P., Gulbis, B., Eleftheriou, A., et al. (2014). The ENERCA white book - recommendations

for centres of expertise in rare anaemias. *Thalassemia Rep.* 4, 86–90. doi: 10. 4081/thal.2014.4878

Weidmann, M., Faye, O., Faye, O., Abd El Wahed, A., Patel, P., Batejat, C., et al. (2018). Development of mobile laboratory for viral hemorrhagic fever detection in Africa. *J. Infect. Dis.* 218, 1622–1630. doi: 10.1093/infdis/jiy362

Zaninoni, A., Fermo, E., Vercellati, C., Consonni, D., Marcello, A. P., Zanella, A., et al. (2018). Use of laser assisted optical rotational cell analyzer (LoRRca MaxSis) in the diagnosis of RBC membrane disorders, enzyme defects, and congenital dyserythropoietic anemias: a monocentric study on 202 patients. *Front. Physiol.* 9:451. doi: 10.3389/fphys.2018. 00451

19

Rapid Gardos Hereditary Xerocytosis Diagnosis in 8 Families using Reticulocyte Indices

Véronique Picard[1,2], Corinne Guitton[3], Lamisse Mansour-Hendili[4], Bernard Jondeau[1], Laurence Bendélac[1], Maha Denguir[5], Julien Demagny[6], Valérie Proulle[1], Frédéric Galactéros[7] and Loic Garçon[1,6]*

[1] Service d'Hématologie Biologique, Assistance Publique-Hôpitaux de Paris, Hôpital Bicêtre, Hôpitaux Universitaires Paris-Saclay, Le Kremlin-Bicêtre, France, [2] Faculté de Pharmacie, Université Paris-Saclay, Châtenay-Malabry, France, [3] Service de Pédiatrie Générale, Assistance Publique-Hôpitaux de Paris, Hôpital Bicêtre, Filière MCGRE, Hôpitaux Universitaires Paris-Saclay, Le Kremlin-Bicêtre, France, [4] Department of Molecular Genetics, Assistance Publique-Hôpitaux de Paris, Henri Mondor University Hospital, Créteil, France, [5] Service de Biochimie, Assistance Publique-Hôpitaux de Paris, Bicêtre Hospital, Le Kremlin-Bicêtre, France, [6] Service d'Hématologie Biologique, CHU Amiens, EA 4666 HEMATIM-UPJV, Amiens, France, [7] Centre de Référence des Syndromes Drépanocytaires Majeurs, Hôpital Henri-Mondor, AP-HP, Créteil, France

***Correspondence:**
Véronique Picard
veronique.picard@aphp.fr

Gardos channelopathy (Gardos-HX) or type 2 stomatocytosis/xerocytosis is a hereditary hemolytic anemia due to mutations in the *KCNN4* gene. It is rarer than inherited type 1 xerocytosis due to *PIEZO1* mutations (Piezo1-HX) and its diagnosis is difficult given the absence of a specific clinical or biological phenotype. We report here that this diagnosis can be sped up using red blood cell (RBC) indices performed on an ADVIA 2120 (Siemens®) analyzer, which measures reticulocyte mean corpuscular volume (rMCV) and mean corpuscular hemoglobin concentration (rMCHC). We studied reticulocyte indices in 3 new and 12 described patients (8 families) with Gardos-HX, 11 subjects presented the recurrent p.Arg352His mutation, 4 cases (two families) carried a private *KCNN4* mutation. They were compared to 79 described patients (49 families) with Piezo1-HX. Surprisingly, in Gardos-HX cases, rMCV revealed to be smaller than MCV and rMCHC higher than MCHC, in contrast with normal or Piezo1-HX RBC. Consequently, ΔMCV (rMCV-MCV) was -0.9 ± 5 fL vs. 19.8 ± 3 fL ($p < 0.001$) in Gardos compared with Piezo1-HX and ΔMCHC (rMCHC-MCHC) was 18.7 ± 13 vs. -50 ± 8.7 g/L ($p < 0.001$). A threshold of 8.6 fL for ΔMCV and -5.5 g/L for ΔMCHC could discriminate between Gardos and Piezo1-HX with 100% sensitivity and specificity, regardless of age, mutation or splenectomy status. Consequently, we showed that reticulocytes indices are useful to suggest Gardos-HX on blood count results, allowing to rapidly target these patients for gene analysis. In addition, these parameters may prove useful as a 'functional tool' in interpreting new *KCNN4* variants.

Keywords: reticulocytes, Piezo1, xerocytosis, Gardos, red cell indices

Gardos hereditary xerocytosis (Gardos-HX) is the most recently described hereditary hemolysis, also known as dehydrated stomatocytosis type II, or Gardos channelopathy (Glogowska et al., 2015; Rapetti-Mauss et al., 2015). Hereditary xerocytosis (HX) are rare dominant red cell membrane disorders initially characterized by K^+ leak leading to decreased intracellular cationic content, loss of water and red cell dehydration (Delaunay, 2004). In most cases, HX is associated with gain of function heterozygous mutations in *PIEZO1* (Piezo1-HX). Piezo1 is a mechanosensitive cation channel that translates a mechanical force into a biological signal, mainly through a Ca^{2+} influx into red cells (Zarychanski et al., 2012; Albuisson et al., 2013; Andolfo et al., 2013). Gardos-HX is caused by heterozygous activating mutations in *KCNN4*, encoding the Ca^{2+}-dependent K^+ exporter Gardos channel. Functional studies have suggested that Gardos channel activation is the common effector of red cell dehydration, triggered either directly by *KCNN4* activating mutation or indirectly by Piezo1-dependant intracellular Ca^{2+} increase (Rapetti-Mauss et al., 2017; Caulier et al., 2018).

Gardos-HX diagnosis is difficult and often delayed because of the absence of typical clinical and biological phenotype. Indeed, it presents as a chronic hemolysis with negative red cell phenotypic investigations including osmolar gradient ektacytometry, which is normal or not specific. Diagnosis is made by genetic analysis, often performed after ruling out many other causes of hemolysis (King et al., 2015). It is important to distinguish Piezo1- and Gardos-HX because of several distinct clinical issues: non-spherocytic chronic hemolysis and risk of iron overload are common to both disorders, however, post-splenectomy thrombotic events and perinatal edemas without anemia are observed in Piezo1-HX, whereas anemia –including pre/neonatal anemia – are more severe in Gardos-HX (Fermo et al., 2017; Andolfo et al., 2018; Picard et al., 2019).

We report here that diagnosis of Gardos-channelopathy can be substantially sped up using reticulocyte indices. In a previous retrospective study including 12 Gardos and 91 Piezo1-HX cases, we have already described red cell parameters – Hb level, reticulocytes count, mean cell volume (MCV), mean cell hemoglobin concentration (MCHC) – in Gardos and Piezo1-HX (Picard et al., 2019). Specifically, we showed that MCV was not significantly different in both disorders, however, MCHC, which is in the normal range in subjects with Gardos channelopathy, was significantly higher in Piezo1-HX (Picard et al., 2019). These observations were also reported by others, they are consistent with data from osmotic gradient ektacytometry assays since these curves show a characteristic left-shifted dehydrated profile in Piezo1-HX but not in Gardos-HX indicating the absence of clear red cell dehydration features (Rapetti-Mauss et al., 2015; Picard et al., 2019). Here, we have extended these observations by analyzing reticulocyte indices in an enlarged series of 15 KCNN4-mutated cases (three new subjects, 12 already described) aged 1–59 years from eight families (two new families), that were compared to 79 Piezo1-HX subjects (49 families) with complete blood count records from the described cohort (Picard et al., 2019). Our primary objective was to better define how red cell and/or reticulocyte indices could be used in diagnosing HX. All patients gave their informed consent according to the Helsinki protocol, and this report followed the French regulations in terms of non-interventional retrospective study. HX diagnosis was based on clinical and biological data and a typical ektacytometry curve for the 79 Piezo1-HX patients, all of them carried at least one rare (MAF < 0.01) *PIEZO1* mutation, this mutation may be already described as associated with HX or not. In the 15 Gardos-HX patients, the ektacytometric profile was normal ($n = 10$) or atypical ($n = 5$) and therefore not contributive for diagnosis,

genetic testing identified a *KCNN4* mutation that segregated with the disease in the eight families. Six Gardos-HX families (12 patients) were already described, one patient was found to carry a new KCNN4 mutation (c.965C > T, p.Ala322Val), in addition, three unreported subjects from two novel families were found to carry the recurrent p.Arg352His mutation. Overall, 11 cases from six families carried the recurrent p.Arg352His substitution and 4 cases from two families carried private KCNN4 mutations. The clinical, biological and genetic characteristics of KCNN4-mutated cases are summarized in **Supplementary Table 1**. All subjects had a complete blood count performed in the same lab on EDTA-blood samples within a 24 h/4°C delay

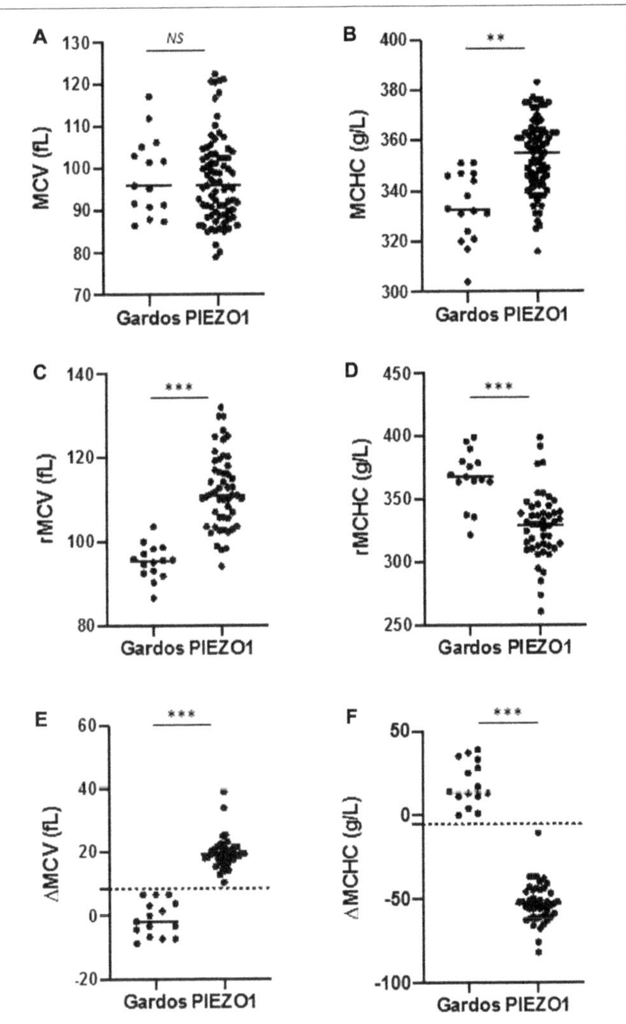

FIGURE 1 | Red cell and reticulocytes indices in Piezo1- and Gardos-HX. **(A,B)** Comparison of MCV (fL) and MCHC (g/L) of all red blood cells (mature + reticulocytes) in Piezo1-HX ($n = 79$) and Gardos-HX ($n = 15$). **(C,D)** Comparison of MCV (fL) and MCHC (g/L) of reticulocytes in Piezo1-HX ($n = 79$) and Gardos-HX ($n = 15$). **(E,F)** ΔMCV (rMCV – mMCV) and ΔMCHC (rMCHC – mMCHC) parameters could discriminate between Gardos and Piezo1-HX, with cut-off values of 8.6 fL and –5.5 g/L, respectively (dashed line). *NS*, non-significant; **$p < 0.01$, ***$p < 0.001$ as analyzed by Student t test. Normal range: MCV: 80–100 fL, MCHC: 310–360 g/L, rMCV: 92–120 fL, rMCHC: 270–330 g/L.

TABLE 1 | Comparison of red cell, reticulocyte and differential indices in splenectomized vs. non-splenectomized Gardos and Piezo1-HX cases (mean ± SEM, p: student T test).

	Normal range	Gardos-HX (n = 15)			Piezo1-HX (n = 79)		
		Splenectomy			Splenectomy		
		Yes	No	p	Yes	No	p
N		7	8		8	71	
Hb (g/L)	120–160 (female) 130–180 (male)	91 ± 8.7	112 ± 1.1	< 0.05	123.7 ± 17	124.8 ± 12	0.24
Reticulocyte (G/L)	20–120	341 ± 15	174 ± 62	0.07	284 ± 135	307 ± 100	0.78
MCV	80–100	107 ± 4.2	92 ± 2.7	< 0.001	103 ± 11	96 ± 7.1	0.25
rMCV (fL)	92–120	96.8 ± 5.3	93.6 ± 3.8	0.27	114 ± 10	111 ± 7	0.66
mMCHC (g/L)	310–360	343 ± 12	353 ± 23	0.5	365 ± 37	370 ± 21	0.84
rMCHC (g/L)	270–330	372 ± 13	364 ± 17	0.1	305 ± 29	334 ± 18	0.1
ΔMCV (fL)	–	−5.8 ± 1.9	3.5 ± 2.5	< 0.001	16.3 ± 6	19.6 ± 6	0.05
ΔMCHC (g/L)	–	28 ± 9	10 ± 6	< 0.001	−65.4 ± 14	−52.4 ± 8.6	0.21

In Gardos-HX, ΔMCV was significantly lower and MCV and ΔMCHC were significantly higher in splenectomized subjects, in splenectomized or non-splenectomized cases, the differential indices remained discriminant for Gardos or Piezo1-HX.

after blood harvesting using an ADVIA2120 (Siemens®) analyzer, that provided specialized red cell and reticulocyte parameters. The ADVIA2120 analyzer measures MCV and MCHC using single cell analysis by dual angle laser scattering cytometry. On one channel, MCV and MCHC indices are measured on red cells, including reticulocytes as well as mature cells. On a second channel using a different set of reagents, reticulocyte counts and indices (rMCV, rMCHC) are distinguished from mature red cells (mMCV, mMCHC) by using oxazine 750 staining. All statistical analyses were performed using two-tailed p value and parametric tests. Statistical significance used was α = 0.05. For quantitative variables, we used Student's t-test or one-way ANOVA test and Tukey *post hoc* analysis for multiparametric analysis. All numeric values were expressed as mean values ± SEM.

As already described, mean hemoglobin level was lower in Gardos vs. Piezo1-HX (103 ± 16 vs. 134 ± 19 g/L, respectively), MCV was not significantly different (98.8 ± 9 vs. 97.7 ± 8 fL), and MCHC was lower in Gardos compared to Piezo1-HX subjects (332 ± 13 vs. 354 ± 24 g/L, Student's t-test, p < 0.01) (**Figures 1A,B**; Picard et al., 2019). Now, a focus on reticulocytes indices indicated that, quite surprisingly, Gardos-HX reticulocytes appeared smaller than expected compared with total red cells (95.1 ± 3 vs. 98.8 ± 9 fL) and their MCHC was higher (367 ± 16 vs. 332 ± 13 g/L). This was not the case in Piezo1-HX, where, as expected, reticulocytes were larger than red cells with a lower MCHC (113.7 ± 7 vs. 97.7 ± 8 fL and 329 ± 20 vs. 354 ± 24 g/L). Therefore, Gardos-HX reticulocytes appeared significantly smaller and had a significantly higher MCHC compared with Piezo1-HX (95.1 ± 3 vs. 113.7 ± 7 fL and 367 ± 16 vs. 329 ± 20 g/L, respectively) (**Figures 1C,D**). As a consequence, the differential MCV between reticulocytes and mature red cells ΔMCV (rMCV – mMCV) was −0.9 ± 5 in Gardos vs. 19.8 ± 3 fL in Piezo1-HX (p < 0.001), and the ΔMCHC (rMCHC – mMCHC) was 18.7 ± 3 in Gardos vs. −50 ± 8.7 g/L in Piezo1-HX (p < 0.001) (**Figures 1E,F**). We then used ROC curves to analyze the performance of these reticulocyte parameters in differentiating Gardos or Piezo1-mutated cases.

The ΔMCHC and ΔMCV were found to be the best parameters (**Supplementary Figure 1**). Indeed, a value of 8.6 fL for ΔMCV and −5.5 g/L for ΔMCHC could discriminate both genotypes with a 100% sensitivity and 100% specificity with no overlap in this small series (**Figures 1E,F**). Because of the low number of Gardos-HX cases, these parameters should be further tested and validated in other labs before use in routine testing. Interestingly, these observations revealed correct for each individual subject, whatever the age, whether they carry the p.Arg352His recurrent mutation or a private mutation. We have not tested subjects carrying the p.Val282Met or p.Val282Glu substitutions, both associated with Gardos-HX, reticulocyte indices in these patients deserve to be investigated (Glogowska et al., 2015).

Then, since 7/15 Gardos vs. 8/79 Piezo1-HX cases were splenectomized, we asked whether splenectomy might influence red cell indices. As shown in **Table 1**, there was no

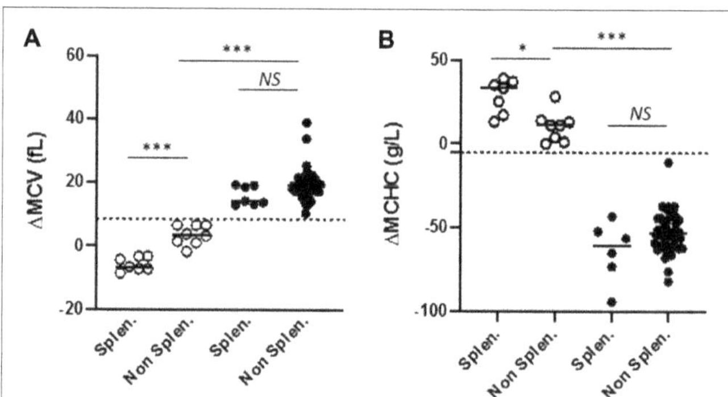

FIGURE 2 | Comparison of ΔMCV **(A)** and ΔMCHC **(B)** in splenectomized and non-splenectomized patients in Gardos-HX (o) and Piezo1-HX (•). The cut-off value of 8.6 fL and –5.5 g/L, respectively are indicated (dashed line). These differential indices could discriminate between Gardos and Piezo1-HX whether subjects were splenectomized or not. NS, non-significant; *p < 0.05, ***p < 0.001 as measured by one-way ANOVA test and Tukey *post hoc* analysis for multiparametric analysis.

significant difference in terms of rMCV and rMCHC between splenectomized and non-splenectomized subjects in both disorders. We observed that Gardos-HX splenectomized subjects had an increased MCV compared to non-splenectomized subjects (MCV: 107 ± 4 vs. 92 ± 3 fL, $p < 0.001$). Nonetheless, ΔMCV and ΔMCHC remained highly suggestive of Gardos-HX whether patients were splenectomized or not (**Table 1** and **Figure 2**). In our setting, these new parameters revealed highly useful and reliable in order to target patients for *KCNN4* gene analysis as soon as the blood count was performed.

Reticulocyte indices are not widely used in diagnosis, although reticulocytes hemoglobin content measured by most hematology analyzers has proved to be helpful in diagnosis of iron deficient anemia (Brugnara et al., 2006; David et al., 2006). Reticulocytes are immature cells, their volume is larger compared with mature red cells and their hemoglobin concentration lower, consistent with rMCV and rMCHC reference range provided by the manufacturer (rMCV: 92.0–120.0 fL, rMCHC: 270–330 g/L in adults) (Thomas, 2008). Currently, most blood cell analyzers do not measure rMCV and rMCHC. Whether red cell indices obtained on other systems using different optic or impedance-based technologies might allow to target Gardos-HX should be evaluated. The ADVIA® technology measures individual red cell volume and hemoglobin content by cell light scattering after isovolumetric spherization. In this setting, Gardos-HX reticulocytes appeared small and dehydrated, they react differently from reticulocytes in Piezo1-HX or in other hemolytic anemia, acquired or hereditary, whether they are related to hemoglobin, enzyme or other membrane defects. This specific behavior of Gardos-HX reticulocytes might be due to an ionic imbalance induced by the *KCNN4* mutation *per se* or due to other differences in the membrane structure or composition. A difference in Gardos channel concentration at the membrane in reticulocytes compared with mature red cells might also account for this observation, to our knowledge, this point has not been studied before. Alternately, the interaction of the Gardos channel with proteins present at a different level in reticulocytes may be involved. Finally, we cannot rule out a possible *in vitro*

artifact due to the ADVIA® technology. Although we cannot infer that this *in vitro* observation does reflect reticulocytes properties *in vivo*, we hypothesize that dehydration predominates in immature cells, i.e., reticulocytes in Gardos-HX while mild or absent in mature red cells. Because reticulocytes represent a minor fraction of circulating red cells, this would account for the normal MCHC reported by several teams in Gardos-HX as well as the non-dehydrated ektacytometry profile. In Piezo1-HX, we propose that dehydration predominates in mature red cells and accounts for the observed high MCHC and dehydrated ektacytometric profile.

In conclusion, these data revealed a new feature distinguishing Gardos and Piezo1-HX. That a simple analysis of red cell and reticulocyte indices on a blood count were able to target each individual Gardos-HX subject was quite unexpected. Although the cause of this specific behavior of Gardos-HX remains to be established, it provides a very useful and cost-free tool for labs using ADVIA analyzers in order to suggest Gardos-HX and speed up diagnosis by *KCNN4* gene analysis. In addition, ΔMCV and ΔMCHC may prove useful as a "functional tool" in interpreting new *KCNN4* variants when Gardos-HX is suspected.

AUTHOR CONTRIBUTIONS

VPi and LG designed the study and wrote the manuscript. LB and LM-H performed genetic studies. VPi, MD, VPr, BJ, and JD analyzed the data. LG, CG, and FG followed patients. All authors contributed to the article and approved the submitted version.

ACKNOWLEDGMENTS

We wish to thank the patients, their family, the clinicians, and the French Cohort of Hereditary Stomatocytosis, as well as the technicians and biologists of the Hematology Laboratory of Hôpital Bicêtre and Mme Hélène Ponsin, for excellent resource management, and ARFH (Association Recherche et Formation en Hematopathologie) for financial support.

REFERENCES

Albuisson, J., Murthy, S. E., Bandell, M., Coste, B., Louis-Dit-Picard, H., Mathur, J., et al. (2013). Dehydrated hereditary stomatocytosis linked to gain-of-function mutations in mechanically activated PIEZO1 ion channels. *Nat. Commun.* 4:1884.

Andolfo, I., Alper, S. L., De Franceschi, L., Auriemma, C., Russo, R., De Falco, L., et al. (2013). Multiple clinical forms of dehydrated hereditary stomatocytosis arise from mutations in PIEZO1. *Blood* 121, 3925–3935. doi: 10.1182/blood-2013-02-482489

Andolfo, I., Russo, R., Rosato, B. E., Manna, F., Gambale, A., Brugnara, C., et al. (2018). Genotype-phenotype correlation and risk stratification in a cohort of 123 hereditary stomatocytosis patients. *Am. J. Hematol.* 93, 1509–1517. doi: 10.1002/ajh.25276

Brugnara, C., Schiller, B., and Moran, J. (2006). Reticulocyte hemoglobin equivalent (Ret He) and assessment of iron-deficient states. *Clin. Lab. Haematol.* 28, 303–308. doi: 10.1111/j.1365-2257.2006.00812.x

Caulier, A., Rapetti-Mauss, R., Guizouarn, H., Picard, V., Garçon, L., and Badens, C. (2018). Primary red cell hydration disorders: pathogenesis and diagnosis. *Int. J. Lab. Hematol.* 40, 68–73. doi: 10.1111/ijlh.12820

David, O., Grillo, A., Ceoloni, B., Cavallo, F., Podda, G., Biancotti, P. P., et al. (2006). Analysis of red cell parameters on the Sysmex XE 2100 and ADVIA 120 in iron deficiency and in uraemic chronic disease. *Scand. J. Clin. Lab. Invest.* 66, 113–120. doi: 10.1080/00365510500406910

Delaunay, J. (2004). The hereditary stomatocytoses : disorders of the red cell membrane permeability to monovalent cations. *Semin. Hematol.* 41, 165–172. doi: 10.1053/j.seminhematol.2004.02.005

Fermo, E., Bogdanova, A., Petkova-Kirova, P., Zaninoni, A., Marcello, A. P., Makhro, A., et al. (2017). 'Gardos Channelopathy': a variant of hereditary Stomatocytosis with complex molecular regulation. *Sci. Rep.* 7:1744.

Glogowska, E., Lezon-Geyda, K., Maksimova, Y., Schulz, V. P., and Gallagher, P. G. (2015). Mutations in the Gardos channel (KCNN4) are associated with hereditary xerocytosis. *Blood* 126, 1281–1284. doi: 10.1182/blood-2015-07-657957

King, M. J., Garçon, L., Hoyer, J. D., Iolascon, A., Picard, V., Stewart, G., et al. (2015). International Council for Standardization in Haematology. ICSH guidelines for the laboratory diagnosis of nonimmune hereditary red cell membrane disorders. *Int. J. Lab. Hematol.* 37, 304–325. doi: 10.1111/ijlh.12335

Picard, V., Guitton, C., Thuret, I., Rose, C., Bendelac, L., Ghazal, K., et al. (2019). Clinical and biological features in *PIEZO1* -hereditary xerocytosis and

Gardos channelopathy: a retrospective series of 126 patients. *Haematologica* 104, 1554–1564. doi: 10.3324/haematol.2018.205328

Rapetti-Mauss, R., Lacoste, C., Picard, V., Guitton, C., Lombard, E., Loosveld, M., et al. (2015). A mutation in the Gardos channel is associated with hereditary xerocytosis. *Blood* 126, 1273–1280. doi: 10.1182/blood-2015-04-642496

Rapetti-Mauss, R., Picard, V., Guitton, C., Ghazal, K., Proulle, V., Badens, C., et al. (2017). Red blood cell Gardos channel (KCNN4): the essential determinant of erythrocyte dehydration in hereditary xerocytosis. *Haematologica* 102, e415–e418.

Thomas, L. (2008). *Labor und Diagnose, 7. Auflage.* Frankfurt: TH-books Verlagsgesellschaft mbH.

Zarychanski, R., Schulz, V. P., Houston, B. L., Maksimova, Y., Houston, D. S., Smith, B., et al. (2012). Mutations in the mechanotransduction protein PIEZO1 are associated with hereditary xerocytosis. *Blood* 120, 1908–1915. doi: 10.1182/blood-2012-04-422253

20

Heterogeneity of Red Blood Cells: Causes and Consequences

Anna Bogdanova[1]*, Lars Kaestner[2,3], Greta Simionato[2,4], Amittha Wickrema[5] and Asya Makhro[1]

[1] Red Blood Cell Research Group, Vetsuisse Faculty, The Zurich Center for Integrative Human Physiology (ZHIP), Institute of Veterinary Physiology, University of Zurich, Zurich, Switzerland, [2] Experimental Physics, Dynamics of Fluids, Faculty of Natural Sciences and Technology, Saarland University, Saarbrücken, Germany, [3] Theoretical Medicine and Biosciences, Medical Faculty, Saarland University, Homburg, Germany, [4] Institute for Clinical and Experimental Surgery, Saarland University, Homburg, Germany, [5] Section of Hematology/Oncology, Department of Medicine, University of Chicago, Chicago, IL, United States

*Correspondence:
Anna Bogdanova
annab@access.uzh.ch

Mean values of hematological parameters are currently used in the clinical laboratory settings to characterize red blood cell properties. Those include red blood cell indices, osmotic fragility test, eosin 5-maleimide (EMA) test, and deformability assessment using ektacytometry to name a few. Diagnosis of hereditary red blood cell disorders is complemented by identification of mutations in distinct genes that are recognized "molecular causes of disease." The power of these measurements is clinically well-established. However, the evidence is growing that the available information is not enough to understand the determinants of severity of diseases and heterogeneity in manifestation of pathologies such as hereditary hemolytic anemias. This review focuses on an alternative approach to assess red blood cell properties based on heterogeneity of red blood cells and characterization of fractions of cells with similar properties such as density, hydration, membrane loss, redox state, Ca^{2+} levels, and morphology. Methodological approaches to detect variance of red blood cell properties will be presented. Causes of red blood cell heterogeneity include cell age, environmental stress as well as shear and metabolic stress, and multiple other factors. Heterogeneity of red blood cell properties is also promoted by pathological conditions that are not limited to the red blood cells disorders, but inflammatory state, metabolic diseases and cancer. Therapeutic interventions such as splenectomy and transfusion as well as drug administration also impact the variance in red blood cell properties. Based on the overview of the studies in this area, the possible applications of heterogeneity in red blood cell properties as prognostic and diagnostic marker commenting on the power and selectivity of such markers are discussed.

Keywords: red blood cells, heterogeneity, morphology, erythroid precursor cells, age

INTRODUCTION

Our understanding of red blood cells (RBCs) evolved from acknowledgment of the basic and fundamental role of these cells as key players in gas exchange to the state where we assign multiple complex functions related to sensing and signaling, maintenance of homeostasis of pH and redox state and participation in control of vascular tone, clotting (Andrews and Low, 1999; Bernhardt et al., 2019), and other processes (Helms et al., 2018; Pernow et al., 2019).

Broadening of RBC functions was accompanied with our awareness of complexity of the cellular architecture and biochemistry. Spatial compartmentalization of processes and resources in RBCs was discovered (Hoffman et al., 2009; Chu et al., 2012). Complex dynamics precise orchestration of processes occurring in the circulating RBCs in response to the changes in micro- and macro-environment (hormonal and mechanical stimulation, changes in local or ambient oxygen availability, temperature, circadian rhythm-related processes and others) is becoming evident (e.g., O'Neill and Reddy, 2011; Cahalan et al., 2015; Zhou et al., 2019).

With time it became clear that these changes and responses do not necessarily involve all the circulating cells. As our knowledge of these cells accumulates more and more reports mention the presence of "responding" and "non-responding" cells in the circulation (e.g., Kaestner et al., 2012; Makhro et al., 2013; Wang et al., 2013; Rotordam et al., 2019). As we recognize the existence of multiple fractions of RBCs that are functionally different from each other, we feel a growing need to unravel the nature of these differences, their causes and the potential information hidden in RBC heterogeneity on systemic distress and pathology. In this review we aimed to summarize the current state of knowledge in this rapidly developing research area. We focus on RBCs of healthy humans and give only a few examples of how RBC heterogeneity may be used to predict RBC disease nature and severity. Heterogeneity of stored or transfused RBCs is a broad topic also out of the scope of this review.

INTER-INDIVIDUAL HETEROGENEITY

Inter-individual variation in properties of circulating RBCs of healthy donors reflects genetic and epigenetic variance as well as the state in which the organism resides over the past 3–4 months during which the cells undergo transitions from erythroid progenitors to young, mature, and senescent state. Variance spreads to the number of copies of proteins per cell, activity of enzymes and ion transporters, shapes, differences in density, deformability, membrane stability, redox state, and the collection of Hb variants in a given cell. Most of the studies for healthy humans were performed on stored blood to assess its quality and identify a cohort of best donors (Sparrow, 2017).

The possible causes of this inter-individual variation originate at the level of erythroid precursor cells (as in case of ineffective erythropoiesis (Oikonomidou and Rivella, 2018), for more details see section Cellular Heterogeneity During Erythropoiesis) or emerge later on as the cells enter the circulation and get exposed to a variety of microenvironments (osmolarity gradients in the kidneys, shear in capillaries and spleen, changes in oxygen availability and pH within peripheral tissues, changes in redox state next to the inflammatory side or to the exercising muscle). Development of heterogeneous RBC populations may be an intrinsic property of blood (e.g., RBC aging), or be triggered by the changes in life style or environmental conditions (e.g., hypoxia, microgravity) or state of the organism (e.g., stress, inflammation, changes in dietary preferences and blood metabolites). Finally, it may result from hereditary diseases

that destabilize the RBC membrane or perturb its rheological properties, redox or metabolic state. In this review we focus on the possible physiological causes of heterogeneity.

PARAMETERS SHOWING INTER-CELLULAR HETEROGENEITY AND METHODS TO DETECT THEM

Table 1 summarizes the information on the parameters displaying inter-individual and inter-cellular heterogeneity, as well as methodological approaches for detection of heterogeneity.

Shape and Size

First descriptions of RBCs as "red corpuscles" given by Jan Swammerdam and dates back to 1658 (Swammerdam, 1737; Bessis and Delpech, 1981; Hajdu, 2003). Since then, substantial progress was made in imaging equipment as well as in fixation and staining of RBC. Blood smears still remain a part of common diagnostic practice in most of the clinical laboratories (Bain, 2005) despite the fact that smear preparation results in distortions of RBC morphology and lysis of the most fragile of them (Wenk, 1976). This technique allows to discriminate between numerous shapes from discocytes to a broad variety of "static" shapes such as echinocytes and stomatocytes, for healthy humans. The list of shape types will extend manifolds for patients with hereditary or acute disorders.

The biggest drawback of the whole approach with smears is that it provides an immediate snapshot of the shape distribution, whereas living RBCs are very dynamic entities. So are their shapes, and, rather than discussing their "absolute shape," it would be feasible to assign them a probability to be observed in one of the shape types. The first attempts to address RBC shapes in terms of probability density distribution are recently undertaken (Reichel et al., 2019). Each cell has its "static shape" that is preferred over the other ones if no force is applied to it. There are also several preferred shape types caused by shear stress in flow. The probability to observe one of those depends on the shear rates and flow dynamics (Abkarian et al., 2008; Dupire et al., 2015; Lanotte et al., 2016; Kihm et al., 2018; Mauer et al., 2018; Reichel et al., 2019). The restoration of the initial shape of the cells as soon as the flow stops got the name of "shape memory" (Fischer, 2004; Cordasco and Bagchi, 2017). Acute shape changes associated with the ion movements across the cell membrane (dehydration or overhydration) are often reversible (Brugnara, 1997; Cossins and Gibson, 1997; Zhu et al., 2018) whereas shape alterations related to the permanent damage of the cytoskeleton or membrane loss are irreversible (Gallagher, 2005; Perrotta et al., 2008).

Preferred shapes reflect the optimal cytoskeletal conformation, hemoglobin concentration, redox state and metabolic balance and free Ca^{2+} levels that, in turn, define the activity of ion transporters, hydration state and phosphorylation state of proteins. Some of these variables will be addressed below.

Parameters to describe dynamics of RBCs morphology are currently in development. Former classifications of shapes performed by eye (Bessis and Lessin, 1970; Bessis and Delpech,

TABLE 1 | Overview of parameters showing inter-cellular heterogeneity as well as basic principle and methodological approaches of their detection in single cells and sub-populations.

Parameter	Indicator	Method	References
Shape/size	**Direct:** Shape classification Projected area Perimeter/roughness Sphericity/elongation	Microscopy: Blood smears, images of living cells (snapshots, time series in flow, microfluidics), Imaging flow cytometers	Gonzalez-Hidalgo et al., 2015; Quint et al., 2018; Herold-Garcia and Fernandes, 2019
	Volume	Confocal microscopy + 3D deconvolution	Sadafi et al., 2019
		Scanning probe microscopy (semi quantitative)	Kihm et al., 2018
	Indirect: Forward and side light scatter Impedance (coulter principle)	Flow cytometry Coulter counters Multiple Blood analyzers	
Density	**Direct:** Separation according to RBC density	Fractionation in Percoll-, Stractan or similar density gradients	Lutz et al., 1992
		Lab-on-a-chip approaches	Catarino et al., 2019
	Indirect: Swelling- or shrinkage- resistance (e.g., the changes in SS and FS within the swelling test) Single cell rheology	Flow cytometry Cell-flow properties analyzer	Kaul et al., 2008; Barshtein et al., 2016; Fermo et al., 2017
	Membrane surface/EMA test	Flow cytometry	
Free Ca^{2+}/channel activity	Fluorescent dyes for Ca^{2+} Detection of ionic currents across the membranes of single cells	Flow cytometry and fluorescence microscopy Patch-clamp incl. automated planar chips	Kaestner et al., 2006; Makhro et al., 2013; Wang et al., 2013; Fermo et al., 2017; Rotordam et al., 2019
Redox state and metabolism	Fluorescent dyes for reduced thiols (e.g., thiol tracker, monobromobimane), Fluorescent dyes for N_2O_3 (DAF-DA), Dyes for detection of H_2O_2, ONOO, HO* (e.g., H_2DCF-DA) Single cell metabolomics (not yet used for red blood cells)	Flow cytometry Fluorescence microscopy Mass-spectrometry	Jemaa et al., 2017; Gilmore et al., 2019
Hb levels and variance	Antibodies with fluorescent tags Chromicity Sodium metabisulfite (Na_2S_2O5) and similar deoxygenation-based sickling tests Hemoglobin Distribution Width (HDW)	Flow cytometry, Fluorescence microscopy Microscopy	Kunicka et al., 2001; Darrow et al., 2016; Jung et al., 2016
Age	Labeling of cells (biotin conjugated with fluorescent tag or staining with PKH dyes) Reticulocyte count RNA-positive or Transferrin receptor-positive	Flow cytometry, microscopy	Mock et al., 1999; Piva et al., 2010

1982) are non-numerical and cannot be reliably translated into the algorithms for automated segmentation and classification of smears and images of living cells. New approaches are currently developing (Tomari et al., 2014). Roundness, roughness, projected areas are among such numeric descriptors of RBC shapes. 3D volume reconstruction of, e.g., confocal recordings are more informative than 2-dimentional images. High resolution 3D-imaging was performed for fixed RBC (Abay et al., 2019). First attempts to get the 3D imaging working for RBCs in flow are undertaken but is not yet available as a high-throughput mode (Quint et al., 2018). Cell shape recognition and classification involving artificial intelligence (AI) algorithms based on artificial neural networks (Kihm et al., 2018). New optical concepts using optofluidic microlenses-like behavior of RBCs (Mugnano et al., 2018) and indirect adaptive optics as well as label-free quantitative phase imaging (Miccio et al., 2015) enables assessment of cell volume of individual cells, and monitoring of morphometric features (e.g., label-free optical markers) that make high throughput reliable quantification of cell phenotypes possible. It allows to stay unbiased, omit "human factor," and allocate RBC shapes to a continuous scale with high throughput and precision. The challenge is that artificial neural networks need to be set up, customized and most notably trained. This type of analysis will become available routinely in the nearest future.

Hydration State and Density

The best method to visualize the variance in RBC density is fractionation on a Percoll (**Figure 1**), Ficoll, Stractan, or phtalate density gradient (Danon and Marikovsky, 1964; Corry et al., 1982; Salvo et al., 1982; Mosca et al., 1991; Lutz et al., 1992). Upon centrifugation in isotonic solution of any of these materials forming continuous or discontinuous gradients, RBCs

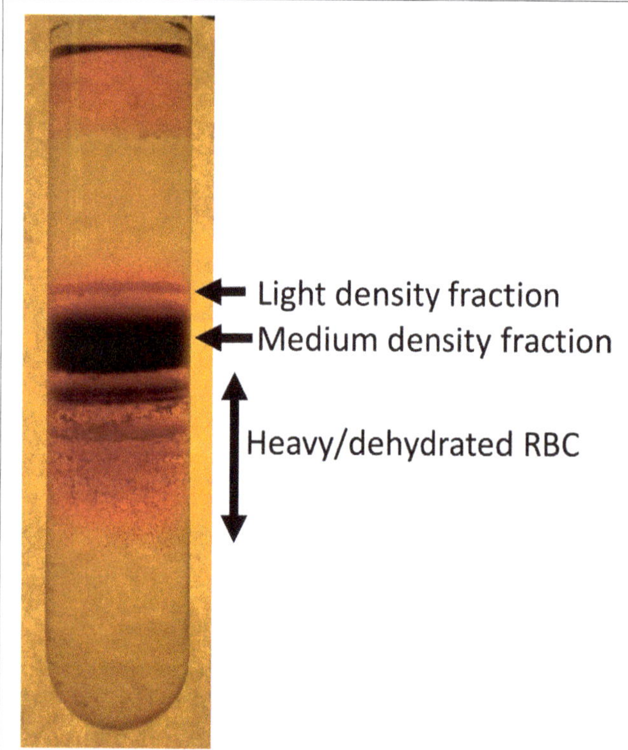

FIGURE 1 | An example of heterogeneity of RBC density revealed by fractionation of RBCs on a self-forming Percoll density gradient. Composition of light, medium and high-density fractions vary depending on human health and environmental stress. About 15–20% of RBCs of healthy human donors forming low density are reticulocytes. However, along with young cells this fraction is "contaminated" with swollen RBCs at the terminal senescence stage (Lew and Tiffert, 2013). Medium fraction is formed by mature RBCs, and heavy dehydrated cells are those with senescent phenotype.

Factors defining RBC density include changes in water and ion content and membrane loss. During the density fractionation RBCs experience shear stress during centrifugation as they move through the isotonic Percoll solution containing micromolar concentrations of Ca^{2+} in the absence of EGTA. Shear forces may activate mechano-sensitive channels such as PIEZO1 channels (Cahalan et al., 2015) and NMDA receptors (Hanggi et al., 2014) that are permeable for Ca^{2+}. Uptake of Ca^{2+} via these receptors triggers loss of K^+ mediated by opening of Ca^{2+}-dependent Gardos channels. Thus, fractionation of RBCs on Percoll should be viewed as a functional test in which distribution of the cells is not only driven by the steady state density, but also by their mechano-sensitivity.

Indirect methods to assess heterogeneity in RBC density include detection of hypo- and hyperchromic cells in blood smears, HDW as well as the shape of the curve in osmotic fragility test of the right arm of the osmoscan curve obtained by ektacytometry (Clark et al., 1983; Lutz et al., 1992). Hight throughput devices for evaluation of RBC density using functional tests at the single cell level are being developed.

Ca^{2+} Levels (Static and Dynamic Tests) and Electrophysiological Properties

Heterogeneity in basal free Ca^{2+} levels was recorded in RBCs of healthy humans (Kaestner et al., 2006; Makhro et al., 2013; Fermo et al., 2017). Stimulation of Ca^{2+} uptake by treatment of healthy human RBCs with PGE_2 (Danielczok et al., 2017), lysophosphatidic acid (Steffen et al., 2011; Kaestner et al., 2012; Wang et al., 2013; Wesseling et al., 2016) or glutamate (Makhro et al., 2013; Hanggi et al., 2014; Makhro et al., 2016a; Petkova-Kirova et al., 2019) increases variance in the intracellular Ca^{2+}. Not all cells respond to shear stress or pro-oxidative condition with an increase in Ca^{2+}.

Molecular causes for this heterogeneity in responses to various stressors are poorly understood. It is obvious, that they relate to the differences in abundance of either Ca^{2+} channels (Kaestner et al., 1999; Makhro et al., 2013; Kaestner and Egee, 2018; Rotordam et al., 2019) or of the primary receptors responding to the stressor (such as LPA or prostaglandin receptors; Wang et al., 2013; Danielczok et al., 2017). In human RBCs several ion channels are known to mediate Ca^{2+} uptake including PIEZO1, TRPC6, NMDA receptors, $Ca_V2.1$ and several others (for a recent review see Kaestner et al., 2020). As a result of stochastic distribution and opening probability, Ca^{2+} entry into individual RBCs varies in response to stimulation by individual Ca^{2+} channels substantially giving rise to "responders" and "non-responders" cellular sup-populations. This uneven behavior may be further amplified due to the existence of feedback loops supporting Ca^{2+}-dependent Ca^{2+} uptake (Kaestner et al., 2018).

Most documented is inter-cellular variance in distribution of the Ca^{2+}-dependent K^+ (Gardos) channel in RBCs. However, majority of the recordings for this best-studied channel in RBCs were performed as mean values for the unseparated populations, using radioactive tracer kinetics technique or single channel recordings. Reports based on whole-cell recordings for this channel are still sparse (Kucherenko et al., 2005; Kucherenko

distribute within them according to their densities. As RBCs of healthy human donors are fractionated on a self-forming Percoll gradient, three to five fractions may be collected. A small fraction of cells with lower density bands as the top layer, followed by one or several RBC populations with a medium density and a minor fraction of cells is presented with the highest density (Lutz et al., 1992; Makhro et al., 2013; Makhro et al., 2016b). RBCs of patients with hereditary hemolytic anemias are generally characterized with a broader variance in densities. Often this diversity may contain clinically relevant information on the severity of disease state. For sickle cell disease the abundance of dense cells was suggested to be a predictor of severity of disease manifestation due to the increased probability of irreversible aggregation of HbS (Kaul et al., 1983). For hereditary spherocytosis severity is associated with an increase in abundance of well-hydrated cells that are lost before they have time to mature and lose some of their membrane (Huisjes et al., 2019). Increase in heterogeneity is high in patients with cryohydrocytosis (Bogdanova et al., 2010), Gardos channelopathy (Fermo et al., 2017) beta-thalassemia, G6PD, and pyruvate kinase (PK) deficiency (Mosca et al., 1991).

et al., 2013; Fermo et al., 2017). A further factor that may amplify heterogeneity of Gardos channel recordings in RBCs is its inactivation upon hypoxic exercises (Mao et al., 2011).

Redox State and Metabolism

Staining of individual cells with fluorescent probes sensitive to pro-oxidative free radicals such as dicarbofluorescein (Amer et al., 2003; Grinberg et al., 2005) and monobromobimane (Kosower and Kosower, 1995) provide a possibility to follow the changes in redox balance in individual cells. One more approach to record redox state in sub-fractions of RBCs is based on pre-fractionation of cells into low, medium and high density fractions before assessment of reduced and oxidized glutathione (GSH and GSSG) and NAD(P)H (Piccinini et al., 1995; D'Alessandro et al., 2013) in these sub-populations. Dense cells were shown to be deprived of GSH and enriched with GSSG compared to the mature RBCs of medium density. Accumulation of GSSG and reduction in GSH was not associated with any substantial changes in the intracellular ATP or NADPH (Sass et al., 1965; D'Alessandro et al., 2013). Finally, redox state of RBCs may be expressed as the ability to tolerate oxidative challenge (Lisovskaya et al., 2008; Sinha et al., 2015) which differs between individual RBCs as well.

Shifts in redox equilibrium in RBCs of healthy donors are associated with age-dependent decrease in pyruvate kinase, hexokinase, glucose-phosphate dehydrogenase, aldolase activities (Salvo et al., 1982; Suzuki and Dale, 1988). Oxidative stress is a hallmark of RBCs of patients with hereditary hemolytic anemias presented with one or two alleles of mutated glucose-6 phosphate dehydrogenase (G6PD). The resulting in acute hemolytic condition known as favism is associated with depletion in NADPH in favor of $NADP^+$ (Mason et al., 2007; Peters and van Noorden, 2017). Furthermore, systemic oxidative stress caused by inflammatory processes, infection and other causes may result in release of reduced glutathione from RBCs and temporary increase in oxidative load and aggravate the differences in redox state between the cells of different ages (Giustarini et al., 2008).

Hb Levels and Variants

Inter-cellular heterogeneity in intracellular hemoglobin content in clinical settings is reflected by the abundance of hypochromic and hyperchromic cells in blood smears. The abundance of hypochrome RBCs for healthy humans should not exceed 2.5% of circulating RBCs (Macdougall et al., 1992; Schaefer and Schaefer, 1995; Braun et al., 1997), dropping below 1% in patients with iron overload, and increasing to 20% and more in patients with iron deficiency. Higher levels of hyperchromic cells was also reported for patients with hereditary spherocytosis (Conway et al., 2002) and sickle cell disease (Ballas and Kocher, 1988).

Even more intercellular heterogeneity is introduced by a pronounced variance in the presence of fetal hemoglobin in a small fraction of cells (F-cells) in healthy humans (Boyer et al., 1975; Thein and Craig, 1998). The abundance of F-cells increases during high altitude exposure (Narayan et al., 2005). Pregnancy has an impact on this parameter (Prus and Fibach, 2013). Moreover, the abundance of F-cells as well as the amount of HbF in them may differ from cell to cell in patients with

beta-thalassemia (Narayan et al., 2005). In sickle cell disease, HbF abundance furthermore strongly depends on the haplotype (Menzel and Thein, 2019). Sickle cell trait results in uneven distribution of HbS between the cells, and the pattern for such variance seems to be hereditary (Anyaibe et al., 1985).

CAUSES OF HETEROGENEITY

If we want to make extensive use of RBC heterogeneity as a diagnostic and prognostic marker, we have to understand the origin of the observed variance in RBC properties.

This may stem from the different pools of erythroid precursor cells that equip the resulting reticulocytes with various sets of proteins that may only be produced as long as the synthesis machinery is active before the enucleation.

The other cause of heterogeneity are the age-dependent differences between the young, mature and senescent RBCs. The third cause occurring at the systemic level originates from the alteration in the micro- and macro-environmental conditions (changes in hormonal and metabolic levels, inflammation, shear stress load, hyperthermia and others).

These three sources of heterogeneity will be reviewed below.

Cellular Heterogeneity During Erythropoiesis

Accumulated evidence over the last 25 years has demonstrated the existence of heterogeneity within the erythroid compartment schematically shown in **Figure 2**. Although it is quite expected to have a heterogenic population within the mature RBC population due to the long- life span of mature RBCs (100–120 days in human) representing cells of various ages, it is less clear and more intriguing the reasons for erythroid precursors/progenitors to be heterogeneous in multiple facets of their form and function.

One of the most established and well-explained aspect of erythroid precursor heterogeneity pertains to erythroid precursors possessing differing sensitivities to erythropoietin (EPO). Soon after discovering the precise molecular function of EPO to be a cell survival function (Koury and Bondurant, 1988, 1990), studies revealed that even within a highly homogenous population in terms of the differentiation stage (operationally defined as colony-forming unit-erythroid; CFU-E), erythroid precursors underwent apoptosis following EPO withdrawal in an asynchronous manner (Kelley et al., 1993). These studies demonstrated a dose response effect as reflected by increasing numbers of CFU-Es undergoing apoptosis as EPO concentrations were gradually decreased. These observations clearly highlighted the built-in heterogeneity within the developing erythroid cell compartment with respect to the biochemical nature of each cell within an otherwise "homogenous" precursor pool as defined by morphological characteristics. One of the possible causes supporting heterogeneity are the gradients in various signaling messengers, growth factors, chemokines, oxygen levels and the resulting reactive oxygen species, and other factors (e.g., Thompson et al., 2010; Spencer et al., 2014; Itkin et al., 2016) making conditions in which precursor cells differentiate unique and dependent on their location within the bone marrow

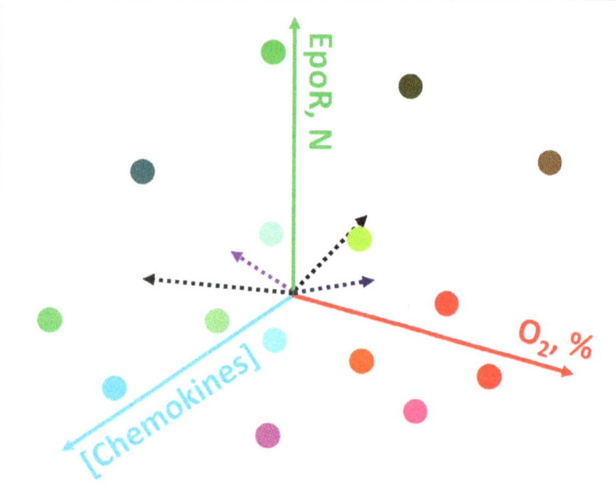

FIGURE 2 | Schematic representation of the possible causes of heterogeneity for erythroid precursor cells within bone marrow. Gradients in oxygen availability, chemokines and other signaling messengers create a plethora of conditions in which cells find themselves during differentiation. For details see the text.

(**Figure 2**). An elegant model proposed by Koury and Bondurant (1992), explained the basis of differing EPO sensitivities as a built-in mechanism to prevent all erythroid precursors undergoing apoptosis during low EPO levels in circulation such as in patients with renal failure. The work by several other groups (Miura et al., 1991; Landschulz et al., 1992; Nakamura et al., 1992; Kelley et al., 1993) had shown that heterogeneic EPO response within the same precursor population cannot be attributed to the numbers of EPO-receptors, affinity or structure, thereby suggesting differences in signal transduction as the likely mechanism for the existence of heterogeneity in EPO response. Based on these findings one can appreciate the existence of signaling heterogeneity within the erythroid precursor compartment as a necessary component during the development process to yield mature red blood cells. Recently developed single-cell intracellular flow cytometry approaches (Liu et al., 2019) are bound to further uncover previously unrecognized levels of regulatory heterogeneity during erythroid cell development.

Besides the existence of biochemical/signaling heterogeneity within the developing erythroid precursors other aspects of erythroid precursor heterogeneity have been observed especially most recently due to the advancement of single-cell technologies at both trascriptomic and phenotypic levels (Woll et al., 2014; La Manno et al., 2018; Brierley and Mead, 2019). Within the erythroid compartment especially during the early stages of erythropoiesis a significant level of transcriptomic variability and heterogeneity seem to exist at least based on mouse bone marrow erythroid precursors (Tusi et al., 2018). The same study also found that cell cycle in erythroid precursors are continuously remodeled during the differentiation program but consistent with very early studies using bulk erythroid precursors (CFU-E), the vast majority of cells were in the S-phase of the cell cycle (Iscove, 1977). These results demonstrate that an individual cell, especially during development, has the ability to program itself to act not in concert each other with respect to signal transduction, gene transcription, cell cycle and many other aspects even though morphologically a cell population may look alike at a particular stage of differentiation.

Overall, accumulated data suggests that heterogeneity during erythroid development may not be evenly spread during the entire development cascade. Most data points to greatest level of inter-cellular heterogeneity during the early phases of development when these cells are responsive to various growth factors. Beyond the late polychromatic stage, when the cells have exited the cell cycle one observes less heterogeneity and most cells undergo dramatic reduction in cell size, chromatin condensation and enucleation. However, it is conceivable even in the bone marrow niche within the blood island not all erythroblasts undergo enucleation adding another layer of heterogeneity. It is also conceivable that due to differing levels of chemokine receptors on these cells the progenitors also exhibit varying degrees of migration within the bone marrow niche. Overall, it may seem the inter-cellular heterogeneity during erythroid precursor development. Each cell is possessing different sensitivity to EPO, and as a result a vast majority of precursors die due to apoptosis, the strategy that seems quite wasteful. However, we speculate that such heterogeneity is critical in order to respond to rapid changes in the micro and macro environment such as changes in oxygen concentration due to changes in altitude, pro-inflammatory and oxidative stress conditions as well as sudden blood loss due to trauma and onset of anemia due to renal failure.

Age of RBCs

Most of the findings for the age-related variance for RBCs of healthy humans were obtained for the fractions of cells of low, medium and high density, that were enriched with young, mature and senescent cells, respectively (Mueller et al., 1985, 1987; Lutz et al., 1992; **Figure 1**). Gradual changes occurring with cell aging were described in several reviews (Lutz and Bogdanova, 2013; Lew and Tiffert, 2017; Badior and Casey, 2018; Minetti et al., 2018) and article collections (Beutler, 1988; Mangani, 1991), and schematically represented in **Figure 3**.

Recent studies of the age-dependent changes in RBCs involve single cell approaches such as flow cytometry and microscopy as well as proteomics (D'Alessandro et al., 2013; Minetti et al., 2013).

Deamidation of asparagine residue 502 of the band 4.1 protein was shown to occur gradually with RBC age as the deamidation rate is an exclusive function of temperature and time (Inaba and Maede, 1988, 1992). Deamidation is manifested as an appearance of a double band on the gels as the native and deamidated form of the protein differ in electrophoretic mobility of the protein. Fractionation of RBCs of healthy humans according to their density has shown that young cells have lower density than mature cells. Senescence is associated with further increase in RBC density and mean corpuscular hemoglobin concentration, and reduction in RBC volume. Using the changes in deamidation of band 4.1 protein or direct labeling of RBCs and monitoring of their aging (Luthra et al., 1979), increase in density were revealed as an intrinsic feature of *in vivo* aging of RBCs of healthy humans. Dense cells obtained by fractionation of leukodepleted RBCs on

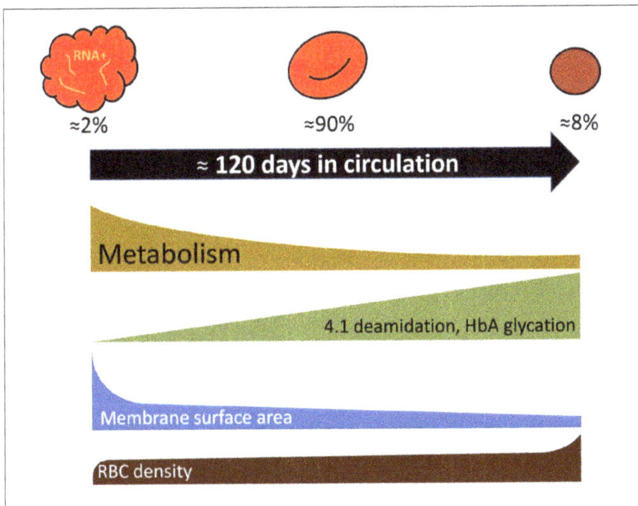

FIGURE 3 | Selected parameters that change during RBC aging following exponential or linear kinetics as cells turn from reticulocytes to mature cells and finally enter the senescent stage. Percentage of reticulocytes, mature cells, and dense senescent cells shown in the scheme correspond to those in adult healthy donors. For more details see Lutz and Bogdanova (2013) and the text.

Percoll density gradient were presented with substantially lower GSH levels and GSSG levels that were doubled compared to the mature RBCs, whereas ATP and NADPH levels were only slightly reduced in the densest cell fractions (Sass et al., 1965; D'Alessandro et al., 2013). These changes were associated with the age-driven decrease in pyruvate kinase, hexokinase, glucose-6-phosphate dehydrogenase, aldolase activities (Salvo et al., 1982; Suzuki and Dale, 1988). Some of the terminally senescent RBCs, that lose control over their Na^+ gradients and volume regulation due to the reduction in Na,K-ATPase activity, were reported to swell and lyse (Lew and Tiffert, 2013, 2017).

Reports on the changes in free Ca^{2+} levels are controversial and depend on the techniques used for assessment of these parameters (Romero and Romero, 1997, 1999; Makhro et al., 2013; Lew and Tiffert, 2017). Both Ca^{2+}-permeable channel activity and that of plasma membrane Ca^{2+} pumps decreases with cellular aging (Romero et al., 2002; Makhro et al., 2013). Despite this inconsistency, changes in the intracellular free Ca^{2+} and the ability to maintain low levels of Ca^{2+} are the factors in control of RBC longevity (Bogdanova et al., 2013; Lew and Tiffert, 2017).

Further hallmarks of RBC aging include the changes in phosphorylation pattern (Fairbanks et al., 1983) and membrane loss (Mohandas and Groner, 1989).

Physical Activity, High Altitude, and Other Stress Conditions

How substantial would the change be at the level of circulating RBCs if the gene expression reprogramming occurs at the level of precursor cells? Simple calculations assuming that the RBC longevity is not affected by these changes and all cells are equally affected by this change, gives a rough estimate of ~0.82% of RBC population changing per day for the "normal" production rate

of 2.4×10^6 cells/s. If erythropoiesis is boosted to its maximum (10-fold increase, 8.2% of new cells will appear daily (Elliott and Molineux, 2009). This means that acute reversible changes at the bone marrow level will hardly be noticed if stress conditions persist for just 24 h. On the contrary, when stress conditions boosting erythropoiesis persist for a week, 5.7–57% of cells will get a new feature.

Such kinetics does not favor *de novo* production as an efficient strategy for acute adaptation to hypoxia or single endurance sport exercise bout, dietary changes, or to pathological conditions such as infection or sepsis, cancer, diabetes, or cardiovascular diseases (**Figure 4**). These changes in turn translate into the changes in shear stress, oxygen availability, pH, hormones and proinflammatory cytokines and other microenvironmental factors sensed by RBC directly. Species that undergo such acute changes from hyperoxygenation to severe hypoxia, such as Rainbow trout (*Oncorhynchus mykiss*) (Fago et al., 2001) or Rüppell's griffon vulture (*Gyps rueppelli*). Rüppell's griffon vulture was spotted at 37,000 feet (11277.6 m) when colliding with the plane (Laybourne, 1974) permanently possess several hemoglobin variants. Hemoglobin A and D chains are present in RBC vulture producing high and low affinity hemoglobin variants and allowing these unique birds to fly above 10,000 m with no need to engage any complex adaptive processes as they land (Weber et al., 1988; Hiebl et al., 1989).

Adult humans have by far lower adaptive capacity, possessing generally one Hb variant, HbA with some minor additions of HbF. However, plasticity of O_2 delivery, and its fast on-demand optimization upon the changes in environmental O_2 availability may be associated with other types of heterogeneity in RBC structure and function. Potential adaptive role of variance in RBC properties has to be further explored.

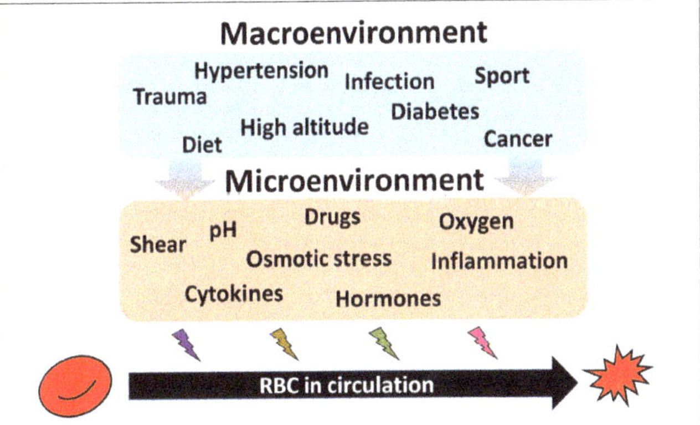

FIGURE 4 | Summary on the environmental causes imposing heterogeneity of circulating RBCs. Exposure of the organism to high altitude or practicing endurance sport as well as dietary preferences cause durable or acute impact on the RBC properties. Along with RBC diseases (anemia, polycythemia), pathologies such as hypertension, diabetes, infection, trauma, cancer, and further systemic diseases are influencing both erythropoietic niche and the circulating cells. All these macroenvironmental stresses translated into the changes in microenvironment for erythroid precursors and circulating RBCs. Shear, alterations in pH and oxygen levels, proinflammatory cytokines, and hormones, as well as drugs work to shape the features of each individual RBC resulting in an increase in the inter-cellular heterogeneity.

It is largely accepted that multiple forms of pathologies, both related to abnormal structure RBC membrane or cytosolic proteins and lipids, as well as systemic disorders such as cancer, diabetes, cardiovascular diseases, sepsis and other diseases of inflammation are associated with anemia, RBC damage and their premature removal from the circulation and increase in their heterogeneity (e.g., Salvagno et al., 2015; Feng et al., 2017; Ahmad et al., 2018; Ko et al., 2018; Yin et al., 2018; Parizadeh et al., 2019; Wang et al., 2019). The causes and consequences as well as predictive power of this increase in variability of RBC properties is out of the scope of this review but deserve special attention.

SUMMARY AND THE STANDING CHALLENGES

The present collection of information on the possible causes and consequences of inter-cellular heterogeneity justifies the increasing attention of researchers to the RBC sub-populations and individual cells. It appears that vast amount of information on the near and distant (within months) past is lost when RBC properties are reduced to a set of single "mean" values. This information appears to be of substantial importance when severity of disease or efficacy of therapy are to be assessed for individual patients. At present we do not have the commercially available and standardized methodologies and machines to be able to compare the data obtained of the single cell features in different labs. These challenges are already addressed by some researchers and will drive the transformation of our understanding of red blood cell biology in the nearest future.

AUTHOR CONTRIBUTIONS

AB and AW have composed the text. All authors contributed to editing and proofreading of the text.

REFERENCES

Abay, A., Simionato, G., Chachanidze, R., Bogdanova, A., Hertz, L., Bianchi, P., et al. (2019). Glutaraldehyde – a subtle tool in the investigation of healthy and pathologic red blood cells. *Front. Physiol.* 10:514. doi: 10.3389/fphys.2019.00514

Abkarian, M., Faivre, M., Horton, R., Smistrup, K., Best-Popescu, C. A., and Stone, H. A. (2008). Cellular-scale hydrodynamics. *Biomed. Mater.* 3:034011. doi: 10.1088/1748-6041/3/3/034011

Ahmad, H., Khan, M., Laugle, M., Jackson, D. A., Burant, C., Malemud, C. J., et al. (2018). Red cell distribution width is positively correlated with atherosclerotic cardiovascular disease 10-year risk score, age, and CRP in Spondyloarthritis with axial or peripheral disease. *Int. J. Rheumatol.* 2018:2476239. doi: 10.1155/2018/2476239

Amer, J., Goldfarb, A., and Fibach, E. (2003). Flow cytometric measurement of reactive oxygen species production by normal and thalassaemic red blood cells. *Eur. J. Haematol.* 70, 84–90. doi: 10.1034/j.1600-0609.2003.00011.x

Andrews, D. A., and Low, P. S. (1999). Role of red blood cells in thrombosis. *Curr. Opin. Hematol.* 6, 76–82.

Anyaibe, S., Castro, O., and Headings, V. (1985). Distributions of hemoglobins A and S among erythrocytes of heterozygotes. *Hemoglobin* 9, 137–155. doi: 10.3109/03630268508996996

Badior, K. E., and Casey, J. R. (2018). Molecular mechanism for the red blood cell senescence clock. *IUBMB Life* 70, 32–40. doi: 10.1002/iub.1703

Bain, B. J. (2005). Diagnosis from the blood smear. *N. Engl. J. Med.* 353, 498–507. doi: 10.1056/nejmra043442

Ballas, S. K., and Kocher, W. (1988). Erythrocytes in Hb SC disease are microcytic and hyperchromic. *Am. J. Hematol.* 28, 37–39. doi: 10.1002/ajh.283028 0108

Barshtein, G., Pries, A. R., Goldschmidt, N., Zukerman, A., Orbach, A., Zelig, O., et al. (2016). Deformability of transfused red blood cells is a potent determinant of transfusion-induced change in recipient's blood flow. *Microcirculation* 23, 479–486. doi: 10.1111/micc.12296

Bernhardt, I., Wesseling, M. C., Nguen, D. B., and Kaestner, L. (2019). "Red blood cells actively contribute to blood coagulation and thrombus formation," in *Erythrocyte*, ed. A. Tombak, (London: IntechOpen).

Bessis, M., and Delpech, G. (1981). Discovery of the red blood cell with notes on priorities and credits of discoveries, past, present and future. *Blood Cells* 7, 447–480.

Bessis, M., and Delpech, G. (1982). Sickle cell shape and structure: images and concepts (1840-1980). *Blood Cells* 8, 359–435.

Bessis, M., and Lessin, L. S. (1970). The discocyte-echinocyte equilibrium of the normal and pathologic red cell. *Blood* 36, 399–403. doi: 10.1182/blood.v36.3.399.399

Beutler, E. (ed.) (1988). *Red Cell Senescence.* Bethesda, MD: American Physiological Society.

Bogdanova, A., Goede, J. S., Weiss, E., Bogdanov, N., Bennekou, P., Bernhardt, I., et al. (2010). Cryohydrocytosis: increased activity of cation carriers in red cells from a patient with a band 3 mutation. *Haematologica* 95, 189–198. doi: 10.3324/haematol.2009.010215

Bogdanova, A., Makhro, A., Wang, J., Lipp, P., and Kaestner, L. (2013). Calcium in red blood cells-a perilous balance. *Int. J. Mol. Sci.* 14, 9848–9872. doi: 10.3390/ijms14059848

Boyer, S. H., Belding, T. K., Margolet, L., and Noyes, A. N. (1975). Fetal hemoglobin restriction to a few erythrocytes (F cells) in normal human adults. *Science* 188, 361–363. doi: 10.1126/science.804182

Braun, J., Lindner, K., Schreiber, M., Heidler, R. A., and Horl, W. H. (1997). Percentage of hypochromic red blood cells as predictor of erythropoietic and iron response after i.v. iron supplementation in maintenance haemodialysis patients. *Nephrol. Dial. Transplant.* 12, 1173–1181. doi: 10.1093/ndt/12.6.1173

Brierley, C. K., and Mead, A. J. (2019). Single-cell sequencing in hematology. *Curr. Opin. Oncol.* 32, 139–145. doi: 10.1097/cco.0000000000000613

Brugnara, C. (1997). Erythrocyte membrane transport physiology. *Curr. Opin. Hematol.* 4, 122–127. doi: 10.1097/00062752-199704020-00008

Cahalan, S. M., Lukacs, V., Ranade, S. S., Chien, S., Bandell, M., and Patapoutian, A. (2015). Piezo1 links mechanical forces to red blood cell volume. *eLife* 4:e07370. doi: 10.7554/eLife.07370

Catarino, S. O., Rodrigues, R. O., Pinho, D., Miranda, J. M., Minas, G., and Lima R. (2019). Blood cells separation and sorting techniques of passive microfluidic devices: from fabrication to applications. *Micromachines (Basel).* 10:593. doi: 10.3390/mi10090593

Chu, H., Puchulu-Campanella, E., Galan, J. A., Tao, W. A., Low, P. S., and Hoffman, J. F. (2012). Identification of cytoskeletal elements enclosing the ATP pools that fuel human red blood cell membrane cation pumps. *Proc. Natl. Acad. Sci. U.S.A.* 109, 12794–12799. doi: 10.1073/pnas.1209014109

Clark, M. R., Mohandas, N., and Shohet, S. B. (1983). Osmotic gradient ektacytometry: comprehensive characterization of red cell volume and surface maintenance. *Blood* 61, 899–910. doi: 10.1182/blood.v61.5.899.bloodjournal615899

Conway, A. M., Vora, A. J., and Hinchliffe, R. F. (2002). The clinical relevance of an isolated increase in the number of circulating hyperchromic red blood cells. *J. Clin. Pathol.* 55, 841–844. doi: 10.1136/jcp.55.11.841

Cordasco, D., and Bagchi, P. (2017). On the shape memory of red blood cells. *Phys. Fluids* 29:041901. doi: 10.1063/1.4979271

Corry, W. D., Bresnahan, P. A., and Seaman, G. V. (1982). Evaluation of density gradient separation methods. *J. Biochem. Biophys. Methods* 7, 71–82. doi: 10.1016/0165-022x(82)90038-0

Cossins, A. R., and Gibson, J. S. (1997). Volume-sensitive transport systems and volume homeostasis in vertebrate red blood cells. *J. Exp. Biol.* 200, 343–352.

D'Alessandro, A., Blasi, B., D'amici, G. M., Marrocco, C., and Zolla, L. (2013). Red blood cell subpopulations in freshly drawn blood: application of proteomics and metabolomics to a decades-long biological issue. *Blood Transfus.* 11, 75–87. doi: 10.2450/2012.0164-11

Danielczok, J., Hertz, L., Ruppenthal, S., Kaiser, E., Petkova-Kirova, P., Bogdanova, A., et al. (2017). Does erythropoietin regulate TRPC channels in red blood cells? *Cell. Physiol. Biochem.* 41, 1219–1228. doi: 10.1159/000464384

Danon, D., and Marikovsky, V. (1964). Determination of density distribution of red cell population. *J. Lab. Clin. Med.* 64, 668–674.

Darrow, M. C., Zhang, Y., Cinquin, B. P., Smith, E. A., Boudreau, R., Rochat, R. H., et al. (2016). Visualizing red blood cell sickling and the effects of inhibition of sphingosine kinase 1 using soft X-ray tomography. *J. Cell Sci.* 129, 3511–3517. doi: 10.1242/jcs.189225

Dupire, J., Abkarian, M., and Viallat, A. (2015). A simple model to understand the effect of membrane shear elasticity and stress-free shape on the motion of red blood cells in shear flow. *Soft Matter* 11, 8372–8382. doi: 10.1039/c5sm01407g

Elliott, S. M. F., and Molineux, G. (eds.) (2009). *Erythropoietins, Erythropoietic Factors, and Erythropoiesis. Molecular, Cellular, Preclinical, and Clinical Biology.* Boston, MA: Birkenhäuser.

Fago, A., Forest, E., and Weber, R. E. (2001). Hemoglobin and subunit multiplicity in the rainbow trout (*Oncorhynchus mykiss*) hemoglobin system. *Fish Physiol. Biochem.* 24, 335–342.

Fairbanks, G., Palek, J., Dino, J. E., and Liu, P. A. (1983). Protein kinases and membrane protein phosphorylation in normal and abnormal human erythrocytes: variation related to mean cell age. *Blood* 61, 850–857. doi: 10. 1182/blood.v61.5.850.850

Feng, G. H., Li, H. P., Li, Q. L., Fu, Y., and Huang, R. B. (2017). Red blood cell distribution width and ischaemic stroke. *Stroke Vasc. Neurol.* 2, 172–175. doi: 10.1136/svn-2017-000071

Fermo, E., Bogdanova, A., Petkova-Kirova, P., Zaninoni, A., Marcello, A. P., Makhro, A., et al. (2017). 'Gardos Channelopathy': a variant of hereditary Stomatocytosis with complex molecular regulation. *Sci. Rep.* 7:1744. doi: 10. 1038/s41598-017-01591-w

Fischer, T. M. (2004). Shape memory of human red blood cells. *Biophys. J.* 86, 3304–3313. doi: 10.1016/s0006-3495(04)74378-7

Gallagher, P. G. (2005). Red cell membrane disorders. *Hematol. Am. Soc. Hematol. Educ. Program* 2005, 13–18.

Gilmore, I. S., Heiles, S., and Pieterse, C. L. (2019). Metabolic imaging at the single-cell scale: recent advances in mass spectrometry imaging. *Annu. Rev. Anal. Chem.* 12, 201–224. doi: 10.1146/annurev-anchem-061318-115516

Giustarini, D., Milzani, A., Dalle-Donne, I., and Rossi, R. (2008). Red blood cells as a physiological source of glutathione for extracellular fluids. *Blood Cells Mol. Dis.* 40, 174–179. doi: 10.1016/j.bcmd.2007.09.001

Gonzalez-Hidalgo, M., Guerrero-Pena, F. A., Herold-Garcia, S., Jaume, I. C. A., and Marrero-Fernandez, P. D. (2015). Red blood cell cluster separation from digital images for use in sickle cell disease. *IEEE J. Biomed. Health Inform.* 19, 1514–1525. doi: 10.1109/JBHI.2014.2356402

Grinberg, L., Fibach, E., Amer, J., and Atlas, D. (2005). N-acetylcysteine amide, a novel cell-permeating thiol, restores cellular glutathione and protects human red blood cells from oxidative stress. *Free Radic. Biol. Med.* 38, 136–145. doi: 10.1016/j.freeradbiomed.2004.09.025

Hajdu, S. I. (2003). A note from history: the discovery of blood cells. *Ann. Clin. Lab. Sci.* 33, 237–238.

Hanggi, P., Makhro, A., Gassmann, M., Schmugge, M., Goede, J. S., Speer, O., et al. (2014). Red blood cells of sickle cell disease patients exhibit abnormally high abundance of N-methyl D-aspartate receptors mediating excessive calcium uptake. *Br. J. Haematol.* 167, 252–264. doi: 10.1111/bjh.13028

Helms, C. C., Gladwin, M. T., and Kim-Shapiro, D. B. (2018). Erythrocytes and vascular function: oxygen and nitric oxide. *Front. Physiol.* 9:125. doi: 10.3389/fphys.2018.00125

Herold-Garcia, S., and Fernandes, L. F. (2019). New Methods for Morphological Erythrocytes Classification. *Conf. Proc. IEEE Eng. Med. Biol. Soc.* 2019, 4068–4071.

Hiebl, I., Weber, R. E., Schneeganss, D., and Braunitzer, G. (1989). High-altitude respiration of falconiformes. The primary structures and functional properties of the major and minor hemoglobin components of the adult White-Headed Vulture (*Trigonoceps occipitalis*, Aegyptinae). *Biol. Chem. Hoppe Seyler* 370, 699–706. doi: 10.1515/bchm3.1989.370.2.699

Hoffman, J. F., Dodson, A., and Proverbio, F. (2009). On the functional use of the membrane compartmentalized pool of ATP by the Na+ and Ca++ pumps in human red blood cell ghosts. *J. Gen. Physiol.* 134, 351–361. doi: 10.1085/jgp. 200910270

Huisjes, R., Makhro, A., Llaudet-Planas, E., Hertz, L., Petkova-Kirova, P., Verhagen, L. P., et al. (2019). Density, heterogeneity and deformability of red cells as markers of clinical severity in hereditary spherocytosis. *Haematologica* 105, 338–347. doi: 10.3324/haematol.2018.188151

Inaba, M., and Maede, Y. (1988). Correlation between protein 4.1a/4.1b ratio and erythrocyte life span. *Biochim. Biophys. Acta* 944, 256–264. doi: 10.1016/0005-2736(88)90439-7

Inaba, M., and Maede, Y. (1992). The critical role of asparagine 502 in post-translational alteration of protein 4.1. *Comp. Biochem. Physiol. B* 103, 523–526. doi: 10.1016/0305-0491(92)90364-w

Iscove, N. N. (1977). The role of erythropoietin in regulation of population size and cell cycling of early and late erythroid precursors in mouse bone marrow. *Cell Tissue Kinet.* 10, 323–334. doi: 10.1111/j.1365-2184.1977.tb00300.x

Itkin, T., Gur-Cohen, S., Spencer, J. A., Schajnovitz, A., Ramasamy, S. K., Kusumbe, A. P., et al. (2016). Distinct bone marrow blood vessels differentially regulate haematopoiesis. *Nature* 532, 323–328. doi: 10.1038/nature17624

Jemaa, M., Fezai, M., Bissinger, R., and Lang, F. (2017). Methods employed in cytofluorometric assessment of eryptosis, the suicidal erythrocyte death. *Cell Physiol. Biochem.* 43, 431–444. doi: 10.1159/000480469

Jung, J., Matemba, L. E., Lee, K., Kazyoba, P. E., Yoon, J., Massaga, J. J., et al. (2016). Optical characterization of red blood cells from individuals with sickle cell trait and disease in Tanzania using quantitative phase imaging. *Sci. Rep.* 6:31698. doi: 10.1038/srep31698

Kaestner, L., Bogdanova, A., and Egee, S. (2020). "Calcium channels and calcium-regulated channels in human red blood cells," in *Calcium Signalling*, ed. S. Islam, (New York, NY: Springer).

Kaestner, L., Bollensdorff, C., and Bernhardt, I. (1999). Non-selective voltage-activated cation channel in the human red blood cell membrane. *Biochim. Biophys. Acta* 1417, 9–15. doi: 10.1016/s0005-2736(98)00240-5

Kaestner, L., and Egee, S. (2018). Commentary: voltage gating of mechanosensitive PIEZO channels. *Front. Physiol.* 9:1565. doi: 10.3389/fphys.2018.01565

Kaestner, L., Steffen, P., Nguyen, D. B., Wang, J., Wagner-Britz, L., Jung, A., et al. (2012). Lysophosphatidic acid induced red blood cell aggregation in vitro. *Bioelectrochemistry* 87, 89–95. doi: 10.1016/j.bioelechem.2011.08.004

Kaestner, L., Tabellion, W., Weiss, E., Bernhardt, I., and Lipp, P. (2006). Calcium imaging of individual erythrocytes: problems and approaches. *Cell Calcium* 39, 13–19. doi: 10.1016/j.ceca.2005.09.004

Kaestner, L., Wang, X., Hertz, L., and Bernhardt, I. (2018). Voltage-activated ion channels in non-excitable cells-a viewpoint regarding their physiological justification. *Front. Physiol.* 9:450. doi: 10.3389/fphys.2018.00450

Kaul, D. K., Fabry, M. E., Windisch, P., Baez, S., and Nagel, R. L. (1983). Erythrocytes in sickle cell anemia are heterogeneous in their rheological and hemodynamic characteristics. *J. Clin. Invest.* 72, 22–31. doi: 10.1172/jci11 0960

Kaul, D. K., Koshkaryev, A., Artmann, G., Barshtein, G., and Yedgar, S. (2008). Additive effect of red blood cell rigidity and adherence to endothelial cells in inducing vascular resistance. *Am. J. Physiol. Heart Circ. Physiol.* 295, H1788–H1793. doi: 10.1152/ajpheart.253.2008

Kelley, L. L., Koury, M. J., Bondurant, M. C., Koury, S. T., Sawyer, S. T., and Wickrema, A. (1993). Survival or death of individual proerythroblasts results from differing erythropoietin sensitivities: a mechanism for controlled rates of erythrocyte production. *Blood* 82, 2340–2352. doi: 10.1182/blood.v82.8.2340. 2340

Kihm, A., Kaestner, L., Wagner, C., and Quint, S. (2018). Classification of red blood cell shapes in flow using outlier tolerant machine learning. *PLoS Comput. Biol.* 14:e1006278. doi: 10.1371/journal.pcbi.1006278

Ko, E., Youn, J. M., Park, H. S., Song, M., Koh, K. H., and Lim, C. H. (2018). Early red blood cell abnormalities as a clinical variable in sepsis diagnosis. *Clin. Hemorheol. Microcirc.* 70, 355–363. doi: 10.3233/CH-180430

Kosower, E. M., and Kosower, N. S. (1995). Bromobimane probes for thiols. *Methods Enzymol.* 251, 133–148. doi: 10.1016/0076-6879(95)51117-2

Koury, M. J., and Bondurant, M. C. (1988). Maintenance by erythropoietin of viability and maturation of murine erythroid precursor cells. *J. Cell. Physiol.* 137, 65–74. doi: 10.1002/jcp.1041370108

Koury, M. J., and Bondurant, M. C. (1990). Erythropoietin retards DNA breakdown and prevents programmed death in erythroid progenitor cells. *Science* 248, 378–381. doi: 10.1126/science.2326648

Koury, M. J., and Bondurant, M. C. (1992). The molecular mechanism of erythropoietin action. *Eur. J. Biochem.* 210, 649–663. doi: 10.1111/j.1432-1033.1992.tb17466.x

Kucherenko, Y., Browning, J., Tattersall, A., Ellory, J. C., and Gibson, J. S. (2005). Effect of peroxynitrite on passive K+ transport in human red blood cells. *Cell. Physiol. Biochem.* 15, 271–280. doi: 10.1159/000087237

Kucherenko, Y. V., Wagner-Britz, L., Bernhardt, I., and Lang, F. (2013). Effect of chloride channel inhibitors on cytosolic Ca2+ levels and Ca2+-activated K+ (Gardos) channel activity in human red blood cells. *J. Membr. Biol.* 246, 315–326. doi: 10.1007/s00232-013-9532-0

Kunicka, J., Malin, M., Zelmanovic, D., Katzenberg, M., Canfield, W., Shapiro, P., et al. (2001). Automated quantitation of hemoglobin-based blood substitutes in whole blood samples. *Am. J. Clin. Pathol.* 116, 913–919.

La Manno, G., Soldatov, R., Zeisel, A., Braun, E., Hochgerner, H., Petukhov, V., et al. (2018). RNA velocity of single cells. *Nature* 560, 494–498. doi: 10.1038/s41586-018-0414-6

Landschulz, K. T., Boyer, S. H., Noyes, A. N., Rogers, O. C., and Frelin, L. P. (1992). Onset of erythropoietin response in murine erythroid colony-forming units: assignment to early S-phase in a specific cell generation. *Blood* 79, 2749–2758. doi: 10.1182/blood.v79.10.2749.bloodjournal79102749

Lanotte, L., Mauer, J., Mendez, S., Fedosov, D. A., Fromental, J. M., Claveria, V., et al. (2016). Red cells' dynamic morphologies govern blood shear thinning under microcirculatory flow conditions. *Proc. Natl. Acad. Sci. U.S.A.* 113, 13289–13294. doi: 10.1073/pnas.1608074113

Laybourne, R. C. (1974). Collision between a vulture and an aircraft at an altitude of 37,000 feet. *Wilson Bull.* 86, 461–462.

Lew, V. L., and Tiffert, T. (2013). The terminal density reversal phenomenon of aging human red blood cells. *Front. Physiol.* 4:171. doi: 10.3389/fphys.2013.00171

Lew, V. L., and Tiffert, T. (2017). On the mechanism of human red blood cell longevity: roles of calcium, the sodium pump, PIEZO1, and gardos channels. *Front. Physiol.* 8:977. doi: 10.3389/fphys.2017.00977

Lisovskaya, I. L., Shcherbachenko, I. M., Volkova, R. I., and Tikhonov, V. P. (2008). Modulation of RBC volume distributions by oxidants (phenazine methosulfate and tert-butyl hydroperoxide): role of Gardos channel activation. *Bioelectrochemistry* 73, 49–54. doi: 10.1016/j.bioelechem.2008.04.008

Liu, L., Liu, C., Quintero, A., Wu, L., Yuan, Y., Wang, M., et al. (2019). Deconvolution of single-cell multi-omics layers reveals regulatory heterogeneity. *Nat. Commun.* 10:470. doi: 10.1038/s41467-018-08205-7

Luthra, M. G., Friedman, J. M., and Sears, D. A. (1979). Studies of density fractions of normal human erythrocytes labeled with iron-59 in vivo. *J. Lab. Clin. Med.* 94, 879–896.

Lutz, H. U., and Bogdanova, A. (2013). Mechanisms tagging senescent red blood cells for clearance in healthy humans. *Front. Physiol.* 4:387. doi: 10.3389/fphys.2013.00387

Lutz, H. U., Stammler, P., Fasler, S., Ingold, M., and Fehr, J. (1992). Density separation of human red blood cells on self forming Percoll gradients: correlation with cell age. *Biochim. Biophys. Acta* 1116, 1–10. doi: 10.1016/0304-4165(92)90120-j

Macdougall, I. C., Cavill, I., Hulme, B., Bain, B., Mcgregor, E., Mckay, P., et al. (1992). Detection of functional iron deficiency during erythropoietin treatment: a new approach. *BMJ* 304, 225–226. doi: 10.1136/bmj.304.6821.225

Makhro, A., Haider, T., Wang, J., Bogdanov, N., Steffen, P., Wagner, C., et al. (2016a). Comparing the impact of an acute exercise bout on plasma amino acid composition, intraerythrocytic Ca(2+) handling, and red cell function in athletes and untrained subjects. *Cell Calcium* 60, 235–244. doi: 10.1016/j.ceca.2016.05.005

Makhro, A., Hanggi, P., Goede, J. S., Wang, J., Bruggemann, A., Gassmann, M., et al. (2013). N-methyl-D-aspartate receptors in human erythroid precursor cells and in circulating red blood cells contribute to the intracellular calcium regulation. *Am. J. Physiol. Cell Physiol.* 305, C1123–C1138. doi: 10.1152/ajpcell.00031.2013

Makhro, A., Huisjes, R., Verhagen, L. P., Manu-Pereira Mdel, M., Llaudet-Planas, E., Petkova-Kirova, P., et al. (2016b). Red cell properties after different modes of blood transportation. *Front. Physiol.* 7:288. doi: 10.3389/fphys.2016.00288

Mangani, M. (ed.) (1991). *Red Blood Cell Aging*. New York, NY: Plenum Press.

Mao, T. Y., Fu, L. L., and Wang, J. S. (2011). Hypoxic exercise training causes erythrocyte senescence and rheological dysfunction by depressed Gardos channel activity. *J. Appl. Physiol.* 111, 382–391. doi: 10.1152/japplphysiol.00096.2011

Mason, P. J., Bautista, J. M., and Gilsanz, F. (2007). G6PD deficiency: the genotype-phenotype association. *Blood Rev.* 21, 267–283. doi: 10.1016/j.blre.2007.05.002

Mauer, J., Mendez, S., Lanotte, L., Nicoud, F., Abkarian, M., Gompper, G., et al. (2018). Flow-induced transitions of red blood cell shapes under shear. *Phys. Rev. Lett.* 121:118103. doi: 10.1103/PhysRevLett.121.118103

Menzel, S., and Thein, S. L. (2019). Genetic modifiers of fetal haemoglobin in sickle cell disease. *Mol. Diagn. Ther.* 23, 235–244. doi: 10.1007/s40291-018-0370-8

Miccio, L., Memmolo, P., Merola, F., Netti, P. A., and Ferraro, P. (2015). Red blood cell as an adaptive optofluidic microlens. *Nat. Commun.* 6:6502. doi: 10.1038/ncomms7502

Minetti, G., Achilli, C., Perotti, C., and Ciana, A. (2018). Continuous change in membrane and membrane-skeleton organization during development from proerythroblast to senescent red blood cell. *Front. Physiol.* 9:286. doi: 10.3389/fphys.2018.00286

Minetti, G., Egee, S., Morsdorf, D., Steffen, P., Makhro, A., Achilli, C., et al. (2013). Red cell investigations: art and artefacts. *Blood Rev.* 27, 91–101. doi: 10.1016/j.blre.2013.02.002

Miura, O., D'andrea, A., Kabat, D., and Ihle, J. N. (1991). Induction of tyrosine phosphorylation by the erythropoietin receptor correlates with mitogenesis. *Mol. Cell. Biol.* 11, 4895–4902. doi: 10.1128/mcb.11.10.4895

Mock, D. M., Lankford, G. L., Widness, J. A., Burmeister, L. F., Kahn, D., and Strauss, R. G. (1999). Measurement of red cell survival using biotin-labeled red cells: validation against 51Cr-labeled red cells. *Transfusion* 39, 156–162.

Mohandas, N., and Groner, W. (1989). Cell membrane and volume changes during red cell development and aging. *Ann. N. Y. Acad. Sci.* 554, 217–224. doi: 10.1111/j.1749-6632.1989.tb22423.x

Mosca, A., Paleari, R., Modenese, A., Rossini, S., Parma, R., Rocco, C., et al. (1991). Clinical utility of fractionating erythrocytes into "Percoll" density gradients. *Adv. Exp. Med. Biol.* 307, 227–238. doi: 10.1007/978-1-4684-5985-2_21

Mueller, T. J., Jackson, C. W., Dockter, M. E., and Morrison, M. (1985). Use of an in vivo enrichment procedure to study membrane skeletal protein changes during red cell aging. *Prog. Clin. Biol. Res.* 195, 227–236.

Mueller, T. J., Jackson, C. W., Dockter, M. E., and Morrison, M. (1987). Membrane skeletal alterations during in vivo mouse red cell aging. Increase in the band 4.1a:4.1b ratio. *J. Clin. Invest.* 79, 492–499. doi: 10.1172/jci112839

Mugnano, M., Memmolo, P., Miccio, L., Merola, F., Bianco, V., Bramanti, A., et al. (2018). Label-free optical marker for red-blood-cell phenotyping of inherited anemias. *Anal. Chem.* 90, 7495–7501. doi: 10.1021/acs.analchem.8b01076

Nakamura, Y., Komatsu, N., and Nakauchi, H. (1992). A truncated erythropoietin receptor that fails to prevent programmed cell death of erythroid cells. *Science* 257, 1138–1141. doi: 10.1126/science.257.5073.1138

Narayan, A. D., Ersek, A., Campbell, T. A., Colon, D. M., Pixley, J. S., and Zanjani, E. D. (2005). The effect of hypoxia and stem cell source on haemoglobin switching. *Br. J. Haematol.* 128, 562–570. doi: 10.1111/j.1365-2141.2004.05336.x

Oikonomidou, P. R., and Rivella, S. (2018). What can we learn from ineffective erythropoiesis in thalassemia? *Blood Rev.* 32, 130–143. doi: 10.1016/j.blre.2017.10.001

O'Neill, J. S., and Reddy, A. B. (2011). Circadian clocks in human red blood cells. *Nature* 469, 498–503. doi: 10.1038/nature09702

Parizadeh, S. M., Jafarzadeh-Esfehani, R., Bahreyni, A., Ghandehari, M., Shafiee, M., Rahmani, F., et al. (2019). The diagnostic and prognostic value of red cell distribution width in cardiovascular disease; current status and prospective. *Biofactors* 45, 507–516. doi: 10.1002/biof.1518

Pernow, J., Mahdi, A., Yang, J., and Zhou, Z. (2019). Red blood cell dysfunction: a new player in cardiovascular disease. *Cardiovasc. Res.* 115, 1596–1605. doi: 10.1093/cvr/cvz156

Perrotta, S., Gallagher, P. G., and Mohandas, N. (2008). Hereditary spherocytosis. *Lancet* 372, 1411–1426. doi: 10.1016/S0140-6736(08)61588-3

Peters, A. L., and van Noorden, C. J. (2017). Single cell cytochemistry illustrated by the demonstration of Glucose-6-phosphate dehydrogenase deficiency in erythrocytes. *Methods Mol. Biol.* 1560, 3–13. doi: 10.1007/978-1-4939-6788-9_1

Petkova-Kirova, P., Hertz, L., Danielczok, J., Huisjes, R., Makhro, A., Bogdanova, A., et al. (2019). Red blood cell membrane conductance in hereditary haemolytic anaemias. *Front. Physiol.* 10:386. doi: 10.3389/fphys.2019.00386

Piccinini, G., Minetti, G., Balduini, C., and Brovelli, A. (1995). Oxidation state of glutathione and membrane proteins in human red cells of different age. *Mech. Ageing Dev.* 78, 15–26. doi: 10.1016/0047-6374(94)01511-j

Piva, E., Brugnara, C., Chiandetti, L., and Plebani, M. (2010). Automated reticulocyte counting: state of the art and clinical applications in the evaluation of erythropoiesis. *Clin. Chem. Lab. Med.* 48, 1369–1380. doi: 10.1515/CCLM. 2010.292

Prus, E., and Fibach, E. (2013). Heterogeneity of F cells in beta-thalassemia. *Transfusion* 53, 499–504. doi: 10.1111/j.1537-2995.2012.03769.x

Quint, S., Christ, A. F., Guckenberger, A., Himbert, S., Kaestner, L., Gekle, S., et al. (2018). 3D tomography of cells in micro-channels. *Appl. Phys. Lett.* 111:103701. doi: 10.1063/1.4986392

Reichel, F., Mauer, J., Nawaz, A. A., Gompper, G., Guck, J., and Fedosov, D. A. (2019). High-throughput microfluidic characterization of erythrocyte shape and mechanical variability. *bioRxiv* [Preprint]. doi: 10.1101/488189

Romero, P. J., and Romero, E. A. (1997). Differences in Ca2+ pumping activity between sub-populations of human red cells. *Cell Calcium* 21, 353–358. doi: 10.1016/s0143-4160(97)90028-2

Romero, P. J., and Romero, E. A. (1999). The role of calcium metabolism in human red blood cell ageing: a proposal. *Blood Cells Mol. Dis.* 25, 9–19.

Romero, P. J., Salas, V., and Hernandez, C. (2002). Calcium pump phosphoenzyme from young and old human red cells. *Cell Biol. Int.* 26, 945–949.

Rotordam, M. G., Fermo, E., Becker, N., Barcellini, W., Bruggemann, A., Fertig, N., et al. (2019). A novel gain-of-function mutation of Piezo1 is functionally affirmed in red blood cells by high-throughput patch clamp. *Haematologica* 104, e179–e183.

Sadafi, A., Koehler, N., Makhro, A., Bogdanova, A., Navab, N., Marr, C., et al. (2019). "Multiclass deep active learning for detecting red blood cell subtypes in brightfield microscopy," in *Medical Image Computing and Computer Assisted Intervention – MICCAI 2019 Lecture Notes in Computer Science*, ed. S. D. E. Al (Berlin: Springer).

Salvagno, G. L., Sanchis-Gomar, F., Picanza, A., and Lippi, G. (2015). Red blood cell distribution width: a simple parameter with multiple clinical applications. *Crit. Rev. Clin. Lab. Sci.* 52, 86–105. doi: 10.3109/10408363.2014.992064

Salvo, G., Caprari, P., Samoggia, P., Mariani, G., and Salvati, A. M. (1982). Human erythrocyte separation according to age on a discontinuous "Percoll" density gradient. *Clin. Chim. Acta* 122, 293–300.

Sass, M. D., Caruso, C. J., and O'connell, D. J. (1965). Decreased glutathione in aging red Cells. *Clin. Chim. Acta* 11, 334–340.

Schaefer, R. M., and Schaefer, L. (1995). The hypochromic red cell: a new parameter for monitoring of iron supplementation during rhEPO therapy. *J. Perinat. Med.* 23, 83–88.

Sinha, A., Chu, T. T., Dao, M., and Chandramohanadas, R. (2015). Single-cell evaluation of red blood cell bio-mechanical and nano-structural alterations upon chemically induced oxidative stress. *Sci. Rep.* 5:9768. doi: 10.1038/srep09768

Sparrow, R. L. (2017). Red blood cell components: time to revisit the sources of variability. *Blood Transfus.* 15, 116–125. doi: 10.2450/2017.0326-16

Spencer, J. A., Ferraro, F., Roussakis, E., Klein, A., Wu, J., Runnels, J. M., et al. (2014). Direct measurement of local oxygen concentration in the bone marrow of live animals. *Nature* 508, 269–273. doi: 10.1038/nature1 3034

Steffen, P., Jung, A., Nguyen, D. B., Muller, T., Bernhardt, I., Kaestner, L., et al. (2011). Stimulation of human red blood cells leads to Ca2+-mediated intercellular adhesion. *Cell Calcium* 50, 54–61. doi: 10.1016/j.ceca.2011.05.002

Suzuki, T., and Dale, G. L. (1988). Senescent erythrocytes: isolation of in vivo aged cells and their biochemical characteristics. *Proc. Natl. Acad. Sci. U.S.A.* 85, 1647–1651.

Swammerdam, J. (1737). *Bybel der Natuure of Historie der Insecten/Biblia Naturae sive Historia Insectorum.* Utrecht: De Banier.

Thein, S. L., and Craig, J. E. (1998). Genetics of Hb F/F cell variance in adults and heterocellular hereditary persistence of fetal hemoglobin. *Hemoglobin* 22, 401–414.

Thompson, C. J., Schilling, T., Howard, M. R., and Genever, P. G. (2010). SNARE-dependent glutamate release in megakaryocytes. *Exp. Hematol.* 38, 504–515. doi: 10.1016/j.exphem.2010.03.011

Tomari, R., Zakaria, W. N. W., Jamil, M. M. A., and Fuad, N. F. N. (2014). Computer aided system for red blood cell classification in blood smear image. *Proc. Comp. Sci.* 42, 206–213.

Tusi, B. K., Wolock, S. L., Weinreb, C., Hwang, Y., Hidalgo, D., Zilionis, R., et al. (2018). Population snapshots predict early haematopoietic and erythroid hierarchies. *Nature* 555, 54–60. doi: 10.1038/nature25741

Wang, J., Wagner-Britz, L., Bogdanova, A., Ruppenthal, S., Wiesen, K., Kaiser, E., et al. (2013). Morphologically homogeneous red blood cells present a heterogeneous response to hormonal stimulation. *PLoS One* 8:e67697. doi: 10.1371/journal.pone.0067697

Wang, P. F., Song, S. Y., Guo, H., Wang, T. J., Liu, N., and Yan, C. X. (2019). Prognostic role of pretreatment red blood cell distribution width in patients with cancer: a meta-analysis of 49 studies. *J. Cancer* 10, 4305–4317. doi: 10. 7150/jca.31598

Weber, R. E., Hiebl, I., and Braunitzer, G. (1988). High altitude and hemoglobin function in the vultures *Gyps rueppellii* and *Aegypius monachus*. *Biol. Chem. Hoppe Seyler* 369, 233–240.

Wenk, R. E. (1976). Comparison of five methods for preparing blood smears. *Am. J. Med. Technol.* 42, 71–78.

Wesseling, M. C., Wagner-Britz, L., Boukhdoud, F., Asanidze, S., Nguyen, D. B., Kaestner, L., et al. (2016). Measurements of intracellular Ca2+ content and phosphatidylserine exposure in human red blood cells: methodological issues. *Cell. Physiol. Biochem.* 38, 2414–2425. doi: 10.1159/000445593

Woll, P. S., Kjallquist, U., Chowdhury, O., Doolittle, H., Wedge, D. C., Thongjuea, S., et al. (2014). Myelodysplastic syndromes are propagated by rare and distinct human cancer stem cells in vivo. *Cancer Cell* 25, 794–808. doi: 10.1016/j.ccr. 2014.03.036

Yin, Y., Ye, S., Wang, H., Li, B., Wang, A., Yan, W., et al. (2018). Red blood cell distribution width and the risk of being in poor glycemic control among patients with established type 2 diabetes. *Ther. Clin. Risk Manag.* 14, 265–273. doi: 10.2147/TCRM.S155753

Zhou, S., Giannetto, M., Decourcey, J., Kang, H., Kang, N., Li, Y., et al. (2019). Oxygen tension-mediated erythrocyte membrane interactions regulate cerebral capillary hyperemia. *Sci. Adv.* 5:eaaw4466. doi: 10.1126/sciadv.aaw4466

Zhu, C., Shi, W., Daleke, D. L., and Baker, L. A. (2018). Monitoring dynamic spiculation in red blood cells with scanning ion conductance microscopy. *Analyst* 143, 1087–1093. doi: 10.1039/c7an01986f

Permissions

All chapters in this book were first published by Frontiers; hereby published with permission under the Creative Commons Attribution License or equivalent. Every chapter published in this book has been scrutinized by our experts. Their significance has been extensively debated. The topics covered herein carry significant findings which will fuel the growth of the discipline. They may even be implemented as practical applications or may be referred to as a beginning point for another development.

The contributors of this book come from diverse backgrounds, making this book a truly international effort. This book will bring forth new frontiers with its revolutionizing research information and detailed analysis of the nascent developments around the world.

We would like to thank all the contributing authors for lending their expertise to make the book truly unique. They have played a crucial role in the development of this book. Without their invaluable contributions this book wouldn't have been possible. They have made vital efforts to compile up to date information on the varied aspects of this subject to make this book a valuable addition to the collection of many professionals and students.

This book was conceptualized with the vision of imparting up-to-date information and advanced data in this field. To ensure the same, a matchless editorial board was set up. Every individual on the board went through rigorous rounds of assessment to prove their worth. After which they invested a large part of their time researching and compiling the most relevant data for our readers.

The editorial board has been involved in producing this book since its inception. They have spent rigorous hours researching and exploring the diverse topics which have resulted in the successful publishing of this book. They have passed on their knowledge of decades through this book. To expedite this challenging task, the publisher supported the team at every step. A small team of assistant editors was also appointed to further simplify the editing procedure and attain best results for the readers.

Apart from the editorial board, the designing team has also invested a significant amount of their time in understanding the subject and creating the most relevant covers. They scrutinized every image to scout for the most suitable representation of the subject and create an appropriate cover for the book.

The publishing team has been an ardent support to the editorial, designing and production team. Their endless efforts to recruit the best for this project, has resulted in the accomplishment of this book. They are a veteran in the field of academics and their pool of knowledge is as vast as their experience in printing. Their expertise and guidance has proved useful at every step. Their uncompromising quality standards have made this book an exceptional effort. Their encouragement from time to time has been an inspiration for everyone.

The publisher and the editorial board hope that this book will prove to be a valuable piece of knowledge for researchers, students, practitioners and scholars across the globe.

List of Contributors

Lorenzo Bertolone, Davide Stefanoni, Evan J. Morrison and Travis Nemkov
Department of Biochemistry and Molecular Genetics, University of Colorado Denver – Anschutz Medical Campus, Aurora, CO, United States

Hye K. Shin, Jin Hyen Baek and Yamei Gao
Center for Biologics Evaluation and Research, Food and Drug Administration, Silver Spring, MD, United States

Tiffany Thomas, Richard O. Francis, Eldad A. Hod, Joseph Schwartz, Krystalyn E. Hudson and Steven L. Spitalnik
Department of Pathology and Cell Biology, Columbia University, New York, United States

James C. Zimring
Department of Pathology, University of Virginia, Charloteseville, VA, United States

Tatsuro Yoshida
Hemanext, Inc., Lexington, MA, United States

Matthew Karafin
Blood Center of Wisconsin, Milwaukee, WI, United States
Department of Pathology and Laboratory Medicine, Milwaukee, WI, United States

Paul W. Buehler
Department of Pathology, University of Maryland School of Medicine, Baltimore, MD, United States
Department of Pediatrics, Center for Blood Oxygen Transport and Hemostasis, University of Maryland School of Medicine, Baltimore, MD, United States

Angelo D'Alessandro
Department of Biochemistry and Molecular Genetics, University of Colorado Denver – Anschutz Medical Campus, Aurora, CO, United States
Division of Hematology, Department of Medicine, University of Colorado Denver – Anschutz Medical Campus, Aurora, CO, United States

Sukanya Suresh, Praveen Kumar Rajvanshi and Constance T. Noguchi
Molecular Medicine Branch, National Institute of Diabetes and Digestive and Kidney Diseases, National Institutes of Health, Bethesda, MD, United States

Anne-Sophie Cloos, Mauriane Maja, Amaury Stommen, Juliette Vanderroost, Patrick Van Der Smissen and Donatienne Tyteca
CELL Unit & PICT Imaging Platform, de Duve Institute, UCLouvain, Brussels, Belgium

Laura G. M. Daenen
Department of Hematology, University Medical Center Utrecht, Utrecht University, Utrecht, Netherlands

Minke Rab and Richard van Wijk
Central Diagnostic Laboratory – Research, University Medical Center Utrecht, Utrecht University, Utrecht, Netherlands

Jan Westerink
Department of Vascular Medicine, University Medical Center Utrecht, Utrecht University, Utrecht, Netherlands

Eric Mignolet, Yvan Larondelle, Andra C. Dumitru and David Alsteens
Louvain Institute of Biomolecular Science and Technology, UCLouvain, Ottignies-Louvain-la-Neuve, Belgium

Romano Terrasi and Giulio G. Muccioli
Bioanalysis and Pharmacology of Bioactive Lipids Research Group, Louvain Drug Research Institute, UCLouvain, Brussels, Belgium

Benedetta Porro, Edoardo Conte, Fabrizio Veglia, Simone Barbieri, Susanna Fiorelli, Sonia Eligini, Alessandro Di Minno, Saima Mushtaq, Elena Tremoli and Viviana Cavalca
Centro Cardiologico Monzino, Istituto di Ricovero e Cura a Carattere Scientifico (IRCCS), Milan, Italy

Anna Zaninoni and Paola Bianchi
Fondazione IRCCS Ca' Granda Ospedale Maggiore Policlinico Milano, Unità Operativa Complessa (UOC) Ematologia, Unità Operativa Semplice (UOS) Fisiopatologia delle Anemie, Milan, Italy

Daniele Andreini
Centro Cardiologico Monzino, Istituto di Ricovero e Cura a Carattere Scientifico (IRCCS), Milan, Italy
Department of Clinical Sciences and Community Health, Cardiovascular Section, University of Milan, Milan, Italy

Elisa Fermo, Cristina Vercellati, Anna Paola Marcello, Anna Zaninoni, Alberto Zanella, Juri A. Giannotta, Wilma Barcellini
UOS Fisiopatologia delle Anemie, UOC Ematologia, Fondazione IRCCS Ca' Granda Ospedale Maggiore Policlinico, Milan, Italy

Ebru Yilmaz Keskin
Department of Pediatric Hematology and Oncology, Suleyman Demirel University, Isparta, Turkey

Silverio Perrotta
Dipartimento della Donna, del Bambino e di Chirurgia Generale e Specialistica, Università degli Studi della Campania "Luigi Vanvitelli," Naples, Italy

Valentina Brancaleoni
UOC Medicina Generale, Fondazione IRCCS Ca' Granda Ospedale Maggiore Policlinico, Milan, Italy

Alessia Pagani, Antonella Nai and Laura Silvestri
Division of Genetics and Cell Biology, San Raffaele Scientific Institute, Milan, Italy
Vita-Salute San Raffaele University, Milan, Italy

Clara Camaschella
Division of Genetics and Cell Biology, San Raffaele Scientific Institute, Milan, Italy
Vita-Salute San Raffaele University, Milan, Italy

Roberta Russo, Roberta Marra, Barbara Eleni Rosato, Achille Iolascon and Immacolata Andolfo
Department of Molecular Medicine and Medical Biotechnologies, University of Naples Federico II, Naples, Italy
CEINGE Biotecnologie Avanzate, Naples, Italy

Maggie L. Kalev-Zylinska and James I. Hearn
Blood and Cancer Biology Laboratory, Department of Molecular Medicine and Pathology, University of Auckland, Auckland, New Zealand
Department of Pathology and Laboratory Medicine, LabPlus Haematology, Auckland City Hospital, Auckland, New Zealand

Asya Makhro and Anna Bogdanova
Red Blood Cell Research Group, Institute of Veterinary Physiology, Vetsuisse Faculty, University of Zurich, Zürich, Switzerland
Zurich Center for Integrative Human Physiology, University of Zurich, Zürich, Switzerland

Francesca Aglialoro, Naomi Hofsink, Menno Hofman, Nicole Brandhorst and Emile van den Akker
Sanquin Research and Landsteiner Laboratory, Department of Haematopoiesis, Amsterdam UMC, University of Amsterdam, Amsterdam, Netherlands

Elie Nader, Sarah Skinner and Philippe Connes
Laboratory LIBM EA7424, Team "Vascular Biology and Red Blood Cell", University of Lyon 1, Lyon, France
Laboratory of Excellence GR-Ex, Paris, France

Marc Romana and Marie-Dominique Hardy-Dessources
Laboratory of Excellence GR-Ex, Paris, France
Biologie Intégrée du Globule Rouge, Université de Paris, UMR_S1134, BIGR, INSERM, F-75015, Paris, France
Biologie Intégrée du Globule Rouge, The Université des Antilles, UMR_S1134, BIGR, F- 97157, Pointe-a-Pitre, France

Romain Fort
Laboratory LIBM EA7424, Team "Vascular Biology and Red Blood Cell", University of Lyon 1, Lyon, France
Laboratory of Excellence GR-Ex, Paris, France
Département de Médecine, Hôpital Edouard Herriot, Hospices Civils de Lyon, Lyon, France

Nathalie Lemonne
Unité Transversale de la Drépanocytose, Hôpital de Pointe-a-Pitre, Hôpital Ricou, Pointe-a-Pitre, France

Nicolas Guillot
Laboratoire Carmen INSERM 1060, INSA Lyon, Université Claude Bernard Lyon 1, Université de Lyon, Villeurbanne, France

Sophie Antoine-Jonville
Laboratoire ACTES EA 3596, The Université des Antilles, Pointe-a-Pitre, France

Céline Renoux
Laboratory LIBM EA7424, Team "Vascular Biology and Red Blood Cell", University of Lyon 1, Lyon, France
Laboratory of Excellence GR-Ex, Paris, France
Laboratoire de Biochimie et de Biologie Moleìculaire, UF de Biochimie des Pathologies Eìrythrocytaires, Centre de Biologie et de Pathologie Est, Hospices Civils de Lyon, Lyon, France

Emeric Stauffer
Laboratory LIBM EA7424, Team "Vascular Biology and Red Blood Cell", University of Lyon 1, Lyon, France
Laboratory of Excellence GR-Ex, Paris, France
Centre de Médecine du Sommeil et des Maladies Respiratoires, Hospices Civils de Lyon, Hôpital de la Croix Rousse, Lyon, France

Philippe Joly
Laboratory LIBM EA7424, Team "Vascular Biology and Red Blood Cell", University of Lyon 1, Lyon, France
Laboratory of Excellence GR-Ex, Paris, France
Laboratoire de Biochimie et de Biologie Moleìculaire, UF de Biochimie des Pathologies Eìrythrocytaires, Centre de Biologie et de Pathologie Est, Hospices Civils de Lyon, Lyon, France

Yves Bertrand
d'Hématologie et d'Oncologie Pédiatrique, Hospices Civils de Lyon, Lyon, France

M. C. Berrevoets, J. Bos, R. Huisjes, T. H. Merkx, B. A. van Oirschot, W. W. van Solinge and R. van Wijk
Central Diagnostic Laboratory-Research, University Medical Center Utrecht, Utrecht University, Utrecht, Netherlands

J. W. Verweij and M. Y. A. Lindeboom
Department of Pediatric Surgery, University Medical Center Utrecht, Utrecht University, Utrecht, Netherlands

E. J. van Beers and M. Bartels
Van Creveldkliniek, University Medical Center Utrecht, Utrecht University, Utrecht, Netherlands

M. A. E. Rab
Central Diagnostic Laboratory-Research, University Medical Center Utrecht, Utrecht University, Utrecht, Netherlands
Van Creveldkliniek, University Medical Center Utrecht, Utrecht University, Utrecht, Netherlands

Iñigo San-Millán
Department of Human Physiology and Nutrition, University of Colorado Colorado Springs, Colorado Springs, CO, United States
Division of Endocrinology, Metabolism and Diabetes, Department of Medicine, University of Colorado Anschutz Medical Campus, Aurora, CO, United States
Department of Research and Development, UAE Team Emirates, Abu Dhabi, United Arab Emirates

Davide Stefanoni, Kirk C. Hansen and Angelo D'Alessandro
Department of Biochemistry and Molecular Genetics, University of Colorado Anschutz Medical Campus, Aurora, CO, United States

Janel L. Martinez
Division of Endocrinology, Metabolism and Diabetes, Department of Medicine, University of Colorado Anschutz Medical Campus, Aurora, CO, United States

Sanu Susan Jacob and Raghavendra K. Rao
Department of Physiology, Kasturba Medical College-Manipal, Manipal Academy of Higher Education, Manipal, India

Aseefhali Bankapur, Surekha Barkur, Mahendra Acharya and Santhosh Chidangil
Department of Atomic and Molecular Physics, Centre of Excellence for Biophotonics, Manipal Academy of Higher Education, Manipal, India

Pragna Rao
Department of Biochemistry, Kasturba Medical College-Manipal, Manipal Academy of Higher Education, Manipal, India

Asha Kamath and R. Vani Lakshmi
Department of Data Science, Prasanna School of Public Health, Manipal Academy of Higher Education, Manipal, India

Prathap M. Baby
Department of Physiology, Melaka Manipal Medical College, Manipal Academy of Higher Education, Manipal, India

Gabriel Salinas Cisneros
Sickle Cell Branch, National Heart Lung and Blood Institute, National Institutes of Health, Bethesda, MD, United States
Division of Hematology and Oncology, Children's National Medical Center, Washington, DC, United States

Swee L. Thein
Sickle Cell Branch, National Heart Lung and Blood Institute, National Institutes of Health, Bethesda, MD, United States

Valeria Rizzuto
Translational Research in Child and Adolescent Cancer – Rare Anemia Disorders Research Laboratory, Vall d'Hebron Research Institute, ERN-EuroBloodNet Member, Barcelona, Spain
Josep Carreras Leukaemia Research Institute, Badalona, Spain
Department of Medicine, Universitat de Barcelona, Barcelona, Spain

Tamara T. Koopmann, Gijs W. E. Santen and Cornelis L. Harteveld
Department of Clinical Genetics, Leiden University Medical Center, ERN-EuroBloodNet Member, Leiden, Netherlands

Adoración Blanco-Álvarez and Barbara Tazón-Vega
Hematologic Molecular Genetics Unit, Hematology Department, Hospital Universitari Vall d'Hebron, ERN-EuroBloodNet Member, Barcelona, Spain

Amira Idrizovic and Maria del Mar Mañú-Pereira
Translational Research in Child and Adolescent Cancer – Rare Anemia Disorders Research Laboratory, Vall d'Hebron Research Institute, ERN-EuroBloodNet Member, Barcelona, Spain

Cristina Díaz de Heredia and Pablo Velasco
Oncohematologic Pediatrics Department, Hospital Universitari Vall d'Hebron, ERN-EuroBloodNet Member, Barcelona, Spain

Rafael Del Orbe and Miriam Vara Pampliega
Hematology Department, Hospital Universitario Cruces, Barakaldo, Spain

David Beneitez
Red Blood Cell Disorders Unit, Hematology Department, Hospital Universitari Vall d'Hebron, ERN-EuroBloodNet Member, Barcelona, Spain

Quinten Waisfisz and Mariet Elting
Department of Clinical Genetics, VU Medical Center, Amsterdam, Netherlands

Frans J. W. Smiers and Anne J. de Pagter
Department of Pediatric Hematology, Leiden University Medical Center, Leiden, Netherlands

Jean-Louis H. Kerkhoffs
Department of Hematology, HAGA City Hospital, The Hague, Netherlands

Saša Svetina
Institute of Biophysics, Faculty of Medicine, University of Ljubljana, Ljubljana, Slovenia
Jožef Stefan Institute, Ljubljana, Slovenia

Magalie Faivre
Université de Lyon, Institut des Nanotechnologies de Lyon INL-UMR 5270 CNRS, Université Lyon 1, Villeurbanne, France

Amel Bessaa
Laboratoire Interuniversitaire de Biologie de la Motricité (LIBM) EA7424, Equipe "Biologie Vasculaire et du Globule Rouge", UCBL1, Villeurbanne, France
Laboratoire d'Excellence (Labex) GR-Ex, Paris, France

Lydie Da Costa
Laboratoire d'Excellence (Labex) GR-Ex, Paris, France
AP-HP, Service d'Hématologie Biologique, Hôpital Robert-Debré, Paris, France
Université Paris Diderot, Université Sorbonne, Paris Cité, Paris, France
INSERM U1149, CRI, Faculté de Médecine Bichat-Claude Bernard, Paris, France

Alexandra Gauthier
Laboratoire Interuniversitaire de Biologie de la Motricité (LIBM) EA7424, Equipe "Biologie Vasculaire et du Globule Rouge", UCBL1, Villeurbanne, France
Laboratoire d'Excellence (Labex) GR-Ex, Paris, France
Institut d'Hématologie et d'Oncologie Pédiatrique (IHOP), Hospices Civils de Lyon, Lyon, France

Philippe Connes
Laboratoire Interuniversitaire de Biologie de la Motricité (LIBM) EA7424, Equipe "Biologie Vasculaire et du Globule Rouge", UCBL1, Villeurbanne, France
Laboratoire d'Excellence (Labex) GR-Ex, Paris, France
Institut Universitaire de France, Paris, France

Paola Bianchi
Fondazione IRCCS Ca' Granda Ospedale Maggiore Policlinico Milano, UOC Ematologia, UOS Fisiopatologia delle Anemie, Milan, Italy

Véronique Picard
Service d'Hématologie Biologique, Assistance Publique-Hôpitaux de Paris, Hôpital Bicêtre, Hôpitaux Universitaires Paris-Saclay, Le Kremlin-Bicêtre, France
Faculté de Pharmacie, Université Paris-Saclay, Châtenay-Malabry, France

Corinne Guitton
Service de Pédiatrie Générale, Assistance Publique-Hôpitaux de Paris, Hôpital Bicêtre, Filière MCGRE, Hôpitaux Universitaires Paris-Saclay, Le Kremlin-Bicêtre, France

Lamisse Mansour-Hendili
Department of Molecular Genetics, Assistance Publique-Hôpitaux de Paris, Henri Mondor University Hospital, Créteil, France

Bernard Jondeau, Laurence Bendélac and Valérie Proulle
Service d'Hématologie Biologique, Assistance Publique-Hôpitaux de Paris, Hôpital Bicêtre, Hôpitaux Universitaires Paris-Saclay, Le Kremlin-Bicêtre, France

Maha Denguir
Service de Biochimie, Assistance Publique-Hôpitaux de Paris, Bicêtre Hospital, Le Kremlin-Bicêtre, France

Julien Demagny
Service d'Hématologie Biologique, CHU Amiens, EA 4666 HEMATIM-UPJV, Amiens, France

Frédéric Galactéros
Centre de Référence des Syndromes Drépanocytaires Majeurs, Hôpital Henri-Mondor, AP-HP, Créteil, France

Loic Garçon
Service d'Hématologie Biologique, Assistance Publique-Hôpitaux de Paris, Hôpital Bicêtre, Hôpitaux Universitaires Paris-Saclay, Le Kremlin-Bicêtre, France
Service d'Hématologie Biologique, CHU Amiens, EA 4666 HEMATIM-UPJV, Amiens, France

Anna Bogdanova and Asya Makhro
Red Blood Cell Research Group, Vetsuisse Faculty, The Zurich Center for Integrative Human Physiology (ZHIP), Institute of Veterinary Physiology, University of Zurich, Zurich, Switzerland

Lars Kaestner
Experimental Physics, Dynamics of Fluids, Faculty of Natural Sciences and Technology, Saarland University, Saarbrücken, Germany
Theoretical Medicine and Biosciences, Medical Faculty, Saarland University, Homburg, Germany

Greta Simionato
Experimental Physics, Dynamics of Fluids, Faculty of Natural Sciences and Technology, Saarland University, Saarbrücken, Germany
Institute for Clinical and Experimental Surgery, Saarland University, Homburg, Germany

Amittha Wickrema
Section of Hematology/Oncology, Department of Medicine, University of Chicago, Chicago, IL, United States

Index

Printed in the USA
CPSIA information can be obtained
at www.ICGtesting.com
LVHW080041211023
761596LV00010BA/1856